T0280492

CAMBRIDGE LIBRARY COLLECTION

Books of enduring scholarly value

Mathematical Sciences

From its pre-historic roots in simple counting to the algorithms powering modern desktop computers, from the genius of Archimedes to the genius of Einstein, advances in mathematical understanding and numerical techniques have been directly responsible for creating the modern world as we know it. This series will provide a library of the most influential publications and writers on mathematics in its broadest sense. As such, it will show not only the deep roots from which modern science and technology have grown, but also the astonishing breadth of application of mathematical techniques in the humanities and social sciences, and in everyday life.

Scientific Papers

This volume includes papers frpm 1887, when Lord Rayleigh became Professor of Natural Philosophy at the Royal Institution in London, to 1892. An 1888 contribution on the densities of hydrogen and oxygen led to a series of experiments on the densities of the atmospheric gases. This resulted in the unsettling discovery that the density of atmospheric nitrogen seemed very slightly to exceed the density of nitrogen derived from its chemical compounds. A substantial 1888 paper, on the wave theory of light, was written for the Encyclopaedia Britannica in the immediate aftermath of the crucial Michelson–Morley experiment in which the speed of light had been measured. In addition, this wide-ranging volume shows Rayleigh's developing interest in the properties of liquid surfaces, with a discourse on foams (1890), and a substantial paper on surface films (1892). It also includes a charming brief appreciation (1890) of James Clerk Maxwell's legacy to science.

Scientific Papers

Volume 3: 1887-1892

Baron John William Strutt Rayleigh

Cambridge University Press

CAMBRIDGE UNIVERSITY PRESS

Cambridge New York Melbourne Madrid Cape Town Singapore São Paolo Delhi

Published in the United States of America by Cambridge University Press, New York

www.cambridge.org
Information on this title: www.cambridge.org/9781108005449

This edition first published 1902
This digitally printed version 2009

ISBN 978-1-108-00544-9

SCIENTIFIC PAPERS

London: C. J. CLAY AND SONS,
CAMBRIDGE UNIVERSITY PRESS WAREHOUSE,
AVE MARIA LANE.
Glasgow: 50, WELLINGTON STREET.

Leipzig: F. A. BROCKHAUS.
New York: THE MACMILLAN COMPANY.

SCIENTIFIC PAPERS

BY

JOHN WILLIAM STRUTT,

BARON RAYLEIGH,

D.Sc., F.R.S.,

HONORARY FELLOW OF TRINITY COLLEGE, CAMBRIDGE,
PROFESSOR OF NATURAL PHILOSOPHY IN THE ROYAL INSTITUTION.

VOL. III.

1887—1892.

CAMBRIDGE:
AT THE UNIVERSITY PRESS.
1902

Cambridge:
PRINTED BY J. AND C. F. CLAY,
AT THE UNIVERSITY PRESS.

PREFACE.

AS in former volumes, the papers here included embrace a wide range of subjects. In Optics, Arts. 149, 150 deal with the reflexion of light at a twin plane of a crystal and, besides revealing unexpected peculiarities respecting polarization, explain some remarkable phenomena observed by Prof. Stokes. Attention may also be called to Art. 185 in which it is shown that the light found by Jamin to be reflected from water at the polarizing angle is to be attributed to a film of grease, and to Art. 157 "On the Limit to Interference when Light is radiated from moving Molecules."

Several papers treat of capillary questions. In Art. 170 Plateau's "Superficial Viscosity" is traced to greasy contamination of water surfaces. The theory of Surface Forces is expounded in Arts. 176, 186, 193, and attention is called to T. Young's remarkable estimates of molecular magnitudes.

The relative densities of Hydrogen and Oxygen and the composition of Water are the subjects of Arts. 146, 153, 187.

In Acoustics the most important paper is probably that on Bells (Art. 164). The modes of vibration and the corresponding partial tones of a large number of bells are there recorded.

The next volume will bring the Collection down to about the present time and, it is hoped, may be ready in about a year.

TERLING PLACE, WITHAM,
December 1901.

The works of the Lord are great,
Sought out of all them that have pleasure therein.

CONTENTS.

ILLUSTRATIONS.

142.

ON THE MAINTENANCE OF VIBRATIONS BY FORCES OF DOUBLE FREQUENCY, AND ON THE PROPAGATION OF WAVES THROUGH A MEDIUM ENDOWED WITH A PERIODIC STRUCTURE.

[*Philosophical Magazine*, XXIV. pp. 145—159; 1887.]

THE nature of the question to be first considered may be best explained by a paragraph from a former paper*, in which the subject was briefly treated. "There is also another kind of maintained vibration which, from one point of view, may be regarded as forced, inasmuch as the period is imposed from without, but which differs from the kind just referred to (ordinary forced vibrations) in that the imposed periodic variations do not tend directly to displace the body from its configuration of equilibrium. Probably the best-known example of this kind of action is that form of Melde's experiment in which a fine string is maintained in transverse vibration by connecting one of its extremities with the vibrating prong of a massive tuning-fork, *the direction of motion of the point of attachment being parallel to the length of the string*†. The effect of the motion is to render the tension of the string periodically variable; and at first sight there is nothing to cause the string to depart from its equilibrium condition of straightness. It is known, however, that under these circumstances the equilibrium position may become unstable, and that the string may settle down into a state of permanent and vigorous vibration *whose period is the double of that of the point of attachment*‡." Other examples of acoustical interest are mentioned in the paper.

* "On Maintained Vibrations." *Phil. Mag.* April, 1883, p. 229. [Vol. II. Art. 97.]

† "When the direction of motion is transverse, the case falls under the head of ordinary forced vibrations."

‡ "See Tyndall's *Sound*, 3rd ed. ch. III. § 7, where will also be found a general explanation of the mode of action."

My attention was recalled to the subject by Mr Glaisher's Address to the Astronomical Society*, in which he gives an interesting account of the treatment of mathematically similar questions in the Lunar Theory by Mr Hill† and by Prof. Adams‡. The analysis of Mr Hill is in many respects incomparably more complete than that which I had attempted; but his devotion to the Lunar Theory leads the author to pass by many points of great interest which arise when his results are applied to other physical questions.

By a suitable choice of the unit of time, the equation of motion of the vibrating body may be put into the form

$$\frac{d^2w}{dt^2} + 2k\frac{dw}{dt} + (\Theta_0 + 2\Theta_1 \cos 2t)\, w = 0\,; \quad\quad (1)$$

where k is a positive quantity, which may usually be treated as small, representing the dissipative forces. $(\Theta_0 + 2\Theta_1 \cos 2t)$ represents the coefficient of restitution, which is here regarded as subject to a small imposed periodic variation of period π. Thus Θ_0 is positive, and Θ_1 is to be treated as relatively small.

The equation to which Mr Hill's researches relate is in one respect less general than (1), and in another more general. It omits the dissipative term proportional to k; but, on the other hand, as the Lunar Theory demands, it includes terms proportional to $\cos 4t$, $\cos 6t$, &c. Thus

$$\frac{d^2w}{dt^2} + (\Theta_0 + 2\Theta_1 \cos 2t + 2\Theta_2 \cos 4t + \ldots)\, w = 0\,; \quad\quad (2)$$

or

$$\frac{d^2w}{dt^2} + \Theta w = 0, \quad\quad (3)$$

where

$$\Theta = \Sigma_n \Theta_n e^{2int}, \quad\quad (4)$$

n being any integer, and i representing $\sqrt{(-1)}$. In the present investigation $\Theta_{-n} = \Theta_n$.

It will be convenient to give here a sketch of Mr Hill's method and results. Remarking that when Θ_1, Θ_2, &c. vanish, the solution of (3) is

$$w = Ke^{ict} + K'e^{-ict}, \quad\quad (5)$$

* *Monthly Notices*, Feb. 1887.

† "On the Part of the Motion of the Lunar Perigee which is a Function of the Mean Motions of the Sun and Moon," *Acta Mathematica*, VIII. 1; 1886. Mr Hill's work was first published in 1877.

‡ "On the Motion of the Moon's Node, in the case when the orbits of the Sun and Moon are supposed to have no Eccentricities, and when their Mutual Inclination is supposed to be indefinitely small." *Monthly Notices*, Nov. 1877.

where K, K' are arbitrary constants, and $c = \sqrt{(\Theta_0)}$, he shows that in the general case we may assume as a particular solution

$$w = \Sigma_n b_n e^{ict + 2int}, \qquad \text{.................................(6)}$$

the value of c being modified by the operation of Θ_1, &c., and the original term $b_0 e^{ict}$ being accompanied by subordinate terms corresponding to the positive and negative integral values of n.

The multiplication by Θ, as given in (4), does not alter the form of (6); and the result of the substitution in the differential equation (3) may be written

$$(c + 2m)^2 b_m - \Sigma_n \Theta_{m-n} b_n = 0, \qquad \text{....................(7)}$$

which holds for all integral values of m, positive and negative. These conditions determine the ratios of all the coefficients b_n to one of them, *e.g.*, b_0, which may then be regarded as the arbitrary constant. They also determine c, the main subject of quest. Mr Hill writes

$$[n] = (c + 2n)^2 - \Theta_0; \qquad \text{........................(8)}$$

so that the equations take the form

$$\left.\begin{array}{l}
\dots\dots\dots\dots\dots\dots\dots\dots\dots\dots\dots\dots\dots\dots\dots \\
\dots + [-2]\, b_{-2} - \Theta_1\, b_{-1} - \Theta_2\, b_0 - \Theta_3\, b_1 - \Theta_4\, b_2 - \dots = 0, \\
\dots - \Theta_1\, b_{-2} + [-1]\, b_{-1} - \Theta_1\, b_0 - \Theta_2\, b_1 - \Theta_3\, b_2 - \dots = 0, \\
\dots - \Theta_2\, b_{-2} - \Theta_1\, b_{-1} + [0]\, b_0 - \Theta_1\, b_1 - \Theta_2\, b_2 - \dots = 0, \\
\dots - \Theta_3\, b_{-2} - \Theta_2\, b_{-1} - \Theta_1\, b_0 + [1]\, b_1 - \Theta_1\, b_2 - \dots = 0, \\
\dots - \Theta_4\, b_{-2} - \Theta_3\, b_{-1} - \Theta_2\, b_0 - \Theta_1\, b_1 + [2]\, b_2 - \dots = 0, \\
\dots\dots\dots\dots\dots\dots\dots\dots\dots\dots\dots\dots\dots\dots\dots
\end{array}\right\} \text{......(9)}$$

The determinant formed by eliminating the b's from these equations is denoted by $\mathfrak{D}(c)$; so that the equation from which c is to be found is

$$\mathfrak{D}(c) = 0. \qquad \text{..............................(10)}$$

The infinite series of values of c determined by (10) cannot give independent solutions of (3),—a differential equation of the second order only. It is evident, in fact, that the system of equations by which c is determined is not altered if we replace c by $c + 2\nu$, where ν is any positive or negative integer. Neither is any change incurred by the substitution of $-c$ for c. "It follows that if (10) is satisfied by a root $c = c_0$, it will also have, as roots, all the quantities contained in the expression

$$\pm c_0 + 2n,$$

where n is any positive or negative integer or zero. And these are all the roots the equation admits of; for each of the expressions denoted by $[n]$ is of two dimensions in c, and may be regarded as introducing into the equation the two roots $2n + c_0$ and $2n - c_0$. Consequently the roots are either all real or all imaginary; and it is impossible that the equation should have any equal root unless all the roots are integral."

On these grounds Mr Hill concludes that $\mathfrak{D}(c)$ must be such that

$$\mathfrak{D}(c) = A\,[\cos(\pi c) - \cos(\pi c_0)] \quad\text{.....................(11)}$$

identically, where A is some constant independent of c; whence on putting $c = 0$,

$$\mathfrak{D}(0) = A\,[1 - \cos(\pi c_0)], \quad\text{...........................(12)}$$

in which, if we please, c_0 may be replaced by c. The value of A may now be determined by comparison with the particular case $\Theta_1 = 0$, $\Theta_2 = 0$, &c., for which of course $c = \sqrt{\Theta_0}$. Thus if $\mathfrak{D}'(0)$ denote the special form then assumed, *i.e.* the simple product of the diagonal constituents,

$$\mathfrak{D}'(0) = A\,[1 - \cos(\pi\sqrt{\Theta_0})], \quad\text{........................(13)}$$

and

$$\frac{1-\cos(\pi c)}{1-\cos(\pi\sqrt{\Theta_0})} = \frac{\sin^2(\frac{1}{2}\pi c)}{\sin^2(\frac{1}{2}\pi\sqrt{\Theta_0})} = \frac{\mathfrak{D}(0)}{\mathfrak{D}'(0)} \quad\text{...............(14)}$$

The fraction $\mathfrak{D}(0) \div \mathfrak{D}'(0)$ is denoted by $\square(0)$. It is the determinant formed from the original one by dividing each row by the constituent in the diagonal, so as to reduce all the diagonal constituents to unity, and by making c vanish. Thus

$$\frac{1-\cos(\pi c)}{1-\cos(\pi\sqrt{\Theta_0})} = \square(0), \quad\text{.......................(15)}$$

where

$$\square(0) = \left|\begin{array}{ccccccc}
 & \cdots & \cdots & \cdots & \cdots & \cdots & \\
\cdots + & 1 & -\dfrac{\Theta_1}{4^2-\Theta_0} & -\dfrac{\Theta_2}{4^2-\Theta_0} & -\dfrac{\Theta_3}{4^2-\Theta_0} & -\dfrac{\Theta_4}{4^2-\Theta_0} & \cdots \\
\cdots & -\dfrac{\Theta_1}{2^2-\Theta_0} & +\;1 & -\dfrac{\Theta_1}{2^2-\Theta_0} & -\dfrac{\Theta_2}{2^2-\Theta_0} & -\dfrac{\Theta_3}{2^2-\Theta_0} & \cdots \\
\cdots & -\dfrac{\Theta_2}{0^2-\Theta_0} & -\dfrac{\Theta_1}{0^2-\Theta_0} & +\;1 & -\dfrac{\Theta_1}{0^2-\Theta_0} & -\dfrac{\Theta_2}{0^2-\Theta_0} & \cdots \\
\cdots & -\dfrac{\Theta_3}{2^2-\Theta_0} & -\dfrac{\Theta_2}{2^2-\Theta_0} & -\dfrac{\Theta_1}{2^2-\Theta_0} & +\;1 & -\dfrac{\Theta_1}{2^2-\Theta_0} & \cdots \\
\cdots & -\dfrac{\Theta_4}{4^2-\Theta_0} & -\dfrac{\Theta_3}{4^2-\Theta_0} & -\dfrac{\Theta_2}{4^2-\Theta_0} & -\dfrac{\Theta_1}{4^2-\Theta_0} & +\;1 & \cdots \\
 & \cdots & \cdots & \cdots & \cdots & \cdots &
\end{array}\right| \quad\text{....(16)}$$

The value of $\square(0)$ is calculated for the purposes of the Lunar Theory to a high order of approximation. It will here suffice to give the part which depends upon the squares of Θ_1, Θ_2, &c. Thus

$$\square(0) = 1 + \frac{\pi\cot(\frac{1}{2}\pi\sqrt{\Theta_0})}{4\sqrt{\Theta_0}}\left[\frac{\Theta_1^2}{1-\Theta_0} + \frac{\Theta_2^2}{4-\Theta_0} + \frac{\Theta_3^2}{9-\Theta_0} + \cdots\right]. \quad (17)$$

Another determinant, $\nabla(0)$, is employed by Mr Hill, the relation of which to $\square(0)$ is expressed by

$$\nabla(0) = 2\sin^2(\tfrac{1}{2}\pi\sqrt{\Theta_0}) \,.\, \square(0)\,; \quad\text{.....................(18)}$$

so that the general solution for c may be written

$$\cos(\pi c) = 1 - \nabla(0). \quad \text{.........................(19)}$$

Mr Hill observes that the reality of c requires that $1 - \nabla(0)$ should lie between -1 and $+1$. In the Lunar Theory this condition is satisfied; but in the application to Acoustics the case of an imaginary c is the one of greater interest, for the vibrations then tend to increase indefinitely.

Cos(πc) being itself always real, let us suppose that πc is complex, so that

$$c = \alpha + i\beta,$$

where α and β are real. Thus

$$\cos \pi c = \cos \pi\alpha \cos i\pi\beta - \sin \pi\alpha \sin i\pi\beta;$$

and the reality of $\cos \pi c$ requires either (1) that $\beta = 0$, or (2) that $\alpha = n$, n being an integer. In the first case c is real. In the second

$$\cos \pi c = \pm \cos i\pi\beta = 1 - \nabla(0), \quad \text{.....................(20)}$$

which gives but one (real) value of β. If $1 - \nabla(0)$ be positive,

$$c = \pm i\beta + 2n; \quad \text{...............................(21)}$$

but if $1 - \nabla(0)$ be negative,

$$\cos \pi c = -\cos i\pi\beta,$$

whence

$$c = \pm i\beta + 2n + 1. \quad \text{.............................(22)}$$

The latter is the case with which we have to do when Θ_0, and therefore c, is nearly equal to unity; and the conclusion that when c is complex, the real part is independent of Θ_1, Θ_2, &c. is of importance. The complete value of w may then be written

$$w = e^{\beta t} \Sigma b_n e^{it(1+2n)} + e^{-\beta t} \Sigma b_n' e^{it(1+2n)}, \quad \text{.................(23)}$$

the ratios of b_n and also of b_n' being determined by (9). After the lapse of a sufficient time, the second set of terms in $e^{-\beta t}$ become insignificant.

In the application of greatest acoustical interest Θ_0 (and c) are nearly equal to unity; so that the free vibrations are performed with a frequency about the half of that introduced by Θ_1. In this case the leading equations in (9) are those which involve the small quantities [0] and [−1]; but for the sake of symmetry, it is advisable to retain also the equation containing [1]. If we now neglect Θ_2, as well as the b's whose suffix is numerically greater than unity, we find

$$\frac{b_{-1}}{\Theta_1 [1]} = \frac{b_0}{[1][-1]} = \frac{b_1}{\Theta_1 [-1]}, \quad \text{....................(24)}$$

and

$$[0][1][-1] - \Theta_1^2 \{[1] + [-1]\} = 0. \quad \text{..................(25)}$$

For the sake of distinctness it will be well to repeat here that

$$[0] = c^2 - \Theta_0, \qquad [-1] = (c-2)^2 - \Theta_0, \qquad [1] = (c+2)^2 - \Theta_0.$$

Substituting these values in (25), Mr Hill obtains

$$(c^2 - \Theta_0)\{(c^2 + 4 - \Theta_0)^2 - 16c^2\} - 2\Theta_1^2\{c^2 + 4 - \Theta_0\} = 0,$$

and neglecting the cube of $(c^2 - \Theta_0)$, as well as its product with Θ_1^2,

$$(c^2 - \Theta_0)^2 + 2(\Theta_0 - 1)(c^2 - \Theta_0) + \Theta_1^2 = 0;$$

and from this again

$$c^2 = 1 + \sqrt{\{(\Theta_0 - 1)^2 - \Theta_1^2\}}. \quad\text{.......................(26)}$$

It appears, therefore, that c is real or imaginary according as $(\Theta_0 - 1)^2$ is greater or less than Θ_1^2. In the problem of the Moon's apse, treated by Mr Hill,

$$\Theta_0 = 1{\cdot}1588439, \qquad \Theta_1 = -0{\cdot}0570440;$$

and in the corresponding problem of the node, investigated by Prof. Adams,

$$\Theta_0 = 1{\cdot}17804,44973,149,$$
$$\Theta_1 = 0{\cdot}01261,68354,6.$$

In both these cases the value of c is real, though of course not to be accurately determined by (26).

Mr Hill's results are not immediately applicable to the acoustical problem embodied in (1), in consequence of the omission of k, representing the dissipation to which all actual vibrations are subject. The inclusion of this term leads, however, merely to the substitution for $(c + 2n)^2 - \Theta_0$ in (8) of

$$(c + 2n)^2 - 2ik(c + 2n) - \Theta_0;$$

so that the whole operation of k is represented if we write $(c - ik)$ in place of c, and $(\Theta_0 - k^2)$ in place of Θ_0. Accordingly

$$\cos\pi(c - ik) = 1 - \nabla'(0), \quad\text{.......................(27)}$$

$\nabla'(0)$ differing from $\nabla(0)$ only by the substitution of $\Theta_0 - k^2$ for Θ_0.

If $1 - \nabla'(0)$ lies between ± 1, $(c - ik)$ is real, so that

$$c = ik \pm \alpha + 2n. \quad\text{.............................(28)}$$

In this case both solutions are affected with the factor e^{-kt}, indicating that whatever the initial circumstances may be, the motion dies away.

It may be otherwise when $1 - \nabla'(0)$ lies beyond the limits ± 1. In the case of most importance, when Θ_0 is nearly equal to unity, $1 - \nabla'(0)$ is algebraically less than -1. If

$$\cos i\pi\beta = -1 + \nabla'(0), \quad\text{.......................(29)}$$

we may write

$$c = 1 + i(k \pm \beta) + 2n. \quad\text{.......................(30)}$$

Here again both motions die down unless β is numerically greater than k, in which case one motion dies down, while the other increases without limit. The critical relation may be written

$$\cos(i\pi k) = -1 + \nabla'(0). \qquad (31)$$

From (30) we see that, whatever may be the value of k, the vibrations (considered apart from the rise or subsidence indicated by the exponential factors) have the same frequency as if k, as well as Θ_1, Θ_2, &c. vanished.

Before leaving the general theory it may be worth while to point out that Mr Hill's method may be applied when the coefficients of d^2w/dt^2 and dw/dt, as well as of w, are subject to given periodic variations. We may write

$$\Phi \frac{d^2w}{dt^2} + \Psi \frac{dw}{dt} + \Theta w = 0, \qquad (32)$$

where $\Phi = \Sigma\Phi_n e^{2int}, \quad \Psi = \Sigma\Psi_n e^{2int}, \quad \Theta = \Sigma\Theta_n e^{2int}$. .... (33)

Assuming, as before, $w = \Sigma_n b_n e^{ict+2int}$, ... (34)

we obtain, on substitution, as the coefficient of $e^{ict+2imt}$,

$$-\Sigma_n b_n (c+2n)^2 \Phi_{m-n} + i\Sigma_n b_n (c+2n)\Psi_{m-n} + \Sigma b_n \Theta_{m-n},$$

which is to be equated to zero. The equation for c may still be written

$$\mathfrak{D}(c) = 0, \qquad (35)$$

where

$$\mathfrak{D}(c) = \left| \begin{matrix} \cdots & \cdots & \cdots & \cdots & \cdots & \cdots & \cdots \\ \ldots & [-2, 0], & [-1, -1], & [0, -2], & [1, -3], & [2, -4], & \ldots \\ \ldots & [-2, 1], & [-1, 0], & [0, -1], & [1, -2], & [2, -3], & \ldots \\ \ldots & [-2, 2], & [-1, 1], & [0, 0], & [1, -1], & [2, -2], & \ldots \\ \ldots & [-2, 3], & [-1, 2], & [0, 1], & [1, 0], & [2, -1], & \ldots \\ \ldots & [-2, 4], & [-1, 3], & [0, 2], & [1, 1], & [2, 0], & \ldots \\ \cdots & \cdots & \cdots & \cdots & \cdots & \cdots & \cdots \end{matrix} \right| \qquad (36)$$

and $[n, r] = (c+2n)^2 \Phi_r - i(c+2n)\Psi_r - \Theta_r$. .... (37)

By similar reasoning to that employed by Mr Hill we may show that

$$\mathfrak{D}(c) = A(\cos\pi c - \cos\pi c_0) + B(\sin\pi c - \sin\pi c_0)\ldots,$$

where A and B are constants independent of c; and, further, that

$$\mathfrak{D}(0) = A(1 - \cos\pi c) - B\sin\pi c. \qquad (38)$$

If all the quantities Φ_r, Ψ_r, Θ_r vanish except Φ_0, Ψ_0, Θ_0, $\mathfrak{D}(0)$ reduces to the diagonal row simply, say $\mathfrak{D}'(0)$. Let c_1, c_2 be the roots of

$$\Phi_0 \frac{d^2w}{dt^2} + \Psi_0 \frac{dw}{dt} + \Theta_0 w = 0, \qquad (39)$$

then $\mathfrak{D}'(0) = A(1 - \cos\pi c_1) - B\sin\pi c_1 = A(1 - \cos\pi c_2) - B\sin\pi c_2$;

so that the equation for c may be written

$$\begin{vmatrix} \mathfrak{D}(0), & 1-\cos\pi c, & \sin\pi c, \\ \mathfrak{D}'(0), & 1-\cos\pi c_1, & \sin\pi c_1, \\ \mathfrak{D}'(0), & 1-\cos\pi c_2, & \sin\pi c_2, \end{vmatrix} = 0. \qquad (40)$$

In this equation $\mathfrak{D}(0) \div \mathfrak{D}'(0)$ is the determinant derived from $\mathfrak{D}(0)$ by dividing each row so as to make the diagonal constituent unity.

If $\ldots\Psi_{-1}, \Psi_0, \Psi_1\ldots$ vanish (even though $\ldots\Phi_{-1}, \Phi_0, \Phi_1\ldots$ remain finite), $\mathfrak{D}(c)$ is an even function of c, and the coefficient B vanishes in (38). In this case we have simply

$$\frac{1-\cos\pi c}{1-\cos\pi\sqrt{\Theta_0}} = \frac{\mathfrak{D}(0)}{\mathfrak{D}'(0)},$$

exactly as when $\Phi_1, \Phi_{-1}, \Phi_2, \Phi_{-2}\ldots$ vanish.

Reverting to (24), we have as the approximate particular solution, when there is no dissipation,

$$w = \frac{e^{(c-2)it}}{(c-2)^2-\Theta_0} + \frac{e^{cit}}{\Theta_1} + \frac{e^{(c+2)it}}{(c+2)^2-\Theta_0}. \qquad (41)$$

If c be real, the solution may be completed by the addition of a second, found from (41) by changing the sign of c. Each of these solutions is affected with an arbitrary constant multiplier. The realized general solution may be written

$$w = \frac{R\cos(c-2)t + S\sin(c-2)t}{(c-2)^2-\Theta_3} + \frac{R\cos ct + S\sin ct}{\Theta_1} + \frac{R\cos(c+2)t + S\sin(c+2)t}{(c+2)^2-\Theta_0}, \qquad (42)$$

from which the last term may usually be omitted, in consequence of the relative magnitude of its denominator. In this solution c is determined by (26).

When c^2 is imaginary, we take

$$4s^2 = \Theta_1^2 - (\Theta_0 - 1)^2; \qquad (43)$$

so that

$$c^2 = 1 + 2is, \quad c = 1 + is, \quad c - 2 = -1 + is.$$

The particular solution may be written

$$w = e^{-st}\{\Theta_1 e^{-it} + (1-\Theta_0 - 2is)\,e^{it}\}; \qquad (44)$$

or, in virtue of (43),

$$w = e^{-st}\{(1-\Theta_0+\Theta_1)\cos t + 2s\sin t\}; \qquad (45)$$

or, again,

$$w = e^{-st}\{\sqrt{(\Theta_1+1-\Theta_0)}\,.\cos t + \sqrt{(\Theta_1-1+\Theta_0)}\,.\sin t\}. \qquad (46)$$

The general solution is

$$\left. \begin{array}{l} w = R\, e^{-st} \{(1 - \Theta_0 + \Theta_1) \cos t + 2s \sin t\} \\ \quad + S\, e^{+st} \{(1 - \Theta_0 + \Theta_1) \cos t - 2s \sin t\} \end{array} \right\}, \quad \text{.........(47)}$$

R, S being arbitrary multipliers.

One or two particular cases may be noticed. If $\Theta_0 = 1$, $2s = \Theta_1$, and

$$w = R' e^{-st} \{\cos t + \sin t\} + S' e^{st} \{\cos t - \sin t\}. \quad \text{.........(48)}$$

Again, suppose that

$$\Theta_1^2 = (\Theta_0 - 1)^2, \quad \text{.........(49)}$$

so that s vanishes, giving the transition between the real and imaginary values of c. Of the two terms in (46), one or other preponderates indefinitely in the two alternatives. Thus, if $\Theta_1 = 1 - \Theta_0$, the solution reduces to $\cos t$; but if $\Theta_1 = -1 + \Theta_0$, it reduces to $\sin t$. The apparent loss of generality by the merging of the two solutions may be repaired in the usual way by supposing s infinitely small.

When there are dissipative forces, we are to replace c by $(c - ik)$, and Θ by $(\Theta_0 - k^2)$; but when k is small the latter substitution may be neglected. Thus, from (26),

$$c = 1 + ik + \tfrac{1}{2} \sqrt{\{(\Theta_0 - 1)^2 - \Theta_1^2\}}. \quad \text{.........(50)}$$

Interest here attaches principally to the case where the radical is imaginary; otherwise the motion necessarily dies down. If, as before,

$$4s^2 = \Theta_1^2 - (\Theta_0 - 1)^2, \quad \text{.........(51)}$$

$$c = 1 + ik + is, \qquad c - 2 = -1 + ik + is, \quad \text{.........(52)}$$

and

$$w = \frac{e^{(c-2)it}}{(c - ik - 2)^2 - \Theta_0} + \frac{e^{cit}}{\Theta_1},$$

or

$$w = e^{-(k+s)t} \{\Theta_1 e^{-it} + (1 - \Theta_0 - 2is) e^{it}\},$$

or

$$w = e^{-(k+s)t} \{(1 - \Theta_0 + \Theta_1) \cos t + 2s \sin t\}. \quad \text{.........(53)}$$

This solution corresponds to a motion which dies away.

The second solution (found by changing the sign of s) is

$$w = e^{(s-k)t} \{(1 - \Theta_0 + \Theta_1) \cos t - 2s \sin t\}. \quad \text{.........(54)}$$

The motion dies away or increases without limit according as s is less or greater than k.

The only case in which the motion is periodic is when $s = k$, or

$$4k^2 = \Theta_1^2 - (\Theta_0 - 1)^2; \quad \text{.........(55)}$$

and then

$$w = (1 - \Theta_0 - \Theta_1) \cos t - 2k \sin t. \quad \text{.........(56)}$$

These results, under a different notation, were given in my former paper*.

If $\Theta_0 = 1$, we have by (51), $2s = \Theta$; and from (53), (54),

$$w = Re^{-(k+s)t}\{\cos t + \sin t\} + Se^{-(k-s)t}\{\cos t - \sin t\}. \quad\text{.........(57)}$$

In the former paper some examples were given drawn from ordinary mechanics and acoustics. To these may be added the case of a stretched wire, whose tension is rendered periodically variable by the passage through it of an intermittent electric current. It is probable that an illustration might be arranged in which the vibrations are themselves electrical. Θ_0 would then represent the stiffness of a condenser, Ψ_0 resistance, and Φ_0 self-induction. The most practicable way of introducing the periodic term would be by rendering the self-induction variable with the time (Φ_1). This could be effected by the rotation of a coil forming part of the circuit.

The discrimination of the real and imaginary values of c is of so much importance, that it is desirable to pursue the approximation beyond the point attained in (26). From (11) we find

$$\frac{\mathfrak{D}(1)}{\mathfrak{D}'(1)} = \frac{1 + \cos(\pi c)}{1 + \cos(\pi\sqrt{\Theta_0})}; \quad\text{......................(58)}$$

from which, or directly, we see that if $c = 1$, corresponding to the transition case between real and imaginary values,

$$\mathfrak{D}(1) = 0. \quad\text{..............................(59)}$$

If, as we shall now suppose, $\Theta_2, \Theta_3 \ldots$ vanish, (59) may be written in the form

$$\begin{vmatrix} \cdots & \cdots & \cdots & \cdots & \cdots & \cdots \\ \ldots 1, & a_2, & 1, & 0, & 0, & 0 \ldots \\ \ldots 0, & 1, & a_1, & 1, & 0, & 0 \ldots \\ \ldots 0, & 0, & 1, & a_1, & 1, & 0 \ldots \\ \ldots 0, & 0, & 0, & 1, & a_2, & 1 \ldots \\ \cdots & \cdots & \cdots & \cdots & \cdots & \cdots \end{vmatrix} = 0, \quad\text{...........(60)}$$

where

$$a_1 = \frac{\Theta_0 - 1}{\Theta_1}, \qquad a_2 = \frac{\Theta_0 - 9}{\Theta_1}, \qquad a_3 = \frac{\Theta_0 - 25}{\Theta_1}. \quad\text{.........(61)}$$

The first approximation, equivalent to (26), is found by considering merely the central determinant of the second order involving only a_1; thus,

$$a_1^2 - 1 = 0. \quad\text{...............................(62)}$$

The second approximation is

$$a_2^2\{(a_1 - 1/a_2)^2 - 1\} = 0. \quad\text{.......................(63)}$$

* In consequence of an error of sign, the result for a second approximation there stated is incorrect [rectified in reprint Art. 97].

The third is

$$a_3^2\{a_2-1/a_3\}^2\left\{\left(a_1-\frac{1}{a_2-1/a_3}\right)^2-1\right\}=0, \quad\text{..............(64)}$$

and so on. The equation (60) is thus equivalent to

$$a_1-\frac{1}{a_2-}\,\frac{1}{a_3-}\,\frac{1}{a_4-}\dots=\pm 1; \quad\text{....................(65)}$$

and the successive approximations are

$$N_1=\pm D_1, \qquad N_2=\pm D_2, \quad \text{\&c.}, \quad\text{..................(66)}$$

where

$$N_1/D_1, \qquad N_2/D_2, \qquad \text{\&c.}$$

are the corresponding convergents to the infinite continued fraction*

In terms of Θ_0, Θ_1, the second approximation to the equation discriminating the real and imaginary values of c is

$$(\Theta_0-1)(\Theta_0-9)-\Theta_1^2=\pm\,\Theta_1(\Theta_0-9). \quad\text{..............(67)}$$

One of the most interesting applications of the foregoing analysis is to the case of a laminated medium in which the mechanical properties are periodic functions of one of the coordinates. I was led to the consideration of this problem in connexion with the theory of the colours of thin plates. It is known that old, superficially decomposed, glass presents reflected tints much brighter, and transmitted tints much purer, than any of which a single transparent film is capable. The laminated structure was proved by Brewster; and it is easy to see how the effect may be produced by the occurrence of nearly similar laminæ at nearly equal intervals. Perhaps the simplest case of the kind that can be suggested is that of a stretched string, periodically loaded, and propagating transverse vibrations. We may imagine similar small loads to be disposed at equal intervals. If, then, the wave-length of a train of progressive waves be approximately equal to the *double* interval between the loads, the partial reflexions from the various loads will all concur in phase, and the result must be a powerful aggregate reflexion, even though the effect of an individual load may be insignificant.

The general equation of vibration for a stretched string of periodic density is

$$\left(\rho_0+\rho_1\cos\frac{2\pi x}{l}+\rho_1'\sin\frac{2\pi x}{l}+\rho_2\cos\frac{4\pi x}{l}\right.$$
$$\left.+\,\rho_2'\sin\frac{4\pi x}{l}+\dots\right)\frac{d^2w}{dt^2}=T\frac{d^2w}{dx^2}, \quad\text{........(68)}$$

* The relations of determinants of this kind to continued fractions has been studied by Muir (*Edinb. Proc.* vol. VIII.).

l being the distance in which the density is periodic. We shall suppose that $\rho_1', \rho_2', \ldots$ vanish, so that the sines disappear, a supposition which involves no loss of generality when we restrict ourselves to a simple harmonic variation of density. If we now assume that $w \propto e^{ipt}$, or $\propto \cos pt$, we obtain

$$\frac{d^2w}{d\xi^2} + (\Theta_0 + 2\Theta_1 \cos 2\xi + 2\Theta_2 \cos 4\xi + \ldots) w = 0, \quad \ldots\ldots\ldots(69)$$

where $\xi = \pi x/l$, and

$$\Theta_0 = \frac{p^2 l^2 \rho_0}{\pi^2 T}, \qquad 2\Theta_1 = \frac{p^2 l^2 \rho_1}{\pi^2 T}, \qquad \text{\&c.}; \ldots\ldots\ldots\ldots\ldots\ldots(70)$$

and this is of the form of Mr Hill's equation (2).

When c is real, we may employ the approximate solutions (41), (44). The latter (with ξ written for t) gives, when multiplied by $\cos pt$ or $\sin pt$, the stationary vibrations of the system. From (41) we get

$$w = \frac{\cos [pt + (c-2)\xi]}{(c-2)^2 - \Theta_0} + \frac{\cos [pt + c\xi]}{\Theta_1}, \ldots\ldots\ldots\ldots\ldots(71)$$

in which, if $c = 1$ nearly, the two terms represent waves progressing with nearly equal velocities in the two directions. Neither term gains permanently in relative importance as x is increased or diminished indefinitely.

It is otherwise when the relation of Θ_0 to Θ_1 is such that c is imaginary. By (44) the solution for w, assumed to be proportional to e^{ipt}, now takes the form

$$w = Re^{-s\xi} \{\Theta_1 e^{i(pt-\xi)} + (1 - \Theta_0 - 2is) e^{i(pt+\xi)}\}$$
$$+ Se^{+s\xi} \{\Theta_1 e^{i(pt-\xi)} + (1 - \Theta_0 + 2is) e^{i(pt+\xi)}\}. \ldots\ldots\ldots\ldots(72)$$

Whatever may be the relative values of R and S, the first solution preponderates when x is large and negative, and the second preponderates when x is large and positive. In either extreme case the motion is composed of two progressive waves moving in opposite directions, *whose amplitudes are equal in virtue of* (43).

The meaning of this is that a wave travelling in either direction is ultimately totally reflected. For example, we may so choose the values of R and S that at the origin of x there is a wave (of given strength) in the positive direction only, and we may imagine that it here passes into a uniform medium, and so is propagated on indefinitely without change. But, in order to maintain this state of things, we have to suppose on the negative side the coexistence of positive and negative waves, which at sufficient distances from the origin are of nearly equal and ever-increasing amplitudes. In order therefore that a small wave may emerge at $x = 0$, we have to cause intense waves to be incident upon a face of the medium corresponding to a large negative x, of which nearly the whole are reflected.

It is important to observe that the ultimate totality of reflexion does not require a special adjustment between the frequency of the waves and the linear period of the lamination. The condition that c should be imaginary is merely that Θ_1 should numerically exceed $(1 - \Theta_0)$. If λ be the wave-length of the vibration corresponding to e^{ipt} and to density ρ_0,

$$\frac{p^2\rho_0}{\pi^2 T} = \frac{4}{\lambda^2}; \quad \text{.................................(73)}$$

and thus the limits between real and imaginary values of c are given by

$$\frac{\lambda^2}{4l^2} - 1 = \pm \frac{\rho_1}{2\rho_0}. \quad \text{.................................(74)}$$

If ρ_1 exceeds these limits a train of waves is ultimately totally reflected, in spite of the finite difference between $\frac{1}{2}\lambda$ and l*.

In conclusion, it may be worth while to point out the application to such a problem as the stationary vibrations of a string of variable density fixed at two points. A distribution of density,

$$\rho_0 + \rho_1 \cos\frac{2\pi x}{l} + \rho_2 \cos\frac{4\pi x}{l} + \ldots \quad \text{..................(75)}$$

is symmetrical with respect to the points $x = 0$ and $x = \frac{1}{2}l$, and between those limits is arbitrary. It is therefore possible for a string of this density to vibrate with the points in question undisturbed, and the law of displacement will be

$$w = \cos pt \left\{A_1 \sin\frac{2\pi x}{l} + A_2 \sin\frac{4\pi x}{l} + A_3 \sin\frac{6\pi x}{l} + \ldots\right\}. \quad \text{...(76)}$$

When, therefore, the problem is attacked by the method of Mr Hill, the value of c obtained by the solution of (69) must be equal to 2. By (15) this requires

$$\square\,(0) = 0. \quad \text{.................................(77)}$$

* A detailed experimental examination of various cases in which a laminated structure leads to a powerful but highly selected reflexion would be of value. The most frequent examples are met with in the organic world. It has occurred to me that Becquerel's reproduction of the spectrum in natural colours upon silver plates may perhaps be explicable in this manner. The various parts of the film of subchloride of silver with which the metal is coated may be conceived to be subjected, during exposure, to *stationary* luminous waves of nearly definite wave-length, the effect of which might be to impress upon the substance a periodic structure recurring at intervals equal to *half* the wave-length of the light; just as a sensitive flame exposed to stationary sonorous waves is influenced at the loops but not at the nodes (*Phil. Mag.* March, 1879, p. 153). [Vol. I. p. 406.] In this way the operation of any kind of light would be to produce just such a modification of the film as would cause it to reflect copiously that particular kind of light. I abstain at present from developing this suggestion, in the hope of soon finding an opportunity of making myself experimentally acquainted with the subject. [1900. I need hardly remind the reader of the beautiful coloured photographs which M. Lippmann has since obtained by this method.]

This equation gives a relation between the quantities Θ_0, Θ_1, Θ_2, ... ; and this again, by (70), determines p, or the frequency $(p/2\pi)$ of vibration.

Since $\Theta_0 = 4$ nearly, the most important term in (17) is that involving Θ_2^2. The first approximation to (77) gives

$$\Theta_0 = 4 + \Theta_2;$$

whence, by (70),

$$\left(\frac{2\pi}{p}\right)^2 = \frac{l^2(\rho_0 - \frac{1}{2}\rho_2)}{T}. \qquad (78)$$

To this order of approximation the solution may be obtained with far greater readiness by the method given in my work on Sound*; but it is probable that, if the solution were required in a case where the variation of density is very considerable, advantage might be taken of Mr Hill's determinant □ (0). There are doubtless other physical problems to which a similar remark would be applicable.

* *Theory of Sound*, vol. I. § 140. In comparing the results, it must be borne in mind that the length of the string in (78) is denoted by $\frac{1}{2}l$.

143.

ON THE EXISTENCE OF REFLECTION WHEN THE RELATIVE REFRACTIVE INDEX IS UNITY.

[*British Association Report*, pp. 585, 586; 1887.]

THE copious undisturbed transmission of light by glass powder when surrounded by liquid of the same index, as in Christiansen's experiment [vol. II. p. 433], suggests the question whether the reflection of any particular ray is really annihilated when the relative refractive index is unity for that ray. Such would be the case according to Fresnel's formulæ, but these are known to be in some respects imperfect. Mechanical theory would indicate that when there is dispersion, reflection would cease to be merely a function of the index or ratio of wave-velocities. We may imagine a stretched string vibrating transversely under the influence of tension, and in a subordinate degree of stiffness, to be composed of two parts so related to one another in respect of mass and stiffness that the wave-velocity is the same in both parts for a specified wave-length. But, as it is easy to see, this adjustment will not secure the complete transmission of a train of progressive waves incident upon the junction, even when the wave-length is precisely that for which the velocities are the same.

The experiments that I have tried have been upon plate glass immersed in a mixture of bisulphide of carbon and benzole, of which the first is more refractive and the second less refractive than the glass; and it was found that the reflection of a candle-flame from a carefully cleaned plate remained pretty strong at moderate angles of incidence, in whatever proportions the liquids were mixed.

For a closer examination the plate was roughened behind (to destroy the second reflection), and was mounted in a bottle prism in such a manner that the incidence could be rendered grazing. When the adjustment of indices was for the yellow, the appearances observed were as follows: if the incidence is pretty oblique, the reflection is total for the violet and blue; scanty, but not evanescent, for the yellow; more copious again in the red. As the incidence becomes more and more nearly grazing, the region of total reflection advances from the blue end closer and closer upon the ray of equal index, and ultimately there is a very sharp transition between this region and the band which now looks very dark. On the other side the reflection revives, but more gradually, and becomes very copious in the orange and red. On this side the reflection is not technically total. If the prism be now turned so that the angle of incidence is moderate, it is found that, in spite of the equality of index for the most luminous part of the spectrum, there is a pretty strong reflection of a candle-flame, and apparently without colour. With the aid of sunlight it was proved that in the reflection at moderate incidences there was no marked chromatic selection, and in all probability the blackness of the band in the yellow at grazing incidences is a matter of contrast only.

Indeed calculation shows that, according to Fresnel's formulæ, the reflection would be nearly insensible at all parts of the spectrum when the index is adjusted for the yellow. The outstanding reflection is not due to a difference of wave-velocities, but to some other cause not usually taken into account.

Such a cause might be found in the presence of a film upon the surface of the glass, of index differing from that of the interior, and not removable by mere cleaning. The glass plate was accordingly repolished with putty powder, after which the reflection was very decidedly diminished. But neither by this nor by any other treatment (*e.g.* with hydrofluoric acid) has it been found possible to render the reflection of a candle-flame at moderate incidences even difficult of observation although the adjustment of indices was as good as could be.

It would, however, be hardly safe to conclude that no sufficient film was operative; and I do not see how the question is to be decided unless an experiment can be made upon a surface freshly obtained by fracture.

[1899. At the suggestion of Lord Kelvin I have lately repeated these observations. The residual light reflected at 45° incidence is polarised in the usual way, *i.e.* as if it were reflected from an interface between two media of slightly differing indices.]

144.

ON THE STABILITY OR INSTABILITY OF CERTAIN FLUID MOTIONS, II.

[*Proceedings of the London Mathematical Society*, XIX. pp. 67—74; 1887.]

As the question of the stability, or otherwise, of fluid motions is attracting attention in consequence of Sir W. Thomson's recent work, I think it advisable to point out an error in the solution which I gave some years ago* of one of the problems relating to this subject; and I will take the opportunity to treat the problem with greater generality.

In the steady laminated motion, the velocity (U) is a function of y only. In the disturbed motion $U+u$, v, the small quantities u, v are supposed to be periodic functions of x, proportional to e^{ikx}, and, as dependent upon the time, to be proportional to e^{int}, where n is a constant, real or imaginary. Under these circumstances the equation determining v (51) is

$$\left(\frac{n}{k}+U\right)\left(\frac{d^2v}{dy^2}-k^2v\right)-\frac{d^2U}{dy^2}v=0. \qquad (1)$$

The vorticity (Z) of the steady motion is $\frac{1}{2}dU/dy$. If throughout any layer Z be constant, d^2U/dy^2 vanishes, and wherever $n+kU$ does not also vanish

$$\frac{d^2v}{dy^2}-k^2v=0, \qquad (2)$$

or

$$v=Ae^{ky}+Be^{-ky}. \qquad (3)$$

If there are several layers in each of which Z is constant, the various solutions of the form (3) are to be fitted together, the arbitrary constants being so chosen as to satisfy certain boundary conditions. The first of these conditions is evidently

$$\Delta v=0. \qquad (4)\dagger$$

* *Math. Soc. Proc.* XI. p. 57; 1880. [Vol. I. Art. 66.]

† [1900. Δ being the symbol of finite differences.]

The second may be obtained by integrating (1) across the boundary. Thus

$$\left(\frac{n}{k} + U\right) . \Delta\left(\frac{dv}{dy}\right) - \Delta\left(\frac{dU}{dy}\right) . v = 0. \quad\text{.................. (5)}$$

At a fixed wall $v = 0$.

In the special problem to which attention is here directed, the laminated motion is supposed to take place between two fixed walls, at $y = 0$ and $y = b_1 + b' + b_2$; and the vorticity is supposed to be constant throughout each of the three layers bounded by

$$y = 0, \qquad y = b_1;$$

$$y = b_1, \qquad y = b_1 + b';$$

$$y = b_1 + b', \qquad y = b_1 + b' + b_2.$$

Fig. 1.

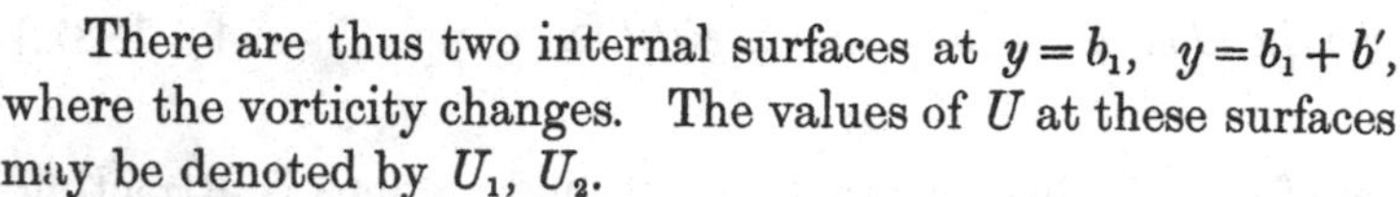

There are thus two internal surfaces at $y = b_1$, $y = b_1 + b'$, where the vorticity changes. The values of U at these surfaces may be denoted by U_1, U_2.

In conformity with (4) and with the condition that $v = 0$ when $y = 0$, we may take in the first layer

$$v = v_1 = \sinh ky; \quad\text{...(6)}$$

in the second layer

$$v = v_2 = v_1 + M_1 \sinh k(y - b_1); \quad\text{........................(7)}$$

in the third layer

$$v = v_3 = v_2 + M_2 \sinh k(y - b' - b_1). \quad\text{......................(8)}$$

The condition that $v = 0$, when $y = b_1 + b' + b_2$, now gives

$$0 = M_2 \sinh kb_2 + M_1 \sinh k(b_2 + b') + \sinh k(b_2 + b' + b_1). \quad\text{......(9)}$$

We have still to express the two other conditions (5) at the surfaces of transition. At the first surface,

$$v = \sinh kb_1, \qquad \Delta(dv/dy) = kM_1;$$

at the second surface,

$$v = M_1 \sinh kb' + \sinh k(b_1 + b'), \qquad \Delta(dv/dy) = kM_2.$$

If we denote the values of $\Delta(dU/dy)$ at the two surfaces respectively by Δ_1, Δ_2, our conditions become

$$\left.\begin{aligned} (n + kU_1) M_1 - \Delta_1 \sinh kb_1 = 0 \\ (n + kU_2) M_2 - \Delta_2 \{M_1 \sinh kb' + \sinh k(b_1 + b')\} = 0 \end{aligned}\right\}. \quad\text{......(10)}$$

By (9) and (10) the values of M_1, M_2, n are determined.

The equation for n is found by equating to zero the determinant

$$\begin{vmatrix} \sinh kb_2, & \sinh k(b_2+b'), & \sinh k(b_2+b'+b_1) \\ n+kU_2, & -\Delta_2 \sinh kb', & -\Delta_2 \sinh k(b_1+b') \\ 0, & n+kU_1, & -\Delta_1 \sinh kb_1 \end{vmatrix};$$

so that n has the values determined by the quadratic

$$An^2+Bn+C=0, \quad \text{.........................(11)}$$

where

$$A=\sinh k(b_2+b'+b_1), \quad \text{..................(12)}$$

$$B=k(U_1+U_2)\sinh k(b_2+b'+b_1)+\Delta_2 \sinh kb_2 \sinh k(b_1+b') + \Delta_1 \sinh kb_1 \sinh k(b_2+b'), \quad \text{......(13)}$$

$$C=k^2U_1U_2\sinh k(b_2+b'+b_1)+kU_1\Delta_2 \sinh kb_2 \sinh k(b_1+b') + kU_2\Delta_1 \sinh kb_1 \sinh k(b_2+b')+\Delta_1\Delta_2 \sinh kb_1 \sinh kb_2 \sinh kb'. \quad \text{...(14)}$$

To find the character of the roots, we have to form the expression for B^2-4AC. Having regard to

$$\sinh k(b_2+b')\sinh k(b_1+b')-\sinh k(b_2+b'+b_1)\sinh kb' = \sinh kb_1 \sinh kb_2,$$

we find

$$B^2-4AC=\{k(U_1-U_2)\sinh k(b_2+b'+b_1) + \Delta_1 \sinh kb_1 \sinh k(b_2+b') - \Delta_2 \sinh kb_2 \sinh k(b_1+b')\}^2 + 4\Delta_1\Delta_2 \sinh^2 kb_1 \sinh^2 kb_2. \quad \text{......(15)}$$

Hence, if Δ_1, Δ_2 have the same sign, that is, if the curve expressing U as a function of y be of one curvature throughout, B^2-4AC is positive, and the two values of n are real. Under these circumstances the disturbance is stable.

We will now suppose that the surfaces at which the vorticity changes are symmetrically situated, so that

$$b_1=b_2=b.$$

In this case we find

$$A=\sinh k(2b+b'), \quad \text{..................(16)}$$

$$B=k(U_1+U_2)\sinh k(2b+b')+(\Delta_1+\Delta_2)\sinh kb \sinh k(b+b'), \quad \text{........(17)}$$

$$C=k^2U_1U_2\sinh k(2b+b')+k(U_1\Delta_2+U_2\Delta_1)\sinh kb \sinh k(b+b') + \Delta_1\Delta_2 \sinh^2 kb \sinh kb', \quad \text{......(18)}$$

$$B^2-4AC=\{k(U_1-U_2)\sinh k(2b+b') + (\Delta_1-\Delta_2)\sinh kb \sinh k(b+b')\}^2 + 4\Delta_1\Delta_2 \sinh^4 kb. \quad \text{......(19)}$$

Under this head there are two sub-cases which may be especially noted. The first is that in which the values of U are the same on both sides of the median plane, so that the middle layer is a region of constant velocity without vorticity, and the velocity curve is that shown in Fig. 2. We may suppose that $U=V$ in the middle layer, and that $U=0$ at the walls, without loss of generality, since any constant velocity (U_0) superposed upon this system merely alters n by the corresponding quantity $-kU_0$, as is evident from (1). Thus

Fig. 2.

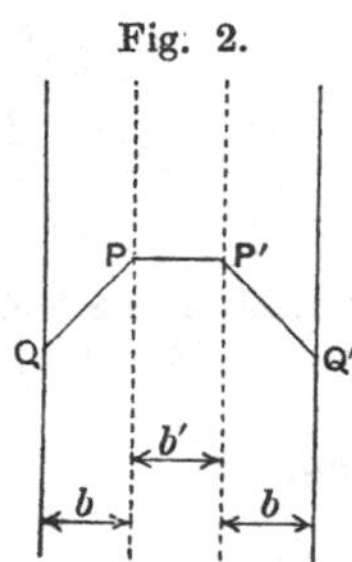

$$U_1 = U_2 = V, \qquad \Delta_2 = \Delta_1 = \Delta = -V/b;$$

and
$$B^2 - 4AC = 4\Delta^2 \sinh^4 kb.$$

Hence
$$n + kV = \frac{V}{b}\,\frac{\sinh kb \sinh k(b+b') \pm \sinh^2 kb}{\sinh k(2b+b')} \text{ ....(20)}$$

If the middle layer be absent, $b'=0$, and

$$n + kV = \frac{V}{b}\,\frac{2\sinh^2 kb}{\sinh 2kb} = \frac{V}{b}\tanh kb, \text{ ....(21)}$$

in conformity with (44) of the former paper; but the more general result (20) does not agree with (46).

The other case which we shall consider is that in which the velocities U on the two sides of the median plane are opposite to one another; so that

$$U_1 = -U_2 = V, \qquad \Delta_2 = -\Delta_1 = -\mu V. \text{ ....(22)}$$

Here $B=0$, and

$$C = -k^2V^2 \sinh k(2b+b') - 2k\mu V^2 \sinh kb \sinh k(b+b') - \mu^2 V^2 \sinh^2 kb \sinh kb'.$$

Thus

$$\frac{n^2}{k^2V^2} = \frac{k^2 \sinh k(2b+b') + 2k\mu \sinh kb \sinh k(b+b') + \mu^2 \sinh^2 kb \sinh kb'}{k^2 \sinh k(2b+b')} \text{ ....(23)}$$

Here the two values of n are equal and opposite; and, since Δ_1, Δ_2 are of opposite signs, the question is open as to whether n is real or imaginary.

Fig. 3.

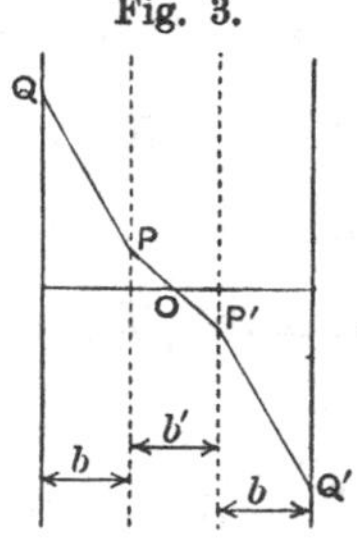

It is at once evident that n is real if μ be positive, that is, if Δ_1 and V are of the same sign, as in Fig. 3.

Even when μ is negative, n^2 is necessarily positive for great values of k, that is, for small wave-lengths. For we have ultimately, from (23),

$$n = \pm kV.$$

We will now inquire for what values of μ n^2 may be negative when k is very small, that is, when the wave-length is very great. Equating the numerator of (23) to zero, and expanding the hyperbolic sines, we get as a quadratic in μ,

$$\mu^2 b^2 b' + 2\mu b (b + b') + 2b + b' = 0,$$

whence
$$\mu = -\frac{1}{b}, \text{ or } -\frac{1}{b} - \frac{2}{b'}. \quad \text{.......................(24)}$$

When μ lies between these limits (and then only), n^2 is negative, and the disturbance (of great wave-length) increases exponentially with the time.

We may express these results by means of the velocity V_0 at the wall where $y = 0$. We have

$$V_0 = V\frac{b + \frac{1}{2}b'}{\frac{1}{2}b'} + \Delta_1 b = V\left(\frac{b + \frac{1}{2}b'}{\frac{1}{2}b'} + \mu b\right).$$

The limiting values of V_0 are therefore

$$\frac{bV}{\frac{1}{2}b'} \text{ and } 0.$$

The velocity curve corresponding to the first limit is shown in Fig. 4 by the line $QPOP'Q'$, the point Q being found by drawing a line AQ parallel to OP to meet the wall in Q. If $b' = 2b$, QP is parallel to OA, or the velocity is constant in each of the extreme layers.

At the second limit $V_0 = 0$, and the velocity-curve is that shown in Fig. 5.

Fig. 4.

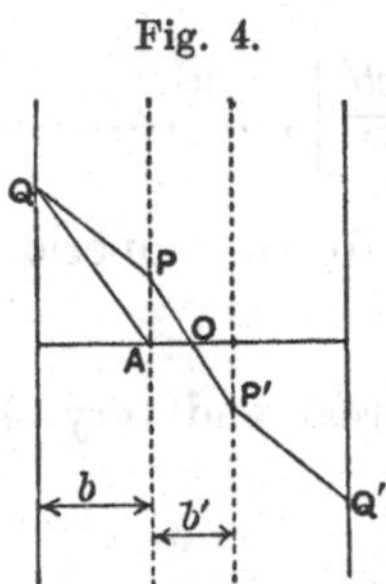

Fig. 5.

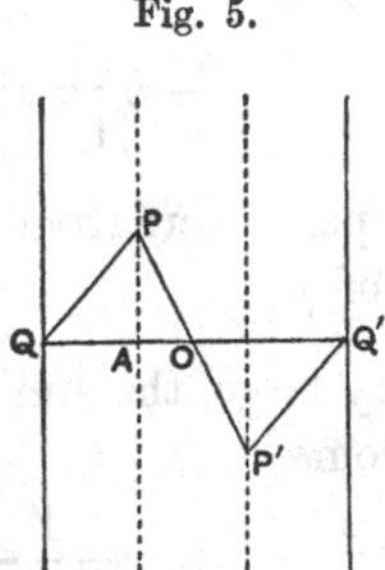

It is important to notice that motions represented by velocity-curves intermediate between these limits are unstable in a manner not possible to motions in which the velocity-curve, as in Fig. 2, is of one curvature throughout.

According to the first approximation, the motion of Fig. 5 is on the border-land between stability and instability for disturbances of great wave-length; but, if we pursue the calculation, we find that it is really unstable. Taking, in (23),

$$\mu = -\frac{1}{b} - \frac{2}{b'},$$

and writing for brevity $kb=x$, $kb'=x'$, we get

$$\frac{n^2}{k^2V^2}=\frac{\left\{\begin{array}{r}x^2x'^2\sinh(2x+x')-2xx'(2x+x')\sinh x\sinh(x+x')\\ +(2x+x')^2\sinh^2 x\sinh x'\end{array}\right\}}{x^2x'^2\sinh(2x+x')};$$

from which, on expanding the hyperbolic sines and retaining two terms, we get, after reduction,

$$\frac{x^2}{k^2V^2}=-\frac{x^2}{3}=-\frac{k^2b^2}{3},\qquad(25)$$

indicating instability.

[January, 1888*. According to (23), we may always, with a prescribed wave-length, determine two values of μ (or V_0), V being regarded as given, between which n^2 will be negative, and the motion unstable. But, if these values of μ were imaginary, the result would be of no significance in the present problem. We may, however, write (23) in the form

$$\frac{n^2}{k^2V^2}=\frac{\{\mu\sinh kb\sinh kb'+k\sinh k(b+b')\}^2-k^2\sinh^2 kb}{k^2\sinh kb'\sinh k(2b+b')},$$

from which we see that, whatever be the value of k, it is possible so to determine μ that the disturbance shall be unstable. The condition is simply that μ must lie between the limits

$$-k\frac{\sinh k(b+b')\pm\sinh kb}{\sinh kb\sinh kb'},$$

or

$$-k\left[\coth kb+\left.\begin{array}{c}\coth\\ \tanh\end{array}\right\}\frac{kb'}{2}\right],\qquad(26)$$

in which the upper alternative corresponds to the superior limit to the *numerical value* of μ.

When k is very large, the limits are very great and very close. When k is small, they become

$$-\frac{1}{b}-\frac{2}{b'}\quad\text{and}\quad-\frac{1}{b},$$

as has already been proved. As k increases from 0 to ∞, the numerical value of the upper limit increases continuously from $1/b+2/b'$ to ∞, and in like manner that of the inferior limit from $1/b$ to ∞. The motion therefore cannot be stable *for all values of k*, if μ (being negative) exceed numerically $1/b$. The final condition of complete stability is therefore that algebraically

$$\mu>-\frac{1}{b}.$$

* This paragraph is re-written, and embodies an improvement suggested in a report communicated to me by the Secretary.

In the transition case

$$V_0 = \left(\mu + \frac{1}{b} + \frac{2}{b'}\right) Vb = \frac{2Vb}{b'};$$

it is that represented in Fig. 4. If PQ be bent more downwards than is there shown, as for example in Fig. 5, the steady motion is certainly unstable.

It would be of interest, in some particular case of instability (such as that of Fig. 5), to calculate for what value of k the instability, measured by in, is greatest, and to ascertain the degree of this instability.]

Reverting to the general equations (11), (12), (13), (14), (15), let us suppose that $\Delta_2 = 0$, amounting to the abolition of the corresponding surface of discontinuity. We get

$$B = k(U_1 + U_2)\sinh k(b_2 + b' + b_1) + \Delta_1 \sinh kb_1 \sinh k(b_2 + b'),$$

$$B^2 - 4AC = \{k(U_1 - U_2)\sinh k(b_2 + b' + b_1) + \Delta_1 \sinh kb_1 \sinh k(b_2 + b')\}^2;$$

so that
$$n = -kU_2, \qquad (27)$$

or
$$n = -kU_1 - \frac{\Delta_1 \sinh kb_1 \sinh k(b_2 + b')}{\sinh k(b_1 + b' + b_2)}. \qquad (28)$$

The latter is the general solution for two layers of constant vorticity of breadths b_1 and $b' + b_2$. An equivalent result may be obtained by supposing in (11), &c., that $b' = 0$, or that $b_1 = 0$.

The occurrence of (27) suggests that any value of $-kU$ is admissible as a value of n, and the meaning of this is apparent from (1). For, at the place where $n + kU = 0$, (2) need not be satisfied, or the arbitrary constants in (3) may change their values. It is evident that, with the prescribed values of n and k, a solution may be found satisfying the required conditions at the walls and at the surfaces where dU/dy changes value, as well as equation (4) at the plane where $n + kU = 0$. Equation (5) is there satisfied independently of the value of v. In this motion an additional vorticity is supposed to be communicated at the plane in question, and moves with the fluid at velocity U.

145.

DIFFRACTION OF SOUND.

[*Royal Institution Proceedings*, XII. pp. 187—198, 1888;
Nature, XXXVIII. pp. 208—211, 1888.]

THE interest of the subject which I propose to bring before you this evening turns principally upon the connection or analogy between light and sound. It has been known for a very long time that sound is a vibration; and every one here knows that light is a vibration also. The last piece of knowledge, however, was not arrived at so easily as the first; and one of the difficulties which retarded the acceptance of the view that light is a vibration was that in some respects the analogy between light and sound seemed to be less perfect than it should be. At the present time many of the students at our schools and universities can tell glibly all about it; yet this difficulty is one not to be despised, for it exercised a determining influence over the great mind of Newton. Newton, it would seem, definitely rejected the wave theory of light on the ground that according to such a theory light would turn round the corners of obstacles, and so abolish shadows, in the way that sound is generally supposed to do. The fact that this difficulty seemed to Newton to be insuperable is, from the point of view of the advancement of science, very encouraging. The difficulty which stopped Newton two centuries ago is no difficulty now. It is well known that the question depends upon the relative wave-lengths in the two cases. Light-shadows are sharp under ordinary circumstances, because the wave-length of light is so small: sound-shadows are usually of a diffused character, because the wave-length of sound is so great. The gap between the two is enormous. I need hardly remind you that the wave-length of C in the middle of the musical scale is about 4 feet. The wave-length of the light with which we are usually concerned, the light towards the middle of the spectrum, is about the forty-thousandth of an inch. The result is that an obstacle which is immensely large for light may be very small for sound, and will therefore behave in a different manner.

That light-shadows are sharp is a familiar fact, but as I can prove it in a moment I will do so. We have here light from the electric arc thrown on the screen; and if I hold up my hand thus we have a sharp shadow at any moderate distance, which shadow can be made sharper still by diminishing the source of light. Sound-shadows, as I have said, are not often sharp; but I believe that they are sharper than is usually supposed, the reason being that when we pass into a sound-shadow—when, for example, we pass into the shade of a large obstacle, such as a building—it requires some little time to effect the transition, and the consequence is that we cannot make a very ready comparison between the intensity of the sound before we enter and its diminution afterwards. When the comparison is made under more favourable conditions, the result is often better than would have been expected. It is, of course, impossible to perform experiments with such obstacles before an audience, and the shadows which I propose to show you to-night are on a much smaller scale. I shall take advantage of the sensitiveness of a flame such as Professor Tyndall has often used here—a flame sensitive to the waves produced by notes so exceedingly high as to be inaudible to the human ear. In fact, all the sounds with which I shall deal to-night will be inaudible to the audience. I hope that no quibbler will object that they are therefore not sounds: they are in every respect analogous to the vibrations which produce the ordinary sensations of hearing.

I will now start the sensitive flame. We must adjust it to a reasonable degree of sensitiveness. I need scarcely explain the mechanism of these flames, which you know are fed from a special gasholder supplying gas at a high pressure. When the pressure is too high, the flame flares on its own account (as this one is doing now), independently of external sound. When the pressure is somewhat diminished, but not too much so—when the flame "stands on the brink of the precipice," were, I think, Tyndall's words—the sound pushes it over, and causes it to flare; whereas, in the absence of such sound, it would remain erect and unaffected. Now, I believe, the flame is flaring under the action of a very high note that I am producing here. That can be tested in a moment by stopping the sound, and seeing whether the flame recovers or not. It recovers now. What I want to show you, however, is that the sound-shadows may be very sharp. I will put my hand between the flame and the source of sound, and you will see the difference. The flame is at present flaring; if I put my hand here, the flame recovers. When the adjustment is correct, my hand is a sufficient obstacle to throw a most conspicuous shadow. The flame is now in the shadow of my hand, and it recovers its steadiness: I move my hand up, the sound comes to the flame again, and it flares. When the conditions are at their best, a very small obstacle is sufficient to make the entire difference, and a sound-shadow may be thrown across several feet from an obstacle as small as the hand. The reason of the divergence from ordinary experience here met with is, that while the hand is a fairly large obstacle in comparison with the wave-length of the sound I am here using, it would not be a sufficiently large obstacle in comparison with the wave-lengths with which we have to do in ordinary life and in music.

Everything then turns upon the question of the wave-length. The wave-length of the sound that I am using now is about half an inch. That is its complete length, and it corresponds to a note that would be very high indeed on the musical scale. The wave-length of middle C being four feet, the C one octave above that is two feet; two octaves above, one foot; three octaves above, six inches; four octaves, three inches; five octaves, one and a half inch; six octaves, three-quarters of an inch; between that and the next octave, that is to say, between six and seven octaves above middle C, is the pitch of the note that I was just now using. There is no difficulty in determining what the wave-length is. The method depends upon the properties of what are known as stationary sonorous waves as opposed to progressive waves. If a train of progressive waves are caused to impinge upon a reflecting wall, there will be sent back or reflected in the reverse direction a second set of waves, and the co-operation of these two sets of waves produces one set or system of stationary waves, the distinction being that, whereas in the one set the places of greatest condensation are continually changing and passing through every point, in the stationary waves there are definite points for the places of greatest condensation (nodes), and others distinct and definite (loops) for the places of greatest motion. The places of greatest variation of density are the places of no motion: the places of greatest motion are places of no variation of density. By the operation of a reflector, such as this board, we obtain a system of stationary waves, in which the nodes and loops occupy given positions relatively to the board.

You will observe that as I hold the board at different distances behind, the flame rises and falls—I can hardly hold it still enough. In one position the flame rises, further off it falls again; and as I move the board back the flame passes continually from the position of the node—the place of no motion—to the loop or place of greatest motion and no variation of pressure. As I move back the aspect of the flame changes; and all these changes are due to the reflection of the sound-waves by the reflector which I am holding. The flame alternately ducks and rises, its behaviour depending upon the different action of the nodes and loops. The nodes occur at distances from the reflecting wall, which are even multiples of the quarter of a wave-length; the loops are, on the other hand, at distances from the reflector which are odd multiples, bisecting therefore the intervals between the nodes. I will now show you that a very slight body is capable of acting as a reflector. This is a screen of tissue paper, and the effect will be apparent when it is held behind the flame and the distances are caused to vary. The flame goes up and down, showing that a considerable proportion of the sonorous intensity incident upon the paper screen is reflected back upon the flame; otherwise the exact position of the reflector would be of no moment. I have here, however, a different sort of reflector. This is a glass plate—I use glass so that those behind may see through it—and it will slide upon a stand here arranged for it. When put in this position the flame is very little affected; the place is what I call a node--a place where there is great pressure variation, but no vibratory velocity. If I move the glass back, the flame becomes vigorously excited; that position is a loop. Move it back still more and the flame becomes fairly quiet; but you see that as the plate travels gradually along, the flame goes through these evolutions as it occupies in succession the position of a node or the position of a loop. The interest of this experiment for our present purpose depends upon this—that the distances through which the glass plate, acting as a reflector, must be successively moved in order to pass the flame from a loop to the next loop, or from a node to the consecutive node, is in each case half the wave-length; so that by measuring the space through which the plate is thus withdrawn one has at once a measurement of the wave-length, and consequently of the pitch of the sound, though one cannot hear it.

The question of whether the flame is excited at the nodes or at the loops,—whether at the places where the pressure varies most or at those where there is no variation of pressure, but considerable motion of air—is one of considerable interest from the point of view of the theory of these flames. The experiment could be made well enough with such a source of sound as I am now using; but it is made rather better by using sounds of a lower pitch and therefore of greater wave-length, the discrimination being then more easy. Here is a table of the distances which the screen must be from the flame in order to give the maximum and the minimum effect, the minimum being practically nothing at all.

TABLE OF MAXIMA AND MINIMA.

Max.	Min.
1·1	
	3·0
4·5	
	5·9
7·5	
	8·9
10·3	
	11·7
13·0	
	14·7
15·9	

The distance between successive maxima or successive minima is very nearly 3 (centims.), and this is accordingly half the length of the wave.

But there is a further question behind. Is it at the loops or is it at the nodes that the flame is most excited? The table shows what the answer must be, because the nodes occur at distances from the screen which are even multiples, and the loops at distances which are odd multiples; and the numbers in the table can be explained in only one way—that the flame is excited at the loops corresponding to the odd multiples, and remains quiescent at the nodes corresponding to the even multiples. This result is especially remarkable, because the ear, when substituted for the flame, behaves in the exactly opposite manner, being excited at the nodes and not at the loops. The experiment may be tried with the aid of a tube, one end of which is placed in the ear, while the other is held close to the burner. It is then found the ear is excited the most when the flame is excited least, and *vice versâ*. The result of the experiment shows, moreover, that the manner in which the flame is disintegrated under the action of sound is not, as might be expected, symmetrical in regard to the axis of the flame. If it were symmetrical, it would be most affected by the symmetrical cause, namely, the variation of pressure. The fact being that it is most excited at the loop, where there is the greatest vibratory velocity, shows that the method of disintegration is unsymmetrical, the velocity being a directed quantity. In that respect the theory of these flames is different from the theory of the water-jets investigated by Savart, which resolve themselves into detached drops under the influence of sonorous vibration. The analogy fails at this point, and it has been pressed too far by some experimenters on the subject. Another simple proof of the correctness of the result of our experiment is that it makes all the difference which way the burner is turned in respect of the direction in which the sound-waves are impinging upon it. If the phenomenon were symmetrical, it would make no difference if the flame were turned round upon its vertical axis. But we find that it does make a difference. This is the way in which I was using the flame, and you see that it is flaring strongly. If I now turn the burner round through a right angle, the flame stops flaring. I have done nothing more than turn the burner round and the flame with it, showing that the sound-waves may impinge in one direction with great effect, and in another direction with no effect. The sensitiveness occurs again when the burner is turned through another right angle; after three right angles there is another place of no effect; and after a complete revolution of the flame the original sensitiveness recurs. So that if the flame were stationary, and the sound-waves came, say, from the north or south, the phenomena would be exhibited; but if they came from the east or west, the flame would make no response.

This is of convenience in experimenting, because, by turning the burner round, I make the flame almost insensitive to a sound, and I am now free to show the effect of any sound that may be brought to it in the perpendicular direction. I am going to use a very small reflector—a small piece of looking-glass. Wood would do as well; but looking-glass facilitates the adjustment, because my assistant, by seeing the reflection, will be able to tell me when I am holding it in the best position. Now, the sound is being reflected from the bit of glass, and is causing the flame to flare, though the same sound, travelling a shorter distance and impinging in another direction, is incompetent to produce the result (Fig. 1).

I am now going to move the reflector to and fro along the line perpendicular to that joining the source and the burner, all the while maintaining the adjustment, so that from the position of the source of sound the image of the flame is seen in the centre of the mirror. Seen from the source, it is still as central as before; but it has lost its effect, and as I move

Fig. 1.

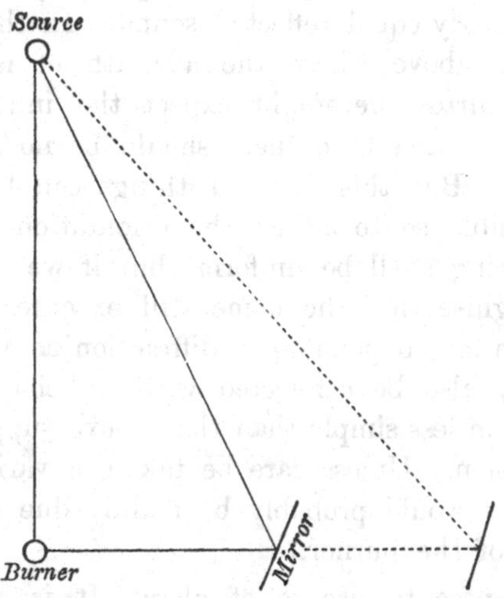

it to and fro I produce cycles of effect and no effect. What is the cause of this? The question depends upon something different from what I have been speaking of hitherto; and the explanation is, that we are here dealing with a diffraction phenomenon. The mirror is a small one, and the sound-waves which it reflects are not big enough to act in the normal manner. We are really dealing with the same sort of phenomena as arise in optics when we use small pin-holes for the entrance of our light. It is not very easy to make the experiment in the present form quite simple, because the mirror would have to be withdrawn, all the while maintaining a somewhat complicated adjustment. In order to raise the question of diffraction in its simplest shape, we must have a direct course for the sound between its origin and the place of observation, and interpose in the path a screen perforated with such holes as we desire to try.

[1900. Further experiments with the arrangement of Fig. 1 have recently been made. When the gas pressure is carefully adjusted, the positions of the mirror corresponding to recovery of the flame may be very well defined, but they depend upon the orientation of the burner. If for example the burner is so turned round its axis that the azimuth of maximum sensibility bisects internally the angle subtended by the source and the mirror, the positions of the mirror for minimum effect are well defined, and they are so spaced along the line of motion that the sum of the distances from the mirror to the source and to the burner increases at each step by one complete wave-length. But if the burner be again turned upon its axis through a right angle, the positions of minimum effect are shifted so as to bisect the intervals between the former ones. In other words the positions of maximum and minimum are interchanged. These effects are just what might have been expected, and they clearly depend upon the co-operation of the direct and the nearly equal reflected sound. In the orientation of the burner contemplated above where the azimuth of maximum sensibility passes through the mirror, we might expect the influence of the direct sound to be eliminated, and then there should be no alternation of effect as the mirror moves. But this state of things can be attained only imperfectly. It is possible so to adjust the orientation of the burner that the sound of the flaring shall be uniform; but if we use our eyes instead of our ears, we recognise that the flame still executes periodic evolutions. The residual variation may depend upon diffraction as above suggested; but I think that it may also be connected with a behaviour of the burner in respect of orientation less simple than that above supposed and applicable as a first approximation. Unless care be taken, a variation of effect with position of the mirror would probably be mainly due to imperfect adjustment of orientation of the burner.]

The screen I propose to use is of glass. It is a practically perfect obstacle for such sounds as we are dealing with; but it is perforated here with a hole (20 cm. in diameter), rendered more evident to those at a distance by means of a circle of paper pasted round it. The edge of the hole corresponds to the inner circumference of the paper. We shall thus be able to try the effect of different sized apertures, all the other circumstances remaining unchanged. The experiment is rather a difficult one before an audience, because everything turns on getting the exact adjustment of distances relatively to the wave-length. At present the sound is passing through this comparatively large hole in the glass screen, and is producing, as you see, scarcely any effect upon the flame situated opposite to its centre. But if (Fig. 2) I diminish the size of the hole by holding this circle of zinc (perforated with a hole 14 cm. in diameter) in front of it, it is seen that, although the hole is smaller, we get a far greater effect. That is a fundamental phenomenon in diffraction. Now I reopen the larger hole, and the flame becomes quiet. So that it is evident that in this case the sound produces a greater effect in passing through a small hole than in passing

Fig. 2.

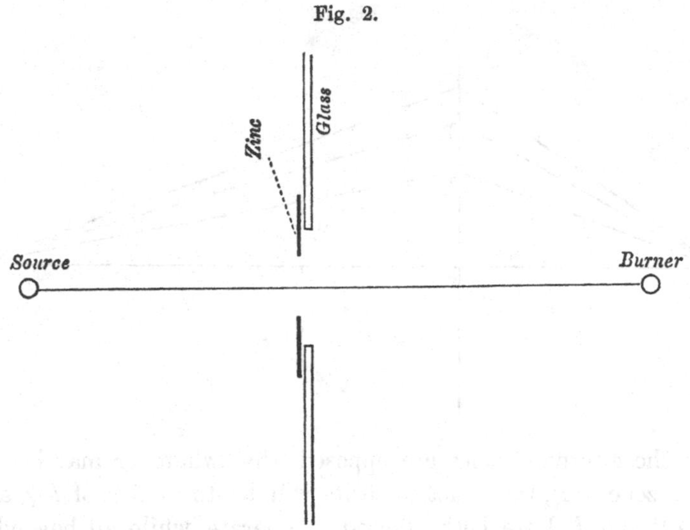

through a larger one. The experiment may be made in another way, by obstructing the central in place of the marginal part of the aperture in the glass. When I hold this unperforated disc of zinc (14 cm. in diameter) centrically in front, we get a greater effect than when the sound is allowed to pass through both parts of the aperture. The flame is now flaring vigorously under the action of the sonorous waves passing the marginal part of the aperture, whereas it will scarcely flare at all under the action of waves passing through both the marginal and the central hole.

This is a point which I should like to dwell upon a little, for it lies at the root of the whole matter. The principle upon which it depends is one that was first formulated by Huygens, one of the leading names in the development of the undulatory theory of light. In this diagram (Fig. 3) is represented in section the different parts of the obstacle. C represents the source of sound, B represents the flame, and APQ is the screen. If we choose a point P on the screen, so that the whole distance from B to C, reckoned through P, viz. BPC, exceeds the shortest distance BAC by exactly half the wave-length of the sound, then the circular area, whose radius is AP, is the first zone. We take next another point, Q, so that the whole distance BQC exceeds the previous one by half a wave-length. Thus we get the second zone represented by PQ. In like manner, by taking different points in succession such that the last distance taken exceeds the previous one every time by half a wave-length, we may map out the whole of the obstructing screen into a series of zones called Huygens' zones. I have here a material embodiment of that motion, in which the zones are actually cut out of a piece of zinc. It is easy to prove that the effects of the parts of the wave

Fig. 3.

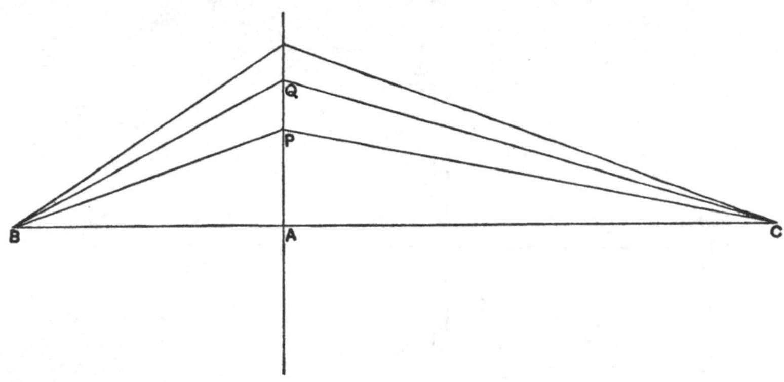

traversing the alternate zones are opposed, that whatever may be the effect of the first zone, AP, the exact opposite will be the effect of PQ, and so on. Thus, if AP and PQ are both allowed to operate, while all beyond Q is cut off, the waves will neutralise one another, and the effect will be immensely less than if AP or PQ operated alone. And that is what you saw just now. When I used the inner aperture only, a comparatively loud sound acted upon the flame. When I added to that inner aperture the additional aperture PQ, the sound disappeared, showing that the effect of the latter was equal and opposite to that of AP, and that the two neutralised each other.

If $AC = a$, $AB = b$, $AP = x$, wave-length $= \lambda$, the value of x for the external radius of the nth zone is

$$x^2 = n\lambda \frac{ab}{a+b},$$

or, if $a = b$,

$$x^2 = \tfrac{1}{2} n\lambda a.$$

With the apertures used above, $x^2 = 49$ for $n = 1$; $x^2 = 100$ for $n = 2$; so that

$$\lambda a = 100,$$

the measurements being in centimetres. This gives the suitable distances, when λ is known. In the present case $\lambda = 1{\cdot}2$, $a = 83$.

Closely connected with this there is another very interesting experiment, which can easily be tried, and which has also an important optical analogy. I mean the experiment of the shadow thrown by a circular disc. If a very small source of light be taken—such a source as would be produced by perforating a thin plate in the shutter of the window of a dark room with a pin and causing the rays of the sun to enter horizontally—and if we interpose in the path of the light a small circular obstacle and then observe the shadow thrown in the rear of that obstacle, a very remarkable peculiarity manifests itself. It is found that in the centre of the shadow of the obstacle, where the darkness might be expected to be greatest, there is, on the contrary, no darkness at all, but a bright spot, a spot as bright as if no obstacle intervened in the course of the light. The history of this subject is curious. The fact was first observed by Delisle in the early part of the eighteenth century, but the observation fell into oblivion. When Fresnel began his important investigations, his memoir on diffraction was communicated to the French Academy and was reported on by the great mathematician Poisson. Poisson was not favourably impressed by Fresnel's theoretical views. Like most mathematicians of the day, he did not take kindly to the wave theory; and in his report on Fresnel's memoir, he made the objection that if the method were applied, as Fresnel had not then done, to investigate what should happen in the shadow of a circular obstacle, it brought out this paradoxical result, that in the centre there would be a bright point. This was regarded as a *reductio ad absurdum* of the theory. All the time, as I have mentioned, the record of Delisle's observations was in existence. The remarks of Poisson were brought to the notice of Fresnel, the experiment was tried, and the bright point was rediscovered, to the gratification of Fresnel and the confirmation of his theoretical views. I don't propose to attempt the optical experiment now, but it can easily be tried in one's own laboratory. A long room or passage must be darkened: a fourpenny bit may be used as the obstacle, strung up by three hairs attached by sealing-wax. When the shadow of the obstacle is received on a piece of ground glass, and examined from behind with a magnifying lens, the bright spot will be seen without much difficulty. But what I propose to show you is the corresponding phenomenon in the case of sound. Fresnel's reasoning is applicable, word for word, to the phenomena we are considering just as much as to that which he, or rather Poisson, had in view. The disc (Fig. 4), which I shall hang up now between the source of sound and the flame, is of glass. It is about

Fig. 4.

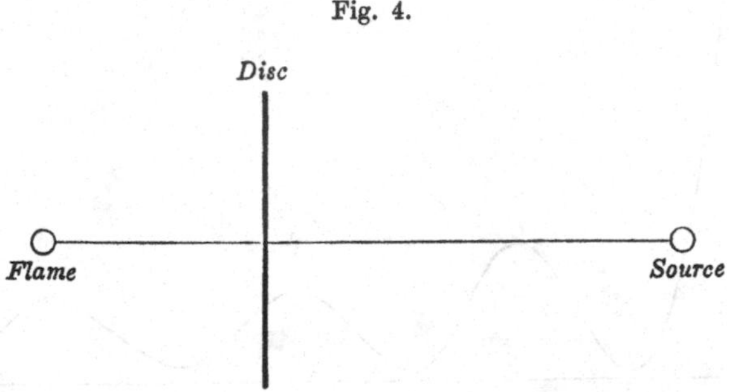

15 inches in diameter. I believe the flame is flaring now from being in the bright spot. If I make a small motion of the disc I shall move the bright

spot and the effect will disappear. I am pushing the disc away now, and the flaring has stopped. The flame is still in the shadow of the disc, but not at the centre. I bring the disc back, and when the flame comes into the centre it flares again vigorously. That is the phenomenon which was discovered by Delisle and confirmed by Arago and Fresnel, but mathematically it was suggested by Poisson.

Poisson's calculation related only to the very central point in the axis of the disc. More recently the theory of this experiment has been very thoroughly examined by a German mathematician, Lommel; and I have exhibited here one of the curves given by him embodying the results of his calculations on the subject (Fig. 5).

The abscissæ, measured horizontally, represent distances drawn outwards from the centre of the shadow O; the ordinates measure the intensity of the light at the various points. The maximum intensity OA is at the centre. A little way outwards at B the intensity falls almost, but not quite, to zero. At C there is a revival of intensity, indicating a bright ring; and further out there is a succession of subordinate fluctuations. The curve on the other side of OA would of course be similar. This curve corresponds to the

Fig. 5.

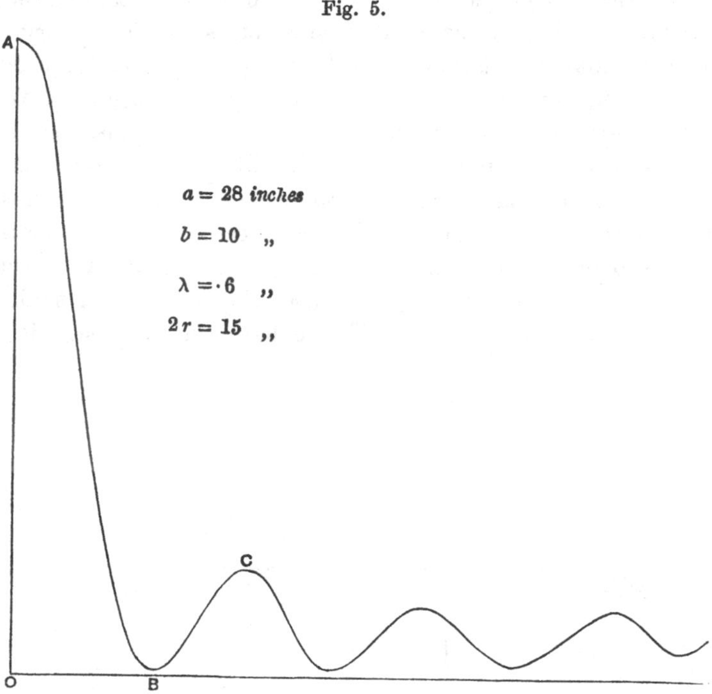

distances and proportions indicated. a is the distance between the source of sound and the disc; b is the distance between the disc and the flame, the place where the intensity is observed. The numbers given are taken from the notes of an experiment which went well. If we can get our flame to the right point of sensitiveness we may succeed in bringing into view not only the central spot, but the revived sound which occurs after we have got away from the central point and have passed through the ring of silence. There is the loud central point. If I push the disc a little we enter the ring of silence B*; a little further, and the flame flares again, being now at C.

Although we have thus imitated the optical experiment, I must not leave you under the idea that we are working under the same conditions that prevail in optics. You see the diameter of my disc is 15 inches, and the length of my sound-wave is about half an inch. My disc is therefore about 30 wave-lengths in diameter, whereas the diameter of a disc representing 30 wave-lengths of light would be only about $\frac{1}{1000}$ inch. Still the conditions are sufficiently alike to get corresponding effects, and to obtain this bright point in the centre of the shadow conspicuously developed.

I will now make an experiment illustrating still further the principle of Huygens' zones, which I have already roughly sketched. I indicated that the effect of contiguous zones was equal and opposite, so that the effect of each of the odd zones is one thing, and of the even zones the opposite thing. If we can succeed in so preparing a screen as to fit the system of zones, allowing the one set to pass, and at the same time intercepting the other set, then we shall get a great effect at the central point, because we shall have removed those parts which, if they remained, would have neutralised the remaining parts. Such a system has been cut out of zinc, and is now hanging before you. When the adjustments are correct there will be produced, under the action of that circular grating, an effect much greater than would result if the sound-waves were allowed to pass on without any obstruction. The only point difficult of explanation is as to what happens when the system of zones is complete, and extends to infinity, viz. when there is no obstruction at all. In that case it may be proved that the aggregate effect of all the zones is, in ordinary cases, half the effect that would be produced by any one zone alone, whereas if we succeed in stopping out a number of the alternate zones, we may expect a large multiple of the effect of one zone. The grating is now in the right position, and you see the flame flaring strongly, under the action of the sound-waves transmitted through these alternate zones, the action of the other zones being stopped by the interposition of the zinc. But the interest of the experiment is principally in this, that the flame is flaring *more* than it would do if the grating were removed altogether. There is now, without the grating, a very trivial flaring†; but when the grating is in position again—though a great

* With the data given above the diameter of the silent ring is two-thirds of an inch.

† Under the best conditions the flame is absolutely unaffected.

part of the sound is thereby stopped out—the effect is far more powerful than when no obstruction intervened. The grating acts, in fact, the part of a lens. It concentrates the sound upon the flame, and so produces the intense magnification of effect which we have seen.

The exterior radius of the nth zone being x, we have, from the formula given above:

$$\frac{1}{a}+\frac{1}{b}=\frac{n\lambda}{x^2};$$

so that if a and b be the distances of the source and image from the grating, the relation required to maintain the focus is as usual,

$$\frac{1}{a}+\frac{1}{b}=\frac{1}{f},$$

where f, the *focal length*, is given by

$$f=\frac{x^2}{n\lambda}.$$

In the actual grating, eight zones (the first, third, fifth, &c.) are occupied by metal. The radius of the first zone, or central circle, is 3 inches, so that $x^2/n=9$. The focal length is necessarily a function of λ. In the present case $\lambda=\frac{1}{2}$ inch nearly, and therefore $f=18$ inches. If a and b are the same, each must be made equal to 36 inches.

146.

ON THE RELATIVE DENSITIES OF HYDROGEN AND OXYGEN. (PRELIMINARY NOTICE.)

[*Proceedings of the Royal Society*, XLIII. pp. 356—363, 1888.]

THE appearance of Professor Cooke's important memoir upon the atomic weights of hydrogen and oxygen* induces me to communicate to the Royal Society a notice of the results that I have obtained with respect to the relative *densities* of these gases. My motive for undertaking this investigation, planned in 1882†, was the same as that which animated Professor Cooke, namely, the desire to examine whether the relative atomic weights of the two bodies really deviated from the simple ratio 1 : 16, demanded by Prout's Law. For this purpose a knowledge of the densities is not of itself sufficient; but it appeared to me that the other factor involved, viz., the relative atomic *volumes* of the two gases, could be measured with great accuracy by eudiometric methods, and I was aware that Mr Scott had in view a redetermination of this number, since in great part carried out‡. If both investigations are conducted with gases under the normal atmospheric conditions as to temperature and pressure, any small departures from the laws of Boyle and Charles will be practically without influence upon the final number representing the ratio of atomic weights.

In weighing the gas the procedure of Regnault was adopted, the working globe being compensated by a similar closed globe of the same external volume, made of the same kind of glass, and of nearly the same weight. In this way the weighings are rendered independent of the atmospheric

* "The Relative Values of the Atomic Weights of Hydrogen and Oxygen," by J. P. Cooke and T. W. Richards, *Amer. Acad. Proc.* Vol. XXIII., 1887.

† Address to Section A, British Association *Report*, 1882 [Vol. II. p. 124].

‡ "On the Composition of Water by Volume," by A. Scott, *Roy. Soc. Proc.*, June 16, 1887 (Vol. XLII. p. 396).

conditions, and only small weights are required. The weight of the globe used in the experiments here to be described was about 200 grams, and the contents were about 1800 c.c.

The balance is by Oertling, and readings with successive releasements of the beam and pans, but without removal of the globes, usually agreed to $\frac{1}{10}$ mg. Each recorded weighing is the mean of the results of several releasements.

The balance was situated in a cellar, where temperature was very constant, but at certain times the air currents, described by Professor Cooke, were very plainly noticeable. The beam left swinging over night would be found still in motion when the weighings were commenced on the following morning. At other times these currents were absent, and the beam would settle down to almost absolute rest. This difference of behaviour was found to depend upon the distribution of temperature at various levels in the room. A delicate thermopile with reflecting cones was arranged so that one cone pointed towards the ceiling and the other to the floor. When the galvanometer indicated that the ceiling was the warmer, the balance behaved well, and *vice versâ*. The reason is of course that air is stable when the temperature increases upwards, and unstable when heat is communicated below. During the winter months the ground was usually warmer than the rest of the room, and air currents developed themselves in the weighing closet. During the summer the air cooled by contact with the ground remained as a layer below, and the balance was undisturbed.

The principal difference to be noted between my arrangements and those of Professor Cooke is that in my case no desiccators were used within the weighing closet. The general air of the room was prevented from getting too damp by means of a large blanket, occasionally removed and dried before a fire*.

In Regnault's experiments the globe was filled with gas to the atmospheric pressure (determined by an independent barometer), and the temperature was maintained at zero by a bath of ice. The use of ice is no doubt to be recommended in the case of the heavier gases; but it involves a cleaning of the globe, and therefore diminishes somewhat the comparability of the weighings, vacuous and full, on which everything depends. Hydrogen is so light that, except perhaps in the mean of a long series, the error of weighing is likely to be more serious than the uncertainty of temperature. I have therefore contented myself with enclosing the body of the globe during the process of filling in a wooden box, into which passed the bulbs of two thermometers, reading to tenths of a degree centigrade. It seems probable that the mean

* I can strongly recommend this method. In twenty-four hours the blanket will frequently absorb two pounds of moisture.

of the readings represents the temperature of the gas to about $\frac{1}{10}$th degree, or at any rate that the differences of temperature on various occasions and with various gases will be given to at least this degree of accuracy. Indeed the results obtained with oxygen exclude a greater uncertainty.

Under these conditions the alternate full and empty weighings can be effected with the minimum of interference with the surface of the globe. The stalk and tap were only touched with a glove, and the body of the globe was scarcely touched at all. To make the symmetry as complete as possible, the counterpoising globe was provided with a similar case, and was carried backwards and forwards between the balance room and the laboratory exactly as was necessary for the working globe.

In my earliest experiments (1885) hydrogen and oxygen were prepared simultaneously in a U-shaped voltameter containing dilute sulphuric acid. Since the same quantity of acid can be used indefinitely, I hoped in this way to eliminate all extraneous impurity, and to obtain hydrogen contaminated only by small quantities of oxygen, and *vice versâ*. The final purification of the gases was to be effected by passing them through red-hot tubes, and subsequent desiccation with phosphoric anhydride. In a few trials I did not succeed in obtaining good hydrogen, a result which I was inclined to attribute to the inadequacy of a red heat to effect the combination of the small residue of oxygen*. Meeting this difficulty, I abandoned the method for a time, purposing to recur to it after I had obtained experience with the more usual methods of preparing the gases. In this part of the investigation my experience runs nearly parallel with that of Professor Cooke. The difficulty of getting quit of the dissolved air when, as in the ordinary preparation of hydrogen, the acid is fed in slowly at the time of working, induced me to design an apparatus whose action can be suspended by breaking an external electrical contact. It may be regarded as a Smee cell thoroughly enclosed. Two points of difference may be noted between this apparatus and that of Professor Cooke. In my manner of working it was necessary that the generator should stand an internal vacuum. To guard more thoroughly against the penetration of external air, every cemented joint was completely covered with vaseline, and the vaseline again with water. Again, the zincs were in the form of solid sheets, closely surrounding the platinised plate on which the hydrogen was liberated, and standing in mercury. It was found far better to work these cells by their own electromotive force, without stimulation by an external battery. If the plates are close, and the contact wires thick, the evolution of gas may be made more rapid than is necessary, or indeed desirable.

* From Professor Cooke's experience it appears not improbable that the impurity may have been sulphurous acid. Is it certain that in his combustions no hydrogen (towards the close largely diluted with nitrogen) escapes the action of the cupric oxide?

Tubes, closed by drowned stopcocks, are provided, in order to allow the acid to be renewed without breaking joints; but one charge is sufficient for a set of experiments (three to five fillings), and during the whole of the time occupied (10 to 14 days) there is no access of atmospheric air. The removal of dissolved air (and other volatile impurity) proved, however, not to be so easy as had been expected, even when assisted by repeated exhaustions with intermittent evolution of hydrogen; and the results often showed a progressive improvement in the hydrogen, even after a somewhat prolonged preliminary treatment. In subsequent experiments greater precautions will be taken*. Experience showed that good hydrogen could not thus be obtained from zinc and ordinary "pure" sulphuric acid, or phosphoric acid, without the aid of purifying agents. The best results so far have been from sulphuric and hydrochloric acid, when the gas is passed in succession over liquid potash, through powdered corrosive sublimate, and then through powdered caustic potash. All the joints of the purifying tubes are connected by fusion, and a tap separates the damp from the dry side of the apparatus. The latter includes a large and long tube charged with phosphoric anhydride, a cotton-wool filter, a blow-off tube sealed with mercury until the filling is completed, besides the globe itself and the Töpler pump. A detailed description is postponed until the experiments are complete. It may be sufficient to mention that there is but one india-rubber connexion,—that between the globe and the rest of the apparatus, and that the leakage through this was usually measured by the Töpler before commencing a filling or an evacuation.

The object of giving a considerable capacity to the phosphoric tube was to provide against the danger of a too rapid passage of gas through the purifying tubes at the commencement of a filling. Suppose the gas to be blowing off, all the apparatus except the globe (and the Töpler) being at a pressure somewhat above the atmospheric. The tap between the damp and dry sides is then closed, and that into the globe is opened. The gas which now enters somewhat rapidly is thoroughly dry, having been in good contact with the phosphoric anhydride. In this way the pressure on the dry side is reduced to about 2 inches of mercury, but this residue is sufficient to allow the damp side of the apparatus to be exhausted to a still lower pressure before the tap between the two sides of the apparatus is re-opened. When this is done, the first movement of the gas is retrograde; and there is no danger at any stage of imperfect purification. The generator is then re-started until the gas (after from two to five hours) begins to blow off again.

In closing the globe some precaution is required to secure that the pressure therein shall really be that measured by the barometer. The mercury seal is at some distance from, and at a lower level than, the rest of the

* Spectrum analysis appears to be incapable of indicating the presence of comparatively large quantities of nitrogen.

apparatus. After removal of the mercury the flow of gas is continued for about one minute, and then the tap between the dry and damp sides is closed. From three to five minutes more were usually allowed for the complete establishment of equilibrium before the tap of the globe was turned off. Experiments on oxygen appeared to show that two minutes was sufficient. For measuring the atmospheric pressure *two* standard mercury barometers were employed.

The evacuations were effected by the Töpler to at least $\frac{1}{20000}$, so that the residual gas (at any rate after one filling with hydrogen) could be neglected.

I will now give some examples of actual results. Those in the following tables relate to gas prepared from *sulphuric* acid, with subsequent purification, as already described:—

Globe (14), empty.

Date	Left	Right	Balance reading
1887			
Oct. 27—Nov. 5	$G_{14}+0·394$	G_{11}	22·66
Nov. 7—Nov. 8	. .	. .	22·89
Nov. 9—Nov. 10	. .	. .	23·00
Nov. 11—Nov. 12......	. .	. .	21·72

Globe (14), full.

Date	Left	Right	Balance reading	Barometer	Temperature
1887				in.	C.
Nov. 5— 7 . .	$G_{14}+0·2400$	G_{11}	20·52	29·416	14·7°
Nov. 8— 9 . .	$G_{14}+0·2364$	G_{11}	19·77	29·830	12·3
Nov. 10—11 . .	$G_{14}+0·2360$	G_{11}	19·18	22·807	11·2
Nov. 12—14 . .	$G_{14}+0·2340$	G_{11}	19·51	30·135	10·3

The second column shows that globe (14) and certain platinum weights were suspended from the left end of the beam, and the third column that (in this series) only the counterpoising globe (11) was hung from the right end. The fourth column gives the mean balance reading in divisions of the scale, each of which (at the time of the above experiments) represented 0·000187 gram. The degree of agreement of these numbers in the first part of the table gives an idea of the errors due to the balance, and to uncertainties in the condition of the exteriors of the globes. A minute and unsystematic correction depending upon imperfect compensation of volumes (to the extent of about 2 c.c.) need not here be regarded.

The weight of the hydrogen at each filling is deduced, whenever possible, by comparison of the "full" reading with the mean of the immediately preceding and following "empty" readings. The difference, interpreted in grams, is taken provisionally as the weight of the gas. Thus for the filling of Nov. 5—

$$H = 0{\cdot}154 - 2{\cdot}25 \times 0{\cdot}000187 = 0{\cdot}15358.$$

The weights thus obtained depend of course upon the temperature and pressure at the time of filling. Reduced to correspond with a temperature of 12°, and to a barometric height of 30 inches (but without a minute correction for varying temperature of the mercury) they stand thus—

November	5.........	0·15811
„	8.........	0·15807
„	10.........	0·15798
„	12.........	0·15792
	Mean......	0·15802

The hydrogen obtained hitherto with similar apparatus and purifying tubes from hydrochloric acid is not quite so light, the mean of two accordant series being 0·15812.

The weighing of oxygen is of course a much easier operation than in the case of hydrogen. The gas was prepared from chlorate of potash, and from a mixture of the chlorates of potash and soda. The discrepancies between the individual weighings were no more than might fairly be attributed to thermometric and manometric errors. The result reduced so as to correspond in all respects with the numbers for hydrogen is 2·5186*.

But before these numbers can be compared with the object of obtaining the relative densities, a correction of some importance is required, which appears to have been overlooked by Professor Cooke, as it was by Regnault. The weight of the gas is *not* to be found by merely taking the difference of the full and empty weighings, unless indeed the weighings are conducted *in vacuo*. The external volume of the globe is larger when it is full than when it is empty, and the weight of the air corresponding to this difference of volume must be *added* to the apparent weight of the gas.

By filling the globe with carefully boiled water, it is not difficult to determine experimentally the expansion per atmosphere. In the case of globe (14) it appears that under normal atmospheric conditions the quantity to be added to the apparent weights of the hydrogen and oxygen is 0·00056 gram.

* An examination of the weights revealed no error worth taking into account at present.

The actually observed alteration of volume (regard being had to the compressibility of water) agrees very nearly with an *à priori* estimate, founded upon the theory of thin spherical elastic shells and the known properties of glass. The proportional value of the required correction, in my case about $\frac{4}{1000}$ of the weight of the hydrogen, will be for spherical globes proportional to a/t, where a is the radius of the globe, and t the thickness of the shell, or to V/W, if V be the contents, and W the weight of the glass. This ratio is nearly the same for Professor Cooke's globe and for mine; but the much greater departure of his globe from the spherical form may increase the amount of the correction which ought to be introduced.

In the estimates now to be given, which must be regarded as provisional, the apparent weight of the hydrogen is taken at 0·15804, so that the real weight is 0·15860. The weight of the same volume of oxygen under the same conditions is $2{\cdot}5186 + 0{\cdot}0006 = 2{\cdot}5192$. The ratio of these numbers is 15·884.

The ratio of densities found by Regnault was 15·964, but the greater part of the difference may well be accounted for by the omission of the correction just now considered.

In order to interpret our result as a ratio of atomic weights, we need to know accurately the ratio of atomic volumes. The number given as most probable by Mr Scott in May, 1887 *, was 1·994, but he informs me that more recent experiments under improved conditions give 1·9965. Combining this with the ratio of densities, we obtain as the ratio of atomic weights—

$$\frac{2 \times 15{\cdot}884}{1{\cdot}9965} = 15{\cdot}912.$$

It is not improbable that experiments conducted on the same lines, but with still greater precautions, may raise the final number by one or even two thousandths of its value.

The ratio obtained by Professor Cooke is 15·953; but the difference between this number and that above obtained may be more than accounted for, if I am right in my suggestion that his gas weighings require correction for the diminished buoyancy of the globe when the internal pressure is removed.

[1901. Further work upon this subject is recorded in *Proc. Roy. Soc.* Vol. L. p. 449, 1892.]

* *Loc. cit.* [1901. Dr Scott's final number (*Proc. Roy. Soc.* Vol. LIII. p. 133, 1893) was 2·00245.]

147.

ON POINT-, LINE-, AND PLANE-SOURCES OF SOUND.

[*Proceedings of the London Mathematical Society*, XIX. pp. 504—507, 1888.]

THE velocity-potential at a distance ρ from a simple source of sound is*

$$\phi = \frac{\Phi_1 e^{ik(at-\rho)}}{4\pi a^2 \rho}, \quad \text{.............................(1)}$$

where $-a^{-2}\Phi_1 e^{ikat}$ represents the rate at which fluid is being introduced at the source at time t. In order to apply this to a linear source of unit intensity, coincident with the axis of y, we have to imagine that the introduction of fluid along the element dy is equal to $dy\, e^{ikat}$; so that, if for the sake of brevity we omit the time factor e^{ikat}, we may take as the velocity potential

$$\phi = -\frac{1}{4\pi}\int_{-\infty}^{+\infty} \frac{e^{-ik\rho}\,dy}{\rho}. \quad \text{......................(2)}$$

If r be the distance of the point at which ϕ is to be estimated from the axis of y,

$$\rho^2 = r^2 + y^2,$$

and

$$\phi = -\frac{1}{2\pi}\int_r^{\infty} \frac{e^{-ik\rho}\,d\rho}{\sqrt{(\rho^2 - r^2)}} = -\frac{1}{2\pi}\int_1^{\infty}\frac{e^{-ikrv}\,dv}{\sqrt{(v^2-1)}}, \quad \text{...........(3)}$$

if $\rho = rv$.

The relation of (3) to Bessel's functions is best studied by the method of Lipschitz†. Consider the integral $\int \frac{e^{-rw}\,dw}{\sqrt{(1+w^2)}}$, where w is a complex variable of the form $u + iv$. If we represent, as usual, simultaneous pairs of values of u and v by the coordinates of a point, the integral will vanish when taken round any closed circuit not including the points $w = \pm i$. The first circuit we have to consider is that enclosed by the axes of u and v, and the quadrant of a circle whose centre is the origin and whose radius is infinite. It is easy

* *Theory of Sound*, § 277.

† *Crelle*, Bd. LVI., 1859.

to see that along this quadrant the integral ultimately vanishes, so that the result is the same whether we integrate from 0 to ∞ along the axis of u or from 0 to $i\infty$ along the axis of v. Thus

$$\int_0^\infty \frac{e^{-ru}\,du}{\sqrt{(1+u^2)}} = \int_0^{i\infty} \frac{e^{-r(iv)}\,d(iv)}{\sqrt{(1+i^2v^2)}} = i\int_0^1 \frac{e^{-irv}\,dv}{\sqrt{(1-v^2)}} + \int_1^\infty \frac{e^{-irv}\,dv}{\sqrt{(v^2-1)}}. \quad \text{...(4)}$$

In like manner, the integral along the axis of u from 0 to ∞ is equal to that along the course from 0 to i along the axis of v, and then to infinity along a line through i parallel to u. Thus

$$\int_0^\infty \frac{e^{-ru}\,du}{\sqrt{(1+u^2)}} = \int_0^i \frac{e^{-irv}\,d(iv)}{\sqrt{(1-v^2)}} + \int_0^\infty \frac{e^{-r(u+i)}\,du}{\sqrt{\{1+(u+i)^2\}}}$$

$$= i\int_0^1 \frac{e^{-irv}\,dv}{\sqrt{(1-v^2)}} + \int_0^\infty \frac{e^{-ir}e^{-ru}\,du}{\sqrt{(2iu+u^2)}}. \quad \text{......................(5)}$$

By comparison of (4), (5), or at once by equating the results of integrating from the point i to $i\infty$, and to $\infty + i$, we get

$$\int_1^\infty \frac{e^{-irv}\,dv}{\sqrt{(v^2-1)}} = \int_0^\infty \frac{e^{-ir}e^{-ru}\,du}{\sqrt{(2iu+u^2)}} = \frac{e^{-ir}}{\sqrt{(2ir)}}\int_0^\infty \frac{e^{-\beta}\beta^{-\frac{1}{2}}\,d\beta}{\sqrt{\left(1+\dfrac{\beta}{2ir}\right)}}$$

$$= \left(\frac{\pi}{2ir}\right)^{\frac{1}{2}} e^{-ir}\left\{1 - \frac{1^2}{1\,.\,8ir} + \frac{1^2\,.\,3^2}{1\,.\,2\,.\,(8ir)^2} - \frac{1^2\,.\,3^2\,.\,5^2}{1\,.\,2\,.\,3\,(8ir)^3} + \ldots\right\}. \quad \text{...(6)}$$

This is the series in descending powers of r by which is expressed the effect of a linear source at a great distance.

Equation (4) may be written in the form

$$\int_0^\infty \frac{e^{-\beta}\,d\beta}{\sqrt{(\beta^2+r^2)}} = i\int_0^{\frac{1}{2}\pi} e^{-ir\cos\theta}\,d\theta + \int_0^1 \frac{e^{-irv}\,dv}{\sqrt{(v^2-1)}};$$

or, if we put, as usual,

$$\frac{2}{\pi}\int_0^{\frac{1}{2}\pi} \cos(r\cos\theta)\,d\theta = J_0(r), \quad \text{......................(7)}$$

$$\frac{2}{\pi}\int_0^{\frac{1}{2}\pi} \sin(r\cos\theta)\,d\theta = K_0(r), \quad \text{......................(8)}$$

and separate the real and imaginary parts,

$$\int_0^1 \frac{\cos(rv)\,dv}{\sqrt{(v^2-1)}} = \int_0^\infty \frac{e^{-\beta}\,d\beta}{\sqrt{(\beta^2+r^2)}} - \frac{\pi}{2}\,.\,K_0(r), \quad \text{..............(9)}$$

$$\int_0^1 \frac{\sin(rv)\,dv}{\sqrt{(v^2-1)}} = \frac{\pi}{2}\,.\,J_0(r), \quad \text{......................(10)}$$

the latter giving Mehler's integral expressive of the Bessel's function of order zero*.

* *Math. Ann.* v. p. 141.

By integrating the effect of a linear source, parallel to y, with respect to a perpendicular coordinate x, we may obtain the effect of a source uniformly distributed over a plane. If the rate of introduction of fluid over the area $dxdy$ be $dxdy\,e^{ikat}$, the value of ϕ at a point distant z from the plane, will be found by integrating (3) with respect to x, connected with r and z by the relation

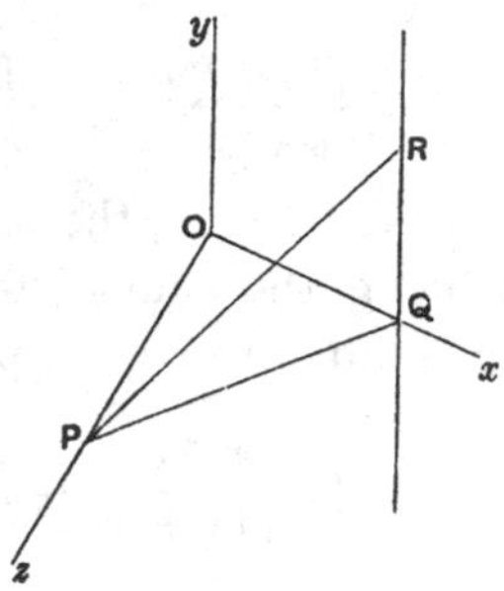

$$r^2 = z^2 + x^2;$$

see Fig. 1, in which

$$RQ = y, \quad PR = \rho, \quad OQ = x,$$
$$PQ = r, \quad OP = z.$$

Fig. 1.

Thus $$\phi = -\frac{1}{\pi}\int_0^\infty dx \int_1^\infty \frac{e^{-ikrv}\,dv}{\sqrt{(v^2-1)}} = -\frac{1}{\pi}\int_z^\infty \frac{r\,dr}{\sqrt{(r^2-z^2)}}\int_1^\infty \frac{e^{-ikrv}\,dv}{\sqrt{(v^2-1)}}. \quad \text{...(11)}$$

The result of a uniform plane source is of course a train of plane waves issuing from it symmetrically in both directions. On the positive side $\phi = Ae^{-ikz}$, where A is a constant readily determined. For $\frac{d\phi}{dz}(z=0) = -ikA$; and this, representing the half of the rate of introduction of fluid per unit area, is by supposition equal to $\frac{1}{2}$. Thus

$$\phi = \frac{i}{2k}e^{-ikz} = \frac{i}{2k}\cos kz + \frac{1}{2k}\sin kz. \quad \text{..................(12)}$$

Comparing the two expressions for ϕ, and having regard to (9) and (10), we see that

$$\int_z^\infty \frac{J_0(kr)\,r\,dr}{\sqrt{(r^2-z^2)}} = \frac{\cos kz}{k}, \quad \text{.......................(13)}$$

$$\int_z^\infty \frac{r\,dr}{\sqrt{(r^2-z^2)}}\left\{\int_0^\infty \frac{e^{-\beta}\,d\beta}{\sqrt{(\beta^2+k^2r^2)}} - \frac{\pi}{2}\,.\,K_0(kr)\right\} = \frac{\sin kz}{k}. \quad \text{......(14)}$$

If we use the series (6), the identity may be written

$$-\frac{1}{\pi}\int_z^\infty \frac{r\,dr}{\sqrt{(r^2-z^2)}}\left(\frac{\pi}{2ikr}\right)^{\frac{1}{2}} e^{-ikr}\left\{1 - \frac{1^2}{1\,.\,8\,.\,ikr} + \ldots\right\} = \frac{i}{2k}e^{-ikz}. \quad \text{...(15)}$$

This equation is easily verified when kz (and therefore kr) is great. Under these circumstances the series may be replaced by its first term; also with sufficient approximation

$$\frac{\sqrt{r}}{\sqrt{(r^2-z^2)}} = \frac{1}{\sqrt{2}\,.\,\sqrt{(r-z)}},$$

since only those elements for which r differs little from z contribute sensibly to the integral.

148.

WAVE THEORY OF LIGHT.

[*Encyclopædia Britannica*, XXIV., 1888.]

§ 1. A GENERAL statement of the principles of the undulatory theory, with elementary explanations, has already been given under Light [*Enc. Brit.* Vol. XIV.], and in the article on Ether the arguments which point to the existence of an all-pervading medium, susceptible in its various parts of an alternating change of state, have been traced by a master hand; but the subject is of such great importance, and is so intimately involved in recent optical investigation and discovery, that a more detailed exposition of the theory, with application to the leading phenomena, was reserved for a special article. That the subject is one of difficulty may be at once admitted. Even in the theory of sound, as conveyed by aerial vibrations, where we are well acquainted with the nature and properties of the vehicle, the fundamental conceptions are not very easy to grasp, and their development makes heavy demands upon our mathematical resources. That the situation is not improved when the medium is hypothetical will be easily understood. For, although the evidence is overwhelming in favour of the conclusion that light is propagated as a vibration, we are almost entirely in the dark as to what it is that vibrates and the manner of vibration. This ignorance entails an appearance of vagueness even in those parts of the subject the treatment of which would not really be modified by the acquisition of a more precise knowledge, *e.g.*, the theory of the colours of thin plates, and of the resolving power of optical instruments. But in other parts of the subject, such as the explanation of the laws of double refraction and of the intensity of light reflected at the surface of a transparent medium, the vagueness is not merely one of language; and if we wish to reach definite results by the *à priori* road we must admit a hypothetical element, for which little justification can be given. The distinction here indicated should be borne clearly in mind. Many optical phenomena must necessarily agree with any kind of wave theory that can be proposed; others may agree or disagree with a particular form of it. In the latter case we may regard the special form as disproved, but the undulatory theory in the proper wider sense remains untouched.

Of such special forms of the wave theory the most famous is that which assimilates light to the transverse vibrations of an elastic solid. *Transverse* they must be in order to give room for the phenomena of polarization. This theory is a great help to the imagination, and allows of the deduction of definite results which are at any rate mechanically possible. An isotropic solid has in general two elastic properties—one relating to the recovery from an alteration of volume, and the other to the recovery from a state of shear, in which the strata are caused to slide over one another. It has been shown by Green that it would be necessary to suppose the luminiferous medium to be incompressible, and thus the only admissible differences between one isotropic medium and another are those of *rigidity* and of *density*. Between these we are in the first instance free to choose. The slower propagation of light in glass than in air may be equally well explained by supposing the rigidity the same in both cases while the density is greater in glass, or by supposing that the density is the same in both cases while the rigidity is greater in air. Indeed there is nothing, so far, to exclude a more complicated condition of things, in which both the density and rigidity vary in passing from one medium to another, subject to the one condition only of making the ratio of velocities of propagation equal to the known refractive index between the media.

When we come to apply this theory to investigate the intensity of light reflected from (say) a glass surface, and to the diffraction of light by very small particles (as in the sky), we find that a reasonable agreement with the facts can be brought about only upon the supposition that the rigidity is the same (approximately, at any rate) in various media, and that the density alone varies. At the same time we have to suppose that the vibration is perpendicular to the plane of polarization.

Up to this point the accordance may be regarded as fairly satisfactory; but, when we extend the investigation to crystalline media in the hope of explaining the observed laws of double refraction, we find that the suppositions which would suit best here are inconsistent with the conclusions we have already arrived at. In the first place, and so long as we hold strictly to the analogy of an elastic solid, we can only explain double refraction as depending upon anisotropic rigidity, and this can hardly be reconciled with the view that the rigidity is the same in different isotropic media. And if we pass over this difficulty, and inquire what kind of double refraction a crystalline solid would admit of, we find no such correspondence with observation as would lead us to think that we are upon the right track.

The theory of anisotropic solids, with its twenty-one elastic constants, seems to be too wide for optical double refraction, which is of a much simpler character*.

For these and other reasons, especially the awkwardness with which it lends itself to the explanation of dispersion, the elastic solid theory, valuable as a piece of purely dynamical reasoning, and probably not without mathematical analogy to the truth, can in Optics be regarded only as an illustration.

In recent years a theory has been received with much favour in which light is regarded as an electromagnetic phenomenon. The dielectric medium is conceived to be subject to a rapidly periodic "electric displacement," the variations of which have the magnetic properties of an electric current. On the basis of purely electrical observations Maxwell calculated the velocity of propagation of such disturbances, and obtained a value not certainly distinguishable from the velocity of light. Such an agreement is very striking; and a further deduction from the theory, that the specific inductive capacity of a transparent medium is equal to the square of the refractive index, is supported to some extent by observation. The foundations of the electrical theory are not as yet quite cleared of more or less arbitrary hypothesis; but, when it becomes certain that a dielectric medium is susceptible of vibrations propagated with the velocity of light, there will be no hesitation in accepting the identity of such vibrations with those to which optical phenomena are due. In the meantime, and apart altogether from the question of its probable truth, the electromagnetic theory is very instructive, in showing us how careful we must be to avoid limiting our ideas too much as to the nature of the luminous vibrations.

§ 2. *Plane Waves of Simple Type.*

Whatever may be the character of the medium and of its vibration, the analytical expression for an infinite train of plane waves is

$$A \cos \left\{ \frac{2\pi}{\lambda} (Vt - x) + \alpha \right\}, \quad \text{.......................(1)}$$

in which λ represents the wave-length, and V the corresponding velocity of propagation. The coefficient A is called the amplitude, and its nature depends upon the medium, and must therefore here be left an open question. The phase of the wave at a given time and place is represented by α. The expression retains the same value whatever integral number of wave-lengths

* See Stokes, "Report on Double Refraction," *Brit. Assoc. Report*, 1862, p. 253.

be added to or subtracted from x. It is also periodic with respect to t, and the period is

$$\tau = \lambda / V. \quad \text{.................................(2)}$$

In experimenting upon sound we are able to determine independently τ, λ, and V; but, on account of its smallness, the periodic time of luminous vibrations eludes altogether our means of observation and is only known indirectly from λ and V by means of (2).

There is nothing arbitrary in the use of a circular function to represent the waves. As a general rule this is the only kind of wave which can be propagated without a change of form; and, even in the exceptional cases where the velocity is independent of wave-length, no generality is really lost by this procedure, because in accordance with Fourier's theorem any kind of periodic wave may be regarded as compounded of a series of such as (1), with wave-lengths in harmonical progression.

A well-known characteristic of waves of type (1) is that any number of trains of various amplitudes and phases, but of the *same wave-length*, are equivalent to a single train of the same type. Thus

$$\Sigma A \cos\left\{\frac{2\pi}{\lambda}(Vt - x) + \alpha\right\}$$

$$= \Sigma A \cos\alpha \,.\, \cos\frac{2\pi}{\lambda}(Vt - x) - \Sigma A \sin\alpha \,.\, \sin\frac{2\pi}{\lambda}(Vt - x)$$

$$= P \cos\left\{\frac{2\pi}{\lambda}(Vt - x) + \phi\right\}, \quad \text{....................(3)}$$

where

$$P^2 = (\Sigma A \cos\alpha)^2 + \Sigma(A \sin\alpha)^2, \qquad \tan\phi = \frac{\Sigma(A\sin\alpha)}{\Sigma(A\cos\alpha)}. \quad \text{...(4, 5)}$$

An important particular case is that of two component trains only.

$$A \cos\left\{\frac{2\pi}{\lambda}(Vt - x) + \alpha\right\} + A' \cos\left\{\frac{2\pi}{\lambda}(Vt - x) + \alpha'\right\} = P \cos\left\{\frac{2\pi}{\lambda}(Vt - x) + \phi\right\},$$

where

$$P^2 = A^2 + A'^2 + 2AA' \cos(\alpha - \alpha'). \quad \text{....................(6)}$$

The composition of vibrations of the same period is precisely analogous, as was pointed out by Fresnel, to the composition of forces, or indeed of any other two-dimensional vector quantities. The magnitude of the force corresponds to the amplitude of the vibration, and the inclination of the force corresponds to the phase. A group of forces of equal intensity, represented by lines drawn from the centre to the angular points of a regular polygon, constitute a system in equilibrium. Consequently, a system of vibrations of equal amplitude and of phases symmetrically distributed round the period has a zero resultant.

According to the phase-relation, determined by $(\alpha - \alpha')$, the amplitude of the resultant may vary from $(A - A')$ to $(A + A')$. If A' and A are equal, the minimum resultant is zero, showing that two equal trains of waves may neutralize one another. This happens when the phases are opposite, or differ by half a (complete) period, and the effect is usually spoken of as the *interference* of light. From a purely dynamical point of view the word is not very appropriate, the vibrations being simply *superposed* with as little interference as can be imagined.

§ 3. *Intensity.*

The intensity of light of given wave-length must depend upon the amplitude, but the precise nature of the relation is not at once apparent. We are not able to appreciate by simple inspection the relative intensities of two unequal lights; and when we say, for example, that one candle is twice as bright as another, we mean that two of the latter burning independently would give us the same light as one of the former. This may be regarded as the definition; and then experiment may be appealed to to prove that the intensity of light from a given source varies inversely as the square of the distance. But our conviction of the truth of the law is perhaps founded quite as much upon the idea that something not liable to loss is radiated outwards, and is distributed in succession over the surfaces of spheres concentric with the source, whose areas are as the squares of the radii. The something can only be energy; and thus we are led to regard the rate at which energy is propagated across a given area parallel to the waves as the measure of intensity; and this is proportional, not to the first power, but to the *square* of the amplitude.

Practical photometry is usually founded upon the law of inverse squares (*Enc. Brit.* Vol. XIV. p. 583); and it should be remembered that the method involves essentially the use of a diffusing screen, the illumination of which, seen in a certain direction, is assumed to be independent of the precise direction in which the light falls upon it; for the distance of a candle, for example, cannot be altered without introducing at the same time a change in the apparent magnitude, and therefore in the incidence of some part at any rate of the light.

With this objection is connected another which is often of greater importance, the necessary enfeeblement of the light by the process of diffusion. And, if to maintain the brilliancy we substitute regular reflectors for diffusing screens, the method breaks down altogether by the apparent illumination becoming independent of the distance of the source of light.

The use of a revolving disk with transparent and opaque sectors in order to control the brightness, as proposed by Fox Talbot*, may often be recom-

* *Phil. Mag.* Vol. v. p. 331, 1834.

mended in scientific photometry, when a great loss of light is inadmissible. The law that, when the frequency of intermittence is sufficient to give a steady appearance, the brightness is proportional to the angular magnitude of the open sectors appears to be well established.

§ 4. *Resultant of a Large Number of Vibrations of Arbitrary Phase.*

We have seen that the resultant of two vibrations of equal amplitude is wholly dependent upon their phase-relation, and it is of interest to inquire what we are to expect from the composition of a large number (n) of equal vibrations of amplitude unity, and of arbitrary phases. The intensity of the resultant will of course depend upon the precise manner in which the phases are distributed, and may vary from n^2 to zero. But is there a definite intensity which becomes more and more probable as n is increased without limit?

The nature of the question here raised is well illustrated by the special case in which the possible phases are restricted to two *opposite* phases. We may then conveniently discard the idea of phase, and regard the amplitudes as at random *positive or negative.* If all the signs are the same, the intensity is n^2; if, on the other hand, there are as many positive as negative, the result is zero. But, although the intensity may range from 0 to n^2, the smaller values are much more probable than the greater.

The simplest part of the problem relates to what is called in the theory of probabilities the "expectation" of intensity, that is, the mean intensity to be expected after a great number of trials, in each of which the phases are taken at random. The chance that all the vibrations are positive is 2^{-n}, and thus the expectation of intensity corresponding to this contingency is $2^{-n} . n^2$. In like manner the expectation corresponding to the number of positive vibrations being $(n-1)$ is

$$2^{-n} . n . (n-2)^2,$$

and so on. The whole expectation of intensity is thus

$$\frac{1}{2^n}\left\{1 . n^2 + n . (n-2)^2 + \frac{n(n-1)}{1.2}(n-4)^2 + \frac{n(n-1)(n-2)}{1.2.3}(n-6)^2 + \ldots\right\}. \quad \ldots\ldots\ldots(1)$$

Now the sum of the $(n+1)$ terms of this series is simply n, as may be proved by comparison of coefficients of x^2 in the equivalent forms

$$(e^x + e^{-x})^n = 2^n(1 + \tfrac{1}{2}x^2 + \ldots)^n$$

$$= e^{nx} + ne^{(n-2)x} + \frac{n(n-1)}{1.2}e^{(n-4)x} + \ldots .$$

The expectation of intensity is therefore n, and this whether n be great or small.

The same conclusion holds good when the phases are unrestricted. From (4), § 2, if $A = 1$,

$$P^2 = n + 2\Sigma \cos(\alpha_2 - \alpha_1), \quad \text{.......................(2)}$$

where under the sign of summation are to be included the cosines of the $\frac{1}{2}n(n-1)$ differences of phase. When the phases are arbitrary, this sum is as likely to be positive as negative, and thus the mean value of P^2 is n.

The reader must be on his guard here against a fallacy which has misled some high authorities. We have not proved that when n is large there is any tendency for a single combination to give the intensity equal to n, but the quite different proposition that in a large number of trials, in each of which the phases are rearranged arbitrarily, the *mean* intensity will tend more and more to the value n. It is true that even in a single combination there is no reason why any of the cosines in (2) should be positive rather than negative, and from this we may infer that when n is increased the sum of the terms tends to vanish in comparison with the number of terms. But, the number of terms being of the order n^2, we can infer nothing as to the value of the sum of the series in comparison with n.

Indeed it is not true that the intensity in a single combination approximates to n, when n is large. It can be proved* that the probability of a resultant intermediate in amplitude between r and $r + dr$ is

$$\frac{2}{n} e^{-r^2/n} r\,dr. \quad \text{.............................(3)}$$

The probability of an amplitude less than r is thus

$$\frac{2}{n}\int_0^r e^{-r^2/n} r\,dr = 1 - e^{-r^2/n}, \quad \text{.......................(4)}$$

or, which is the same thing, the probability of an amplitude greater than r is

$$e^{-r^2/n}. \quad \text{...................................(5)}$$

The accompanying table gives the probabilities of intensities less than the fractions of n named in the first column. For example, the probability of intensity less than n is ·6321.

·05	·0488	·80	·5506
·10	·0952	1·00	·6321
·20	·1813	1·50	·7768
·40	·3296	2·00	·8647
·60	·4512	3·00	·9502

It will be seen that, however great n may be, there is a fair chance of considerable relative fluctuations of intensity in consecutive combinations.

* *Phil. Mag.* Aug. 1880 [Vol. I. p. 491].

The *mean* intensity, expressed by

$$\frac{2}{n}\int_0^\infty e^{-r^2/n}\,.\,r^2\,.\,r\,dr,$$

is, as we have already seen, equal to n.

It is with this mean intensity only that we are concerned in ordinary photometry. A source of light, such as a candle or even a soda flame, may be regarded as composed of a very large number of luminous centres disposed throughout a very sensible space; and, even though it be true that the intensity at a particular point of a screen illuminated by it and at a particular moment of time is a matter of chance, further processes of averaging must be gone through before anything is arrived at of which our senses could ordinarily take cognizance. In the smallest interval of time during which the eye could be impressed, there would be opportunity for any number of rearrangements of phase, due either to motions of the particles or to irregularities in their modes of vibration. And even if we supposed that each luminous centre was fixed, and emitted perfectly regular vibrations, the manner of composition and consequent intensity would vary rapidly from point to point of the screen, and in ordinary cases the mean illumination over the smallest appreciable area would correspond to a thorough averaging of the phase-relationships. In this way the idea of the intensity of a luminous source, independently of any questions of phase, is seen to be justified, and we may properly say that two candles are twice as bright as one.

§ 5. *Propagation of Waves in General.*

It has been shown under Optics [Vol. II. p. 387], that a system of rays, however many reflexions or refractions they may have undergone, are always normal to a certain surface, or rather system of surfaces. From our present point of view these surfaces are to be regarded as wave-surfaces, that is, surfaces of constant phase. It is evident that, so long as the radius of curvature is very large in comparison with λ, each small part of a wave-surface propagates itself just as an infinite plane wave coincident with the tangent plane would do. If we start at time t with a given surface, the corresponding wave-surface at time $t+dt$ is to be found by prolonging every normal by the length Vdt, where V denotes the velocity of propagation at the place in question. If the medium be uniform, so that V is constant, the new surface is *parallel* to the old one, and this property is retained however many short intervals of time be considered in succession. A wave-surface thus propagates itself *normally*, and the corresponding parts of successive surfaces are those which lie upon the same normal. In this sense the normal may be regarded as a *ray*, but the idea must not be pushed to streams of light limited to pass through small apertures. The manner in which the phase is determined by the length of the ray, and the conditions under which energy may be regarded as travelling along a ray, will be better treated under the head of Huygens's principle, and the theory of shadows (§ 10).

From the law of propagation, according to which the wave-surfaces are always as far advanced as possible, it follows that the course of a ray is that for which the time, represented by $\int V^{-1}ds$, is a minimum. This is Fermat's principle of least time. Since the refractive index (μ) varies as V^{-1}, we may take $\int \mu ds$ as the measure of the retardation between one wave-surface and another; and it is the same along whichever ray it may be measured.

The principle that $\int \mu ds$ is a minimum along a ray lends itself readily to the investigation of optical laws. As an example, we will consider the very important theory of magnifying power. Let A_0, B_0 be two points upon a wave-surface before the light enters the object-glass of a telescope, A, B the corresponding points upon a wave-surface after emergence from the eye-piece, both surfaces being plane. The value of $\int \mu ds$ is the same along the ray A_0A as along B_0B; and, if from any cause B_0 be slightly retarded relatively to A_0, then B will be retarded to the same amount relatively to A. Suppose now that the retardation in question is due to a small rotation (θ) of the wave-surface A_0B_0 about an axis in its own plane perpendicular to AB. The retardation of B_0 relatively to A_0 is then $A_0B_0 . \theta$; and in like manner, if ϕ be the corresponding rotation of AB, the retardation is $AB . \phi$. Since these retardations are the same, we have

$$\frac{\phi}{\theta} = \frac{A_0B_0}{AB},$$

or *the magnifying power is equal to the ratio of the widths of the stream of light before and after passing the telescope.*

The magnifying power is not necessarily the same in all directions. Consider the case of a prism arranged as for spectrum work. Passage through the prism does not alter the vertical width of the stream of light; hence there is no magnifying power in this direction. What happens in a horizontal direction depends upon circumstances. A single prism in the position of minimum deviation does not alter the horizontal width of the beam. The same is true of a sequence of any number of prisms each in the position of minimum deviation, or of the combination called by Thollon a couple, when the deviation is the least that can be obtained by rotating the couple as a *rigid system*, although a further diminution might be arrived at by violating this tie. In all these cases there is neither horizontal nor vertical magnification, and the instrument behaves as a telescope of power unity. If, however, a prism be so placed that the angle of emergence differs from the angle of incidence, the horizontal width of the beam undergoes a change. If the emergence be nearly grazing, there will be a high magnifying power in the horizontal direction; and, whatever may be the character of the system of prisms, the horizontal magnifying power is represented by the ratio of widths. Brewster suggested that, by combining two prisms with refracting edges at right angles, it would be possible to secure equal magnifying power in the two directions, and thus to imitate the action of an ordinary telescope.

The theory of magnifying power is intimately connected with that of apparent brightness. By the use of a telescope in regarding a bright body, such, for example, as the moon, there is a concentration of light upon the pupil in proportion to the ratio of the area of the object-glass to that of the pupil*. But the apparent brightness remains unaltered, the apparent superficial magnitude of the object being changed in precisely the same proportion, in accordance with the law just established.

These fundamental propositions were proved a long while since by Cotes and Smith; and a complete exposition of them, from the point of view of geometrical optics, is to be found in Smith's treatise†.

§ 6. *Waves Approximately Plane or Spherical.*

A plane wave of course remains plane after reflexion from a truly plane surface; but any irregularities in the surface impress themselves upon the wave. In the simplest case, that of perpendicular incidence, the irregularities are *doubled*, any depressed portion of the surface giving rise to a retardation in the wave-front of twice its own amount. It is assumed that the lateral dimensions of the depressed or elevated parts are large multiples of the wave-length; otherwise the assimilation of the various parts to plane waves is not legitimate.

In like manner, if a plane wave passes perpendicularly through a parallel plate of refracting material, a small elevation t at any part of one of the surfaces introduces a retardation $(\mu - 1)t$ in the corresponding part of the wave-surface. An error in a glass surface is thus of only one-quarter of the importance of an equal error in a reflecting surface. Further, if a plate, otherwise true, be distorted by bending, the errors introduced at the two surfaces are approximately opposite, and neutralize one another‡.

* It is here assumed that the object-glass is large enough to fill the whole of the pupil with light; also that the glasses are perfectly transparent, and that there is no loss of light by reflexion. For theoretical purposes the latter requirement may be satisfied by supposing the transition between one optical medium and another to be gradual in all cases.

† Smith, *Compleat System of Optics*, Cambridge, 1738. The reader may be referred to a paper entitled "Notes, chiefly Historical, on some Fundamental Propositions in Optics" (*Phil. Mag.* June 1886 [Vol. II. Art. 137]), in which some account is given of Smith's work, and its relation to modern investigations.

‡ On this principle Grubb has explained the observation that the effects of bending stress are nearly as prejudicial in the case of thick object-glasses as in the case of thin ones.

In practical applications it is of importance to recognize the effects of a small departure of the wave-surface from its ideal plane or spherical form. Let the surface be referred to a system of rectangular coordinates, the axis of z being normal at the centre of the section of the beam, and the origin being the point of contact of the tangent plane. If, as happens in many cases, the surface be one of symmetry round OZ, the equation of the surface may be represented approximately by

$$z = r^2/2\rho + Ar^4 + \ldots, \qquad \text{.........................(1)}$$

in which ρ is the radius of curvature, or focal length, and $r^2 = x^2 + y^2$. If the surface be truly spherical, $A = 1/8\rho^3$, and any deviation of A from this value indicates ordinary symmetrical spherical aberration.

If, however, the surface be not symmetrical, we may have to encounter aberration of a lower order of small quantities, and therefore presumably of higher importance. By taking the axis of x and y coincident with the directions of principal curvature at O, we may write the equation of the surface

$$z = \frac{x^2}{2\rho} + \frac{y^2}{2\rho'} + \alpha x^3 + \beta x^2 y + \gamma xy^2 + \delta y^3, \qquad \text{..............(2)}$$

ρ, ρ' being the principal radii of curvature, or focal lengths. The most important example of unsymmetrical aberration is in the spectroscope, where (if the faces of the prisms may be regarded as at any rate surfaces of revolution) the wave-surface may by suitable adjustments be rendered symmetrical with respect to the horizontal plane $y = 0$. This plane may then be regarded as primary, ρ being the primary focal length, at which distance the spectrum is formed. Under these circumstances β and δ may be omitted from (2), which thus takes the form

$$z = \frac{x^2}{2\rho} + \frac{y^2}{2\rho'} + \alpha x^3 + \gamma xy^2. \qquad \text{......................(3)}$$

The constants α and γ in (3) may be interpreted in terms of the differential coefficients of the principal radii of curvature. By the usual formula the radius of curvature at the point x of the intersection of (3) with the plane $y = 0$ is approximately $\rho(1 - 6\alpha\rho x)$. Since $y = 0$ is a principal plane throughout, this radius of curvature is a principal radius of the surface; so that, denoting it by ρ, we have

$$\alpha = \frac{1}{6}\frac{d\rho^{-1}}{dx}. \qquad \text{............................(4)}$$

Again, in the neighbourhood of the origin, the approximate value of the product of the principal curvatures is

$$\frac{1}{\rho\rho'} + \frac{6\alpha x}{\rho'} + \frac{2\gamma x}{\rho}.$$

Thus

$$d\left(\frac{1}{\rho\rho'}\right) = -\frac{d\rho}{\rho^2\rho'} - \frac{d\rho'}{\rho'^2\rho} = \frac{6\alpha x}{\rho'} + \frac{2\gamma x}{\rho};$$

whence by (4)

$$\gamma = \tfrac{1}{2}\frac{d\rho'^{-1}}{dx}. \quad \text{.............................(5)}$$

The equation of the normal at the point x, y, z is

$$\frac{\zeta - z}{-1} = \frac{\xi - x}{\rho^{-1}x + 3\alpha x^2 + \gamma y^2} = \frac{\eta - y}{\rho'^{-1}y + 2\gamma xy}; \quad \text{............(6)}$$

and its intersection with the plane $\zeta = \rho$ occurs at the point determined approximately by

$$\xi = -\rho\,(3\alpha x^2 + \gamma y^2), \qquad \eta = \frac{(\rho' - \rho)\,y}{\rho'} - 2\rho\gamma xy, \quad \text{...........(7)}$$

terms of the third order being omitted.

According to geometrical optics, the thickness of the image of a luminous line at the primary focus is determined by the extreme value of ξ; and for good definition in the spectroscope it is necessary to reduce this thickness as much as possible. One way of attaining the desired result would be to narrow the aperture; but, as we shall see later, to narrow the horizontal aperture is really to throw away the peculiar advantage of large instruments. The same objection, however, does not apply to narrowing the *vertical* aperture; and in many spectroscopes a great improvement in definition may be thus secured. In general, it is necessary that both γ and α be small. Since the value of ξ does not depend on ρ', it would seem that in respect of definition there is no advantage in avoiding astigmatism.

The width of the image when $\eta = 0$ (corresponding to $y = 0$) is $3\alpha\rho x^2$, and vanishes when $\alpha = 0$, *i.e.*, when there is no aberration for rays in the primary plane. In this case the image reduces to a linear arc. If further $\gamma = 0$, this arc becomes straight, and then the image at the primary focus is perfect to this order of approximation. As an example where $\alpha = 0$, the image of a luminous point, formed at an equal distance on the further side of a sloped equi-convex lens, may be mentioned.

At the secondary focus, $\zeta = \rho'$, and from (6)

$$\xi = x\,\frac{\rho - \rho'}{\rho}, \qquad \eta = -2\rho'\gamma xy. \quad \text{...................(8)}$$

If $\gamma = 0$, the secondary focal line is formed without aberration, but not otherwise. Both focal lines are well formed when parallel rays fall upon a plano-convex lens, sloped at about 30°, the curved side of the lens being turned towards the parallel rays.

[1900. A plane reflecting plate, which reflects approximately plane waves perpendicularly, may be *bent* so as to eliminate all the errors of wave-front represented in (2). For the solution of the equation of bending, viz.

$$\frac{d^4w}{dx^4} + 2\frac{d^4w}{dx^2dy^2} + \frac{d^4w}{dy^4} = 0,$$

includes arbitrary terms in all powers of x, y below the fourth. Errors of the same nature in the (unbent) surface of the reflector are eliminated at the same time.]

§ 7. *Interference Fringes.*

We have seen (§ 2) that, when two trains of parallel waves of equal wave-length are superposed, the intensity of the resultant depends upon the phase-relation of the components; but it is necessarily the same at all points of the wave-front. It not unfrequently happens that the parallelism of the component trains is approximate only, and there then arises the phenomenon known as interference fringes. If the two directions of propagation be inclined on opposite sides to the axis of x at small angles α, the expressions for two components of equal amplitudes are

$$\cos\frac{2\pi}{\lambda}\left\{Vt - x\cos\alpha - y\sin\alpha\right\}, \quad \text{and} \quad \cos\frac{2\pi}{\lambda}\left\{Vt - x\cos\alpha + y\sin\alpha\right\};$$

so that the resultant is expressed by

$$2\cos\frac{2\pi y\sin\alpha}{\lambda}\,.\,\cos\frac{2\pi}{\lambda}(Vt - x\cos\alpha);\;\ldots\ldots\ldots\ldots\ldots(1)$$

from which it appears that the vibrations advance parallel to the axis of x, unchanged in type, and with a uniform velocity $V/\cos\alpha$. Considered as depending on y, the vibration is a maximum when $y\sin\alpha$ is equal to 0, λ, 2λ, 3λ, etc., corresponding to the centres of the bright bands, while for the intermediate values $\frac{1}{2}\lambda$, $\frac{3}{2}\lambda$, &c., there is no vibration. This is the interference of light proceeding from two similar homogeneous and very distant sources.

In the form of experiment adopted by Fresnel the sources O_1, O_2* are situated at a finite distance D from the place of observation (*Enc. Brit.* Vol. XIV. p. 606). If A be the point of the screen equidistant from O_1, O_2, and P a neighbouring point, then approximately

$$O_1P - O_2P = \sqrt{\{D^2 + (u + \tfrac{1}{2}b)^2\}} - \sqrt{\{D^2 + (u - \tfrac{1}{2}b)^2\}} = ub/D,$$

where $O_1O_2 = b$, $AP = u$.

* It is scarcely necessary to say that O_1, O_2 must not be distinct sources of light; otherwise there could be no fixed phase-relation and consequently no regular interference. In Fresnel's experiment O_1, O_2 are virtual images of one real source O, obtained by reflexion in two mirrors. The mirrors may be replaced by a bi-prism. Or, as in Lloyd's arrangement, O_1 may be identical with O, and O_2 obtained by a grazing reflexion from a single mirror.

Thus, if λ be the wave-length, the places where the phases are accordant are given by

$$u = n\lambda D/b, \quad \text{.................................(2)}$$

n being an integer.

If the light were really homogeneous, the successive fringes would be similar to one another and unlimited in number; moreover there would be no place that could be picked out by inspection as the centre of the system. In practice λ varies, and the only place of complete accordance for all kinds of light is at A, where $u = 0$. Theoretically, there is no place of complete discordance for all kinds of light, and consequently no complete blackness. In consequence, however, of the fact that the range of sensitiveness of the eye is limited to less than an "octave," the centre of the first dark band (on either side) is sensibly black, even when white light is employed; but it should be carefully remarked that the existence of even one band is due to selection, and that the formation of several visible bands is favoured by the capability of the retina to make chromatic distinctions within the visible range.

The number of perceptible bands increases *pari passu* with the approach of the light to homogeneity. For this purpose there are two methods that may be used.

We may employ light, such as that from the soda flame, which possesses *ab initio* a high degree of homogeneity. If the range of wave-length included be $\frac{1}{50000}$, a corresponding number of interference fringes may be made visible. The above is the number obtained by Fizeau, and Michelson has recently gone as far as 200,000. The narrowness of the bright line of light seen in the spectroscope, and the possibility of a large number of Fresnel's bands, depend upon precisely the same conditions; the one is in truth as much an interference phenomenon as the other.

In the second method the original light may be highly composite, and homogeneity is brought about with the aid of a spectroscope. The analogy with the first method is closest if we use the spectroscope to give us a line of homogeneous light in simple substitution for the artificial flame. Or, following Foucault and Fizeau, we may allow the white light to pass, and subsequently analyse the mixture transmitted by a narrow slit in the screen upon which the interference bands are thrown. In the latter case we observe a channelled spectrum, with maxima of brightness corresponding to the wave-lengths $bu/(nD)$. In either case the number of bands observable is limited solely by the resolving power of the spectroscope (§ 13), and proves nothing with respect to the regularity, or otherwise, of the vibrations of the original light.

The truth of this remark is strikingly illustrated by the possible formation, with white light, of a large number of achromatic bands. The unequal widths of the bands for the various colours, and consequent overlapping and obliteration, met with in the usual form of the experiment, depend upon the constancy of b (the mutual distance of the two sources) while λ varies. It is obvious that, if b were proportional to λ, the widths of the bands would be independent of λ, and that the various systems would fit together perfectly. To carry out the idea in its entirety, it would be necessary to use a diffraction spectrum as a source, and to duplicate this by Lloyd's method with a single reflector placed so that $b=0$ when $\lambda=0$. [*Phil. Mag.* XXVIII. p. 77, 1889.] In practice a sufficiently good result could doubtless be obtained with a prismatic spectrum (especially if the red and violet were removed by absorbing agents) under the condition that $d(b/\lambda)=0$ in the yellow-green. It is remarkable that, in spite of the achromatic character of the bands, their possible number is limited still by the resolving power of the instrument used to form the spectrum.

If a system of Fresnel's bands be examined through a prism, the central white band undergoes an abnormal displacement, which has been supposed to be inconsistent with theory. The explanation has been shown by Airy* to depend upon the peculiar manner in which the white band is in general formed.

"Any one of the kinds of homogeneous light composing the incident heterogeneous light will produce a series of bright and dark bars, unlimited in number as far as the mixture of light from the two pencils extends, and undistinguishable in quality. The consideration, therefore, of homogeneous light will never enable us to determine which is the point that the eye immediately turns to as the centre of the fringes. What then is the physical circumstance that determines the centre of the fringes?

"The answer is very easy. For different colours the bars have different breadths. If then the bars of all colours coincide at one part of the mixture of light, they will not coincide at any other part; but at equal distances on both sides from that place of coincidence they will be equally far from a state of coincidence. If then we can find where the bars of all colours coincide, that point is the centre of the fringes.

"It appears then that the centre of the fringes is *not* necessarily the point where the two pencils of light have described equal paths, but is determined by considerations of a perfectly different kind......The distinction is important in this and in other experiments."

The effect in question depends upon the dispersive power of the prism. If v be the linear shifting due to the prism of the originally central band, v must be regarded as a function of λ. Measured from the original centre, the position of the nth bar is now

$$v + n\lambda D/b.$$

* "Remarks on Mr Potter's Experiment on Interference," *Phil. Mag.* Vol. II. p. 161, 1833.

The coincidence of the various bright bands occurs when this quantity is as independent as possible of λ, that is, when n is the nearest integer to

$$n = -\frac{b}{D}\frac{dv}{d\lambda}; \quad \text{.................................(3)}$$

or, as Airy expresses it in terms of the width of a band (h), $n = -dv/dh$.

The apparent displacement of the white band is thus not v simply, but

$$v - h\frac{dv}{dh}. \quad \text{.....................................(4)}$$

The signs of dv and dh being opposite, the abnormal displacement is in addition to the normal effect of the prism. But, since dv/dh, or $dv/d\lambda$, is not constant, the achromatism of the white band is less perfect than when no prism is used.

If a grating were substituted for the prism, v would vary as h, and (4) would vanish, so that in all orders of spectra the white band would be seen undisplaced.

The theoretical error, dependent upon the dispersive power, involved in the method of determining the refractive index of a plate by means of the displacement of a system of interference fringes (*Enc. Brit.* Vol. XIV. p. 607) has been discussed by Stokes*. In the absence of dispersion the retardation R due to the plate would be independent of λ, and therefore completely compensated at the point determined by $u = DR/b$; but when there is dispersion it is accompanied by a fictitious displacement of the fringes on the principle explained by Airy.

More recently the matter has engaged the attention of Cornu†, who thus formulates the general principle:—"*Dans un système de franges d'interférences produites à l'aide d'une lumière hétérogène ayant un spectre continu, il existe toujours une frange achromatique qui joue le rôle de frange centrale et qui se trouve au point de champ où les radiations les plus intenses présentent une différence de phase maximum ou minimum.*"

In Fresnel's experiment, if the retardation of phase due to an interposed plate, or to any other cause, be $F(\lambda)$, the whole relative retardation of the two pencils at the point u is

$$\phi = F(\lambda) + \frac{bu}{\lambda D}; \quad \text{..............................(5)}$$

and the situation of the central, or achromatic, band is determined, not by $\phi = 0$, but by $d\phi/d\lambda = 0$, or

$$u = \lambda^2 DF'(\lambda)/b. \quad \text{..............................(6)}$$

* *Brit. Assoc. Rep.*, 1850.

† *Jour. de Physique*, I. p. 293, 1882.

In the theoretical statement we have supposed the source of light to be limited to a mathematical point, or to be extended only in the vertical direction (parallel to the bands). Such a vertical extension, while it increases illumination, has no prejudicial effect upon distinctness, the various systems due to different points of the luminous line being sensibly superposed. On the other hand, the horizontal dimension of the source must be confined within narrow limits, the condition obviously being that the displacement of the centre of the system incurred by using in succession the two edges only of the slit should be small in comparison with the width of an interference band.

Before quitting this subject it is proper to remark that Fresnel's bands are more influenced by diffraction than their discoverer supposed. On this account the fringes are often unequally broad and undergo fluctuations of brightness. A more precise calculation has been given by H. F. Weber* and by H. Struve†, but the matter is too complicated to be further considered here. The observations of Struve appear to agree well with the corrected theory.

§ 8. *Colours of Thin Plates.*

When plane waves of homogeneous light (λ) fall upon a parallel plate of index μ, the resultant reflected wave is made up of an infinite number of components, of which the most important are the first, reflected at the upper surface of the plate, and the second, transmitted at the upper surface, reflected at the under surface, and then transmitted at the upper surface. It is readily proved (*Enc. Brit.* Vol. XIV. p. 608) that so far as it depends

[Fig. 0.]

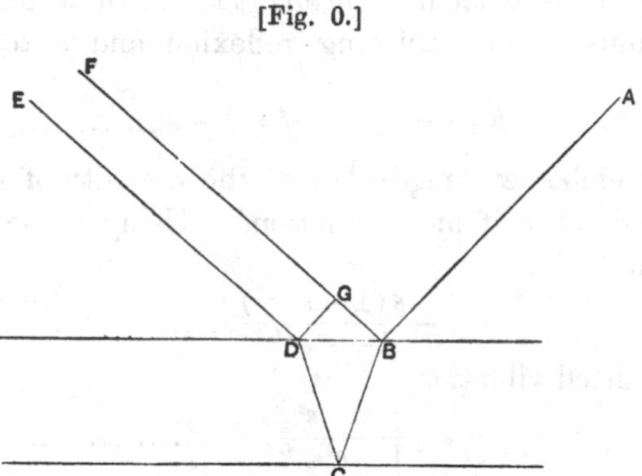

upon the distances to be travelled in the plate and in air the retardation (δ) of the second wave relatively to the first is given by

$$\delta = 2\mu l \cos \alpha', \quad \ldots\ldots\ldots\ldots\ldots\ldots\ldots\ldots\ldots\ldots(1)$$

* *Wied. Ann.* VIII. p. 407.

† *Wied. Ann.* XV. p. 49.

where t denotes the thickness of the plate, and α' the angle of refraction corresponding to the first entrance. [1900. $ABF = 2\alpha$, $BCD = 2\alpha'$,

$$\begin{aligned}\delta &= \mu(BC + CD) - BG\\ &= 2\mu BC - 2BC \sin\alpha' \sin\alpha\\ &= 2\mu BC(1 - \sin^2\alpha') = 2\mu t \cos\alpha'.]\end{aligned}$$

If we represent all the vibrations by complex quantities, from which finally the imaginary parts are to be rejected, the retardation δ may be expressed by the introduction of the factor $\epsilon^{-i\kappa\delta}$, where $i = \sqrt{(-1)}$, and $\kappa = 2\pi/\lambda$.

At each reflexion or refraction the amplitude of the incident wave must be supposed to be altered by a certain factor. When the light proceeds from the surrounding medium to the plate, the factor for reflexion will be supposed to be b, and for refraction c; the corresponding quantities when the progress is from the plate to the surrounding medium will be denoted by e, f. Denoting the incident vibration by unity, we have then for the first component of the reflected wave b, for the second $cef\epsilon^{-i\kappa\delta}$, for the third $ce^3f\epsilon^{-2i\kappa\delta}$, and so on. Adding these together, and summing the geometric series, we find

$$b + \frac{cef\epsilon^{-i\kappa\delta}}{1 - e^2\epsilon^{-i\kappa\delta}}. \qquad \text{(2)}$$

In like manner for the wave transmitted through the plate we get

$$\frac{cf}{1 - e^2\epsilon^{-i\kappa\delta}}. \qquad \text{(3)}$$

The quantities b, c, e, f are not independent. The simplest way to find the relations between them is to trace the consequences of supposing $\delta = 0$ in (2) and (3). For it is evident *à priori* that with a plate of vanishing thickness there must be a vanishing reflexion and a total transmission. Accordingly,

$$b + e = 0, \qquad cf = 1 - e^2, \qquad \text{(4)}$$

the first of which embodies Arago's law of the equality of reflexions, as well as the famous "loss of half an undulation." Using these we find for the reflected vibration,

$$-\frac{e(1 - \epsilon^{-i\kappa\delta})}{1 - e^2\epsilon^{-i\kappa\delta}}, \qquad \text{(5)}$$

and for the transmitted vibration

$$\frac{1 - e^2}{1 - e^2\epsilon^{-i\kappa\delta}}. \qquad \text{(6)}$$

The intensities of the reflected and transmitted lights are the squares of the moduli of these expressions. Thus

Intensity of reflected light

$$= e^2\frac{(1 - \cos\kappa\delta)^2 + \sin^2\kappa\delta}{(1 - e^2\cos\kappa\delta)^2 + e^4\sin^2\kappa\delta} = \frac{4e^2\sin^2(\tfrac{1}{2}\kappa\delta)}{1 - 2e^2\cos\kappa\delta + e^4}; \qquad \text{(7)}$$

Intensity of transmitted light

$$= \frac{(1-e^2)^2}{1-2e^2\cos\kappa\delta+e^4}, \quad \text{.........................(8)}$$

the sum of the two expressions being unity.

According to (7) not only does the reflected light vanish completely when $\delta = 0$, but also whenever $\frac{1}{2}\kappa\delta = n\pi$, n being an integer, that is, whenever $\delta = n\lambda$. When the first and third medium are the same, as we have here supposed, the central spot in the system of Newton's rings is *black*, even though the original light contain a mixture of all wave-lengths. The general explanation of the colours of Newton's rings is given under "Light" [*Enc. Brit.* Vol. XIV.], to which reference must be made. If the light reflected from a plate of any thickness be examined with a spectroscope of sufficient resolving power (§ 13), the spectrum will be traversed by dark bands, of which the centres correspond to those wave-lengths which the plate is incompetent to reflect. It is obvious that there is no limit to the fineness of the bands which may be thus impressed upon a spectrum, whatever may be the character of the original mixed light.

[1900. As ordinarily observed, Newton's rings depend upon the variable thickness of the thin plate, which is seen in focus. This disposition implies that the rays which proceeding from a given part of the plate and filling the aperture of the eye are ultimately brought to a point upon the retina, are incident at *various* obliquities. The confusion is least when the incidence is approximately perpendicular, and it is usually of no importance when the whole retardation is small, as when coloured bands are formed from white light. But when we proceed to high interference the difficulty arising from variable obliquity increases, and it becomes necessary to pay great attention to the perpendicularity of the incidence, and perhaps to contract the aperture of the eye. A stage is soon reached at which it is better to abandon this procedure altogether and to focus the eye, not upon the plate, but for an infinite distance, so as to combine at one point of the retina rays which are incident in a *given direction*. If the surfaces of the plate are absolutely parallel, an ideal ring system is then formed, the centre of the system corresponding to perpendicular incidence, and each ring to a definite degree of obliquity. Accurately parallel surfaces may be obtained very simply from a layer of water resting upon mercury (*Nature*, XLVIII. p. 212, 1893). In this method no slit, or limitation of the beam, otherwise than in the pupil of the eye, is anywhere required.

The illumination depends upon the intensity of the monochromatic source and upon the reflecting power of the surfaces. If R denote the intensity of reflected light, as given in (7),

$$\frac{1}{R} = 1 + \frac{(1-e^2)^2}{4e^2\sin^2(\frac{1}{2}\kappa\delta)}. \quad \text{.........................(7')}$$

If $e=1$ absolutely,

$$1/R = R = 1$$

for all values of δ. If $e=1$ very nearly, $R=1$ nearly for all values of δ for which $\sin(\frac{1}{2}\kappa\delta)$ is not very small. The field will be of the full brightness corresponding to the source, but will be traversed by *narrow* black lines.

This condition of things may be approximated to in the case of the layer of water over mercury by making the reflexion very oblique. The experiment in this form succeeds, but the high obliquity is inconvenient. In the researches of MM. Fabry and Perot the *transmitted* light is employed with an incidence approximately perpendicular. If a transparent plate could be composed of material for which $e=1$ nearly, the transmitted light $(1-R)$ would nearly vanish except when $\sin(\frac{1}{2}\kappa\delta)$ is close to zero. The field would be *dark* in general, but be traversed by *narrow bright* lines. Unfortunately there is no transparent material giving nearly complete reflexion at perpendicular incidence, but MM. Fabry and Perot have obtained very interesting results by the use of lightly silvered glass surfaces. The silvered surfaces may include a plate of air, of which the thickness can then be regulated, or they may be the external surfaces of a plate of glass, which needs to be very accurately formed. This arrangement constitutes a *spectroscope,* inasmuch as it allows the structure of a complex spectrum line to be directly observed. If for example we look at a soda flame, we see in general two distinct systems of narrow bright circles corresponding to the two *D*-lines. With particular values of the thickness of the plate of air the two systems may coincide so as to be seen as a single system, but a slight alteration of thickness will cause a separation. One peculiarity of the light from a soda flame will at once strike the observer more conspicuously than with any other form of spectroscope. If the flame contains but little soda, the lines of the two systems are very unequal in brightness, but the difference greatly diminishes when the supply of soda is increased, as would be necessary from the first in other methods of observation. In using this apparatus the eye of the observer must be focused for infinity, and the adjustment of the reflecting surfaces to parallelism must be very exact. A small movement of the eye in any direction should not entail an expansion or contraction of the rings.

In Michelson's apparatus the colours reflected from a thin plate are obtained without actual approximation of the reflecting surfaces. By means of it Michelson has made a very thorough and successful comparison of the standards of length and the wave-lengths of the radiation obtained by electric discharge from cadmium vapour in a vacuum tube.]

The relations between the factors *b, c, e, f* have been proved, independently of the theory of thin plates, in a general manner by Stokes*, who called to his

* "On the Perfect Blackness of the Central Spot in Newton's Rings, and on the Verification of Fresnel's Formulæ for the Intensities of Reflected and Refracted Rays." *Camb. and Dub. Math. Jour.* Vol. IV. p. 1, 1849; reprint Vol. II. p. 89.

aid the general mechanical principle of *reversibility*. If the motions constituting the reflected and refracted rays to which an incident ray gives rise be supposed to be reversed, they will reconstitute a reversed incident ray. This gives one relation; and another is obtained from the consideration that there is no ray in the second medium, such as would be generated by the operation alone of either the reversed reflected or refracted rays. Space does not allow of the reproduction of the argument at length, but a few words may perhaps give the reader an idea of how the conclusions are arrived at. The incident ray (IA) being 1, the reflected (AR) and refracted (AF) rays are denoted by b and c. When b is reversed, it gives rise to a reflected ray b^2 along AI, and a refracted ray bc along AG (say). When c is reversed, it gives rise to cf along AI, and ce along AG. Hence $bc + ce = 0$, $b^2 + cf = 1$, which agree with (4).

Fig. 1.

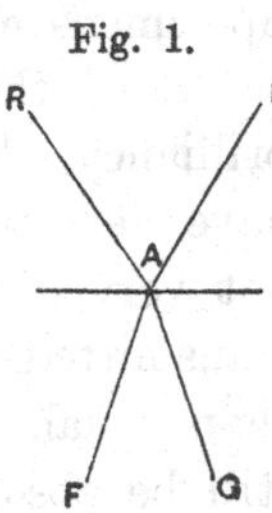

It is here assumed that there is no change of phase in the act of reflexion or refraction, except such as can be represented by a change of sign. Professor Stokes has, however, pushed the application of his method to the case where changes of phase are admitted, and arrives at the conclusion that "the sum of the accelerations of phase at the two reflexions is equal to the sum of the accelerations at the two refractions, and the accelerations of the two refractions are equal to each other." The accelerations are supposed to be so measured as to give like signs to c and f, and unlike to b and e. The same relations as before obtain between the factors b, c, e, f, expressing the ratios of amplitudes*.

When the third medium differs from the first, the theory of thin plates is more complicated, and need not here be discussed. One particular case, however, may be mentioned. When a thin transparent film is backed by a perfect reflector, no colours should be visible, all the light being ultimately reflected, whatever the wave-length may be. The experiment may be tried with a thin layer of gelatin on a polished silver plate. In other cases where

* It would appear, however, that these laws cannot be properly applied to the calculation of reflexion from a thin plate. This is sufficiently proved by the fact that the resultant expression for the intensity founded upon them does not vanish with the thickness. The truth is that the method of deducing the aggregate reflexion from the consideration of the successive partial reflexions and refractions is applicable only when the disturbance in the interior of the plate is fully represented by the transverse waves considered in the argument, whereas the occurrence of a change of phase is probably connected with the existence of additional superficial waves (§ 27). The existence of these superficial waves may be ignored when the reflected and refracted waves are to be considered only at distances from the surface exceeding a few wave-lengths, but in the application to thin plates this limitation is violated. If indeed the method of calculating the aggregate reflexion from a thin plate were sound when a change of phase occurs, we could still use the expressions (2) and (3), merely understanding by b, c, e, f, factors which may be complex; and the same formal relations (4) would still hold good. These do not agree with those found by Stokes by the method of reversion; and the discrepancy indicates that, when there are changes of phase, the action of a thin plate cannot be calculated in the usual way.

a different result is observed, the inference is that either the metal does not reflect perfectly, or else that the material of which the film is composed is not sufficiently transparent.

Theory and observation alike show that the transmitted colours of a thin plate, *e.g.*, a soap film or a layer of air, are very inferior to those reflected. Specimens of ancient glass, which have undergone superficial decomposition, on the other hand, sometimes show transmitted colours of remarkable brilliancy. The probable explanation, suggested by Brewster, is that we have here to deal not merely with one, but with a series of thin plates of not very different thicknesses. It is evident that with such a series the transmitted colours would be much purer, and the reflected much brighter, than usual. If the thicknesses are strictly equal, certain wave-lengths must still be absolutely missing in the reflected light; while on the other hand a constancy of the interval between the plates will in general lead to a special preponderance of light of some other wave-length for which all the component parts as they ultimately emerge are in agreement as to phase*.

All that can be expected from a physical theory is the determination of the composition of the light reflected from or transmitted by a thin plate in terms of the composition of the incident light. The further question of the chromatic character of the mixtures thus obtained belongs rather to physiological optics, and cannot be answered without a complete knowledge of the chromatic relations of the spectral colours themselves. Experiments upon this subject have been made by various observers, and especially by Maxwell†, who has exhibited his results on a colour diagram as used by Newton. A calculation of the colours of thin plates, based upon Maxwell's data, and accompanied by a drawing showing the curve representative of the entire series up to the fifth order, has recently been published‡; and to this the reader who desires further information must be referred, with the remark that the true colours are not seen in the usual manner of operating with a plate of air enclosed between glass surfaces, on account of the contamination with white light reflected at the other surfaces of the glasses. This objection is avoided when a soap film is employed, to the manifest advantage of the darker colours, such as the red of the first order. The colours of Newton's scale are met with also in the light transmitted by a somewhat thin plate of doubly-refracting material, such as mica, the plane of analysis being perpendicular to that of primitive polarization.

* The analytical investigations and formulæ given by Stokes for a pile of plates (*Proc. Roy. Soc.* Vol. XI. p. 545, 1860) may be applied to this question, provided that we understand the quantities r, t, ϕ, ψ, &c., to be complex, so as to express the luminous displacement in phase as well as in amplitude, instead of real quantities relating merely to *intensities*.

† Maxwell, "Theory of Compound Colours," *Phil. Trans.*, 1860.

‡ *Edin. Trans.*, 1887 [Vol. II. p. 498].

The same series of colours occur also in other optical experiments, *e.g.*, at the centre of the illuminated area when light issuing from a point passes through a small round aperture in an otherwise opaque screen (§ 10).

The colours of which we have been speaking are those formed at nearly perpendicular incidence, so that the retardation (reckoned as a distance), viz., $2\mu t \cos\alpha'$, is sensibly independent of λ. This state of things may be greatly departed from when the thin plate is rarer than its surroundings, and the incidence is such that α' is nearly equal to 90°, for then, in consequence of the powerful dispersion, $\cos\alpha'$ may vary greatly as we pass from one colour to another. Under these circumstances the series of colours entirely alters its character, and the bands (corresponding to a graduated thickness) may even lose their coloration, becoming sensibly black and white through many alternations*. The general explanation of this remarkable phenomenon was suggested by Newton, but it does not appear to have been followed out in accordance with the wave theory.

Let us suppose that plane waves of white light travelling in glass are incident at angle α upon a plate of air, which is bounded again on the other side by glass. If μ be the index of the *glass*, α' the angle of refraction, then $\sin\alpha' = \mu \sin\alpha$; and the retardation, expressed by the equivalent distance in air, is

$$2t \sec\alpha' - \mu \,.\, 2t \tan\alpha' \sin\alpha = 2t \cos\alpha';$$

and the retardation in *phase* is $2t \cos\alpha'/\lambda$, λ being as usual the wave-length in air.

The first thing to be noticed is that, when α approaches the critical angle, $\cos\alpha'$ becomes as small as we please, and that consequently the retardation corresponding to a given thickness is very much less than at perpendicular incidence. Hence the glass surfaces need not be so close as usual.

A second feature is the increased brilliancy of the light. According to (7) the intensity of the reflected light when at a maximum ($\sin\frac{1}{2}\kappa\delta = 1$) is $4e^2/(1+e^2)^2$. At perpendicular incidence e is about $\frac{1}{5}$, and the intensity is somewhat small; but, as $\cos\alpha'$ approaches zero, e approaches unity (§ 26), and the brilliancy is much increased.

But the peculiarity which most demands attention is the lessened influence of a variation in λ upon the phase-retardation. A diminution of λ of itself increases the retardation of phase, but, since waves of shorter wave-length are more refrangible, this effect may be more or less perfectly compensated by the greater obliquity and consequent diminution in the value of $\cos\alpha'$. We will investigate the conditions under which the retardation of phase is stationary in spite of a variation of λ.

* Newton's *Optics*, bk. II.; Fox Talbot, *Phil. Mag.* Vol. IX. p. 401, 1836.

In order that $\lambda^{-1}\cos\alpha'$ may be stationary, we must have

$$\lambda \sin\alpha' d\alpha' + \cos\alpha' d\lambda = 0,$$

where (α being constant)

$$\cos\alpha' d\alpha' = \sin\alpha\, d\mu.$$

Thus

$$\cot^2\alpha' = -\frac{\lambda}{\mu}\frac{d\mu}{d\lambda}, \qquad (9)$$

giving α' when the relation between μ and λ is known.

According to Cauchy's formula, which represents the facts very well throughout most of the visible spectrum,

$$\mu = A + B\lambda^{-2}, \qquad (10)$$

so that

$$\cot^2\alpha' = \frac{2B}{\lambda^2\mu} = \frac{2(\mu - A)}{\mu}. \qquad (11)$$

If we take, as for Chance's "extra-dense flint," $B = \cdot 984 \times 10^{-10}$, and as for the soda lines, $\mu = 1\cdot 65$, $\lambda = 5\cdot 89 \times 10^{-5}$, we get

$$\alpha' = 79° \, 30'.$$

At this angle of refraction, and with this kind of glass, the retardation of phase is accordingly nearly independent of wave-length, and therefore the bands formed, as the thickness varies, are approximately achromatic. Perfect achromatism would be possible only under a law of dispersion

$$\mu^2 = A' - B'\lambda^2.$$

If the source of light be distant and very small, the black bands are wonderfully fine and numerous. The experiment is best made (after Newton) with a right-angled prism, whose hypothenusal surface may be brought into approximate contact with a plate of black glass. The bands should be observed with a convex lens, of about 8 inches focus. If the eye be at twice this distance from the prism, and the lens be held midway between, the advantages are combined of a large field and of maximum distinctness.

If Newton's rings are examined through a prism, some very remarkable phenomena are exhibited, described in his twenty-fourth observation*: "When the two object-glasses are laid upon one another, so as to make the rings of the colours appear, though with my naked eye I could not discern above eight or nine of those rings, yet by viewing them through a prism I could see a far greater multitude, insomuch that I could number more than forty......And I believe that the experiment may be improved to the discovery of far greater numbers.. ...But it was but one side of these rings, namely, that towards which the refraction was made, which by the

* Newton's *Optics*. See also Place, *Pogg. Ann.* CXIV. p. 504, 1861.

refraction was rendered distinct, and the other side became more confused than when viewed with the naked eye......

"I have sometimes so laid one object-glass upon the other that to the naked eye they have all over seemed uniformly white, without the least appearance of any of the coloured rings; and yet by viewing them through a prism great multitudes of those rings have discovered themselves."

Newton was evidently much struck with these "so odd circumstances"; and he explains the occurrence of the rings at unusual thicknesses as due to the dispersing power of the prism. The blue system being more refracted than the red, it is possible under certain conditions that the nth blue ring may be so much displaced relatively to the corresponding red ring as *at one part of the circumference* to compensate for the different diameters. A white stripe may thus be formed in a situation where without the prism the mixture of colours would be complete, so far as could be judged by the eye.

The simplest case that can be considered is when the "thin plate" is bounded by plane surfaces inclined to one another at a small angle. By drawing back the prism (whose edge is parallel to the intersection of the above-mentioned planes) it will always be possible so to adjust the effective dispersing power as to bring the nth bars to coincidence for any two assigned colours, and therefore approximately for the entire spectrum. The formation of the achromatic band, or rather central black band, depends indeed upon the same principles as the fictitious shifting of the centre of a system of Fresnel's bands when viewed through a prism.

But neither Newton nor, as would appear, any of his successors has explained why the bands should be more numerous than usual, and under certain conditions sensibly achromatic for a large number of alternations. It is evident that, in the particular case of the wedge-shaped plate above specified, such a result would not occur. The width of the bands for any colour would be proportional to λ, as well after the displacement by the prism as before; and the succession of colours formed in white light and the number of perceptible bands would be much as usual.

The peculiarity to be explained appears to depend upon the *curvature* of the surfaces bounding the plate. For simplicity suppose that the lower surface is plane ($y = 0$), and that the approximate equation of the upper surface is $y = a + bx^2$, a being thus the least distance between the plates. The black of the nth order for wave-length λ occurs when

$$\tfrac{1}{2}n\lambda = a + bx^2; \qquad \text{.............................(12)}$$

and thus the width (δx) at this place of the band is given by

$$\tfrac{1}{2}\lambda = 2bx\,\delta x, \qquad \text{.................................(13)}$$

or

$$\delta x = \frac{\lambda}{4bx} = \frac{\lambda}{4\sqrt{b}\,.\,\sqrt{(\frac{1}{2}n\lambda - a)}}. \qquad \text{.....................(14)}$$

If the glasses be in contact, as is usually supposed in the theory of Newton's rings, $a = 0$, and $\delta x \propto \lambda^{\frac{1}{2}}$, or the width of the band of the nth order varies as the square root of the wave-length, instead of as the first power. Even in this case the overlapping and subsequent obliteration of the bands is greatly retarded by the use of the prism, but the full development of the phenomenon requires that a should be finite. Let us inquire what is the condition in order that the width of the band of the nth order may be stationary, as λ varies. By (14) it is necessary that the variation of $\lambda^2/(\frac{1}{2}n\lambda - a)$ should vanish. Hence $a = \frac{1}{4}n\lambda$, so that the interval between the surfaces at the place where the nth band is formed should be half due to curvature and half to imperfect contact at the place of closest approach. If this condition be satisfied, the achromatism of the nth band, effected by the prism, carries with it the achromatism of a large number of neighbouring bands, and thus gives rise to the remarkable effects described by Newton. [1901. For further developments see *Phil. Mag.* Vol. XXVIII. p. 200, 1889.]

§ 9. *Newton's Diffusion Rings.*

In the fourth part of the second book of his *Optics* Newton investigates another series of rings, usually (though not very appropriately) known as the colours of thick plates. The fundamental experiment is as follows. At the centre of curvature of a concave looking-glass, quicksilvered behind, is placed an opaque card, perforated by a small hole through which sunlight is admitted. The main body of the light returns through the aperture; but a series of concentric rings are seen upon the card, the formation of which was proved by Newton to require the co-operation of the two surfaces of the mirror. Thus the diameters of the rings depend upon the thickness of the glass, and none are formed when the glass is replaced by a metallic speculum. The brilliancy of the rings depends upon imperfect polish of the anterior surface of the glass, and may be augmented by a coat of diluted milk, a device used by the Duc de Chaulnes. The rings may also be well observed without a screen in the manner recommended by Stokes. For this purpose all that is required is to place a *small* flame at the centre of curvature of the prepared glass, so as to coincide with its image. The rings are then seen surrounding the flame and occupying a definite position in space.

The explanation of the rings, suggested by Young and developed by Herschel, refers them to interference between one portion of light scattered or diffracted by a particle of dust and then regularly refracted and reflected, and another portion first regularly refracted and reflected and then diffracted at emergence by the same particle. It has been shown by Stokes* that no

* *Camb. Trans.* Vol. IX. p. 147, 1851.

regular interference is to be expected between portions of light diffracted by different particles of dust.

In the memoir of Stokes will be found a very complete discussion of the whole subject, and to this the reader must be referred who desires a fuller knowledge. Our limits will not allow us to do more than touch upon one or two points. The condition of fixity of the rings when observed in air, and of distinctness when a screen is used, is that the systems due to all parts of the diffusing surface should coincide; and it is fulfilled only when, as in Newton's experiments, the source and screen are in the plane passing through the centre of curvature of the glass.

As the simplest for actual calculation, we will consider a little further the case where the glass is plane and parallel, of thickness t and index μ, and is supplemented by a lens at whose focus the source of light is placed. This lens acts both as collimator and as object-glass, so that the combination of lens and plane mirror replaces the concave mirror of Newton's experiment. The retardation is calculated in the same way as for thin plates. In Fig. 2 the diffracting particle is situated at B, and we have to find the relative retardation of the two rays which emerge finally at inclination θ, the one diffracted at emergence following the path $ABDBIE$, and the other diffracted at entrance and following the path $ABFGH$. The retardation of the former from B to I is $2\mu t + BI$, and of the latter from B to the equivalent place G is $2\mu BF$. Now $FB = t \sec \theta'$, θ' being the angle of refraction; $BI = 2t \tan \theta' \sin \theta$; so that the relative retardation R is given by

Fig. 2.

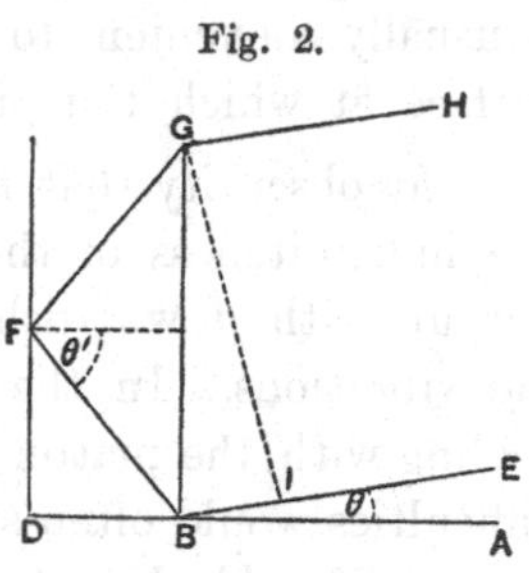

$$R = 2\mu t \{1 + \mu^{-1} \tan \theta' \sin \theta - \sec \theta'\} = 2\mu t (1 - \cos \theta').$$

If θ, θ' be small, we may take

$$R = 2t\theta^2/\mu, \quad \text{..............................(1)}$$

as sufficiently approximate.

The condition of distinctness is here satisfied, since R is the same for every ray emergent parallel to a given one. The rays of one parallel system are collected by the lens to a focus at a definite point in the neighbourhood of the original source.

The formula (1) was discussed by Herschel, and shown to agree with Newton's measures. The law of formation of the rings follows immediately from the expression for the retardation, the radius of the ring of nth order being proportional to n and to the square root of the wave-length.

§ 10. *Huygens's Principle. Theory of Shadows.*

The objection most frequently brought against the undulatory theory in its infancy was the difficulty of explaining in accordance with it the existence of shadows. Thanks to Fresnel and his followers, this department of Optics is now precisely the one in which the theory has secured its greatest triumphs.

The principle employed in these investigations is due to Huygens, and may be thus formulated. If round the origin of waves an ideal closed surface be drawn, the whole action of the waves in the region beyond may be regarded as due to the motion continually propagated across the various elements of this surface. The wave motion due to any element of the surface is called a *secondary* wave, and in estimating the total effect regard must be paid to the phases as well as the amplitudes of the components. It is usually convenient to choose as the surface of resolution a *wave-front, i.e.*, a surface at which the primary vibrations are in one phase.

Any obscurity that may hang over Huygens's principle is due mainly to the indefiniteness of thought and expression which we must be content to put up with if we wish to avoid pledging ourselves as to the character of the vibrations. In the application to sound, where we know what we are dealing with, the matter is simple enough in principle, although mathematical difficulties would often stand in the way of the calculations we might wish to make. The ideal surface of resolution may be there regarded as a flexible lamina; and we know that, if by forces locally applied every element of the lamina be made to move normally to itself exactly as the air at that place does, the external aerial motion is fully determined. By the principle of superposition the whole effect may be found by integration of the partial effects due to each element of the surface, the other elements remaining at rest.

We will now consider in detail the important case in which uniform plane waves are resolved at a surface coincident with a wave-front (OQ). We imagine the wave-front divided into elementary rings or zones, called Huygens's zones, by spheres described round P (the point at which the aggregate effect is to be estimated), the first sphere, touching the plane at O, with a radius equal to PO, and the succeeding spheres with radii increasing at each step by $\frac{1}{2}\lambda$. There are thus marked out a series of circles, whose radii x are given by $x^2 + r^2 = (r + \frac{1}{2}n\lambda)^2$, or $x^2 = n\lambda r$ nearly; so that the rings are at first of nearly equal area. Now the effect upon P of each element of the plane is proportional to its area; but it depends also upon

Fig. 3.

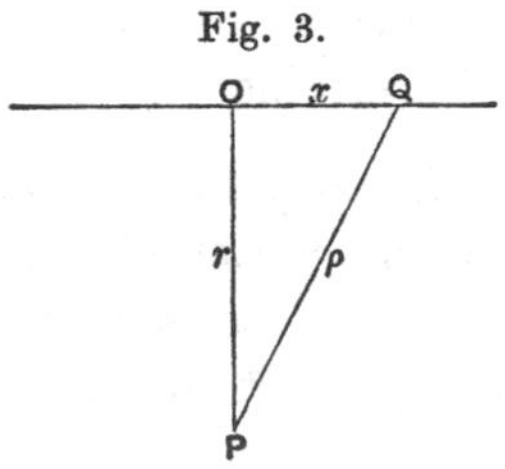

the distance from P, and possibly upon the inclination of the secondary ray to the direction of vibration and to the wave-front. These questions will be further considered in connexion with the dynamical theory; but under all ordinary circumstances the result is independent of the precise answer that may be given. All that it is necessary to assume is that the effects of the successive zones gradually diminish, whether from the increasing obliquity of the secondary ray or because (on account of the limitation of the region of integration) the zones become at last more and more incomplete. The component vibrations at P due to the successive zones are thus nearly equal in amplitude and opposite in phase (the phase of each corresponding to that of the infinitesimal circle midway between the boundaries), and the series which we have to sum is one in which the terms are alternately opposite in sign and, while at first nearly constant in numerical magnitude, gradually diminish to zero. In such a series each term may be regarded as very nearly indeed destroyed by the halves of its immediate neighbours, and thus the sum of the whole series is represented by half the first term, which stands over uncompensated. The question is thus reduced to that of finding the effect of the first zone, or central circle, of which the area is $\pi\lambda r$.

We have seen that the problem before us is independent of the law of the secondary wave as regards obliquity; but the result of the integration necessarily involves the law of the intensity and phase of a secondary wave as a function of r, the distance from the origin. And we may in fact, as was done by A. Smith*, determine the law of the secondary wave, by comparing the result of the integration with that obtained by supposing the primary wave to pass on to P without resolution.

Now as to the phase of the secondary wave, it might appear natural to suppose that it starts from any point Q with the phase of the primary wave, so that on arrival at P it is retarded by the amount corresponding to QP. But a little consideration will prove that in that case the series of secondary waves could not reconstitute the primary wave. For the aggregate effect of the secondary waves is the half of that of the first Huygens zone, and it is the central element only of that zone for which the distance to be travelled is equal to r. Let us conceive the zone in question to be divided into infinitesimal rings of equal area. The effects due to each of these rings are equal in amplitude and of phase ranging uniformly over half a complete period. The phase of the resultant is midway between those of the extreme elements, that is to say, a quarter of a period behind that due to the element at the centre of the circle. It is accordingly necessary to suppose that the secondary waves start with a phase one-quarter of a period in advance of that of the primary wave at the surface of resolution.

* *Camb. Math. Journ.* Vol. III. p. 46, 1843.

Further, it is evident that account must be taken of the variation of phase in estimating the magnitude of the effect at P of the first zone. The middle element alone contributes without deduction; the effect of every other must be found by introduction of a resolving factor, equal to $\cos\theta$, if θ represent the difference of phase between this element and the resultant. Accordingly, the amplitude of the resultant will be less than if all its components had the same phase, in the ratio

$$\int_{-\frac{1}{2}\pi}^{+\frac{1}{2}\pi} \cos\theta \, d\theta : \pi,$$

or $2 : \pi$. Now $2 \text{ area}/\pi = 2\lambda r$; so that, in order to reconcile the amplitude of the primary wave (taken as unity) with the half effect of the first zone, the amplitude, at distance r, of the secondary wave emitted from the element of area dS must be taken to be

$$\frac{dS}{\lambda r} \quad \text{..(1)}$$

By this expression, in conjunction with the quarter-period acceleration of phase, the law of the secondary wave is determined.

That the amplitude of the secondary wave should vary as r^{-1} was to be expected from considerations respecting energy; but the occurrence of the factor λ^{-1}, and the acceleration of phase, have sometimes been regarded as mysterious. It may be well therefore to remember that precisely these laws apply to a secondary wave of sound, which can be investigated upon the strictest mechanical principles.

The recomposition of the secondary waves may also be treated analytically. If the primary wave at O be $\cos kat$, the effect of the secondary wave proceeding from the element dS at Q is

$$\frac{dS}{\lambda\rho}\cos k\,(at-\rho+\tfrac{1}{4}\lambda) = -\frac{dS}{\lambda\rho}\sin k\,(at-\rho).$$

If $dS = 2\pi x\,dx$, we have for the whole effect

$$-\frac{2\pi}{\lambda}\int_0^\infty \frac{\sin k\,(at-\rho)\,x\,dx}{\rho},$$

or, since $x\,dx = \rho\,d\rho$, $k = 2\pi/\lambda$,

$$-k\int_r^\infty \sin k\,(at-\rho)\,d\rho = \Big[-\cos k\,(at-\rho)\Big]_r^\infty.$$

In order to obtain the effect of the primary wave, as retarded by traversing the distance r, viz. $\cos k\,(at-r)$, it is necessary to suppose that the integrated term vanishes at the upper limit. And it is important to notice that without some further understanding the integral is really ambiguous. According to the assumed law of the secondary wave, the result must actually depend upon the precise radius of the outer boundary of the region of integration, supposed to be exactly circular. This case is, however, at most very special and exceptional. We may usually suppose that a large number of the outer rings are incomplete, so that the integrated term at the upper limit may properly be taken to vanish. If a formal proof be desired, it may be obtained by introducing into the integral a factor such as $e^{-h\rho}$, in which h is ultimately made to diminish without limit.

When the primary wave is plane, the area of the first Huygens zone is $\pi\lambda r$, and, since the secondary waves vary as r^{-1}, the intensity is independent of r, as of course it should be. If, however, the primary wave be spherical, and of radius a at the wave-front of resolution, then we know that at a distance r further on the amplitude of the primary wave will be diminished in the ratio $a:(r+a)$. This may be regarded as a consequence of the altered area of the first Huygens zone. For, if x be its radius, we have

$$\sqrt{\{(r+\tfrac{1}{2}\lambda)^2-x^2\}}+\sqrt{\{a^2-x^2\}}=r+a,$$

so that

$$x^2=\frac{\lambda ar}{a+r}\text{ nearly.}$$

Since the distance to be travelled by the secondary waves is still r, we see how the effect of the first zone, and therefore of the whole series, is proportional to $a/(a+r)$. In like manner may be treated other cases, such as that of a primary wave-front of unequal principal curvatures.

The general explanation of the formation of shadows may also be conveniently based upon Huygens's zones. If the point under consideration be so far away from the geometrical shadow that a large number of the earlier zones are complete, then the illumination, determined sensibly by the first zone, is the same as if there were no obstruction at all. If, on the other hand, the point be well immersed in the geometrical shadow, the earlier zones are altogether missing, and, instead of a series of terms beginning with finite numerical magnitude and gradually diminishing to zero, we have now to deal with one of which the terms diminish to zero *at both ends*. The sum of such a series is very approximately zero, each term being neutralized by the halves of its immediate neighbours, which are of the opposite sign. The question of light or darkness then depends upon whether the series begins or ends abruptly. With few exceptions, abruptness can occur only in the presence of the first term, viz. when the secondary wave of least retardation is unobstructed, or when a *ray* passes through the point under consideration. According to the undulatory theory the light cannot be regarded strictly as travelling along a ray; but the existence of an unobstructed ray implies that the system of Huygens's zones can be commenced, and, if a large number of these zones are fully developed and do not terminate abruptly, the illumination is unaffected by the neighbourhood of obstacles. Intermediate cases in which a few zones only are formed belong especially to the province of diffraction.

An interesting exception to the general rule that full brightness requires the existence of the first zone occurs when the obstacle assumes the form of a small circular disk parallel to the plane of the incident waves. In the earlier half of the 18th century* Delisle found that the centre of the circular shadow was occupied by a bright point of light, but the observation passed into oblivion until Poisson brought forward as an objection to Fresnel's theory that it required at the centre of a circular shadow a point as bright as if no obstacle were intervening. If we conceive the primary wave to be broken up at the plane of the disk, a system of Huygens's zones can be constructed which begin from the circumference; and the first zone external to the disk plays the part ordinarily taken by the centre of the entire system. The whole effect is the half of that of the first existing zone, and this is sensibly the same as if there were no obstruction.

When light passes through a small circular or annular aperture, the illumination at any point along the axis depends upon the precise relation between the aperture and the distance from it at which the point is taken. If, as in the last paragraph, we imagine a system of zones to be drawn commencing from the inner circular boundary of the aperture, the question turns upon the manner in which the series terminates at the outer boundary. If the aperture be such as to fit exactly an integral number of zones, the aggregate effect may be regarded as the half of those due to the first and last zones. If the number of zones be even, the action of the first and last zones are antagonistic, and there is complete darkness at the point. If on the other hand the number of zones be odd, the effects conspire; and the illumination (proportional to the square of the amplitude) is four times as great as if there were no obstruction at all.

The process of augmenting the resultant illumination at a particular point by stopping some of the secondary rays may be carried much further†. By the aid of photography it is easy to prepare a plate, transparent where the zones of odd order fall, and opaque where those of even order fall. Such a plate has the power of a condensing lens, and gives an illumination out of all proportion to what could be obtained without it. An even greater effect (fourfold) would be attained if it were possible to provide that the stoppage of the light from the alternate zones were replaced by a phase-reversal without loss of amplitude.

In such experiments the narrowness of the zones renders necessary a pretty close approximation to the geometrical conditions. Thus in the case of the circular disk, equidistant (r) from the source of light and from the

* Verdet, *Leçons d'Optique Physique*, I. § 66.

† Soret, *Pogg. Ann.* CLVI. p. 99, 1875.

screen upon which the shadow is observed, the width of the first exterior zone is given by

$$dx = \frac{\lambda (2r)}{4(2x)},$$

$2x$ being the diameter of the disk. If $2r=1000$ cm., $2x=1$ cm., $\lambda=6\times10^{-5}$ cm., then $dx=\cdot0015$ cm. Hence, in order that this zone may be perfectly formed, there should be no error in the circumference of the order of ·001 cm.* The experiment succeeds in a dark room of the length above mentioned, with a threepenny bit (supported by three threads) as obstacle, the origin of light being a small needle-hole in a plate of tin, through which the sun's rays shine horizontally after reflexion from an external mirror. In the absence of a heliostat it is more convenient to obtain a point of light with the aid of a lens of short focus.

The amplitude of the light at any point in the axis, when plane waves are incident perpendicularly upon an annular aperture, is, as above,

$$\cos k(at - r_1) - \cos k(at - r_2) = 2 \sin kat \,.\, \sin k(r_1 - r_2),$$

r_2, r_1 being the distances of the outer and inner boundaries from the point in question. It is scarcely necessary to remark that in all such cases the calculation applies in the first instance to homogeneous light, and that, in accordance with Fourier's theorem, each homogeneous component of a mixture may be treated separately. When the original light is white, the presence of some components and the absence of others will usually give rise to coloured effects, variable with the precise circumstances of the case.

Although what we have to say upon the subject is better postponed until we consider the dynamical theory, it is proper to point out at once that there is an element of assumption in the application of Huygens's principle to the calculation of the effects produced by opaque screens of limited extent. Properly applied, the principle could not fail; but, as may readily be proved in the case of sonorous waves, it is not in strictness sufficient to assume the expression for a secondary wave suitable when the primary wave is undisturbed, with mere limitation of the integration to the transparent parts of the screen. But, except perhaps in the case of very fine gratings, it is probable that the error thus caused is insignificant; for the incorrect estimation of the secondary waves will be limited to distances of a few wave-lengths only from the boundary of opaque and transparent parts.

§ 11. *Fraunhofer's Diffraction Phenomena.*

A very general problem in diffraction is the investigation of the distribution of light over a screen upon which impinge divergent or convergent spherical waves after passage through various diffracting apertures. When

* It is easy to see that the radius of the bright spot is of the same order of magnitude.

the waves are convergent and the recipient screen is placed so as to contain the centre of convergency—the image of the original radiant point, the calculation assumes a less complicated form. This class of phenomena was investigated by Fraunhofer (upon principles laid down by Fresnel), and are sometimes called after his name. We may conveniently commence with them on account of their simplicity and great importance in respect to the theory of optical instruments.

Fig. 4.

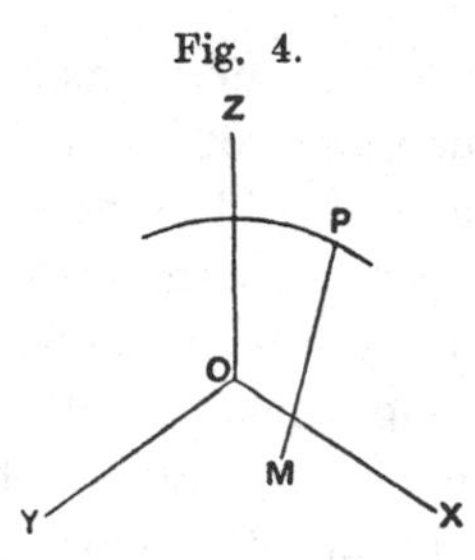

If f be the radius of the spherical wave at the place of resolution, where the vibration is represented by $\cos kat$, then at any point M (Fig. 4) in the recipient screen the vibration due to an element dS of the wave-front is (§ 9)

$$-\frac{dS}{\lambda\rho}\sin k(at-\rho),$$

ρ being the distance between M and the element dS.

Taking coordinates in the plane of the screen with the centre of the wave as origin, let us represent M by ξ, η, and P (where dS is situated) by x, y, z. Then

$$\rho^2=(x-\xi)^2+(y-\eta)^2+z^2, \qquad f^2=x^2+y^2+z^2;$$

so that

$$\rho^2=f^2-2x\xi-2y\eta+\xi^2+\eta^2.$$

In the application with which we are concerned, ξ, η are very small quantities; and we may take

$$\rho=f\left\{1-\frac{x\xi+y\eta}{f^2}\right\}.$$

At the same time dS may be identified with $dxdy$, and in the denominator ρ may be treated as constant and equal to f. Thus the expression for the vibration at M becomes

$$-\frac{1}{\lambda f}\iint\sin k\left\{at-f+\frac{x\xi+y\eta}{f}\right\}dxdy;\ \ldots\ldots\ldots\ldots\ldots\ldots(1)$$

and for the intensity, represented by the square of the amplitude,

$$I^2=\frac{1}{\lambda^2f^2}\left[\iint\sin k\frac{x\xi+y\eta}{f}\,dxdy\right]^2+\frac{1}{\lambda^2f^2}\left[\iint\cos k\frac{x\xi+y\eta}{f}\,dxdy\right]^2.\ \ldots(2)$$

This expression for the intensity becomes rigorously applicable when f is indefinitely great, so that ordinary optical aberration disappears. The incident waves are thus plane, and are limited to a plane aperture coincident with a wave-front. The integrals are then properly functions of the *direction* in which the light is to be estimated.

In experiment under ordinary circumstances it makes no difference whether the collecting lens is in front of or behind the diffracting aperture.

It is usually most convenient to employ a telescope focused upon the radiant point, and to place the diffracting apertures immediately in front of the object-glass. What is seen through the eye-piece in any case is the same as would be depicted upon a screen in the focal plane.

Before proceeding to special cases it may be well to call attention to some general properties of the solution expressed by (2)*.

If, when the aperture is given, the wave-length (proportional to k^{-1}) varies, the composition of the integrals is unaltered, provided ξ and η are taken [directly] proportional to λ. A diminution of λ thus leads to a simple proportional shrinkage of the diffraction pattern, attended by an augmentation of brilliancy in proportion to λ^{-2}.

If the wave-length remains unchanged, similar effects are produced by an increase in the scale of the aperture. The linear dimension of the diffraction pattern is inversely as that of the aperture, and the brightness at corresponding points is as the *square* of the area of aperture.

If the aperture and wave-length increase in the same proportion, the size and shape of the diffraction pattern undergo no change.

We will now apply the integrals (2) to the case of a rectangular aperture of width a parallel to x and of width b parallel to y. The limits of integration for x may thus be taken to be $-\frac{1}{2}a$ and $+\frac{1}{2}a$, and for y to be $-\frac{1}{2}b$, $+\frac{1}{2}b$. We readily find (with substitution for k of $2\pi/\lambda$)

$$I^2 = \frac{a^2b^2}{f^2\lambda^2} \cdot \frac{\sin^2(\pi a\xi/f\lambda)}{\pi^2a^2\xi^2/f^2\lambda^2} \cdot \frac{\sin^2(\pi b\eta/f\lambda)}{\pi^2b^2\eta^2/f^2\lambda^2}, \quad \text{..............(3)}$$

as representing the distribution of light in the image of a mathematical point when the aperture is rectangular, as is often the case in spectroscopes.

The second and third factors of (3) being each of the form $\sin^2 u/u^2$, we have to examine the character of this function. It vanishes when $u = m\pi$, m being any whole number other than zero. When $u = 0$, it takes the value unity. The maxima occur when

$$u = \tan u, \quad \text{....................................(4)}$$

and then

$$\sin^2 u/u^2 = \cos^2 u. \quad \text{...........................(5)}$$

To calculate the roots of (5) we may assume

$$u = (m + \tfrac{1}{2})\pi - y = U - y,$$

where y is a positive quantity which is small when u is large. Substituting this, we find $\cot y = U - y$, whence

$$y = \frac{1}{U}\left(1 + \frac{y}{U} + \frac{y^2}{U^2} + \ldots\right) - \frac{y^3}{3} - \frac{2y^5}{15} - \frac{17y^7}{315}.$$

* Bridge, *Phil. Mag.* Nov. 1858.

This equation is to be solved by successive approximation. It will readily be found that

$$u = U - y = U - U^{-1} - \frac{2}{3}U^{-3} - \frac{13}{15}U^{-5} - \frac{146}{105}U^{-7} - \ldots \quad \ldots\ldots(6)$$

In the first quadrant there is no root after zero, since $\tan u > u$, and in the second quadrant there is none because the signs of u and $\tan u$ are opposite. The first root after zero is thus in the third quadrant, corresponding to $m = 1$. Even in this case the series converges sufficiently to give the value of the root with considerable accuracy, while for higher values of m it is all that could be desired. The actual values of u/π (calculated in another manner by Schwerd) are 1·4303, 2·4590, 3·4709, 4·4747, 5·4818, 6·4844, &c.

Since the maxima occur when $u = (m + \frac{1}{2})\pi$ nearly, the successive values are not very different from

$$\frac{4}{9\pi^2}, \quad \frac{4}{25\pi^2}, \quad \frac{4}{49\pi^2}, \quad \&c.$$

The application of these results to (3) shows that the field is brightest at the centre $\xi = 0$, $\eta = 0$, viz. at the geometrical image. It is traversed by dark lines whose equations are

$$\xi = mf\lambda/a, \quad \eta = mf\lambda/b.$$

Within the rectangle formed by pairs of consecutive dark lines, and not far from its centre, the brightness rises to a maximum; but these subsequent maxima are in all cases much inferior to the brightness at the centre of the entire pattern ($\xi = 0$, $\eta = 0$).

By the principle of energy the illumination over the entire focal plane must be equal to that over the diffracting area; and thus, in accordance with the suppositions by which (3) was obtained, its value when integrated from $\xi = -\infty$ to $\xi = +\infty$, and from $\eta = -\infty$ to $\eta = +\infty$ should be equal to ab. This integration, employed originally by Kelland* to determine the absolute intensity of a secondary wave, may be at once effected by means of the known formula

$$\int_{-\infty}^{+\infty} \frac{\sin^2 u}{u^2}\,du = \int_{-\infty}^{+\infty} \frac{\sin u}{u}\,du = \pi.$$

It will be observed that, while the total intensity is proportional to ab, the intensity at the focal point is proportional to a^2b^2. If the aperture be increased, not only is the total brightness over the focal plane increased with it, but there is also a concentration of the diffraction pattern. The form of (3) shows immediately that, if a and b be altered, the coordinates of any characteristic point in the pattern vary as a^{-1} and b^{-1}.

The contraction of the diffraction pattern with increase of aperture is of fundamental importance with reference to the resolving power of optical

* *Ed. Trans.* xv. 315.

instruments. According to common optics, where images are absolute, the diffraction pattern is supposed to be infinitely small, and two radiant points, however near together, form separated images. This is tantamount to an assumption that λ is infinitely small. The actual finiteness of λ imposes a limit upon the separating or resolving power of an optical instrument.

This indefiniteness of images is sometimes said to be due to diffraction by the edge of the aperture, and proposals have even been made for curing it by causing the transition between the interrupted and transmitted parts of the primary wave to be less abrupt. Such a view of the matter is altogether misleading. What requires explanation is not the imperfection of actual images so much as the possibility of their being as good as we find them.

At the focal point ($\xi = 0$, $\eta = 0$) all the secondary waves agree in phase, and the intensity is easily expressed, whatever be the form of the aperture. From the general formula (2), if A be the *area* of aperture,

$$I_0^2 = A^2/\lambda^2 f^2. \qquad (7)$$

The formation of a sharp image of the radiant point requires that the illumination become insignificant when ξ, η attain small values, and this insignificance can only arise as a consequence of discrepancies of phase among the secondary waves from various parts of the aperture. So long as there is no sensible discrepancy of phase, there can be no sensible diminution of brightness as compared with that to be found at the focal point itself. We may go further, and lay it down that there can be no considerable loss of brightness until the difference of phase of the waves proceeding from the nearest and furthest parts of the aperture amounts to $\frac{1}{4}\lambda$.

When the difference of phase amounts to λ, we may expect the resultant illumination to be very much reduced. In the particular case of a rectangular aperture the course of things can be readily followed, especially if we conceive f to be infinite. In the direction (suppose horizontal) for which $\eta = 0$, $\xi/f = \sin\theta$, the phases of the secondary waves range over a complete period when $\sin\theta = \lambda/a$, and, since all parts of the horizontal aperture are equally effective, there is in this direction a complete compensation and consequent absence of illumination. When $\sin\theta = \frac{3}{2}\lambda/a$, the phases range one and a half periods, and there is revival of illumination. We may compare the brightness with that in the direction $\theta = 0$. The phase of the resultant amplitude is the same as that due to the central secondary wave, and the discrepancies of phase among the components reduce the amplitude in the proportion

$$\frac{1}{3\pi}\int_{-\frac{3}{2}\pi}^{+\frac{3}{2}\pi} \cos\phi \, d\phi : 1,$$

or $-2 : 3\pi$; so that the brightness in this direction is $4/9\pi^2$ of the maximum at $\theta = 0$. In like manner we may find the illumination in any other direction, and it is obvious that it vanishes when $\sin\theta$ is any multiple of λ/a.

The reason of the augmentation of resolving power with aperture will now be evident. The larger the aperture the smaller are the angles through which it is necessary to deviate from the principal direction in order to bring in specified discrepancies of phase—the more concentrated is the image.

In many cases the subject of examination is a luminous line of uniform intensity, the various points of which are to be treated as independent sources of light. If the image of the line be $\xi = 0$, the intensity at any point ξ, η of the diffraction pattern may be represented by

$$\int_{-\infty}^{+\infty} I^2 d\eta = \frac{a^2 b}{\lambda f} \frac{\sin^2(\pi a\xi/\lambda f)}{\pi^2 a^2 \xi^2/\lambda^2 f^2}, \qquad \ldots\ldots\ldots\ldots\ldots\ldots\ldots(8)$$

the same law as obtains for a luminous point when horizontal directions are alone considered. The definition of a fine vertical line, and consequently the resolving power for contiguous vertical lines, is thus *independent of the vertical aperture of the instrument*, a law of great importance in the theory of the spectroscope.

The distribution of illumination in the image of a luminous line is shown by the curve ABC (Fig. 5), representing the value of the function $\sin^2 u/u^2$ from $u = 0$ to $u = 2\pi$. The part corresponding to negative values of u is similar, OA being a line of symmetry.

Let us now consider the distribution of brightness in the image of a double line whose components are of equal strength, and at such an angular interval that the central line in the image of one coincides with the first zero of brightness in the image of the other. In Fig. 5 the curve of brightness for one component is ABC, and for the other $OA'C'$; and the curve representing half the combined brightnesses is $E'BE$. The brightness (corresponding to B) midway between the two central points AA' is ·8106 of the brightness at the central points themselves. We may consider this to be about the limit of closeness at which there could be any decided appearance of resolution, though doubtless an observer accustomed to his instrument would recognize the duplicity with certainty. The obliquity, corresponding to $u = \pi$, is such that the phases of the secondary waves range over a complete period, *i.e.* such that the projection of the horizontal aperture upon this direction is one wave-length. We conclude that *a double line cannot be fairly resolved unless its components subtend an angle exceeding that subtended by the wave-length of light at a distance equal to the horizontal aper-*

Fig. 5.

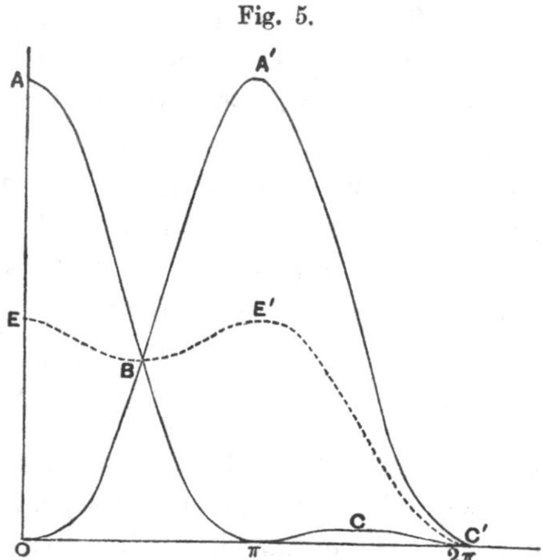

ture. This rule is convenient on account of its simplicity; and it is sufficiently accurate in view of the necessary uncertainty as to what exactly is meant by resolution.

On the experimental confirmation of the theory of the resolving power of rectangular apertures, see Optics, *Enc. Brit.* Vol. XVII. p. 807, [Vol. II. p. 411].

If the angular interval between the components of a double line be half as great again as that supposed in the figure, the brightness midway between is ·1802 as against 1·0450 at the central lines of each image. Such a falling off in the middle must be more than sufficient for resolution. If the angle subtended by the components of a double line be twice that subtended by the wave-length at a distance equal to the horizontal aperture, the central bands are just clear of one another, and there is a line of absolute blackness in the middle of the combined images.

Since the limitation of the width of the central band in the image of a luminous line depends upon discrepancies of phase among the secondary waves, and since the discrepancy is greatest for the waves which come from the edges of the aperture, the question arises how far the operation of the central parts of the aperture is advantageous. If we imagine the aperture reduced to two equal narrow slits bordering its edges, compensation will evidently be complete when the projection on an oblique direction is equal to $\frac{1}{2}\lambda$, instead of λ as for the complete aperture. By this procedure the width of the central band in the diffraction pattern is halved, and so far an advantage is attained. But, as will be evident, the bright bands bordering the central band are now not inferior to it in brightness; in fact, a band similar to the central band is reproduced an indefinite number of times, so long as there is no sensible discrepancy of phase in the secondary waves proceeding from the various parts of the *same* slit. Under these circumstances the narrowing of the band is paid for at a ruinous price, and the arrangement must be condemned altogether.

A more moderate suppression of the central parts is, however, sometimes advantageous. Theory and experiment alike prove that a double line, of which the components are equally strong, is better resolved when, for example, one-sixth of the horizontal aperture is blocked off by a central screen; or the rays quite at the centre may be allowed to pass, while others a little further removed are blocked off. Stops, each occupying one-eighth of the width, and with centres situated at the points of trisection, answer well the required purpose.

It has already been suggested that the principle of energy requires that the general expression for I^2 in (2) when integrated over the whole of the plane ξ, η should be equal to A, where A is the area of the aperture. A general analytical verification has been given by Stokes*. The expression for I^2 may be written in the form

$$I^2 = \frac{1}{\lambda^2 f^2} \iiiint \cos \frac{k}{f} \{\xi (x' - x) + \eta (y' - y)\} \, dx\, dy\, dx'\, dy', \quad \text{.........(9)}$$

the integrations with respect to x', y' as well as those with respect to x, y being over the area of the aperture; and for the present purpose this is to be integrated again with respect to ξ, η over the whole of the focal plane.

In changing the order of integration so as to take first that with respect to ξ, η, it is proper, in order to avoid ambiguity, to introduce under the integral sign the factor $e^{\mp \alpha \xi \mp \beta \eta}$, the + or − being chosen so as to make the elements of the integral vanish at infinity. After the operations have been performed, α and β are to be supposed to vanish.

Thus $\iint I^2 d\xi d\eta =$ Limit of

$$\frac{1}{\lambda^2 f^2} \iiiiiint e^{\mp \alpha \xi \mp \beta \eta} \cos \frac{k}{f} \{\xi (x' - x) + \eta (y' - y)\} \, dx\, dy\, dx'\, dy'\, d\xi\, d\eta.$$

Now
$$\int_{-\infty}^{+\infty} e^{\mp \alpha \xi} \cos (h\xi - H)\, d\xi = \frac{2\alpha \cos H}{\alpha^2 + h^2};$$

and thus

$$\iint_{-\infty}^{+\infty} I^2\, d\xi d\eta = \text{Limit of } \frac{1}{\lambda^2 f^2} \iiiint \frac{4\alpha\beta\, dx\, dy\, dx'\, dy'}{\left\{\alpha^2 + \dfrac{k^2 (x' - x)^2}{f^2}\right\} \left\{\beta^2 + \dfrac{k^2 (y' - y)^2}{f^2}\right\}}$$

Let
$$\frac{k (x' - x)}{f} = \alpha u, \quad dx' = \frac{f\alpha}{k}\, du.$$

The limits for u are ultimately $-\infty$ and $+\infty$, and we have

$$\text{Limit} \int \frac{2\alpha\, dx'}{\left\{\alpha^2 + \dfrac{k^2 (x' - x)^2}{f^2}\right\}} = \frac{2f}{k} \int_{-\infty}^{+\infty} \frac{du}{1 + u^2} = \frac{2f}{k} \cdot \pi = f\lambda.$$

In like manner the integration for y' may be performed; and we find

$$\iint_{-\infty}^{+\infty} I^2\, d\xi d\eta = \iint dx\, dy = A. \quad \text{.....................(10)†}$$

We saw that I_0^2 (the intensity at the focal point) was equal to $A^2/\lambda^2 f^2$. If A' be the area over which the intensity must be I_0^2 in order to give the actual total intensity in accordance with

$$A' I_0^2 = \iint_{-\infty}^{+\infty} I^2 d\xi d\eta,$$

* *Ed. Trans.* xx. p. 317, 1853.

† It is easy to show that this conclusion is not disturbed by the introduction at every point of an arbitrary retardation ρ, a function of x, y. The terms $(\rho' - \rho)$ are then to be added under the cosine in (9); but they are ultimately without effect, since the only elements which contribute are those for which in the limit $x' = x$, $y' = y$, and therefore $\rho' = \rho$.

the relation between A and A' is $AA' = \lambda^2 f^2$. Since A' is in some sense the area of the diffraction pattern, it may be considered to be a rough criterion of the definition, and we infer that the definition of a point depends principally upon the *area* of the aperture, and only in a very secondary degree upon the shape when the area is maintained constant.

§ 12. *Theory of Circular Aperture.*

We will now consider the important case where the form of the aperture is circular. Writing for brevity

$$k\xi/f = p, \quad k\eta/f = q, \quad \text{.............................(1)}$$

we have for the general expression (§ 11) of the intensity

$$\lambda^2 f^2 I^2 = S^2 + C^2, \quad \text{.............................(2)}$$

where $S = \iint \sin(px + qy)\,dx\,dy, \quad C = \iint \cos(px + qy)\,dx\,dy.$ ......(3, 4)

When, as in the application to rectangular or circular apertures, the form is symmetrical with respect to the axes both of x and y, $S = 0$, and C reduces to

$$C = \iint \cos px \cos qy\,dx\,dy. \quad \text{...........................(5)}$$

In the case of the circular aperture the distribution of light is of course symmetrical with respect to the focal point $p = 0$, $q = 0$; and C is a function of p and q only through $\sqrt{(p^2 + q^2)}$. It is thus sufficient to determine the intensity along the axis of p. Putting $q = 0$, we get

$$C = \iint \cos px\,dx\,dy = 2\int_{-R}^{+R} \cos px\,\sqrt{(R^2 - x^2)}\,dx,$$

R being the radius of the aperture. This integral is the Bessel's function of order unity, defined by

$$J_1(z) = \frac{z}{\pi}\int_0^\pi \cos(z\cos\phi)\sin^2\phi\,d\phi. \quad \text{................(6)}$$

Thus, if $x = R\cos\phi$,

$$C = \pi R^2 \frac{2J_1(pR)}{pR}; \quad \text{..........................(7)}$$

and the illumination at distance r from the focal point is

$$I^2 = \frac{\pi^2 R^4}{\lambda^2 f^2} \cdot \frac{4J_1^2(2\pi Rr/f\lambda)}{(2\pi Rr/f\lambda)^2}. \quad \text{........................(8)}$$

The ascending series for $J_1(z)$, used by Airy* in his original investigation of the diffraction of a circular object-glass, and readily obtained from (6), is

$$J_1(z) = \frac{z}{2} - \frac{z^3}{2^2.4} + \frac{z^5}{2^2.4^2.6} - \frac{x^7}{2^2.4^2.6^2.8} + \ldots \quad \text{...........(9)}$$

* "On the Diffraction of an Object-Glass with Circular Aperture," *Camb. Trans.* 1834.

When z is great, we may employ the semi-convergent series

$$J_1(z)=\sqrt{\left(\frac{2}{\pi z}\right)}\sin(z-\tfrac{1}{4}\pi)\left\{1+\frac{3.5.1}{8.16}\left(\frac{1}{z}\right)^2-\frac{3.5.7.9.1.3.5}{8.16.24.32}\left(\frac{1}{z}\right)^4+\dots\right\}$$
$$+\sqrt{\left(\frac{2}{\pi z}\right)}\cos(z-\tfrac{1}{4}\pi)\left\{\frac{3}{8}.\frac{1}{z}-\frac{3.5.7.1.3}{8.16.24}\left(\frac{1}{z}\right)^3\right.$$
$$\left.+\frac{3.5.7.9.11.1.3.5.7}{8.16.24.32.40}\left(\frac{1}{z}\right)^5-\dots\right\}\dots\dots\dots\dots(10)$$

A table of the values of $2z^{-1}J_1(z)$ has been given by Lommel*, to whom is due the first systematic application of Bessel's functions to the diffraction integrals.

The illumination vanishes in correspondence with the roots of the equation $J_1(z)=0$. If these be called z_1, z_2, z_3, ... the radii of the dark rings in the diffraction pattern are

$$\frac{f\lambda z_1}{2\pi R},\quad \frac{f\lambda z_2}{2\pi R},\dots$$

being thus *inversely* proportional to R.

The integrations may also be effected by means of polar coordinates, taking first the integration with respect to ϕ so as to obtain the result for an infinitely thin annular aperture. Thus, if

$$x=\rho\cos\phi,\quad y=\rho\sin\phi,$$

$$C=\iint\cos px\,dx\,dy=\int_0^R\int_0^{2\pi}\cos(p\rho\cos\theta)\,\rho\,d\rho\,d\theta.$$

Now by definition

$$J_0(z)=\frac{2}{\pi}\int_0^{\frac{1}{2}\pi}\cos(z\cos\theta)\,d\theta=1-\frac{z^2}{2^2}+\frac{z^4}{2^2.4^2}-\frac{z^6}{2^2.4^2.6^2}+\dots\dots(11)$$

The value of C for an annular aperture of radius r and width dr is thus

$$dC=2\pi J_0(p\rho)\,\rho\,d\rho.\quad\dots\dots\dots\dots(12)$$

For the complete circle,

$$C=\frac{2\pi}{p^2}\int_0^{pR}J_0(z)\,z\,dz=\frac{2\pi}{p^2}\left\{\frac{p^2R^2}{2}-\frac{p^4R^4}{2^2.4}+\frac{p^6R^6}{2^2.4^2.6}-\dots\right\}$$
$$=\pi R^2.\frac{2J_1(pR)}{pR},$$

as before.

In these expressions we are to replace p by $k\xi/f$, or rather, since the diffraction pattern is symmetrical, by kr/f, where r is the distance of any point in the focal plane from the centre of the system.

* *Schlömilch*, xv. p. 166, 1870.

The roots of $J_0(z)$ after the first may be found from

$$\frac{z}{\pi}=i-\cdot 25+\frac{\cdot 050661}{4i-1}-\frac{\cdot 053041}{(4i-1)^3}+\frac{\cdot 262051}{(4i-1)^5}, \quad\ldots\ldots\ldots\ldots(13)$$

and those of $J_1(z)$ from

$$\frac{z}{\pi}=i+\cdot 25-\frac{\cdot 151982}{4i+1}+\frac{\cdot 015399}{(4i+1)^3}-\frac{\cdot 245835}{(4i+1)^5}, \quad\ldots\ldots\ldots\ldots(14)$$

formulæ derived by Stokes* from the descending series†. The following table gives the actual values:—

i	$\frac{z}{\pi}$ for $J_0(z)=0$	$\frac{z}{\pi}$ for $J_1(z)=0$	i	$\frac{z}{\pi}$ for $J_0(z)=0$	$\frac{z}{\pi}$ for $J_1(z)=0$
1	·7655	1·2197	6	5·7522	6·2439
2	1·7571	2·2330	7	6·7519	7·2448
3	2·7546	3·2383	8	7·7516	8·2454
4	3·7534	4·2411	9	8·7514	9·2459
5	4·7527	5·2428	10	9·7513	10·2463

In both cases the image of a mathematical point is thus a symmetrical ring system. The greatest brightness is at the centre, where

$$dC=2\pi\rho\,d\rho, \quad C=\pi R^2.$$

For a certain distance outwards this remains sensibly unimpaired, and then gradually diminishes to zero, as the secondary waves become discrepant in phase. The subsequent revivals of brightness forming the bright rings are necessarily of inferior brilliancy as compared with the central disk.

The first dark ring in the diffraction pattern of the complete circular aperture occurs when

$$r/f=1\cdot 2197\times\lambda/2R. \quad\ldots\ldots\ldots\ldots\ldots\ldots\ldots\ldots(15)$$

We may compare this with the corresponding result for a rectangular aperture of width a,

$$\xi/f=\lambda/a;$$

and it appears that in consequence of the preponderance of the central parts, the compensation in the case of the circle does not set in at so small an obliquity as when the circle is replaced by a rectangular aperture, whose side is equal to the diameter of the circle.

* *Camb. Trans.* IX. 1850.

† The descending series for $J_0(z)$ appears to have been first given by Sir W. Hamilton in a memoir on "Fluctuating Functions," *Roy. Irish Trans.* 1840.

Again, if we compare the complete circle with a narrow annular aperture of the same radius, we see that in the latter case the first dark ring occurs at a much smaller obliquity, viz.

$$r/f = \cdot 7655 \times \lambda/2R.$$

It has been found by Herschel and others that the definition of a telescope is often improved by stopping off a part of the central area of the object-glass; but the advantage to be obtained in this way is in no case great, and anything like a reduction of the aperture to a narrow annulus is attended by a development of the external luminous rings sufficient to outweigh any improvement due to the diminished diameter of the central area*.

The maximum brightnesses and the places at which they occur are easily determined with the aid of certain properties of the Bessel's functions. It is known† that

$$J_0'(z) = -J_1(z); \quad \text{..........................(16)}$$

$$J_2(z) = \frac{1}{z} J_1(z) - J_1'(z); \qquad J_0(z) + J_2(z) = \frac{2}{z} J_1(z). \quad \text{......(17, 18)}$$

The maxima of C occur when

$$\frac{d}{dz}\left(\frac{J_1(z)}{z}\right) = \frac{J_1'(z)}{z} - \frac{J_1(z)}{z^2} = 0;$$

or by (17) when $J_2(z) = 0$. When z has one of the values thus determined,

$$\frac{2}{z} J_1(z) = J_0(z).$$

The accompanying table is given by Lommel‡, in which the first column gives the roots of $J_2(z) = 0$, and the second and third columns the corresponding values of the functions specified. It appears that the maximum brightness in the first ring is only about $\frac{1}{57}$ of the brightness at the centre.

z	$2z^{-1}J_1(z)$	$4z^{-2}J_1^2(z)$
·000000	+1·000000	1·000000
5·135630	− ·132279	·017498
8·417236	+ ·064482	·004158
11·619857	− ·040008	·001601
14·795938	+ ·027919	·000779
17·959820	− ·020905	·000437

* Airy, *loc. cit.* "Thus the magnitude of the central spot is diminished, and the brightness of the rings increased, by covering the central parts of the object-glass."

† Todhunter's *Laplace's Functions*, ch. XXXI. ‡ *Loc. cit.*

We will now investigate the total illumination distributed over the area of the circle of radius r. We have

$$I^2 = \frac{\pi^2 R^4}{\lambda^2 f^2} \cdot \frac{4J_1^2(z)}{z^2}, \quad \text{.........................(19)}$$

where

$$z = 2\pi Rr/\lambda f. \quad \text{.........................(20)}$$

Thus

$$2\pi \int I^2 r\,dr = \frac{\lambda^2 f^2}{2\pi R^2} \int I^2 z\,dz = \pi R^2 . 2 \int z^{-1} J_1^2(z)\,dz.$$

Now by (17), (18)

$$z^{-1} J_1(z) = J_0(z) - J_1'(z);$$

so that

$$z^{-1} J_1^2(z) = -\tfrac{1}{2} \frac{d}{dz} J_0^2(z) - \tfrac{1}{2} \frac{d}{dz} J_1^2(z),$$

and

$$2 \int_0^z z^{-1} J_1^2(z)\,dz = 1 - J_0^2(z) - J_1^2(z). \quad \text{..............(21)}$$

If r, or z, be infinite, $J_0(z)$, $J_1(z)$ vanish, and the whole illumination is expressed by πR^2, in accordance with the general principle. In any case the proportion of the whole illumination to be found outside the circle of radius r is given by

$$J_0^2(z) + J_1^2(z).$$

For the dark rings $J_1(z) = 0$; so that the fraction of illumination outside any dark ring is simply $J_0^2(z)$. Thus for the first, second, third, and fourth dark rings we get respectively ·161, ·090, ·062, ·047, showing that more than $\frac{9}{10}$ths of the whole light is concentrated within the area of the second dark ring*.

When z is great, the descending series (10) gives

$$\frac{2J_1(z)}{z} = \frac{2}{z} \sqrt{\left(\frac{2}{\pi z}\right)} \sin\left(z - \tfrac{1}{4}\pi\right); \quad \text{.................(22)}$$

so that the places of maxima and minima occur at equal intervals.

The mean brightness varies as z^{-3} (or as r^{-3}), and the integral found by multiplying it by $z\,dz$ and integrating between 0 and ∞ converges.

It may be instructive to contrast this with the case of an infinitely narrow annular aperture, where the brightness is proportional to $J_0^2(z)$. When z is great,

$$J_0(z) = \sqrt{\left(\frac{2}{\pi z}\right)} \cos\left(z - \tfrac{1}{4}\pi\right).$$

The mean brightness varies as z^{-1}; and the integral $\int_0^\infty J_0^2(z)\,z\,dz$ is not convergent.

* *Phil. Mag.* March 1881. [Vol. I. Art. 73.]

The efficiency of a telescope is of course intimately connected with the size of the disk by which it represents a mathematical point. The resolving power upon double stars of telescopes of various apertures has been investigated by Dawes and others (*Enc. Brit.* Vol. XVII. p. 807) [Vol. I. p. 411], with results that agree fairly well with theory.

If we integrate the expression (8) for I^2 with respect to η, we shall obtain a result applicable to a linear luminous source of which the various parts are supposed to act independently.

From (19), (20)

$$d\xi\int_{-\infty}^{+\infty} I^2 d\eta = \frac{2\pi^2 R^4}{\lambda^2 f^2}\, d\xi\int_0^{\infty} 4z^{-2} J_1^2(z)\, d\eta = 2R^2 d\xi \int \frac{J_1^2(z)\, dz}{z\,.\,\eta},$$

since $\eta^2 = r^2 - \xi^2$.

If we write

$$\zeta = 2\pi R\xi/\lambda f, \qquad\qquad (23)$$

we get

$$d\xi\,.\int_{-\infty}^{+\infty} I^2 d\eta = 2R^2 d\zeta\,.\int_{\zeta}^{\infty} \frac{J_1^2(z)\, dz}{z\sqrt{(z^2-\zeta^2)}}. \qquad\qquad (24)$$

This integral has been investigated by H. Struve*, who, calling to his aid various properties of Bessel's functions, shows that

$$\int_{\zeta}^{\infty} \frac{J_1^2(z)\, dz}{z\sqrt{(z^2-\zeta^2)}} = \frac{2}{\pi}\,\frac{1}{\zeta}\int_0^{\frac{1}{2}\pi} \sin(2\zeta\sin\beta)\cos^2\beta\, d\beta, \qquad\qquad (25)$$

of which the right-hand member is readily expanded in powers of ζ. By means of (24) we may verify that

$$\int_{-\infty}^{+\infty} d\xi \int_{-\infty}^{+\infty} I^2 d\eta = \pi R^2.$$

Contrary to what would naturally be expected, the subject is more easily treated without using the results of the integration with respect to x and y, by taking first of all, as in the investigation of Stokes (§ 11), the integration with respect to η. Thus

$$\lambda^2 f^2 \int_{-\infty}^{+\infty} I^2 d\eta = \text{Limit of}$$

$$\iiiint\!\!\int e^{\mp\beta\eta} \cos\frac{k}{f}\{\xi(x'-x)+\eta(y'-y)\}\, dx\, dy\, dx'\, dy'\, d\eta\,; \qquad\qquad (26)$$

and

$$\int_{-\infty}^{+\infty} e^{\mp\beta\eta}\cos\frac{k}{f}\{\xi(x'-x)+\eta(y'-y)\}\, d\eta = \frac{2\beta\cos\dfrac{k\xi}{f}(x'-x)}{\beta^2+\dfrac{k^2(y'-y)^2}{f^2}}. \qquad\qquad (27)$$

* *Wied. Ann.* XVII. 1008, 1882.

We have now to consider

$$\iint \frac{2\beta\, dy\, dy'}{\beta^2 + k^2 (y' - y)^2/f^2} \text{.........................(28)}$$

In the integration with respect to y' every element vanishes in the limit ($\beta = 0$), unless $y' = y$. If the range of integration for y' includes the value y, then

$$\text{Limit} \int \frac{2\beta\, dy'}{\beta^2 + k^2 (y' - y)^2/f^2} = f\lambda;$$

otherwise it vanishes.

The limit of (28) may thus be denoted by $\lambda f Y$, where Y is the *common part* of the ranges of integration for y' and y corresponding to any values of x' and x. Hence

$$\int_{-\infty}^{+\infty} I^2 d\eta = \lambda^{-1} f^{-1} \iint Y \cos \frac{k\xi}{f} (x' - x)\, dx\, dx'$$

$$= \lambda^{-1} f^{-1} \iint Y \cos \frac{k\xi x}{f} \cos \frac{k\xi x'}{f}\, dx\, dx', \text{ ....(29)}$$

if, as for the present purpose, the aperture is symmetrical with respect to the axis of y.

In the application to the circle we may write

$$\int_{-\infty}^{+\infty} I^2 d\eta = 4\lambda^{-1} f^{-1} \int_0^R \int_0^R Y \cos \frac{k\xi x}{f} \cos \frac{k\xi x'}{f}\, dx\, dx',$$

where Y is the smaller of the two quantities $2\sqrt{(R^2 - x'^2)}$, $2\sqrt{(R^2 - x^2)}$, *i.e.*, corresponds to the *larger* of the two abscissæ x', x. If we take $Y = 2\sqrt{(R^2 - x^2)}$, and limit the integration to those values of x' which are less than x, we should obtain exactly the half of the required result. Thus

$$\int_{-\infty}^{+\infty} I^2 d\eta = 16\lambda^{-1} f^{-1} \int_0^R \int_0^x \sqrt{(R^2 - x^2)} \cos \frac{k\xi x}{f} \cos \frac{k\xi x'}{f}\, dx\, dx'$$

$$= \frac{4}{\pi\xi} \int_0^R \sqrt{(R^2 - x^2)} \sin \frac{2k\xi x}{f}\, dx = \frac{4R^2}{\pi\xi} \int_0^{\frac{1}{2}\pi} \cos^2\beta \sin \frac{2k\xi R \sin\beta}{f}\, d\beta.$$

Hence, writing as before $\zeta = 2\pi R\xi/\lambda f$, we get

$$d\xi \int_{-\infty}^{+\infty} I^2 d\eta = \frac{4R^2}{\pi} \cdot \frac{d\xi}{\xi} \cdot \int_0^{\frac{1}{2}\pi} \cos^2\beta \sin(2\zeta\beta)\, d\beta, \text{ ....(30)}$$

in which we may replace $d\xi/\xi$ by $d\zeta/\zeta$, in agreement with the result obtained by Struve.

The integral in (30) may be written in another form. We have

$$\int \zeta \sin(2\zeta\beta) \cos^2\beta\, d\beta = -\tfrac{1}{2} \cos\beta \cos(2\zeta \sin\beta) - \tfrac{1}{2} \int \cos(2\zeta \sin\beta) \sin\beta\, d\beta;$$

and thus

$$\int_0^{\frac{1}{2}\pi} \zeta \sin(2\zeta\beta)\cos^2\beta\,d\beta = \tfrac{1}{2}\int_0^{\frac{1}{2}\pi}\{1-\cos(2\zeta\sin\beta)\}\sin\beta\,d\beta$$

$$= \int_0^{\frac{1}{2}\pi}\sin^2(\zeta\sin\beta)\sin\beta\,d\beta. \quad\text{............(31)}$$

The integral is thus expressible by means of the function K_1,* and we have

$$d\xi\int_{-\infty}^{+\infty} I^2 d\eta = \tfrac{1}{2}R^2\zeta^{-3}d\zeta K_1(2\zeta). \quad\text{....................(32)}$$

The ascending series for $K_1(z)$ is

$$K_1(z) = \frac{2}{\pi}\left\{\frac{z^3}{1^2.3} - \frac{z^5}{1^2.3^2.5} + \frac{z^7}{1^2.3^2.5^2.7} - \ldots\right\}; \quad\text{.........(33)}$$

and this is always convergent. The descending semi-convergent series is

$$K_1(z) = \frac{2}{\pi}\{z + z^{-1} - 3.z^{-3} + 1^2.3^2.5.z^{-5} - 1^2.3^2.5^2.7.z^{-7} + \ldots\}$$

$$-\sqrt{\left(\frac{2z}{\pi}\right)}.\cos(z-\tfrac{1}{4}\pi)\left\{1 - \frac{(1^2-4)(3^2-4)}{1.2.(8z)^2} + \ldots\right\}$$

$$-\sqrt{\left(\frac{2z}{\pi}\right)}.\sin(z-\tfrac{1}{4}\pi)\left\{\frac{1^2-4}{1.8z} - \frac{(1^2-4)(3^2-4)(5^2-4)}{1.2.3.(8z)^3} + \ldots\right\}, \quad\text{...(34)}$$

the series within braces being the same as those which occur in the expression of the function $J_1(z)$.

When ζ (or ξ) is very great,

$$d\xi\int_{-\infty}^{+\infty} I^2 d\eta = \frac{2R^2}{\pi}\zeta^{-2}d\zeta,$$

so that the intensity of the image of a luminous line is ultimately inversely as the square of the distance from the central axis, or geometrical image.

	ζ	Intensity
On the axis itself ...	0·00	1
First minimum	3·55	$\frac{1}{34}$
First maximum	4·65	$\frac{1}{24}$
Second minimum ...	6·80	$\frac{1}{115}$
Second maximum ...	8·00	$\frac{1}{80}$
Third minimum......	9·60	$\frac{1}{419}$
Third maximum......	11·00	$\frac{1}{208}$
Fourth minimum ...	13·20	$\frac{1}{10000}$

* *Theory of Sound*, § 302.

As is evident from its composition, the intensity remains finite for all values of ζ; it is, however, subject to fluctuations presenting maxima and minima, which have been calculated by Ch. André*, using apparently the method of quadratures.

The results are also exhibited by M. André in the form of a curve, of which Fig. 6 is a copy.

It will be seen that the distribution of brightness does not differ greatly from that due to a rectangular aperture whose width (perpendicular to the luminous line) is equal to the diameter of the circular aperture. It will be instructive to examine the image of a double line, whose components present an interval corresponding to $\zeta = \pi$, and to compare the result with that already found for a rectangular aperture (§ 11). We may consider the brightness at distance ζ proportional to

Fig. 6.

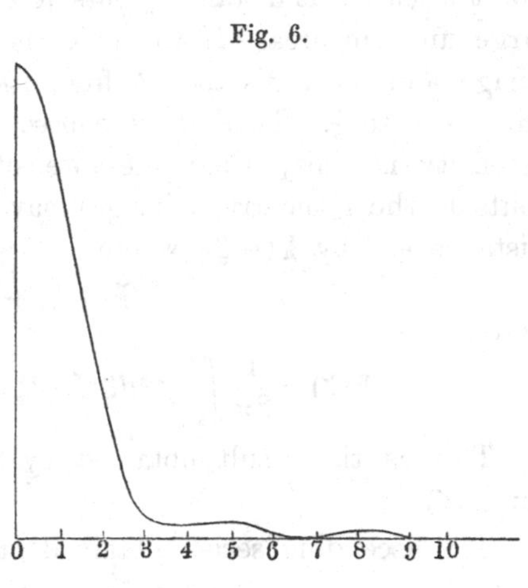

$$L(\zeta) = \frac{1}{1^2 . 3} - \frac{2^2 \zeta^2}{1^2 . 3^2 . 5} + \frac{2^4 \zeta^4}{1^2 . 3^2 . 5^2 . 7} - \dots \dots (35)$$

In the compound image the illumination at the geometrical focus of one of the luminous lines is represented by

$$L(0) + L(\pi);$$

and the illumination midway between the geometrical images of the two lines is

$$2L(\tfrac{1}{2}\pi).$$

We find by actual calculation from the series, $L(\pi) = \cdot 0164$, $L(\frac{1}{2}\pi) = \cdot 1671$, $L(0) = \cdot 3333$, so that

$$L(0) + L(\pi) = \cdot 3497, \quad 2L(\tfrac{1}{2}\pi) = \cdot 3342,$$

and

$$\frac{2L(\frac{1}{2}\pi)}{L(0) + L(\pi)} = \cdot 955.$$

The corresponding number for the rectangular aperture was ·811; so that, as might have been expected, the resolving power of the circular aperture is distinctly less than that of the rectangular aperture of equal width. Hence a telescope will not resolve a double line unless the angular interval between them decidedly exceeds that subtended by the wave-length of light at a distance equal to the diameter of the object-glass. Experiment shows that resolution begins when the angular interval is about a tenth part greater than that mentioned.

* *Ann. d. l'École Normale*, v. p. 310, 1876.

If we integrate (30) with respect to ξ between the limits $-\infty$ and $+\infty$, we obtain πR^2, as has already been remarked. This represents the whole illumination over the focal plane due to a radiant point whose image is at O, or, reciprocally, the illumination at O (the same as at any other point) due to an infinitely extended luminous area. If we take the integration from ξ (supposed positive) to ∞ we get the illumination at O due to a uniform luminous area extending over this region, that is to say, the illumination at a point situated at distance ξ outside the border of the geometrical image of a large uniform area. If the point is supposed to be inside the geometrical image and at a distance ζ from its edge, we are to take the integration from $-\infty$ to ξ. Thus, if we choose the scale of intensities so that the full intensity is unity, then the intensity at a distance corresponding to $+\zeta$ (outside the geometrical image) may be represented by $\mathfrak{I}(+\zeta)$, and that at a distance $-\zeta$ by $\mathfrak{I}(-\zeta)$, where

$$\mathfrak{I}(+\zeta)+\mathfrak{I}(-\zeta)=1,$$

and

$$\mathfrak{I}(\zeta)=\frac{1}{2\pi}\int_\zeta^\infty \zeta^{-3}d\zeta K_1(2\zeta)=\tfrac{1}{2}-\frac{1}{2\pi}\int_0^\zeta \zeta^{-3}d\zeta K_1(2\zeta). \quad\text{.........(36)}$$

This is the result obtained by Struve, who gives the following series for $\mathfrak{I}(\zeta)$.

The ascending series, obtained at once by integration from (33), is

$$\mathfrak{I}(\zeta)=\tfrac{1}{2}-\frac{2}{\pi^2}\sum_{n=1}^{n=\infty}(-1)^{n-1}\frac{2n+1}{2n-1}\,\frac{2^{2n}\zeta^{2n-1}}{1^2.3^2.5^2\ldots(2n+1)^2}. \quad\text{.........(37)}$$

When ζ is great, we have approximately from the descending series

$$\mathfrak{I}(\zeta)=\frac{2}{\pi^2}\left(\frac{1}{\zeta}+\frac{1}{12\zeta^2}\right)-\frac{1}{2\pi^{3/2}}\,\frac{\cos(2\zeta+\frac{1}{4}\pi)}{\zeta^{5/2}}.$$

Thus "at great distances from the edge of the geometrical image the intensity is inversely proportional to the distance, and to the radius of the object-glass."

The following table, abbreviated from that given by Struve, will serve to calculate the enlargement of an image due to diffraction in any case that may arise.

$$\zeta=2\pi R\xi/\lambda f.$$

$$\mathfrak{I}(-\zeta)=1-\mathfrak{I}(+\zeta).$$

ζ	$\mathfrak{I}(\zeta)$	ζ	$\mathfrak{I}(\zeta)$	ζ	$\mathfrak{I}(\zeta)$
0·0	·5000	2·5	·0765	7·0	·0293
0·5	·3678	3·0	·0630	9·0	·0222
1·0	·2521	4·0	·0528	11·0	·0186
1·5	·1642	5·0	·0410	15·0	·0135
2·0	·1073	6·0	·0328		

It may perhaps have struck the reader that there is some want of rigour in our treatment of (30) when we integrate it over the whole focal plane of ξ, η, inasmuch as in the proof of the formulæ ξ and η are supposed to be small. The inconsistency becomes very apparent when we observe that according to the formulæ there is no limit to the relative retardation of secondary waves coming from various parts of the aperture, whereas in reality this retardation could never exceed the longest line capable of being drawn within the aperture. It will be worth while to consider this point a little further, although our limits forbid an extended treatment.

The formula becomes rigorous if we regard it as giving the illumination on the surface of a sphere of very large radius f, in a direction such that

$$\xi = f \sin\theta \cos\phi, \qquad \eta = f \sin\theta \sin\phi;$$

it may then be written

$$I^2 = \lambda^{-2} f^{-2} \iiiint \cos k\{(x'-x)\sin\theta\cos\phi + (y'-y)\sin\theta\sin\phi\}\, dx\, dy\, dx'\, dy'.$$

The whole intensity over the infinite hemisphere is given by

$$\mathfrak{I} = f^2 \int_0^{\frac{1}{2}\pi} \int_0^{2\pi} I^2 \sin\theta\, d\theta\, d\phi. \qquad (38)$$

According to the plan formerly adopted, we postpone the integration with respect to x, y, x', y', and take first that with respect to θ and ϕ. Thus for a single pair of elements of area $dx\, dy$, $dx'\, dy'$ we have to consider

$$\iint \cos k\{(x'-x)\sin\theta\cos\phi + (y'-y)\sin\theta\sin\phi\} \sin\theta\, d\theta\, d\phi,$$

or, if we write

$$x' - x = r\cos\alpha, \quad y' - y = r\sin\alpha,$$

$$\int_0^{\frac{1}{2}\pi} \int_0^{2\pi} \cos(kr\sin\theta\cos\phi)\sin\theta\, d\theta\, d\phi.$$

Now it may be proved (*e.g.*, by expansion in powers of kr) that

$$\int_0^{\frac{1}{2}\pi} \int_0^{2\pi} \cos(kr\sin\theta\cos\phi)\sin\theta\, d\theta\, d\phi = 2\pi \frac{\sin kr}{kr}; \qquad (39)$$

and thus

$$\mathfrak{I} = \frac{2\pi}{\lambda^2} \iiiint \frac{\sin kr}{kr}\, dx\, dy\, dx'\, dy', \qquad (40)$$

r being the distance between the two elements of area $dx\, dy$, $dx'\, dy'$.

In the case of a circular area of radius R, we have*

$$\iiiint \frac{\sin kr}{kr}\, dx\, dy\, dx'\, dy' = \frac{2\pi R^2}{k^2}\left\{1 - \frac{J_1(2kR)}{kR}\right\},$$

* *Theory of Sound*, § 302.

and thus

$$\mathfrak{I} = \pi R^2 \left\{1 - \frac{J_1(2kR)}{kR}\right\}. \qquad \text{.........................(41)}$$

When $kR = \infty$,

$$\mathfrak{I} = \pi R^2, \text{ as before.}$$

It appears therefore that according to the assumed law of the secondary wave the total illumination is proportional to the area of aperture, only under the restriction that the linear dimensions of the aperture are very large in comparison with the wave-length.

A word as to the significance of (39) may not be out of place. We know that

$$\psi = \cos k\{\sin\theta\cos\phi\,.\,x + \sin\theta\sin\phi\,.\,y + \cos\theta\,.\,z\} \qquad \text{.........(42)}$$

satisfies Laplace's extended equation $(\nabla^2 + k^2)\psi = 0$, being of the form $\cos kx'$, where x' is drawn in an oblique direction; and it follows that $\iint\psi\sin\theta d\theta\,d\phi$ satisfies the same equation. Now this, if the integration be taken over the hemisphere $\theta = 0$ to $\theta = \frac{1}{2}\pi$, must become a function of r, or $\sqrt{(x^2 + y^2 + z^2)}$, only.

Hence, putting $x = r$, $y = 0$, $z = 0$, we get

$$\iint\psi\sin\theta d\theta d\phi = \int_0^{\frac{1}{2}\pi}\int_0^{2\pi}\cos(kr\sin\theta\cos\phi)\sin\theta d\theta d\phi.$$

But the only function of r which satisfies Laplace's equation continuously through the origin is $A\sin kr/(kr)$; and that $A = 2\pi$ is proved at once by putting $r = 0$. The truth of the formula may also be established independently of the differential equation by equating the values of

$$\int_0^{\frac{1}{2}\pi}\int_0^{2\pi}\psi\sin\theta d\theta d\phi,$$

when $x = r$, $y = 0$, $z = 0$, and when $x = 0$, $y = 0$, $z = r$. Thus

$$\int_0^{\frac{1}{2}\pi}\int_0^{2\pi}\cos(kr\sin\theta\cos\phi)\sin\theta d\theta d\phi = \int_0^{\frac{1}{2}\pi}\int_0^{2\pi}\cos(kr\cos\theta)\sin\theta d\theta d\phi = 2\pi\frac{\sin kr}{kr}.$$

The formula itself may also be written

$$\int_0^{\frac{1}{2}\pi} J_0(kr\sin\theta)\sin\theta d\theta = \frac{\sin kr}{kr}. \qquad \text{.................(43)}$$

The results of the preceding theory of circular apertures admit of an interesting application to *coronas*, such as are often seen encircling the sun and moon. They are due to the interposition of small spherules of water, which act the part of diffracting obstacles. In order to the formation of a well-defined corona it is essential that the particles be exclusively, or preponderatingly, of one size.

If the origin of light be treated as infinitely small, and be seen in focus, whether with the naked eye or with the aid of a telescope, the whole of the light in the absence of obstacles would be concentrated in the immediate neighbourhood of the focus. At other parts of the field the effect is the same, by Babinet's principle, whether the imaginary screen in front of the object-glass is generally transparent but studded with a number of opaque circular disks, or is generally opaque but perforated with corresponding apertures. Consider now the light diffracted in a direction many times more oblique than any with which we should be concerned, were the whole aperture uninterrupted, and take first the effect of a single small aperture. The light in the proposed direction is that determined by the size of the small aperture in accordance with the laws already investigated, and its phase depends upon the position of the aperture. If we take a direction such that the light (of given wave-length) from a single aperture vanishes, the evanescence continues even when the whole series of apertures is brought into contemplation. Hence, whatever else may happen, there must be a system of dark rings formed, the same as from a single small aperture. In directions other than these it is a more delicate question how the partial effects should be compounded. If we make the extreme suppositions of an infinitely small source and absolutely homogeneous light, there is no escape from the conclusion that the light in a definite direction is arbitrary, that is, dependent upon the chance distribution of apertures. If, however, as in practice, the light be heterogeneous, the source of finite area, the obstacles in motion, and the discrimination of different directions imperfect, we are concerned merely with the mean brightness found by varying the arbitrary phase-relations, and this is obtained by simply multiplying the brightness due to a single aperture by the number of apertures (n)*. The diffraction pattern is therefore that due to a single aperture, merely brightened n times.

In his experiments upon this subject Fraunhofer employed plates of glass dusted over with lycopodium, or studded with small metallic disks of uniform size; and he found that the diameters of the rings were proportional to the length of the waves and inversely as the diameter of the disks.

In another respect the observations of Fraunhofer appear at first sight to be in disaccord with theory; for his measures of the diameters of the red rings, visible when white light was employed, correspond with the law applicable to dark rings, and not to the different law applicable to the luminous maxima. Verdet has, however, pointed out that the observation in this form is essentially different from that in which homogeneous red light is employed, and that the position of the red rings would correspond to the *absence* of blue-green light rather than to the greatest abundance of

* See § 4.

red light. Verdet's own observations, conducted with great care, fully confirm this view, and exhibit a complete agreement with theory.

By measurements of coronas it is possible to infer the size of the particles to which they are due, an application of considerable interest in the case of natural coronas—the general rule being the larger the corona the smaller the water spherules. Young employed this method not only to determine the diameters of cloud particles (*e.g.* $\frac{1}{1000}$ inch), but also those of fibrous material, for which the theory is analogous. His instrument was called the *eriometer**.

§ 13. *Influence of Aberration. Optical Power of Instruments.*

Our investigations and estimates of resolving power have thus far proceeded upon the supposition that there are no optical imperfections, whether of the nature of a regular aberration or dependent upon irregularities of material and workmanship. In practice there will always be a certain aberration, or error of phase, which we may also regard as the deviation of the actual wave-surface from its intended position. In general, we may say that aberration is unimportant, when it nowhere (or at any rate over a relatively small area only) exceeds a small fraction of the wave-length (λ). Thus in estimating the intensity at a focal point, where, in the absence of aberration, all the secondary waves would have exactly the same phase, we see that an aberration nowhere exceeding $\frac{1}{4}\lambda$ can have but little effect.

The only case in which the influence of small aberration upon the entire image has been calculated† is that of a rectangular aperture, traversed by a cylindrical wave with aberration equal to cx^3. The aberration is here unsymmetrical, the wave being in advance of its proper place in one half of the aperture, but behind in the other half. No terms in x or x^2 need be considered. The first would correspond to a general turning of the beam; and the second would imply imperfect focusing of the central parts. The effect of aberration may be considered in two ways. We may suppose the aperture (a) constant, and inquire into the operation of an increasing aberration; or we may take a given value of c (*i.e.* a given wave-surface) and examine the effect of a varying aperture. The results in the second case show that an increase of aperture up to that corresponding to an extreme aberration of half a period has no ill effect upon the central band (§ 11), but it increases unduly the intensity of one of the neighbouring lateral bands; and the practical conclusion is that the best results will be obtained from an aperture giving an extreme aberration of from a quarter to half a period, and that with an increased aperture aberration is not so much a direct cause of

* "Chromatics," in Vol. III. of Supp. to *Enc. Brit.* 1817.

† "Investigations in Optics," *Phil. Mag.* Nov. 1879. [Vol. I. p. 428.]

deterioration as an obstacle to the attainment of that improved definition which should accompany the increase of aperture.

If, on the other hand, we suppose the aperture given, we find that aberration begins to be distinctly mischievous when it amounts to about a quarter period, *i.e.* when the wave-surface deviates at each end by a quarter wave-length from the true plane.

For the focal point itself the calculations are much simpler. We will consider the case of a circular object-glass with a symmetrical aberration proportional to $h\rho^4$. The vibration will be represented by

$$2\int_0^1 \cos(nt - h\rho^4)\,\rho\,d\rho,$$

in which the radius of the aperture is supposed to be unity. The intensity is thus expressed by

$$I_0^2 = \left[2\int_0^1 \cos(h\rho^4)\,\rho\,d\rho\right]^2 + \left[2\int_0^1 \sin(h\rho^4)\,\rho\,d\rho\right]^2, \quad\ldots\ldots\ldots\ldots(1)$$

the scale being such that the intensity is unity when there is no aberration ($h = 0$).

By integration by parts it can be shown that

$$2\int_0^1 e^{ih\rho^4}\rho\,d\rho = e^{ih}\left\{1 - \frac{4ih}{6} + \frac{(4ih)^2}{6.10} - \frac{(4ih)^3}{6.10.14} + \ldots\right\};$$

so that

$$2\int_0^1 \cos(h\rho^4)\,\rho\,d\rho = \cos h\left\{1 - \frac{(4h)^2}{6.10} + \frac{(4h)^4}{6.10.14.18} - \ldots\right\}$$

$$+ \sin h\left\{\frac{4h}{6} - \frac{(4h)^3}{6.10.14} + \ldots\right\}, \quad\ldots\ldots\ldots\ldots\ldots\ldots(2)$$

$$2\int_0^1 \sin(h\rho^4)\,\rho\,d\rho = \sin h\left\{1 - \frac{(4h)^2}{6.10} + \frac{(4h)^4}{6.10.14.18} - \ldots\right\}$$

$$- \cos h\left\{\frac{4h}{6} - \frac{(4h)^3}{6.10.14} + \ldots\right\}. \quad\ldots\ldots\ldots\ldots\ldots\ldots(3)$$

Hence, when $h = \frac{1}{4}\pi$,

$$2\int_0^1 \cos(\tfrac{1}{4}\pi\rho^4)\,\rho\,d\rho = 1{\cdot}32945/\sqrt{2}, \quad 2\int_0^1 \sin(\tfrac{1}{4}\pi\rho^4)\,\rho\,d\rho = {\cdot}35424/\sqrt{2},$$

$$I_0^2 = {\cdot}9464.$$

Similarly, when $h = \frac{1}{2}\pi$,

$$I_0^2 = {\cdot}8003;$$

and when $h = \pi$,

$$I_0^2 = {\cdot}3947.$$

These numbers represent the influence of aberration upon the intensity at the central point, upon the understanding that the focusing is that adapted to a small aperture, for which h might be neglected. If a readjustment of focus be permitted, the numbers will be sensibly raised. The general conclusion is that an aberration between the centre and circumference of a quarter period has but little effect upon the intensity at the central point of the image.

As an application of this result, let us investigate what amount of temperature disturbance in the tube of a telescope may be expected to impair definition. According to Biot and Arago, the index μ for air at $t°$ C. and at atmospheric pressure is given by

$$\mu - 1 = \frac{\cdot 00029}{1 + \cdot 0037\, t}.$$

If we take 0° C. as standard temperature,

$$\delta\mu = -1\cdot 1 t \times 10^{-6}.$$

Thus, on the supposition that the irregularity of temperature t extends through a length l, and produces an acceleration of a quarter of a wave-length,

$$\tfrac{1}{4}\lambda = 1\cdot 1\, lt \times 10^{-6};$$

or, if we take $\lambda = 5\cdot 3 \times 10^{-5}$,

$$lt = 12,$$

the unit of length being the centimetre.

We may infer that, in the case of a telescope tube 12 cm. long, a stratum of air heated 1° C. lying along the top of the tube, and occupying a moderate fraction of the whole volume, would produce a not insensible effect. If the change of temperature progressed uniformly from one side to the other, the result would be a lateral displacement of the image without loss of definition; but in general both effects would be observable. In longer tubes a similar disturbance would be caused by a proportionally less difference of temperature.

We will now consider the application of the principle to the formation of images, unassisted by reflexion or refraction*. The function of a lens in forming an image is to compensate by its variable thickness the differences of phase which would otherwise exist between secondary waves arriving at the focal point from various parts of the aperture (Optics, *Enc. Brit.* Vol. XVII. p. 802 [Vol. II. p. 398]). If we suppose the diameter of the lens to be given ($2R$), and its focal length f gradually to increase, the original differences of phase at the image of an infinitely distant luminous point diminish without limit. When f attains a certain value, say f_1, the extreme error of phase to be compensated falls to $\tfrac{1}{4}\lambda$. But, as we have seen, such an error of phase causes no sensible deterioration in the definition; so that from this point onwards the lens is useless, as only improving an image already sensibly as

* *Phil. Mag.* March 1881. [Vol. I. p. 513.]

perfect as the aperture admits of. Throughout the operation of increasing the focal length, the resolving power of the instrument, which depends only upon the aperture, remains unchanged; and we thus arrive at the rather startling conclusion that a telescope of any degree of resolving power might be constructed without an object-glass, if only there were no limit to the admissible focal length. This last proviso, however, as we shall see, takes away almost all practical importance from the proposition.

To get an idea of the magnitudes of the quantities involved, let us take the case of an aperture of $\frac{1}{5}$ inch, about that of the pupil of the eye. The distance f_1, which the actual focal length must exceed, is given by

$$\sqrt{(f_1^2 + R^2)} - f_1 = \tfrac{1}{4}\lambda;$$

so that

$$f_1 = 2R^2/\lambda. \qquad \text{......................(4)}$$

Thus, if $\lambda = \frac{1}{40000}$, $R = \frac{1}{10}$, we find $f_1 = 800$ inches [inch = 2·54 cm.].

The image of the sun thrown upon a screen at a distance exceeding 66 feet, through a hole $\frac{1}{5}$ inch in diameter, is therefore at least as well defined as that seen direct.

As the minimum focal length increases with the square of the aperture, a quite impracticable distance would be required to rival the resolving power of a modern telescope. Even for an aperture of 4 inches, f_1 would have to be 5 miles.

A similar argument may be applied to find at what point an achromatic lens becomes sensibly superior to a single one. The question is whether, when the adjustment of focus is correct for the central rays of the spectrum, the error of phase for the most extreme rays (which it is necessary to consider) amounts to a quarter of a wave-length. If not, the substitution of an achromatic lens will be of no advantage. Calculation shows that, if the aperture be $\frac{1}{5}$ inch, an achromatic lens has no sensible advantage if the focal length be greater than about 11 inches. If we suppose the focal length to be 66 feet, a single lens is practically perfect up to an aperture of 1·7 inch.

Some estimates of the admissible *aberration* in a spherical lens have already been given under Optics, *Enc. Brit.* Vol. XVII. p. 807 [Vol. II. p. 413]. In a similar manner we may estimate the least visible displacement of the eye-piece of a telescope focused upon a distant object, a question of interest in connexion with range-finders. It appears* that a displacement δf from the true focus will not sensibly impair definition, provided

$$\delta f < f^2\lambda/R^2, \qquad \text{......................(5)}$$

$2R$ being the diameter of aperture. The linear accuracy required is thus a

* *Phil. Mag.* xx. p. 354, 1885. [Vol. II. p. 430.]

function of the *ratio* of aperture to focal length. The formula agrees well with experiment.

The principle gives an instantaneous solution of the question of the ultimate optical efficiency in the method of "mirror-reading," as commonly practised in various physical observations. A rotation by which one edge of the mirror advances $\frac{1}{4}\lambda$ (while the other edge retreats to a like amount) introduces a phase-discrepancy of a whole period where before the rotation there was complete agreement. A rotation of this amount should therefore be easily visible, but the limits of resolving power are being approached; and the conclusion is independent of the focal length of the mirror, and of the employment of a telescope, provided of course that the reflected image is seen in focus, and that the full width of the mirror is utilized.

A comparison with the method of a material pointer, attached to the parts whose rotation is under observation, and viewed through a microscope, is of interest. The limiting efficiency of the microscope is attained when the angular aperture amounts to 180° (Microscope, *Enc. Brit.* Vol. XVI. p. 267; Optics, *Enc. Brit.* Vol. XVII. p. 807 [Vol. II. p. 412]); and it is evident that a lateral displacement of the point under observation through $\frac{1}{2}\lambda$ entails (at the old image) a phase-discrepancy of a whole period, one extreme ray being accelerated and the other retarded by half that amount. We may infer that the limits of efficiency in the two methods are the same when the length of the pointer is equal to the width of the mirror.

An important practical question is the amount of error admissible in optical surfaces. In the case of a mirror, reflecting at nearly perpendicular incidence, there should be no deviation from truth (over any appreciable area) of more than $\frac{1}{8}\lambda$. For glass, $\mu - 1 = \frac{1}{2}$ nearly; and hence the admissible error in a refracting surface of that material is four times as great.

Fig. 7.

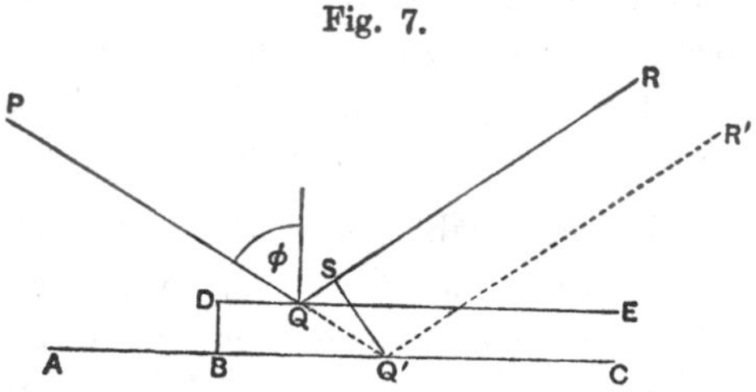

In the case of oblique reflexion at an angle ϕ, the error of retardation due to an elevation BD (Fig. 7) is

$$QQ' - QS = BD \sec\phi\,(1 - \cos SQQ') = BD \sec\phi\,(1 + \cos 2\phi) = 2BD\cos\phi\,;$$

from which it follows that an error of given magnitude in the figure of a surface is less important in oblique than in perpendicular reflexion. It must, however, be borne in mind that errors can sometimes be compensated by altering adjustments. If a surface intended to be flat is affected with a slight general curvature, a remedy may be found in an alteration of focus, and the remedy is the less complete as the reflexion is more oblique.

The formula expressing the optical power of prismatic spectroscopes is given with examples under Optics, *Enc. Brit.* Vol. XVII. p. 807 [Vol. II. p. 412], and may readily be investigated upon the principles of the wave theory. Let A_0B_0 (Fig. 8) be a plane wave-surface of the light before it falls upon the prisms, AB the corresponding wave-surface for a particular part of the

Fig. 8.

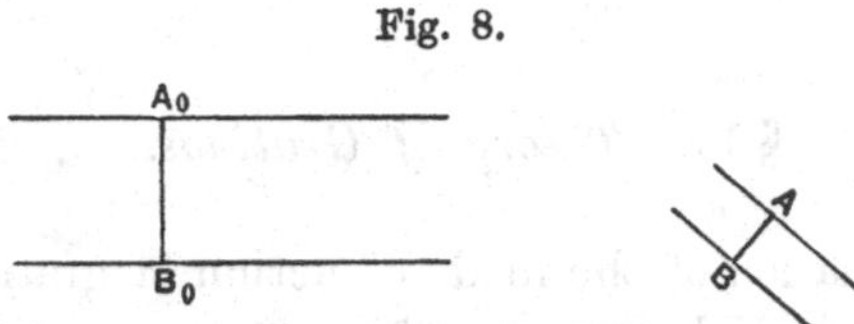

spectrum after the light has passed the prisms, or after it has passed the eye-piece of the observing telescope. The path of a ray from the wave-surface A_0B_0 to A or B is determined by the condition that the optical distance, $\int \mu ds$, is a minimum (Optics, *Enc. Brit.* Vol. XVII. p. 798); and, as AB is by supposition a wave-surface, this optical distance is the same for both points. Thus

$$\int \mu\, ds \text{ (for } A) = \int \mu\, ds \text{ (for } B). \quad\ldots\ldots\ldots\ldots\ldots\ldots(6)$$

We have now to consider the behaviour of light belonging to a neighbouring part of the spectrum. The path of a ray from the wave-surface A_0B_0 to the point A is changed; but in virtue of the minimum property the change may be neglected in calculating the optical distance, as it influences the result by quantities of the second order only in the changes of refrangibility. Accordingly, the optical distance from A_0B_0 to A is represented by $\int(\mu + \delta\mu)\, ds$, the integration being along the original path $A_0 \ldots A$; and similarly the optical distance between A_0B_0 and B is represented by $\int(\mu + \delta\mu)\, ds$, the integration being along $B_0 \ldots B$. In virtue of (6) the difference of the optical distances to A and B is

$$\int \delta\mu\, ds \text{ (along } B_0 \ldots B) - \int \delta\mu\, ds \text{ (along } A_0 \ldots A). \quad\ldots\ldots\ldots(7)$$

The new wave-surface is formed in such a position that the optical distance is constant; and therefore the *dispersion*, or the angle through which the wave-surface is turned by the change of refrangibility, is found simply by dividing (7) by the distance AB. If, as in common flint-glass spectroscopes, there is only one dispersing substance, $\int \delta\mu\, ds = \delta\mu \,.\, s$, where s is simply the thickness traversed by the ray. If t_2 and t_1 be the thicknesses traversed by the extreme rays, and a denote the width of the emergent beam, the dispersion θ is given by

$$\theta = \delta\mu\,(t_2 - t_1)/a,$$

or, if t_1 be negligible,

$$\theta = \delta\mu t/a. \quad\text{.................................}(8)$$

The condition of resolution of a double line whose components subtend an angle θ is that θ must exceed λ/a. Hence, in order that a double line may be resolved whose components have indices μ and $\mu+\delta\mu$, it is necessary that t should exceed the value given by the following equation:—

$$t = \lambda/\delta\mu. \quad\text{.................................}(9)$$

For applications of these results, see Spectroscope (*Enc. Brit.* Vol. XXII. p. 373).

§ 14. *Theory of Gratings.*

The general explanation of the mode of action of gratings has been given under Light (*Enc. Brit.* Vol. XIV. p. 607). If the grating be composed of alternate transparent and opaque parts, the question may be treated by means of the general integrals (§ 11) by merely limiting the integration to the transparent parts of the aperture. For an investigation upon these lines the reader is referred to Airy's *Tracts* and to Verdet's *Leçons.* If, however, we assume the theory of a simple rectangular aperture (§ 11), the results of the ruling can be inferred by elementary methods, which are perhaps more instructive.

Apart from the ruling, we know that the image of a mathematical line will be a series of narrow bands, of which the central one is by far the brightest. At the middle of this band there is complete agreement of phase among the secondary waves. The dark lines which separate the bands are the places at which the phases of the secondary waves range over an integral number of periods. If now we suppose the aperture AB to be covered by a great number of opaque strips or bars of width d, separated by transparent intervals of width a, the condition of things in the directions just spoken of is not materially changed. At the central point there is still complete agreement of phase; but the amplitude is diminished in the ratio of $a : a+d$. In another direction, making a small angle with the last, such that the projection of AB upon it amounts to a few wave-lengths, it is easy to see that the mode of interference is the same as if there were no ruling. For example, when the direction is such that the projection of AB upon it amounts to one wave-length, the elementary components neutralize one another, because their phases are distributed symmetrically, though discontinuously, round the entire period. The only effect of the ruling is to diminish the amplitude in the ratio $a : a+d$; and, except for the difference in illumination, the appearance of a line of light is the same as if the aperture were perfectly free.

The lateral (spectral) images occur in such directions that the projection of the element $(a+d)$ of the grating upon them is an exact multiple of λ. The effect of each of the n elements of the grating is then the same; and, unless this vanishes on account of a particular adjustment of the ratio $a:d$, the resultant amplitude becomes comparatively very great. These directions, in which the retardation between A and B is exactly $mn\lambda$, may be called the principal directions. On either side of any one of them the illumination is distributed according to the same law as for the central image $(m=0)$, vanishing, for example, when the retardation amounts to $(mn \pm 1)\lambda$. In considering the relative brightnesses of the different spectra, it is therefore sufficient to attend merely to the principal directions, provided that the whole deviation be not so great that its cosine differs considerably from unity.

We have now to consider the amplitude due to a single element, which we may conveniently regard as composed of a transparent part a bounded by two opaque parts of width $\frac{1}{2}d$. The phase of the resultant effect is by symmetry that of the component which comes from the middle of a. The fact that the other components have phases differing from this by amounts ranging between $\pm\, am\pi/(a+d)$ causes the resultant amplitude to be less than for the central image (where there is complete phase agreement). If B_m denote the brightness of the mth lateral image, and B_0 that of the central image, we have

$$B_m : B_0 = \left[\int_{-\frac{am\pi}{a+d}}^{+\frac{am\pi}{a+d}} \cos x\, dx \div \frac{2am\pi}{a+d}\right]^2 = \left(\frac{a+d}{am\pi}\right)^2 \sin^2 \frac{am\pi}{a+d}. \quad \text{......(1)}$$

If B denote the brightness of the central image when the whole of the space occupied by the grating is transparent, we have

$$B_0 : B = a^2 : (a+d)^2,$$

and thus

$$B_m : B = \frac{1}{m^2\pi^2} \sin^2 \frac{am\pi}{a+d}. \quad \text{......................(2)}$$

The sine of an angle can never be greater than unity; and consequently under the most favourable circumstances only $1/m^2\pi^2$ of the original light can be obtained in the mth spectrum. We conclude that, with a grating composed of transparent and opaque parts, the utmost light obtainable in any one spectrum is in the first, and there amounts to $1/\pi^2$, or about $\frac{1}{10}$, and that for this purpose a and d must be equal. When $d=a$, the general formula becomes

$$B_m : B = \frac{\sin^2 \frac{1}{2}m\pi}{m^2\pi^2}, \quad \text{.........................(3)}$$

showing that, when m is even, B_m vanishes, and that, when m is odd,

$$B_m : B = 1/m^2\pi^2.$$

The third spectrum has thus only $\frac{1}{9}$ of the brilliancy of the first.

Another particular case of interest is obtained by supposing a small relatively to $(a+d)$. Unless the spectrum be of very high order, we have simply

$$B_m : B = \{a/(a+d)\}^2 \; ; \quad \text{..........................(4)}$$

so that the brightnesses of all the spectra are the same.

The light stopped by the opaque parts of the grating, together with that distributed in the central image and lateral spectra, ought to make up the brightness that would be found in the central image, were all the apertures transparent. Thus, if $a=d$, we should have

$$1 = \frac{1}{2} + \frac{1}{4} + \frac{2}{\pi^2}\left(1 + \frac{1}{9} + \frac{1}{25} + \dots\right),$$

which is true by a known theorem. In the general case

$$\frac{a}{a+d} = \left(\frac{a}{a+d}\right)^2 + \frac{2}{\pi^2} \sum_{m=1}^{m=\infty} \frac{1}{m^2} \sin^2\left(\frac{m\pi a}{a+d}\right),$$

a formula which may be verified by Fourier's theorem.

According to a general principle formulated by Babinet, the brightness of a lateral spectrum is not affected by an interchange of the transparent and opaque parts of the grating. The vibrations corresponding to the two parts are precisely antagonistic, since if both were operative the resultant would be zero. So far as the application to gratings is concerned, the same conclusion may be derived from (2).

From the value of $B_m : B_0$ we see that no lateral spectrum can surpass the central image in brightness; but this result depends upon the hypothesis that the ruling acts by opacity, which is generally very far from being the case in practice. In an engraved glass grating there is no opaque material present by which light could be absorbed, and the effect depends upon a difference of retardation in passing the alternate parts. It is possible to prepare gratings which give a lateral spectrum brighter than the central image, and the explanation is easy. For if the alternate parts were equal and alike transparent, but so constituted as to give a relative retardation of $\frac{1}{2}\lambda$, it is evident that the central image would be entirely extinguished, while the first spectrum would be four times as bright as if the alternate parts were opaque. If it were possible to introduce at every part of the aperture of the grating an arbitrary retardation, all the light might be concentrated in any desired spectrum. By supposing the retardation to vary uniformly and continuously we fall upon the case of an ordinary prism; but there is then no diffraction spectrum in the usual sense. To obtain such it would be necessary that the retardation should gradually alter by a wave-length in passing over any element of the grating, and then fall back to its previous value, thus springing suddenly over a wave-length. It is not likely that such a result will ever be fully attained in practice; but the case is worth stating, in order to show that there is no theoretical limit to the concentration of light of assigned wave-length in one spectrum, and as illustrating the frequently observed unsymmetrical character of the spectra on the two sides of the central image*.

Fig. 9.

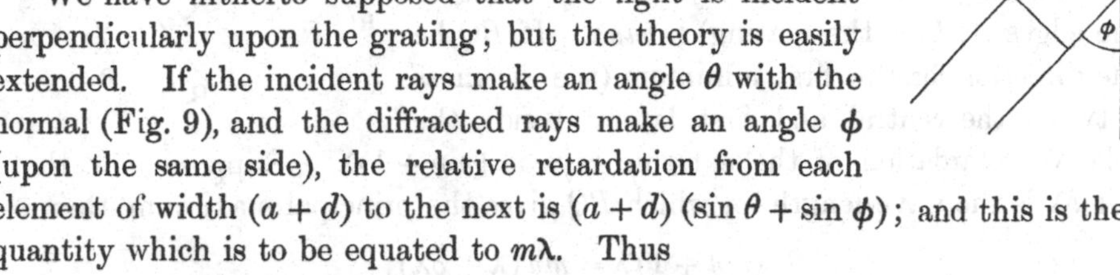

We have hitherto supposed that the light is incident perpendicularly upon the grating; but the theory is easily extended. If the incident rays make an angle θ with the normal (Fig. 9), and the diffracted rays make an angle ϕ (upon the same side), the relative retardation from each element of width $(a+d)$ to the next is $(a+d)(\sin\theta+\sin\phi)$; and this is the quantity which is to be equated to $m\lambda$. Thus

$$\sin\theta+\sin\phi = 2\sin\tfrac{1}{2}(\theta+\phi)\,.\cos\tfrac{1}{2}(\theta-\phi) = m\lambda/(a+d). \quad \text{......(5)}$$

The "deviation" is $(\theta+\phi)$, and is therefore a minimum when $\theta=\phi$, *i.e.* when the grating is so situated that the angles of incidence and diffraction are equal.

In the case of a reflexion grating the same method applies. If θ and ϕ denote the angles with the normal made by the incident and diffracted rays, the formula (5) still holds, and, if the deviation be reckoned from the direction of the regularly reflected rays, it is expressed as before by $(\theta+\phi)$, and is a minimum when $\theta=\phi$, that is, when the diffracted rays return upon the course of the incident rays.

Fig. 10.

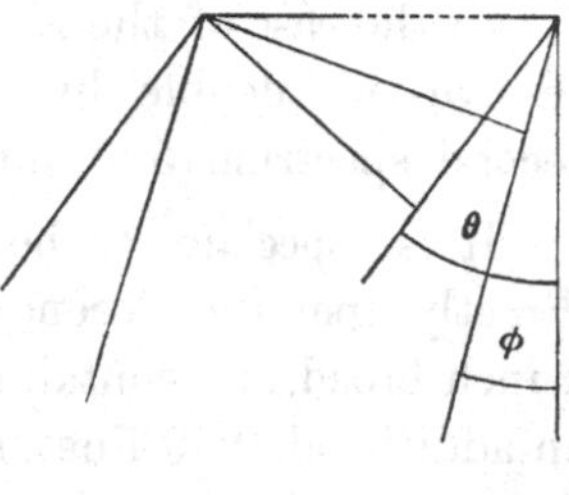

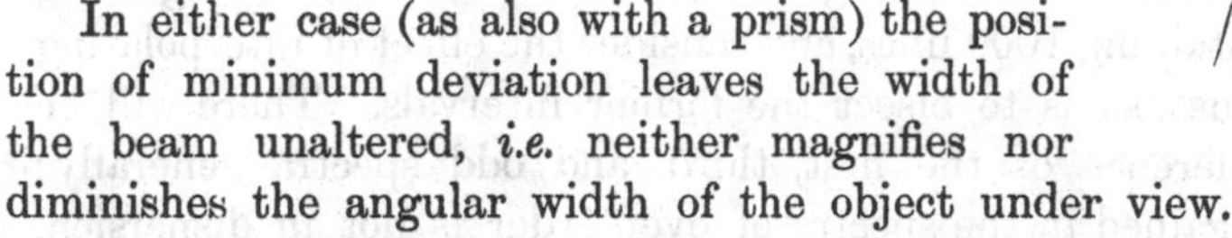

In either case (as also with a prism) the position of minimum deviation leaves the width of the beam unaltered, *i.e.* neither magnifies nor diminishes the angular width of the object under view.

From (5) we see that, when the light falls perpendicularly upon a grating $(\theta=0)$, there is no spectrum formed (the image corresponding to $m=0$ not being counted as a spectrum), if the grating interval σ or $(a+d)$ is less than λ. Under these circumstances, if the material of the grating be completely transparent, the whole of the light must appear in the direct image, and the ruling is not perceptible. From the absence of spectra Fraunhofer argued that there must be a microscopic limit represented by λ; and the inference is plausible, to say the least†. Fraunhofer should, however, have fixed the microscopic limit at $\tfrac{1}{2}\lambda$, as appears from (5), when we suppose

$$\theta=\tfrac{1}{2}\pi, \quad \phi=\tfrac{1}{2}\pi.$$

* *Phil. Mag.* XLVII. 193, 1874. [Vol. I. p. 215.]

† "Notes on some Fundamental Propositions in Optics," *Phil. Mag.* June 1886. [Vol. II. p. 513.]

We will now consider the important subject of the resolving power of gratings, as dependent upon the number of lines (n) and the order of the spectrum observed (m). Let BP (Fig. 11) be the direction of the principal maximum (middle of central band) for the wave-length λ in the mth spectrum. Then the relative retardation of the extreme rays (corresponding to the edges A, B of the grating) is $mn\lambda$. If BQ be the direction for the first minimum (the darkness between the central and first lateral band), the relative retardation of the extreme rays is $(mn+1)\lambda$. Suppose now that $\lambda + \delta\lambda$ is the wave-length for which BQ gives the principal maximum, then

Fig. 11.

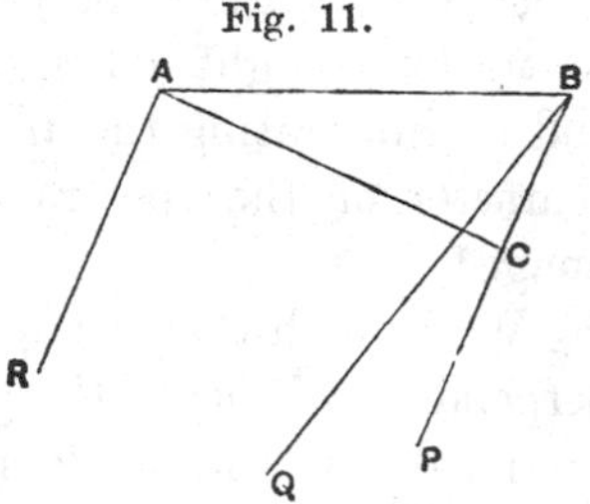

$$(mn+1)\lambda = mn(\lambda + \delta\lambda);$$

whence

$$\delta\lambda/\lambda = 1/mn. \qquad \text{.................................}(6)$$

According to our former standard, this gives the smallest difference of wave-lengths in a double line which can be just resolved; and we conclude that the resolving power of a grating depends only upon the total number of lines, and upon the order of the spectrum, without regard to any other considerations. It is here of course assumed that the n lines are really utilized.

In the case of the D-lines the value of $\delta\lambda/\lambda$ is about 1/1000; so that to resolve this double line in the first spectrum requires 1000 lines, in the second spectrum 500, and so on.

It is especially to be noticed that the resolving power does not depend directly upon the closeness of the ruling. Let us take the case of a grating 1 inch broad, and containing 1000 lines, and consider the effect of interpolating an additional 1000 lines, so as to bisect the former intervals. There will be destruction by interference of the first, third, and odd spectra generally; while the advantage gained in the spectra of even order is not in dispersion, nor in resolving power, but simply in brilliancy, which is increased four times. If we now suppose half the grating cut away, so as to leave 1000 lines in half an inch, the dispersion will not be altered, while the brightness and resolving power are halved.

There is clearly no theoretical limit to the resolving power of gratings, even in spectra of given order. But it is possible that, as suggested by Rowland*, the structure of natural spectra may be too coarse to give opportunity for resolving powers much higher than those now in use. However this may be, it would always be possible, with the aid of a grating of given resolving power, to construct artificially from white light mixtures of

* Compare also Lippich, *Pogg. Ann.* CXXXIX. p. 465, 1870; Rayleigh, *Nature*, Oct. 2, 1873, [Vol. I. p. 183.]

slightly different wave-lengths whose resolution or otherwise would discriminate between powers inferior and superior to the given one*.

If we define as the "dispersion" in a particular part of the spectrum the ratio of the angular interval $d\theta$ to the corresponding increment of wave-length $d\lambda$, we may express it by a very simple formula. For the alteration of wave-length entails, at the two limits of a diffracted wave-front, a relative retardation equal to $mnd\lambda$. Hence, if a be the width of the diffracted beam, and $d\theta$ the angle through which the wave-front is turned,

$$a\,d\theta = mn\,d\lambda,$$

or

$$\text{dispersion} = mn/a. \quad \text{.........................(7)}$$

The resolving power and the width of the emergent beam fix the optical character of the instrument. The latter element must eventually be decreased until less than the diameter of the pupil of the eye. Hence a wide beam demands treatment with further apparatus (usually a telescope) of high magnifying power.

In the above discussion it has been supposed that the ruling is accurate, and we have seen that by increase of m a high resolving power is attainable with a moderate number of lines. But this procedure (apart from the question of illumination) is open to the objection that it makes excessive demands upon accuracy. According to the principle already laid down, it can make but little difference in the principal direction corresponding to the first spectrum, provided each line lie within a quarter of an interval $(a + d)$ from its theoretical position. But, to obtain an equally good result in the mth spectrum, the error must be less than $1/m$ of the above amount†.

There are certain errors of a systematic character which demand special consideration. The spacing is usually effected by means of a screw, to each revolution of which corresponds a large number (*e.g.* one hundred) of lines. In this way it may happen that, although there is almost perfect periodicity with each revolution of the screw after (say) 100 lines, yet the 100 lines themselves are not equally spaced. The "ghosts" thus arising were first described by Quincke‡, and have been elaborately investigated by Peirce§, both theoretically and experimentally. The general nature of the effects to be expected in such a case may be made clear by means of an illustration already

* The power of a grating to construct light of nearly definite wave-length is well illustrated by Young's comparison with the production of a musical note by reflexion of a sudden sound from a row of palings. The objection raised by Herschel (*Light*, § 703) to this comparison depends on a misconception.

† It must not be supposed that errors of this order of magnitude are unobjectionable in all cases. The position of the middle of the bright band representative of a mathematical line can be fixed with a spider-line micrometer within a small fraction of the width of the band, just as the accuracy of astronomical observations far transcends the separating power of the instrument.

‡ *Pogg. Ann.* CXLVI. p. 1, 1872.

§ *Am. Jour. Math.* II. p. 330, 1879.

employed for another purpose. Suppose two similar and accurately ruled transparent gratings to be superposed in such a manner that the lines are parallel. If the one set of lines exactly bisect the intervals between the others, the grating interval is practically halved, and the previously existing spectra of odd order vanish. But a very slight relative displacement will cause the apparition of the odd spectra. In this case there is approximate periodicity in the half interval, but complete periodicity only after the whole interval. The advantage of approximate bisection lies in the superior brilliancy of the surviving spectra; but in any case the compound grating may be considered to be perfect in the longer interval, and the definition is as good as if the bisection were accurate.

The effect of a gradual increase in the interval (Fig. 12) as we pass across the grating has been investigated by Cornu*, who thus explains an anomaly observed by Mascart. The latter found that certain gratings exercised a converging power upon the spectra formed upon one side, and a corresponding

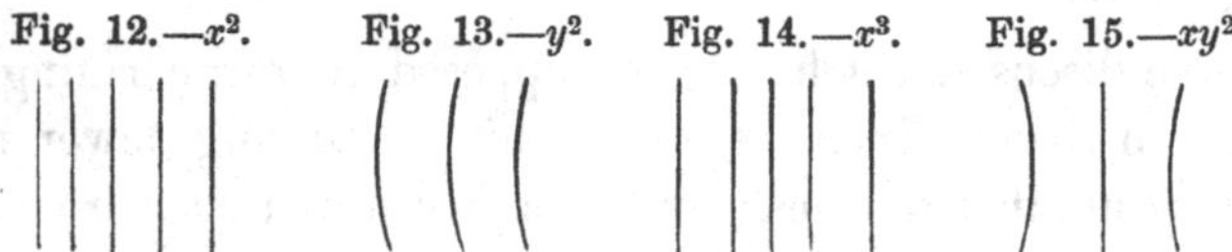
Fig. 12.—x^2. Fig. 13.—y^2. Fig. 14.—x^3. Fig. 15.—xy^2.

diverging power upon the spectra on the other side. Let us suppose that the light is incident perpendicularly, and that the grating interval increases from the centre towards that edge which lies nearest to the spectrum under observation, and decreases towards the hinder edge. It is evident that the waves from *both* halves of the grating are accelerated in an increasing degree, as we pass from the centre outwards, as compared with the phase they would possess were the central value of the grating interval maintained throughout.

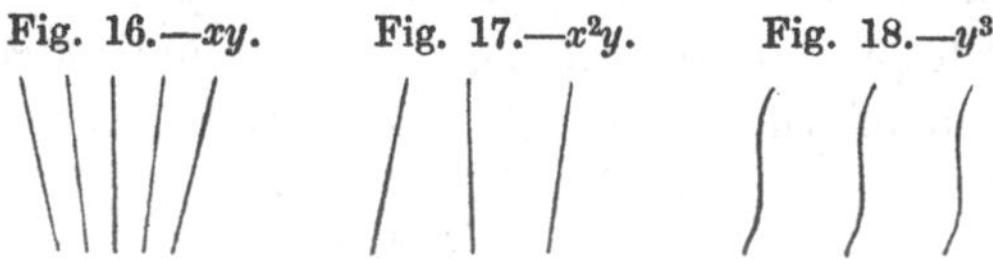
Fig. 16.—xy. Fig. 17.—x^2y. Fig. 18.—y^3.

The irregularity of spacing has thus the effect of a convex lens, which accelerates the marginal relatively to the central rays. On the other side the effect is reversed. This kind of irregularity may clearly be present in a degree surpassing the usual limits, without loss of definition, when the telescope is focused so as to secure the best effect.

It may be worth while to examine further the other variations from correct ruling which correspond to the various terms expressing the deviation of the wave-surface from a perfect plane. If x and y be coordinates in the

* *C. R.* LXXX. p. 645, 1875.

plane of the wave-surface, the axis of y being parallel to the lines of the grating, and the origin corresponding to the centre of the beam, we have as an approximate equation to the wave-surface (§ 6)

$$z = \frac{x^2}{2\rho} + Bxy + \frac{y^2}{2\rho'} + \alpha x^3 + \beta x^2 y + \gamma xy^2 + \delta y^3 + \ldots ; \quad \ldots\ldots\ldots(8)$$

and, as we have just seen, the term in x^2 corresponds to a linear error in the spacing. In like manner, the term in y^2 corresponds to a general *curvature* of the lines (Fig. 13), and does not influence the definition at the (primary) focus, although it may introduce astigmatism*. If we suppose that everything is symmetrical on the two sides of the primary plane $y = 0$, the coefficients B, β, δ vanish. In spite of any inequality between ρ and ρ', the definition will be good to this order of approximation, provided α and γ vanish. The former measures the *thickness* of the primary focal line, and the latter measures its *curvature*. The error of ruling giving rise to α is one in which the intervals increase or decrease in *both* directions from the centre outwards (Fig. 14), and it may often be compensated by a slight rotation in azimuth of the object-glass of the observing telescope. The term in γ corresponds to a *variation* of curvature in crossing the grating (Fig. 15).

When the plane zx is not a plane of symmetry, we have to consider the terms in xy, x^2y, and y^3. The first of these corresponds to a deviation from parallelism, causing the interval to alter gradually as we pass *along* the lines (Fig. 16). The error thus arising may be compensated by a rotation of the object-glass about one of the diameters $y = \pm x$. The term in x^2y corresponds to a deviation from parallelism in the same direction on both sides of the central line (Fig. 17); and that in y^3 would be caused by a curvature such that there is a point of inflexion at the middle of each line (Fig. 18).

All the errors, except that depending on α, and especially those depending on γ and δ, can be diminished, without loss of resolving power, by contracting the *vertical* aperture. A linear error in the spacing, and a general curvature of the lines, are eliminated in the ordinary use of a grating.

The explanation of the difference of focus upon the two sides as due to unequal spacing was verified by Cornu upon gratings purposely constructed with an increasing interval. He has also shown how to rule a plane surface with lines so disposed that the grating shall of itself give well-focused spectra.

* "In the same way we may conclude that in flat gratings any departure from a straight line has the effect of causing the dust in the slit and the spectrum to have different foci—a fact sometimes observed" (Rowland, "On Concave Gratings for Optical Purposes," *Phil. Mag.* September 1883).

A similar idea appears to have guided Rowland to his brilliant invention of concave gratings, by which spectra can be photographed without any further optical appliance. In these instruments the lines are ruled upon a spherical surface of speculum metal, and mark the intersections of the surface by a system of parallel and equidistant planes, of which the middle member passes through the centre of the sphere. If we consider for the present only the primary plane of symmetry, the figure is reduced to two dimensions. Let AP (Fig. 19) represent the surface of the grating, O being the centre of the circle. Then, if Q be any radiant point and Q' its image (primary focus) in the spherical mirror AP, we have

Fig. 19.

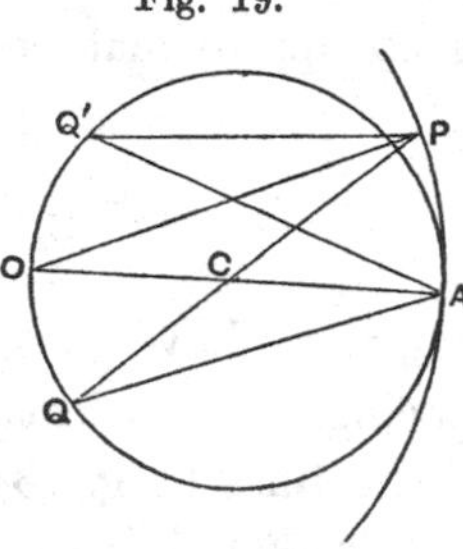

$$\frac{1}{v_1} + \frac{1}{u} = \frac{2\cos\phi}{a}$$

where $v_1 = AQ'$, $u = AQ$, $a = OA$, $\phi =$ angle of incidence QAO, equal to the angle of reflexion $Q'AO$*. If Q be on the circle described upon OA as diameter, so that $u = a\cos\phi$, then Q' lies also upon the same circle; and in this case it follows from the symmetry that the unsymmetrical aberration (depending upon α) vanishes.

This disposition is adopted in Rowland's instrument; only, in addition to the central image formed at the angle $\phi' = \phi$, there are a series of spectra with various values of ϕ', but all disposed upon the same circle. Rowland's investigation is contained in the paper already referred to; but the following account of the theory is in the form adopted by Glazebrook†.

In order to find the difference of optical distances between the courses QAQ', QPQ', we have to express $QP - QA$, $PQ' - AQ'$. To find the former, we have, if $OAQ = \phi$, $AOP = \omega$,

$$QP^2 = u^2 + 4a^2 \sin^2 \tfrac{1}{2}\omega - 4au \sin \tfrac{1}{2}\omega \sin(\tfrac{1}{2}\omega - \phi)$$

$$= (u + a\sin\phi\sin\omega)^2 - a^2\sin^2\phi\sin^2\omega + 4a\sin^2\tfrac{1}{2}\omega\,(a - u\cos\phi).$$

Now as far as ω^4

$$4\sin^2\tfrac{1}{2}\omega = \sin^2\omega + \tfrac{1}{4}\sin^4\omega,$$

and thus to the same order

$$QP^2 = (u + a\sin\phi\sin\omega)^2 - a\cos\phi\,(u - a\cos\phi)\sin^2\omega + \tfrac{1}{4}a\,(a - u\cos\phi)\sin^4\omega.$$

* This formula may be obtained as in Optics, *Enc. Brit.* Vol. XVII. p. 800, equation (3) [Vol. II. p. 390], and may indeed be derived from that equation by writing $\phi' = \phi$, $\mu = -1$.

† *Phil. Mag.* June 1883, Nov. 1883.

But if we now suppose that Q lies on the circle $u = a\cos\phi$, the middle term vanishes, and we get, correct as far as ω^4,

$$QP = (u + a\sin\phi\sin\omega)\sqrt{\left\{1 + \frac{a^2\sin^2\phi\sin^4\omega}{4u}\right\}};$$

so that

$$QP - u = a\sin\phi\sin\omega + \tfrac{1}{8}a\sin\phi\tan\phi\sin^4\omega, \quad\text{..........(9)}$$

in which it is to be noticed that the adjustment necessary to secure the disappearance of $\sin^2\omega$ is sufficient also to destroy the term in $\sin^3\omega$.

A similar expression can be found for $Q'P - Q'A$; and thus, if $Q'A = v$, $Q'AO = \phi'$, where $v = a\cos\phi'$, we get

$$QP + PQ' - QA - AQ' = a\sin\omega(\sin\phi - \sin\phi') + \tfrac{1}{8}a\sin^4\omega(\sin\phi\tan\phi + \sin\phi'\tan\phi'). \quad\text{...(10)}$$

If $\phi' = \phi$, the term of the first order vanishes, and the reduction of the difference of path *via* P and *via* A to a term of the fourth order proves not only that Q and Q' are conjugate foci, but also that the foci are exempt from the most important term in the aberration. In the present application ϕ' is not necessarily equal to ϕ; but if P correspond to a line upon the grating, the difference of retardations for consecutive positions of P, so far as expressed by the term of the first order, will be equal to $\mp m\lambda$ (m integral), and therefore without influence, provided

$$\sigma(\sin\phi - \sin\phi') = \mp m\lambda, \quad\text{......................(11)}$$

where σ denotes the constant interval between the planes containing the lines. This is the ordinary formula for a reflecting plane grating, and it shows that the spectra are formed in the usual directions. They are here focused (so far as the rays in the primary plane are concerned) upon the circle $OQ'A$, and the outstanding aberration is of the fourth order.

In order that a large part of the field of view may be in focus at once, it is desirable that the locus of the focused spectrum should be nearly perpendicular to the line of vision. For this purpose Rowland places the eye-piece at O, so that $\phi = 0$, and then by (11) the value of ϕ' in the mth spectrum is

$$\sigma\sin\phi' = \pm m\lambda. \quad\text{............................(12)}$$

If ω now relate to the edge of the grating, on which there are altogether n lines, $n\sigma = 2a\sin\omega$, and the value of the last term in (10) becomes

$$\tfrac{1}{16}n\sigma\sin^3\omega\sin\phi'\tan\phi',$$

or

$$\tfrac{1}{16}mn\lambda\sin^3\omega\tan\phi'. \quad\text{........................(13)}$$

This expresses the retardation of the extreme relatively to the central ray, and is to be reckoned positive, whatever may be the signs of ω and ϕ'.

If the semi-angular aperture (ω) be $\frac{1}{100}$, and $\tan \phi' = 1$, mn might be as great as four millions before the error of phase would reach $\frac{1}{4}\lambda$. If it were desired to use an angular aperture so large that the aberration according to (13) would be injurious, Rowland points out that on his machine there would be no difficulty in applying a remedy by making σ slightly variable towards the edges. Or, retaining σ constant, we might attain compensation by so polishing the surface as to bring the circumference slightly forward in comparison with the position it would occupy upon a true sphere.

It may be remarked that these calculations apply to the rays in the primary plane only. The image is greatly affected with astigmatism; but this is of little consequence, if γ in (8) be small enough. Curvature of the primary focal line having a very injurious effect upon definition, it may be inferred from the excellent performance of these gratings that γ is in fact small. Its value does not appear to have been calculated. The other coefficients in (8) vanish in virtue of the symmetry.

The mechanical arrangements for maintaining the focus are of great simplicity. The grating at A and the eye-piece at O are rigidly attached to a bar AO, whose ends rest on carriages, moving on rails OQ, AQ at right angles to each other. A tie between C and Q can be used if thought desirable.

The absence of chromatic aberration gives a great advantage in the comparison of overlapping spectra, which Rowland has turned to excellent account in his determinations of the relative wave-lengths of lines in the solar spectrum*.

For absolute determinations of wave-lengths plane gratings are used. It is found† that the angular measurements present less difficulty than the comparison of the grating interval with the standard metre. There is also some uncertainty as to the actual temperature of the grating when in use. In order to minimize the heating action of the light, it might be submitted to a preliminary prismatic analysis before it reaches the slit of the spectrometer, after the manner of Von Helmholtz (Optics, *Enc. Brit.* Vol. XVII. p. 802 [Vol. II. p. 397]).

Bell found further that it is necessary to submit the gratings to calibration, and not to rest satisfied with a knowledge of the number of lines and of the total width. It not unfrequently happens that near the beginning of the ruling the interval is anomalous. If the width of this region be small, it has scarcely any effect upon the angular measurements, and should be left out of account in estimating the effective interval.

* *Phil. Mag.* March 1887.

† Bell, *Phil. Mag.* March 1887.

§ 15. *Theory of Corrugated Waves.*

The theory of gratings is usually given in a form applicable only to the case where the alternate parts are transparent and opaque. Even then it is very improbable that the process of simply including the transparent parts and excluding the opaque parts in the integrations of § 11 gives an accurate result. The condition of things in actual gratings is much more complicated, and all that can with confidence be asserted is the approximate periodicity in the interval σ. The problem thus presents itself—to determine the course of events on the further side of the plane $z=0$ when the amplitude and phase over that plane are periodic functions of x; and the first step in the solution would naturally be to determine the effect corresponding to the infinitesimal strip $y\,dx$ over which the amplitude and phase are constant. In Fig. 20 QQ' represents the strip in question, of which the effect is to be estimated at P, viz. $(0, 0, z)$;

Fig. 20.

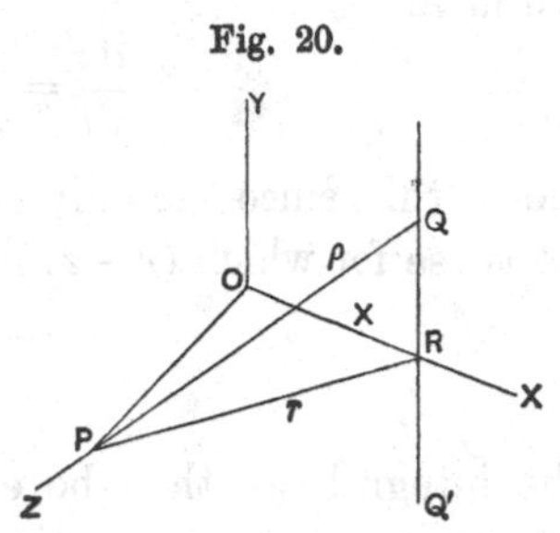

$$QR=y, \quad RP=r, \quad QP=\rho.$$

If we assume the law of secondary wave determined in § 10 so as to suit the resolution of an infinite uniform primary wave, we have, as the effect of QQ',

$$2dx\int_0^\infty \frac{dy}{\lambda\rho}\cos k\left(at-\rho+\tfrac{1}{4}\lambda\right) = -\frac{2dx}{\lambda}\int_r^\infty \frac{d\rho}{\sqrt{(\rho^2-r^2)}}\sin k(at-\rho). \quad \ldots(1)$$

The development of this expression for the operation of a linear source would take us too far*. We must content ourselves with the limiting form assumed when kr is great, as it would almost always be in optics. Under these circumstances the denominator may be simplified by writing

$$\sqrt{(\rho^2-r^2)} = \sqrt{(2r)}\,.\sqrt{(\rho-r)},$$

so that (1) becomes

$$-\frac{2dx}{\lambda\sqrt{(2r)}}\int_0^\infty \frac{d(\rho-r)}{\sqrt{(\rho-r)}}\sin k\{at-r-(\rho-r)\}.$$

Now

$$\int_0^\infty \frac{\sin ku\,du}{\sqrt{u}} = \int_0^\infty \frac{\cos ku\,du}{\sqrt{u}} = \sqrt{\left(\frac{\pi}{2k}\right)} = \tfrac{1}{2}\sqrt{\lambda},$$

and thus we obtain

$$-\frac{dx}{\sqrt{(2\lambda r)}}\{\sin k(at-r)-\cos k(at-r)\} = -\frac{dx}{\sqrt{(\lambda r)}}\sin k\left(at-r-\tfrac{1}{8}\lambda\right), \quad \ldots(2)$$

which gives the effect of a linear source at a great distance. The occurrence of the factor $r^{-\frac{1}{2}}$ is a consequence of the cylindrical expansion of the waves.

* *Theory of Sound*, § 341.

The whole effect is retarded *one-eighth* of a period in comparison with that of the central element, instead of one-quarter of a period as in the case of a uniform wave extending over the whole plane.

The effect of the uniform plane wave can be recovered by integrating (2) with respect to x from $-\infty$ to $+\infty$, on the supposition that kr is great. We have

$$\frac{dx}{\sqrt{r}} = \frac{r\,dr}{\sqrt{r}\,.\,x} = \frac{\sqrt{r}\,.\,d(r-z)}{\sqrt{(r+z)}\,.\,\sqrt{(r-z)}};$$

and in this, since the only elements which contribute sensibly to the integral are those for which $(r-z)$ is small, we may write

$$\frac{\sqrt{r}}{\sqrt{(r+z)}} = \frac{1}{\sqrt{2}}.$$

The integral can then be evaluated by the same formula as before, and we get finally $\cos k(at-z)$, the same as if the primary wave were supposed to advance without resolution. The recomposition of the primary wave by integration with rectangular coordinates is thus verified, but only under the limitation, not really required by the nature of the case, that the point at which the effect is to be estimated is distant by a very great number of wave-lengths from the plane of resolution.

We will now suppose that the amplitude and phase of the primary wave at the plane of resolution $z=0$ are no longer constants, but periodic functions of x. Instead of $\cos kat$ simply, we should have to take in general

$$A\cos(px+f)\cos kat + B\cos(px+g)\sin kat;$$

but it will be sufficient for our purpose to consider the first term only, in which we may further put for simplicity $A=1$, $f=0$. The effect of the linear element at x, 0, upon a point at ξ, z, will be, according to (2),

$$-\frac{dx}{\sqrt{(\lambda r)}}\cos px \sin k(at-r-\tfrac{1}{8}\lambda),$$

where r is the distance, expressed by $r^2=z^2+(x-\xi)^2$.

Thus, if we write $x=\xi+\alpha$, the whole effect is

$$-\int_{-\infty}^{+\infty}\frac{d\alpha}{2\sqrt{(\lambda r)}}\{\sin(kat+p\xi-\tfrac{1}{4}\pi-kr+p\alpha)$$
$$+\sin(kat-p\xi-\tfrac{1}{4}\pi-kr-p\alpha)\}, \quad \ldots(3)$$

where $r^2=z^2+\alpha^2$.

In the two terms of this integral the elements are in general of rapidly fluctuating sign; and the only important part of the range of integration in (for example) the first term is in the neighbourhood of the place where $p\alpha-kr$ is stationary in value, or where

$$p\,d\alpha - k\,dr = 0. \quad \ldots\ldots\ldots\ldots\ldots\ldots(4)$$

In general $\alpha d\alpha - rdr = 0$, so that if the values of α and r corresponding to (4) be called α_0, r_0, we have

$$\frac{\alpha_0}{p} = \frac{r_0}{k} = \frac{z}{\sqrt{(k^2 - p^2)}}. \qquad (5)$$

Now, in the neighbourhood of these values, if $\alpha = \alpha_0 + \alpha_1$,

$$p\alpha - kr = p\alpha_0 - kr_0 + \alpha_1\left(p - k\frac{\alpha_0}{r_0}\right) - \frac{k\alpha_1^2}{2r_0}\left(1 - \frac{p^2}{k^2}\right),$$

in which by (5) the term of the first order vanishes. Using this in (3), we get for the first term

$$-\int_{-\infty}^{+\infty}\frac{d\alpha_1}{2\sqrt{(\lambda r_0)}}\{\sin(kat + p\xi - \tfrac{1}{4}\pi - kr_0 + p\alpha_0)\cos h\alpha_1^2$$
$$- \cos(kat + p\xi - \tfrac{1}{4}\pi - kr_0 + p\alpha_0)\sin h\alpha_1^2\},$$

where for brevity h is written for

$$\frac{k}{2r_0}\left(1 - \frac{p^2}{k^2}\right).$$

The integration is effected by means of the formula

$$\int_{-\infty}^{+\infty}\cos hu^2\,du = \int_{-\infty}^{+\infty}\sin hu^2\,du = \sqrt{\left(\frac{\pi}{2h}\right)};$$

and we find

$$\frac{k}{2\sqrt{(k^2 - p^2)}}\cos(kat + p\xi - kr_0 + p\alpha_0).$$

The other term in (3) gives in like manner

$$\frac{k}{2\sqrt{(k^2 - p^2)}}\cos(kat - p\xi + kr_0 + p\alpha_0);$$

so that the complete value is

$$\frac{k\cos p\xi}{\sqrt{(k^2 - p^2)}}\cos\{kat - \sqrt{(k^2 - p^2)}\,.\,z\}. \qquad (6)$$

When $p = 0$, we fall back on the uniform plane wave travelling with velocity a. In general the velocity is not a, but

$$ka/\sqrt{(k^2 - p^2)}. \qquad (7)$$

The wave represented by (6) is one in which the amplitude at various points of a wave-front is proportional to $\cos p\xi$, or $\cos px$; and, beyond the reversals of phase herein implied, the phase is constant, so that the wave-surfaces are given by z = constant. The wave thus described moves forward at the velocity given by (7), and with type unchanged.

The above investigation may be regarded as applicable to gratings which give spectra of the first order only. Although k vary, there is no separation of colours. Such a separation requires either a limitation in the width of the grating (here supposed to be infinite), or the use of a focusing lens.

It is important to remark that p has been assumed to be less than k, or σ greater than λ; otherwise no part of the range of integration in (3) is exempt from rapid fluctuation of sign, and the result must be considered to be zero. The principle that irregularities in a wave-front of periods less than λ cannot be propagated is of great consequence. Further light will be thrown upon it by a different investigation to be given presently.

The possibility of the wave represented by (6) is perhaps sufficiently established by the preceding method, but the occurrence of the factor $k/\sqrt{(k^2-p^2)}$ shows that the law of the secondary wave (determined originally from a consideration of uniform plane waves) was not rightly assumed.

The correct law applicable in any case may be investigated as follows. Let us assume that the expression for the wave of given periodic time is

$$\psi = e^{ikat}\iint \rho^{-1}\, e^{-ik\rho}\, F(x, y)\, dx\, dy\,; \quad \text{..................(8)}$$

and let us inquire what the value of $F(x, y)$ must be in order that the application of Huygens's principle may give a correct result. From (8)

$$\frac{d\psi}{dz} = e^{ikat}\iint \frac{z}{\rho}\frac{d}{d\rho}\left(\frac{e^{-ik\rho}}{\rho}\right) F(x, y)\, dx\, dy,$$

and
$$\frac{d}{d\rho}\left(\frac{e^{-ik\rho}}{\rho}\right) = -\frac{e^{-ik\rho}(1+k\rho)}{\rho^2}.$$

We propose now to find the limiting value of $d\psi/dz$ when z is very small. The value of the integral will depend upon those elements only for which x and y are very small, so that we replace $F(x, y)$ in the limit by $F(0, 0)$. Also, in the limit,

$$\iint \frac{z}{\rho}\frac{d}{d\rho}\left(\frac{e^{-ik\rho}}{\rho}\right) dx\, dy = \iint \frac{-z}{\rho^3}\, dx\, dy = -2\pi\,;$$

so that

$$\text{Limit}\ \frac{d\psi}{dz} = -2\pi\, e^{ikat}\, F(0, 0).$$

The proper value of $e^{ikat}F(x, y)$ is therefore that of $-d\psi/dz$ at the same point $(x, y, 0)$ divided by 2π, and we have in general

$$\psi = \frac{-1}{2\pi}\iint \left(\frac{d\psi}{dz}\right)\frac{e^{-ik\rho}}{\rho}\, dx\, dy. \quad \text{......................(9)}$$

In the case of the uniform plane wave,

$$\psi = e^{ik(at-z)}, \quad d\psi/dz = -ik\, e^{ik(at-z)}\,;$$

so that

$$\psi = \frac{ie^{ikat}}{\lambda} \iint \frac{e^{-ik\rho}}{\rho} dx\, dy = \iint \frac{e^{ik(at-\rho+\frac{1}{4}\lambda)}}{\lambda\rho} dx\, dy,$$

agreeing with what we have already found for the secondary wave in this case.

But, if $\psi = \cos px \,.\, e^{i[kat - \sqrt{(k^2-p^2)}\,.\,z]}$,

$$\frac{d\psi}{dz}(z=0) = -i\sqrt{(k^2-p^2)} \cos px \; e^{ikat},$$

and

$$\psi = \frac{\sqrt{(k^2-p^2)}}{2\pi} \iint \frac{e^{ik(at-\rho+\frac{1}{4}\lambda)}}{\rho} \cos px \; dx dy.$$

The occurrence of the anomalous factor in (6) is thus explained.

It must be admitted that the present process of investigation is rather artificial; and the cause lies in the attempt to dispense with the differential equation satisfied by ψ, viz.,

$$\frac{d^2\psi}{dx^2} + \frac{d^2\psi}{dy^2} + \frac{d^2\psi}{dz^2} + k^2\psi = 0, \quad \text{.... ....(10)}$$

on which in the case of sound the whole theory is based. It is in fact easy to verify that any value of ψ included under (8), where

$$\rho^2 = (\xi - x)^2 + (\eta - y)^2 + \zeta^2,$$

satisfies the equation

$$\frac{d^2\psi}{d\xi^2} + \frac{d^2\psi}{d\eta^2} + \frac{d^2\psi}{d\zeta^2} + k^2\psi = 0.$$

When there is no question of resolution by Huygens's principle, the distinction between ξ, η and x, y may be dropped.

Starting from the differential equation, we may recover previous results very simply. If ψ be proportional to $\cos px \cos qy$, we have

$$\frac{d^2\psi}{dz^2} + (k^2 - p^2 - q^2)\,\psi = 0. \quad \text{.......................(11)}$$

If $k^2 - p^2 - q^2 = \mu^2$, μ being real, the solution of (11) is

$$\psi = Ae^{i\mu z} + Be^{-i\mu z},$$

where A and B are independent of z. Restoring the factors involving t, x, y, we may write

$$\psi = \cos px \cos qy \,\{Ae^{i(kat+\mu z)} + Be^{i(kat-\mu z)}\}, \quad \text{..............(12)}$$

of which the first term may be dropped when we contemplate waves travelling in the positive direction only. The corresponding realized solution is of the type

$$\psi = \cos px \cos qy \cos \{kat - \sqrt{(k^2 - p^2 - q^2)}\,.\,z\}. \quad \text{...........(13)}$$

When $k^2 > (p^2 + q^2)$, the wave travels without change of type and with velocity

$$V = \frac{ka}{\sqrt{(k^2 - p^2 - q^2)}}. \quad \text{.......................(14)}$$

We have now to consider what occurs when $k^2 < (p^2 + q^2)$. If we write $k^2 - p^2 - q^2 = -\mu^2$, we have in place of (12)

$$\psi = \cos px \cos qy \{Ae^{ikat+\mu z} + Be^{ikat-\mu z}\}; \quad \text{..............(15)}$$

and for the realized solution corresponding to (13)

$$\psi = \cos px \cos qy \, e^{-\mu z} \cos kat. \quad \text{.....................(16)}$$

We conclude that under these circumstances the motion rapidly diminishes as z increases, and that no wave in the usual sense can be propagated at all.

It follows that corrugations of a reflecting surface (no matter how deep) will not disturb the regularity of a perpendicularly reflected wave, provided the wave-length of the corrugation do not exceed that of the vibration. And, whatever the former wave-length may be in relation to the latter, regular reflexion will occur when the incidence is sufficiently oblique.

The first form of solution may be applied to give an explanation of the appearances observed when a plane wave traverses a parallel coarse grating and then impinges upon a screen held at varying distances behind*. As the general expression of the wave periodic with respect to x in distance σ we may take

$$A_0 \cos(kat - kz) + A_1 \cos(px + f_1) \cos(kat - \mu_1 z)$$
$$+ B_1 \cos(px + g_1) \sin(kat - \mu_1 z) + A_2 \cos(2px + f_2) \cos(kat - \mu_2 z) + \ldots,$$

where

$$p = 2\pi/\sigma, \; k = 2\pi/\lambda, \text{ and } \mu_1^2 = k^2 - p^2, \; \mu_2^2 = k^2 - 4p^2, \ldots,$$

the series being continued as long as μ is real. We shall here, however, limit ourselves to the first three terms, and in them suppose A_1 and B_1 to be small relatively to A_0. The intensity may then be represented by

$$A_0^2 + 2A_0A_1 \cos(px + f) \cos(kz - \mu_1 z)$$
$$+ 2A_0B_1 \cos(px + g) \sin(kz - \mu_1 z). \quad \text{...(17)}$$

The stripes thrown upon the screen in various positions are thus periodic functions of z, and the period is

$$z = \frac{2\pi}{k - \sqrt{(k^2 - p^2)}} = \frac{2\sigma^2}{\lambda}, \quad \text{.......................(18)}$$

if λ be supposed small in comparison with σ. It may be noticed that, if the position of the screen be altered by the half of this amount, the effect is equivalent to a shifting parallel to x through the distance $\frac{1}{2}\sigma$. Hence, if the grating consists of alternate transparent and opaque parts of width $\frac{1}{2}\sigma$, the stripes seen upon the screen are *reversed* when the latter is drawn back

* *Phil. Mag.* March 1881, "On Copying Diffraction Gratings and on some Phenomena connected therewith." [Vol. I. p. 504.]

through the distance σ^2/λ. In this case we may suppose B_1 to vanish, and (17) then shows that the field is uniform when the screen occupies positions midway between those which give the most distinct patterns. These results are of interest in connexion with the photographic reproduction of gratings.

§ 16. *Talbot's Bands.*

These very remarkable bands are seen under certain conditions when a tolerably pure spectrum is regarded with the naked eye, or with a telescope, *half the aperture being covered by a thin plate, e.g., of glass or mica.* The view of the matter taken by the discoverer* was that any ray which suffered in traversing the plate a retardation of an odd number of half wave-lengths would be extinguished, and that thus the spectrum would be seen interrupted by a number of dark bars. But this explanation cannot be accepted as it stands, being open to the same objection as Arago's theory of stellar scintillation†. It is as far as possible from being true that a body emitting homogeneous light would disappear on merely covering half the aperture of vision with a half-wave plate. Such a conclusion would be in the face of the principle of energy, which teaches plainly that the retardation in question leaves the aggregate brightness unaltered. The actual formation of the bands comes about in a very curious way, as is shown by a circumstance first observed by Brewster. When the retarding plate is held on the side towards the red of the spectrum, *the bands are not seen.* Even in the contrary case, the thickness of the plate must not exceed a certain limit, however pure the spectrum may be. A satisfactory explanation of these bands was first given by Airy‡, but we shall here follow the investigation of Stokes§, limiting ourselves, however, to the case where the retarded and unretarded beams are contiguous and of equal width. The aperture of the unretarded beam may thus be taken to be limited by $x=-h$, $x=0$, $y=-l$, $y=+l$; and that of the beam retarded by R to be given by $x=0$, $x=h$, $y=-l$, $y=+l$. For the former (1) § 11 gives

$$-\frac{1}{\lambda f}\int_{-h}^{0}\int_{-l}^{+l}\sin k\left\{at-f+\frac{x\xi+y\eta}{f}\right\}dx\,dy$$

$$=-\frac{2lh}{\lambda f}\cdot\frac{f}{k\eta l}\sin\frac{k\eta l}{f}\cdot\frac{2f}{k\xi h}\sin\frac{k\xi h}{2f}\cdot\sin k\left\{at-f-\frac{\xi h}{2f}\right\},\ \ldots\ldots\ldots(1)$$

on integration and reduction.

* *Phil. Mag.* x. p. 364; 1837.

† On account of inequalities in the atmosphere giving a variable refraction, the light from a star would be irregularly distributed over a screen. The experiment is easily made on a laboratory scale, with a small source of light, the rays from which, in their course towards a rather distant screen, are disturbed by the neighbourhood of a heated body. At a moment when the eye, or object-glass of a telescope, occupies a dark position, the star vanishes. A fraction of a second later the aperture occupies a bright place, and the star reappears. According to this view the chromatic effects depend entirely upon atmospheric dispersion.

‡ *Phil. Trans.* 1840, p. 225; 1841, p. 1. § *Ibid.* 1848, p. 227.

For the retarded stream the only difference is that we must subtract R from at, and that the limits of x are 0 and $+h$. We thus get for the disturbance at ξ, η due to this stream

$$-\frac{2lh}{\lambda f}\cdot\frac{f}{k\eta l}\sin\frac{k\eta l}{f}\cdot\frac{2f}{k\xi h}\sin\frac{k\xi h}{2f}\,.\,\sin k\left\{at-f-R+\frac{\xi h}{2f}\right\}.\qquad(2)$$

If we put for shortness τ for the quantity under the last circular function in (1), the expressions (1), (2) may be put under the forms $u\sin\tau$, $v\sin(\tau-\alpha)$ respectively; and, if I be the intensity, I will be measured by the sum of the squares of the coefficients of $\sin\tau$ and $\cos\tau$ in the expression

$$u\sin\tau+v\sin(\tau-\alpha),$$

so that

$$I=u^2+v^2+2uv\cos\alpha,$$

which becomes on putting for u, v, and α their values, and putting

$$\left\{\frac{f}{k\eta l}\sin\frac{k\eta l}{f}\right\}^2=Q,\qquad(3)$$

$$I=Q\,.\,\frac{4l^2}{\pi^2\xi^2}\sin^2\frac{\pi\xi h}{\lambda f}\left\{2+2\cos\left(\frac{2\pi R}{\lambda}-\frac{2\pi\xi h}{\lambda f}\right)\right\}.\qquad(4)$$

If the subject of examination be a luminous line parallel to η, we shall obtain what we require by integrating (4) with respect to η from $-\infty$ to $+\infty$. The constant multiplier is of no especial interest, so that we may take as applicable to the image of a line

$$I=\frac{2}{\xi^2}\sin^2\frac{\pi\xi h}{\lambda f}\left\{1+\cos\left(\frac{2\pi R}{\lambda}-\frac{2\pi\xi h}{\lambda f}\right)\right\}.\qquad(5)$$

If $R=\frac{1}{2}\lambda$, I vanishes at $\xi=0$; but the whole illumination, represented by $\int_{-\infty}^{+\infty}I\,d\xi$, is independent of the value of R. If $R=0$, $I=\frac{1}{\xi^2}\sin^2\frac{2\pi\xi h}{\lambda f}$, in agreement with § 11, where a has the meaning here attached to $2h$.

The expression (5) gives the illumination at ξ due to that part of the complete image whose geometrical focus is at $\xi=0$, the retardation for this component being R. Since we have now to integrate for the whole illumination at a particular point O due to all the components which have their foci in its neighbourhood, we may conveniently regard O as origin. ξ is then the coordinate relatively to O of any focal point O' for which the retardation is R; and the required result is obtained by simply integrating (5) with respect to ξ from $-\infty$ to $+\infty$. To each value of ξ corresponds a different value of λ, and (in consequence of the dispersing power of the plate) of R. The variation of λ may, however, be neglected in the integration, except in $2\pi R/\lambda$, where a small variation of λ entails a comparatively large alteration of phase. If we write

$$\rho=2\pi R/\lambda,\qquad(6)$$

we must regard ρ as a function of ξ, and we may take with sufficient approximation under any ordinary circumstances

$$\rho = \rho' + \varpi\xi, \qquad \text{.................................(7)}$$

where ρ' denotes the value of ρ at O, and ϖ is a constant which is positive when the retarding plate is held at the side on which the blue of the spectrum *is seen.* The possibility of dark bands depends upon ϖ being positive. Only in this case can

$$\cos\{\rho' + (\varpi - 2\pi h/\lambda f)\,\xi\}$$

retain the constant value -1 throughout the integration, and then only when

$$\varpi = 2\pi h/\lambda f, \qquad \text{.................................(8)}$$

and

$$\cos\rho' = -1. \qquad \text{.................................(9)}$$

The first of these equations is the condition for the formation of dark bands, and the second marks their situation, which is the same as that determined by the imperfect theory.

The integration can be effected without much difficulty. For the first term in (5) the evaluation is effected at once by a known formula. In the second term if we observe that

$$\cos\{\rho' + (\varpi - 2\pi h/\lambda f)\,\xi\} = \cos\{\rho' - g_1\xi\} = \cos\rho' \cos g_1\xi + \sin\rho' \sin g_1\xi,$$

we see that the second part vanishes when integrated, and that the remaining integral is of the form

$$w = \int_{-\infty}^{+\infty} \sin^2 h_1\xi \cos g_1\xi \,\frac{d\xi}{\xi^2},$$

where

$$h_1 = \pi h/\lambda f, \qquad g_1 = \varpi - 2\pi h/\lambda f. \qquad \text{....................(10)}$$

By differentiation with respect to g_1 it may be proved that

$$\begin{aligned}
&w = 0 && \text{from } g_1 = -\infty && \text{to } g_1 = -2h_1,\\
&w = \tfrac{1}{2}\pi\,(2h_1 + g_1) && \text{from } g_1 = -2h_1 && \text{to } g_1 = 0,\\
&w = \tfrac{1}{2}\pi\,(2h_1 - g_1) && \text{from } g_1 = 0 && \text{to } g_1 = 2h_1,\\
&w = 0 && \text{from } g_1 = 2h_1 && \text{to } g_1 = \infty.
\end{aligned}$$

The integrated intensity, I', or

$$2\pi h_1 + 2\cos\rho\, w,$$

is thus

$$I' = 2\pi h_1, \qquad \text{.................................(11)}$$

when g_1 numerically exceeds $2h_1$; and, when g_1 lies between $\pm 2h_1$,

$$I = \pi\,\{2h_1 + (2h_1 - \sqrt{g_1^2})\cos\rho'\}. \qquad \text{....................(12)}$$

It appears therefore that there are no bands at all unless ϖ lies between 0 and $+4h_1$, and that within these limits the best bands are formed at the middle of the range when $\varpi = 2h_1$. The formation of bands thus requires that the retarding plate be held upon the side already specified, so that ϖ be positive; and that the thickness of the plate (to which ϖ is proportional) do not exceed a certain limit, which we may call $2T_0$. At the best thickness T_0 the bands are black, and not otherwise.

The linear width of the band (e) is the increment of ξ which alters ρ by 2π, so that

$$e = 2\pi/\varpi. \qquad (13)$$

With the best thickness

$$\varpi = 2\pi h/\lambda f, \qquad (14)$$

so that in this case

$$e = \lambda f/h. \qquad (15)$$

The bands are thus of the same width as those due to two infinitely narrow apertures coincident with the central lines of the retarded and unretarded streams, the subject of examination being itself a fine luminous line.

If it be desired to see a given number of bands in the whole or in any part of the spectrum, the thickness of the retarding plate is thereby determined, independently of all other considerations. But in order that the bands may be really visible, and still more in order that they may be black, another condition must be satisfied. It is necessary that the aperture of the pupil be accommodated to the angular extent of the spectrum, or reciprocally. Black bands will be too fine to be well seen unless the aperture ($2h$) of the pupil be somewhat contracted. One-twentieth to one-fiftieth of an inch is suitable. The aperture and the number of bands being both fixed, the condition of blackness determines the angular magnitude of a band and of the spectrum. The use of a grating is very convenient, for not only are there several spectra in view at the same time, but the dispersion can be varied continuously by sloping the grating. The slits may be cut out of tin-plate, and half covered by mica or "microscopic glass," held in position by a little cement.

If a telescope be employed there is a distinction to be observed, according as the half-covered aperture is between the eye and the ocular, or in front of the object-glass. In the former case the function of the telescope is simply to increase the dispersion, and the formation of the bands is of course independent of the particular manner in which the dispersion arises. If, however, the half-covered aperture be in front of the object-glass, the phenomenon is magnified as a whole, and the desirable relation between the (unmagnified) dispersion and the aperture is the same as without the telescope. There appears to be no further advantage in the use of a telescope than the increased facility of accommodation, and for this of course a very low power suffices.

The original investigation of Stokes, here briefly sketched, extends also to the case where the streams are of unequal widths h, k, and are separated by an interval $2g$. In the case of unequal widths the bands cannot be black; but if $h=k$, the finiteness of $2g$ does not preclude the formation of black bands.

The theory of Talbot's bands with a half-covered *circular* aperture has been treated by H. Struve*.

§ 17. *Diffraction when the Source of Light is not Seen in Focus.*

The phenomena to be considered under this head are of less importance than those investigated by Fraunhofer, and will be treated in less detail; but, in view of their historical interest and of the ease with which many of the experiments may be tried, some account of their theory could not be excluded from such a work as the present. One or two examples have already attracted our attention when considering Huygens's zones, viz., the shadow of a circular disk, and of a screen circularly perforated; but the most famous problem of this class—first solved by Fresnel—relates to the shadow of a screen bounded by a straight edge.

In theoretical investigations these problems are usually treated as of two dimensions only, everything being referred to the plane passing through the luminous point and perpendicular to the diffracting edges, supposed to be straight and parallel. In strictness this idea is appropriate only when the source is a luminous line, emitting cylindrical waves, such as might be obtained from a luminous point with the aid of a cylindrical lens. When, in order to apply Huygens's principle, the wave is supposed to be broken up, the phase is the same at every element of the surface of resolution which lies upon a line perpendicular to the plane of reference, and thus the effect of the whole line, or rather infinitesimal strip, is related in a constant manner (§ 15) to that of the element which lies in the plane of reference, and may be considered to be represented thereby. The same method of representation is applicable to spherical waves, issuing from a *point*, if the radius of curvature be large; for, although there is variation of phase along the length of the infinitesimal strip, the whole effect depends practically upon that of the central parts where the phase is sensibly constant†.

Fig. 21.

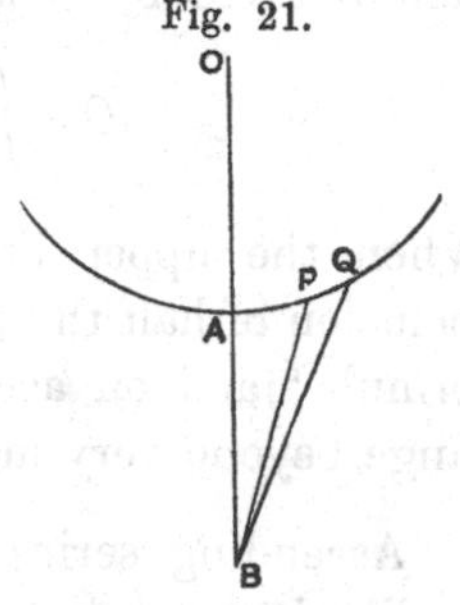

In Fig. 21 APQ is the arc of the circle representative of the wave-front of resolution, the centre being at O, and the radius OA

* *St Petersburg Trans.* XXXI. No. 1, 1883.

† In experiment a line of light is sometimes substituted for a point in order to increase the illumination. The various parts of the line are here *independent* sources, and should be treated accordingly. To assume a cylindrical form of primary wave would be justifiable only when there is synchronism among the secondary waves issuing from the various centres.

being equal to a. B is the point at which the effect is required, distant $a+b$ from O, so that $AB=b$, $AP=s$, $PQ=ds$.

Taking as the standard phase that of the secondary wave from A, we may represent the effect of PQ by

$$\cos 2\pi\left(\frac{t}{\tau}-\frac{\delta}{\lambda}\right).ds,$$

where $\delta=BP-AP$ is the retardation at B of the wave from P relatively to that from A.

Now

$$\delta=(a+b)\,s^2/2ab, \quad \text{..............................(1)}$$

so that, if we write

$$\frac{2\pi\delta}{\lambda}=\frac{\pi(a+b)\,s^2}{ab\lambda}=\frac{\pi}{2}v^2, \quad \text{..............................(2)}$$

the effect at B is

$$\left\{\frac{ab\lambda}{2(a+b)}\right\}^{\frac{1}{2}}\left\{\cos\frac{2\pi t}{\tau}\int\cos\tfrac{1}{2}\pi v^2.\,dv+\sin\frac{2\pi t}{\tau}\int\sin\tfrac{1}{2}\pi v^2.\,dv\right\}, \quad \text{...(3)}$$

the limits of integration depending upon the disposition of the diffracting edges. When a, b, λ are regarded as constant, the first factor may be omitted,—as indeed should be done for consistency's sake, inasmuch as other factors of the same nature have been omitted already.

The intensity I^2, the quantity with which we are principally concerned, may thus be expressed

$$I^2=\{\int\cos\tfrac{1}{2}\pi v^2.\,dv\}^2+\{\int\sin\tfrac{1}{2}\pi v^2.\,dv\}^2. \quad \text{..................(4)}$$

These integrals, taken from $v=0$, are known as Fresnel's integrals; we will denote them by C and S, so that

$$C=\int_0^v\cos\tfrac{1}{2}\pi v^2.\,dv, \qquad S=\int_0^v\sin\tfrac{1}{2}\pi v^2.\,dv. \quad \text{...............(5)}$$

When the upper limit is infinity, so that the limits correspond to the inclusion of half the primary wave, C and S are both equal to $\frac{1}{2}$, by a known formula; and on account of the rapid fluctuation of sign the parts of the range beyond very moderate values of v contribute but little to the result.

Ascending series for C and S were given by Knockenhauer, and are readily investigated. Integrating by parts, we find

$$C+iS=\int_0^v e^{i.\frac{1}{2}\pi v^2}dv=e^{i.\frac{1}{2}\pi v^2}.\,v-\tfrac{1}{3}i\pi\int_0^v e^{i.\frac{1}{2}\pi v^2}dv^3;$$

and, by continuing this process,

$$C+iS=e^{i.\frac{1}{2}\pi v^2}\left\{v-\frac{i\pi}{3}v^3+\frac{i\pi}{3}\frac{i\pi}{5}v^5-\frac{i\pi}{3}\frac{i\pi}{5}\frac{i\pi}{7}v^7+\ldots\right\}.$$

By separation of real and imaginary parts,

$$C = M\cos\tfrac{1}{2}\pi v^2 + N\sin\tfrac{1}{2}\pi v^2, \qquad S = M\sin\tfrac{1}{2}\pi v^2 - N\cos\tfrac{1}{2}\pi v^2, \quad \ldots(6)$$

where

$$M = \frac{v}{1} - \frac{\pi^2 v^5}{3.5} + \frac{\pi^4 v^9}{3.5.7.9} - \ldots, \quad \ldots\ldots(7)$$

$$N = \frac{\pi v^3}{1.3} - \frac{\pi^3 v^7}{1.3.5.7} + \frac{\pi^5 v^{11}}{1.3.5.7.9.11} - \ldots. \quad \ldots\ldots(8)$$

These series are convergent for all values of v, but are practically useful only when v is small.

Expressions suitable for discussion when v is large were obtained by Gilbert*. Taking

$$\tfrac{1}{2}\pi v^2 = u, \quad \ldots\ldots(9)$$

we may write

$$C + iS = \frac{1}{\sqrt{(2\pi)}}\int_0^u \frac{e^{iu}du}{\sqrt{u}}. \quad \ldots\ldots(10)$$

Again, by a known formula,

$$\frac{1}{\sqrt{u}} = \frac{1}{\sqrt{\pi}}\int_0^\infty \frac{e^{-ux}dx}{\sqrt{x}}. \quad \ldots\ldots(11)$$

Substituting this in (10), and inverting the order of integration, we get

$$C + iS = \frac{1}{\pi\sqrt{2}}\int_0^\infty \frac{dx}{\sqrt{x}}\int_0^u e^{u(i-x)}\,dx = \frac{1}{\pi\sqrt{2}}\int_0^\infty \frac{dx}{\sqrt{x}}\,\frac{e^{u(i-x)}-1}{i-x}. \quad \ldots\ldots(12)$$

Thus, if we take

$$G = \frac{1}{\pi\sqrt{2}}\int_0^\infty \frac{e^{-ux}\sqrt{x}\,.\,dx}{1+x^2}, \qquad H = \frac{1}{\pi\sqrt{2}}\int_0^\infty \frac{e^{-ux}dx}{\sqrt{x}\,.\,(1+x^2)}, \quad \ldots(13)$$

$$C = \tfrac{1}{2} - G\cos u + H\sin u, \qquad S = \tfrac{1}{2} - G\sin u - H\cos u. \quad \ldots(14)$$

The constant parts in (14), viz. $\tfrac{1}{2}$, may be determined by direct integration of (12), or from the observation that by their constitution G and H vanish when $u = \infty$, coupled with the fact that C and S then assume the value $\tfrac{1}{2}$.

Comparing the expressions for C, S in terms of M, N, and in terms of G, H, we find that

$$G = \tfrac{1}{2}(\cos u + \sin u) - M, \qquad H = \tfrac{1}{2}(\cos u - \sin u) + N, \ldots\ldots(15)$$

formulæ which may be utilized for the calculation of G, H when u (or v) is small. For example, when $u = 0$, $M = 0$, $N = 0$, and consequently $G = H = \tfrac{1}{2}$.

Descending series of the semi-convergent class, available for numerical calculation when u is moderately large, can be obtained from (12) by writing $x = uy$, and expanding the denominator in powers of y. The integration of the several terms may then be effected by the formula

$$\int_0^\infty e^{-y}y^{q-\frac{1}{2}}\,dy = \Gamma(q+\tfrac{1}{2}) = (q-\tfrac{1}{2})(q-\tfrac{3}{2})\ldots\tfrac{1}{2}\sqrt{\pi};$$

* *Mém. couronnés de l'Acad. de Bruxelles*, XXXI. 1. See also Verdet, *Leçons*, § 86.

and we get in terms of v

$$G = \frac{1}{\pi^2 v^3} - \frac{1.3.5}{\pi^4 v^7} + \frac{1.3.5.7.9}{\pi^6 v^{11}} - \ldots, \quad \text{..............(16)}$$

$$H = \frac{1}{\pi v} - \frac{1.3}{\pi^3 v^5} + \frac{1.3.5.7}{\pi^5 v^9} - \ldots. \quad \text{.......................(17)}$$

The corresponding values of C and S were originally derived by Cauchy, without the use of Gilbert's integrals, by direct integration by parts.

From the series for G and H just obtained it is easy to verify that

$$\frac{dH}{dv} = -\pi v G, \qquad \frac{dG}{dv} = \pi v H - 1. \quad \text{..................(18)}$$

We now proceed to consider more particularly the distribution of light upon a screen PBQ near the shadow of a straight edge A. At a point P within the geometrical shadow of the obstacle, the half of the wave to the right of C (Fig. 22), the nearest point on the wave-front, is wholly intercepted, and on the left the integration is to be taken from $s = CA$ to $s = \infty$. If V be the value of v corresponding to CA, viz.,

$$V = \sqrt{\left\{\frac{2(a+b)}{ab\lambda}\right\}} \, . \, CA, \quad \text{.......................(19)}$$

we may write

$$I^2 = \left(\int_V^\infty \cos \tfrac{1}{2}\pi v^2 \, . \, dv\right)^2 + \left(\int_V^\infty \sin \tfrac{1}{2}\pi v^2 \, . \, dv\right)^2, \quad \text{...........(20)}$$

or, according to our previous notation,

$$I^2 = (\tfrac{1}{2} - C_V)^2 + (\tfrac{1}{2} - S_V)^2 = G^2 + H^2. \quad \text{..............(21)}$$

Now in the integrals represented by G and H every element diminishes as V increases from zero. Hence, as CA increases, viz., as the point P is more and more deeply immersed in the shadow, the illumination *continuously* decreases, and that without limit. It has long been known from observation that there are no bands on the interior side of the shadow of the edge.

Fig. 22.

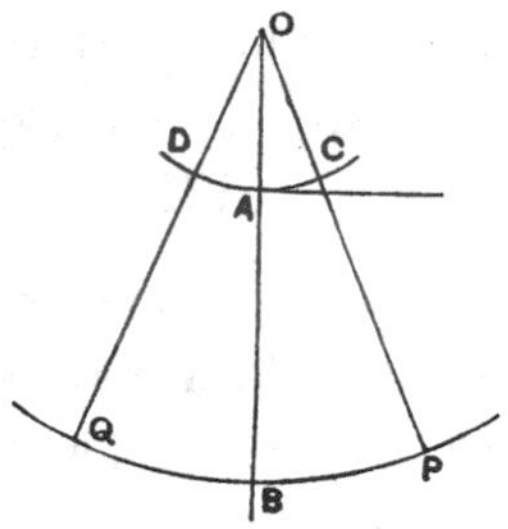

The law of diminution when V is moderately large is easily expressed with the aid of the series (16), (17) for G, H. We have ultimately $G = 0$, $H = (\pi V)^{-1}$, so that

$$I^2 = 1/\pi^2 V^2,$$

or the illumination is inversely as the square of the distance from the shadow of the edge.

For a point Q outside the shadow the integration extends over *more* than half the primary wave. The intensity may be expressed by

$$I^2 = (\tfrac{1}{2} + C_V)^2 + (\tfrac{1}{2} + S_V)^2; \quad \text{....................(22)}$$

and the maxima and minima occur when

$$(\tfrac{1}{2}+C_V)\frac{dC}{dV}+(\tfrac{1}{2}+S_V)\frac{dS}{dV}=0,$$

whence

$$\sin\tfrac{1}{2}\pi V^2+\cos\tfrac{1}{2}\pi V^2=G. \qquad\qquad (23)$$

When $V=0$, viz., at the edge of the shadow, $I^2=\frac{1}{2}$; when $V=\infty$, $I^2=2$, on the scale adopted. The latter is the intensity due to the uninterrupted wave. The quadrupling of the intensity in passing outwards from the edge of the shadow is, however, accompanied by fluctuations giving rise to bright and dark bands. The position of these bands determined by (23) may be very simply expressed when V is large, for then sensibly $G=0$, and

$$\tfrac{1}{2}\pi V^2=\tfrac{3}{4}\pi+n\pi, \qquad\qquad (24)$$

n being an integer. In terms of δ, we have from (2)

$$\delta=(\tfrac{3}{8}+\tfrac{1}{2}n)\lambda. \qquad\qquad (25)$$

The first maximum in fact occurs when $\delta=\frac{3}{8}\lambda-{\cdot}0046\lambda$, and the first minimum when $\delta=\frac{7}{8}\lambda-{\cdot}0016\lambda$*, the corrections being readily obtainable from a table of G by substitution of the approximate value of V.

The position of Q corresponding to a given value of V, that is, to a band of given order, is by (19)

$$BQ=\frac{a+b}{a}AD=V\sqrt{\left\{\frac{b\lambda\,(a+b)}{2a}\right\}}. \qquad\qquad (26)$$

By means of this expression we may trace the locus of a band of given order as b varies. With sufficient approximation we may regard BQ and b as rectangular coordinates of Q. Denoting them by x, y, so that AB is axis of y and a perpendicular through A the axis of x, and rationalizing (26), we have

$$2ax^2-V^2\lambda y^2-V^2a\lambda y=0,$$

which represents a hyperbola with vertices at O and A.

From (24), (26) we see that the width of the bands is of the order $\sqrt{\{b\lambda\,(a+b)/a\}}$. From this we may infer the limitation upon the width of the source of light, in order that the bands may be properly formed. If ω be the apparent magnitude of the source seen from A, ωb should be much smaller than the above quantity, or

$$\omega<\sqrt{\{\lambda\,(a+b)/ab\}}. \qquad\qquad (27)$$

If a be very great in relation to b, the condition becomes

$$\omega<\sqrt{(\lambda/b)}, \qquad\qquad (28)$$

so that if b is to be moderately great (1 metre), the apparent magnitude of the sun must be greatly reduced before it can be used as a source.

* Verdet, *Leçons*, § 90.

The values of V for the maxima and minima of intensity, and the magnitudes of the latter, were calculated by Fresnel. An extract from his results is given in the accompanying table.

	V	I^2
First maximum	1·2172	2·7413
First minimum	1·8726	1·5570
Second maximum ...	2·3449	2·3990
Second minimum ...	2·7392	1·6867
Third maximum	3·0820	2·3022
Third minimum	3·3913	1·7440

A very thorough investigation of this and other related questions, accompanied by fully worked-out tables of the functions concerned, will be found in a recent paper by Lommel*.

When the functions C and S have once been calculated, the discussion of various diffraction problems is much facilitated by the idea, due to Cornu†, of exhibiting as a curve the relationship between C and S, considered as the rectangular coordinates (x, y) of a point. Such a curve is shown in Fig. 23, where, according to the definition (5) of C, S,

$$x = \int_0^v \cos \tfrac{1}{2}\pi v^2 . dv, \qquad y = \int_0^v \sin \tfrac{1}{2}\pi v^2 . dv. \qquad \text{.............. (29)}$$

The origin of coordinates O corresponds to $v = 0$; and the asymptotic points J, J', round which the curve revolves in an ever-closing spiral, correspond to $v = \pm\infty$.

The intrinsic equation, expressing the relation between the arc σ (measured from O) and the inclination ϕ of the tangent at any point to the axis of x, assumes a very simple form. For

$$dx = \cos \tfrac{1}{2}\pi v^2 . dv, \qquad dy = \sin \tfrac{1}{2}\pi v^2 . dv;$$

so that

$$\sigma = \int \sqrt{(dx^2 + dy^2)} = v, \qquad \text{...................... (30)}$$

$$\phi = \tan^{-1}\frac{dy}{dx} = \tfrac{1}{2}\pi v^2. \qquad \text{......................... (31)}$$

Accordingly,

$$\phi = \tfrac{1}{2}\pi\sigma^2; \qquad \text{............................. (32)}$$

* "Die Beugungserscheinungen geradlinig begrenzter Schirme," *Abh. bayer. Akad. der Wiss.* II. Cl. XV. Bd. III. Abth., 1886.

† *Journal de Physique*, III. p. 1, 1874. A similar suggestion has recently been made independently by Fitzgerald.

and for the curvature,

$$\frac{d\phi}{d\sigma} = \pi\sigma. \quad \text{..........................(33)}$$

Cornu remarks that this equation suffices to determine the general character of the curve. For the osculating circle at any point includes the whole of the curve which lies beyond; and the successive convolutions envelop one another without intersection.

The utility of the curve depends upon the fact that the elements of arc represent, in amplitude and phase, the component vibrations due to the corresponding portions of the primary wave-front. For by (30) $d\sigma = dv$, and by (2) dv is proportional to ds. Moreover by (2) and (31) the retardation of phase of the elementary vibration from PQ (Fig. 21) is $2\pi\delta/\lambda$, or ϕ. Hence,

Fig. 23.

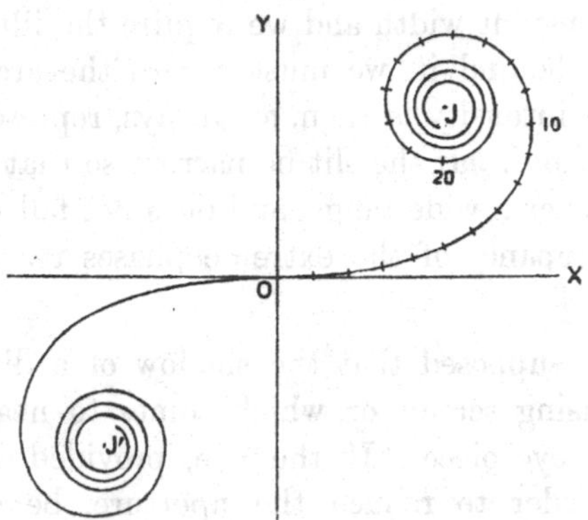

in accordance with the rule for compounding vector quantities, the resultant vibration at B, due to any finite part of the primary wave, is represented in amplitude and phase by the chord joining the extremities of the corresponding arc $(\sigma_2 - \sigma_1)$.

In applying the curve in special cases of diffraction to exhibit the effect at any point P (Fig. 22), the centre of the curve O is to be considered to correspond to that point C of the primary wave-front which lies nearest to P. The operative part, or parts, of the curve are of course those which represent the unobstructed portions of the primary wave.

Let us reconsider, following Cornu, the diffraction of a screen unlimited on one side, and on the other terminated by a straight edge. On the illuminated side, at a distance from the shadow, the vibration is represented by JJ'. The coordinates of J, J' being $(\frac{1}{2}, \frac{1}{2})$, $(-\frac{1}{2}, -\frac{1}{2})$, I^2 is 2; and the phase is $\frac{1}{8}$ period in arrear of that of the element at O. As the point under contemplation is supposed to approach the shadow, the vibration is represented by the chord drawn from J to a point on the other half of the curve, which travels inwards from J' towards O. The amplitude is thus subject to fluctuations, which increase as the shadow is approached. At the point O the intensity is one-quarter of that of the entire wave, and after this point is passed, that is, when we have entered the geometrical shadow, the intensity falls off gradually to zero, *without fluctuations*. The whole progress of the phenomenon is thus exhibited to the eye in a very instructive manner.

We will next suppose that the light is transmitted by a slit, and inquire what is the effect of varying the width of the slit upon the illumination at the projection of its centre. Under these circumstances the arc to be considered is bisected at O, and its length is proportional to the width of the slit. It is easy to see that the length of the chord (which passes in all cases through O) increases to a maximum near the place where the phase-retardation is $\frac{3}{8}$ of a period, then diminishes to a minimum when the retardation is about $\frac{7}{8}$ of a period, and so on.

If the slit is of constant width and we require the illumination at various points on the screen behind it, we must regard the arc of the curve as of *constant length*. The intensity is then, as always, represented by the square of the length of the chord. If the slit be narrow, so that the arc is short, the intensity is constant over a wide range, and does not fall off to an important extent until the discrepancy of the extreme phases reaches about a quarter of a period.

We have hitherto supposed that the shadow of a diffracting obstacle is received upon a diffusing screen, or, which comes to nearly the same thing, is observed with an eye-piece. If the eye, provided if necessary with a perforated plate in order to reduce the aperture, be situated inside the shadow at a place where the illumination is still sensible, and be focused upon the diffracting edge, the light which it receives will appear to come from the neighbourhood of the edge, and will present the effect of a silver lining. This is doubtless the explanation of a "pretty optical phenomenon, seen in Switzerland, when the sun rises from behind distant trees standing on the summit of a mountain*."

§ 18. *Diffraction Symmetrical about an Axis.*

The general problem of the diffraction pattern due to a source of light concentrated in a point, when the system is symmetrical about an axis, has been ably investigated by Lommel†. We must content ourselves here with a very slight sketch of some of his results.

* Necker, *Phil. Mag.* Nov. 1832; Fox Talbot, *Phil. Mag.* June 1833. "When the sun is about to emerge......every branch and leaf is lighted up with a silvery lustre of indescribable beauty......The birds, as Mr Necker very truly describes, appear like flying brilliant sparks." Talbot ascribes the appearance to diffraction; and he recommends the use of a telescope.

† *Abh. der bayer. Akad. der Wiss.* II. Cl. XV. Bd. II. Abth.

Spherical waves, centred upon the axis, of radius a fall upon the diffracting screen; and the illumination is required on a second screen, like the first perpendicular to the axis, at a distance $(a+b)$ from the source. We have first to express the distance (d) between an element dS of the wave-front and a point M in the plane of the second screen. Let ζ denote the distance of M from the axis of symmetry; then, if we take an axis of x to pass through M, the coordinates of M are $(\zeta, 0, 0)$. On the same system the coordinates of dS are

$$a \sin\theta \cos\phi, \qquad a \sin\theta \sin\phi, \qquad a(1-\cos\theta)+b;$$

and the distance is given by

$$d^2 = b^2 + \zeta^2 - 2a\zeta \sin\theta \cos\phi + 4a(a+b)\sin^2 \tfrac{1}{2}\theta.$$

In this expression ζ and θ are to be treated as small quantities. Writing ρ for $a \sin\theta$, we get approximately

$$d = b + \frac{\zeta^2}{2b} - \frac{\zeta \cos\phi}{b}\rho + \frac{a+b}{2ab}\rho^2. \quad\text{.....................(1)}$$

The vibration at the wave-front of resolution being denoted by $a^{-1}\cos 2\pi t/\tau$, the integral expressive of the resultant of the secondary waves is (§ 17)

$$-\frac{1}{ab\lambda}\iint \sin 2\pi\left(\frac{t}{\tau} - \frac{d}{\lambda}\right) dS. \quad\text{.......................(2)}$$

Substituting $\rho\, d\rho\, d\phi$ for dS, and for d its value from (1), we obtain as the expression for the intensity at the point ζ,

$$I^2 = \frac{1}{a^2b^2\lambda^2}(C^2 + S^2), \quad\text{.............................(3)}$$

where

$$C = \iint \cos(\tfrac{1}{2}k\rho^2 - l\rho\cos\phi) \, . \, \rho\, d\rho\, d\phi, \quad\text{.................(4)}$$

$$^*S = \iint \sin(\tfrac{1}{2}k\rho^2 - l\rho\cos\phi) \, . \, \rho\, d\rho\, d\phi, \quad\text{.................(5)}$$

and the following abbreviations have been introduced

$$\frac{2\pi}{\lambda}\frac{a+b}{2ab} = \tfrac{1}{2}k, \qquad \frac{2\pi\zeta}{\lambda b} = l. \quad\text{.......................(6)}$$

The range of integration is for ϕ from 0 to 2π. The limits for ρ depend upon the particular problem in hand; but for the sake of definiteness we will suppose that in the analytical definitions of C and S the limits are 0 and r, so as to apply immediately to the problem of a circular aperture of radius r. If we introduce the notation of Bessel's functions, we have

$$C = 2\pi\int_0^r J_0(l\rho)\cos(\tfrac{1}{2}k\rho^2) \, . \, \rho\, d\rho, \quad S = 2\pi\int_0^r J_0(l\rho)\sin(\tfrac{1}{2}k\rho^2) \, . \, \rho\, d\rho. \quad\text{...(7, 8)}$$

* Used now in an altered sense.

By integration by parts of these expressions Lommel develops series suitable for calculation. Setting

$$kr^2 = y, \qquad lr = z, \quad \text{..............................(9)}$$

he finds in the first place

$$C = \pi r^2 \left\{ \frac{\cos \frac{1}{2}y}{\frac{1}{2}y} U_1 + \frac{\sin \frac{1}{2}y}{\frac{1}{2}y} U_2 \right\}, \quad S = \pi r^2 \left\{ \frac{\sin \frac{1}{2}y}{\frac{1}{2}y} U_1 - \frac{\cos \frac{1}{2}y}{\frac{1}{2}y} U_2 \right\}, \quad (10, 11)$$

where

$$U_1 = \frac{y}{z} J_1(z) - \frac{y^3}{z^3} J_3(z) + \frac{y^5}{z^5} J_5(z) - \ldots, \quad \text{...........(12)}$$

$$U_2 = \frac{y^2}{z^2} J_2(z) - \frac{y^4}{z^4} J_4(z) + \frac{y^6}{z^6} J_6(z) - \ldots. \quad \text{...........(13)}$$

The series are convenient when y is less than z.

The second set of expressions are

$$C = \pi r^2 \left\{ \frac{2}{y} \sin \frac{z^2}{2y} + \frac{\sin \frac{1}{2}y}{\frac{1}{2}y} V_0 - \frac{\cos \frac{1}{2}y}{\frac{1}{2}y} V_1 \right\}, \quad \text{...........(14)}$$

$$S = \pi r^2 \left\{ \frac{2}{y} \cos \frac{z^2}{2y} - \frac{\cos \frac{1}{2}y}{\frac{1}{2}y} V_0 - \frac{\sin \frac{1}{2}y}{\frac{1}{2}y} V_1 \right\}, \quad \text{...........(15)}$$

where

$$V_0 = J_0(z) - \frac{z^2}{y^2} J_2(z) + \frac{z^4}{y^4} J_4(z) - \ldots, \quad \text{..............(16)}$$

$$V_1 = \frac{z}{y} J_1(z) - \frac{z^3}{y^3} J_3(z) + \frac{z^5}{y^5} J_5(z) - \ldots. \quad \text{...........(17)}$$

These series are suitable when z/y is small.

When the primary wave is complete, $r = \infty$, and we have at once from the second set of expressions

$$C_\infty = \frac{2\pi}{k} \sin \frac{l^2}{2k}, \qquad S_\infty = \frac{2\pi}{k} \cos \frac{l^2}{2k}, \quad \text{..................(18)}$$

so that

$$I^2 = \frac{C_\infty{}^2 + S_\infty{}^2}{a^2 b^2 \lambda^2} = \frac{1}{(a+b)^2}, \quad \text{........................(19)}$$

as we know it should be.

In the application to the problem of the shadow of a circular disk the limits of integration are from r to ∞. If these integrals be denoted by C', S', we have

$$C' = C_\infty - C = \pi r^2 \left\{ -\frac{\sin \frac{1}{2}y}{\frac{1}{2}y} V_0 + \frac{\cos \frac{1}{2}y}{\frac{1}{2}y} V_1 \right\}, \quad \text{...........(20)}$$

$$S' = S_\infty - S = \pi r^2 \left\{ \frac{\cos \frac{1}{2}y}{\frac{1}{2}y} V_0 + \frac{\sin \frac{1}{2}y}{\frac{1}{2}y} V_1 \right\}; \quad \text{..............(21)}$$

and

$$C'^2+S'^2=\frac{4\pi^2}{k^2}(V_0^2+V_1^2),\qquad\qquad(22)$$

$$I^2=\frac{V_0^2+V_1^2}{(a+b)^2}.\qquad\qquad(23)$$

When the point where the illumination is required is situated upon the axis, ζ, l, z are zero. Hence $V_0=1$, $V_1=0$, and

$$I^2=\frac{1}{(a+b)^2},$$

the same as if the primary wave had come on unbroken. This is Poisson's theorem, already found (§ 10) by a much simpler method, in which attention is limited from the first to points upon the axis. The distribution of light at other points upon the screen is to be found from (23) by means of the series (16), (17) for V_0 and V_1. Lommel gives curves for the intensity when $y=\pi$, 2π, 3π, ... 6π. The bright central spot is accompanied by rings of varying intensity.

The limit of the geometrical shadow $[\zeta/(a+b)=r/a]$ corresponds to $y=z$. In this case

$$V_0=J_0(z)-J_2(z)+J_4(z)-\ldots=\tfrac{1}{2}\{J_0(z)+\cos z\},\qquad\qquad(24)$$

$$V_1=J_1(z)-J_3(z)+J_5(z)-\ldots=\tfrac{1}{2}\sin z.\qquad\qquad(25)$$

The numbers computed for special values of y and z apply to a whole class of problems. Since

$$y=\frac{2\pi}{\lambda}\frac{a+b}{ab}\cdot r^2,\qquad z=\frac{2\pi}{\lambda}\frac{\zeta}{b}\cdot r,$$

both y and z remain unchanged, even when λ is constant, if we suppose

$$b\propto a,\qquad r\propto\zeta\propto\sqrt{a}.\qquad\qquad(26)$$

We may fall back upon Fraunhofer's phenomena by supposing $a=b=\infty$, or more generally $b=-a$, so that $y=0$.

Under these circumstances

$$C=\pi r^2\frac{J_1(z)}{z},\qquad S=0.$$

But it is unnecessary to add anything further under this head.

§ 19. *Polarization.*

A ray of ordinary light is symmetrical with respect to the direction of propagation. If, for example, this direction be vertical, there is nothing that can be said concerning the north and south sides of the ray that is not equally true concerning the east and west sides. In polarized light this symmetry is lost. Huygens showed that when a ray of such light falls upon a crystal of Iceland spar, which is made to revolve about the ray as an axis, the phenomena vary in a manner not to be represented as a mere revolution with the spar. In Newton's language, the ray itself has *sides*, or is polarized.

Malus discovered that ordinary light may be polarized by reflexion as well as by double refraction; and Brewster proved that the effect is nearly complete when the tangent of the angle of incidence is equal to the refractive index, or (which comes to the same) when the reflected and refracted rays are perpendicular to one another. The light thus obtained is said to be polarized in the plane of reflexion.

Reciprocally, the character of a polarized ray may be revealed by submitting it to the test of reflexion at the appropriate angle. As the normal to the reflecting surface revolves (in a cone) about the ray, there are two azimuths of the plane of incidence, distant 180°, at which the reflexion is a maximum, and two others, distant 90° from the former, at which the reflexion (nearly) vanishes. In the latter case the plane of incidence is perpendicular to that in which the light must be supposed to have been reflected in order to acquire its polarization.

The full statement of the law of double refraction is somewhat complicated, and scarcely to be made intelligible except in terms of the wave theory; but, in order merely to show the relation of double refraction in a uniaxal crystal, such as Iceland spar, to polarized light, we may take the case of a prism so cut that the refracting edge is parallel to the optic axis. By traversing such a prism, in a plane perpendicular to the edge, a ray of ordinary light is divided into two, of equal intensity, each of which is refracted according to the ordinary law of Snell. Whatever may be the angle and setting of the prism, the phenomenon may be represented by supposing half the light to be refracted with one index (1·65), and the other half with the different index (1·48). The rays thus arising are polarized,—the one more refracted in the plane of refraction, and the other in the perpendicular plane. If these rays are now allowed to fall upon a second similar prism, held so that its edge is parallel to that of the first prism, there is no further duplication. The ray first refracted with index 1·65 is refracted again in like manner, and similarly the ray first refracted with index 1·48 is again so refracted. But the case is altered if the second prism be caused to rotate about the incident ray. If the rotation be through an angle of 90°, each ray is indeed refracted singly; but the indices are exchanged. The ray that suffered most refraction at the first prism now suffers least at the second, and *vice versâ*. At intermediate rotations the double refraction reasserts itself, each ray being divided into two, refracted with the above-mentioned indices, and of intensity dependent upon the amount of rotation, but always such that no light is lost (or gained) on the whole by the separation.

The law governing the intensity was formulated by Malus, and has been verified by the measures of Arago and other workers. If θ be the angle of rotation from the position in which one of the rays is at a maximum, while the other vanishes, the intensities are proportional to $\cos^2\theta$ and $\sin^2\theta$. On the same scale, if we neglect the loss by reflexion and absorption, the intensity of the incident light is represented by unity.

A similar law applies to the intensity with which a polarized ray is reflected from a glass surface at the Brewsterian angle. If θ be reckoned from the azimuth of maximum reflexion, the intensity at other angles may be represented by $\cos^2\theta$, vanishing when $\theta = 90°$.

The phenomena here briefly sketched force upon us the view that the vibrations of light are transverse to the direction of propagation. In ordinary light the vibrations are as much in one transverse direction as in another; and when such light falls upon a doubly refracting, or reflecting, medium, the vibrations are ***resolved*** into two definite directions, constituting two rays polarized in perpendicular planes, and differently influenced by the medium. In this case the two rays are necessarily of equal intensity.

Consider, for example, the application of this idea to the reflexion of a ray of ordinary light at the Brewsterian, or polarizing, angle. The incident light may be resolved into two, of equal intensity, and polarized respectively in and perpendicular to the plane of incidence. Now we know that a ray polarized in the plane perpendicular to that of incidence will not be reflected, will in fact be entirely transmitted; and the necessary consequence is that all the light reflected at this angle will be polarized in the plane of incidence. The operation of the plate is thus purely selective, the polarized component, which is missing in the reflected light, being represented in undue proportion in the transmitted light.

If the incident light be polarized, suppose at an angle θ with the plane of incidence, the incident vibration may be resolved into $\cos\theta$ in the one plane and $\sin\theta$ in the other. The latter polarized component is not reflected. The reflected light is thus in all cases polarized in the plane of reflexion; and its ***intensity***, proportional to the square of the vibration, is represented by $h\cos^2\theta$, if h be the intensity in which light is reflected when polarized in the plane of reflexion. The law of Malus is thus a necessary consequence of the principle of resolution.

The idea of transverse vibrations was admitted with reluctance, even by Young and Fresnel themselves. A perfect fluid, such as the ethereal medium was then supposed to be, is essentially incapable of transverse vibrations. But there seems to be no reason *à priori* for preferring one kind of vibration to another; and the phenomena of polarization prove conclusively that, if luminous vibrations are analogous to those of a material medium, it is to solids, and not to fluids, that we must look. An isotropic solid is capable of propagating two distinct kinds of waves,—the first dependent upon *rigidity*, or the force by which shear is resisted, and the second analogous to waves of sound and dependent upon *compressibility*. In the former the vibrations are transverse to the direction of propagation, that is, they may take place in any direction parallel to the wave-front, and they are thus suitable representatives of the vibrations of light. In this theory the luminiferous ether is distinctly assimilated to an elastic solid, and the velocity of light depends upon the *rigidity* and *density* assigned to the medium.

The possibility of longitudinal waves, in which the displacement is perpendicular to the wave-front, is an objection to the elastic solid theory of light, for there is nothing known in optics corresponding thereto. If, however, we suppose with Green that the medium is incompressible, the velocity of longitudinal waves becomes infinite, and the objection is in great degree obviated. Such a supposition is hardly a departure from the original idea, inasmuch as, so far as we know, there is nothing to prevent a solid material possessing these properties, and an approximation is actually presented by such bodies as jelly, in which the velocity of longitudinal vibrations is a large multiple of that of transverse vibrations.

§ 20. *Interference of Polarized Light.*

The conditions of interference of polarized light are most easily deduced from the phenomena of the colours of crystalline plates, if we once admit Young's view that the origin of the colours is to be sought in the interference of the differently refracted rays. Independently of any hypothesis of this kind, the subject was directly investigated by Fresnel and Arago*, who summarized their conclusions thus:—

(1) Under the same conditions in which two rays of ordinary light appear to destroy one another, two rays polarized in contrary (viz., perpendicular) directions are without mutual influence.

(2) Two rays of light polarized in the same direction act upon one another like ordinary rays; so that, with these two kinds of light, the phenomena of interference are identical.

(3) Two rays *originally polarized in opposite directions* may afterwards be brought to the same plane of polarization, *without thereby acquiring the power to influence one another.*

(4) *Two rays polarized in opposite directions, and afterwards brought to similar polarizations*, react in the same manner as natural rays, *if they are derived from a beam originally polarized in one direction.*

* Fresnel's *Works*, Vol. I. p. 521.

The fact that oppositely polarized rays cannot be made to interfere may of itself be regarded as a proof that the vibrations are transverse; and the principle, once admitted, gives an intelligible account of all the varied phenomena in this field of optics. The only points on which any difficulty arises are as to the nature of ordinary unpolarized light, and the rules according to which *intensity* is to be calculated. It will be proper to consider these questions somewhat fully.

In ordinary (plane) polarized light the vibrations are supposed to be in one direction only. If x and y be rectangular coordinates in the plane of the wave, we may take, as representing a regular vibration of plane-polarized light,

$$x = a\cos(\phi - \alpha), \qquad (1)$$

where $\phi = 2\pi t/\tau$, and a, α denote constants. It must be remembered, however, that in optics a regular vibration of this kind never presents itself. In the simplest case of approximately monochromatic light, the amplitude and phase must be regarded (§ 4) as liable to incessant variation, and all that we are able to appreciate is the *mean* intensity, represented by $M(a^2)$. If a number of these irregular streams of light are combined, the intensity of the mixture cannot be calculated from a mere knowledge of the separate intensities, unless we have assurance that the streams are *independent*, that is, without mutual phase-relations of a durable character. For instance, two thoroughly similar streams combine into one of *four-fold* intensity, if the phases are the same; while, if the phases are opposed, the intensity falls to zero. It is only when the streams are independent, so that the phase-relation is arbitrary and variable from moment to moment, that the apparent resultant intensity is necessarily the double of the separate intensities.

If any number of independent vibrations of type (1) be superposed, the resultant is

$$[\Sigma a_1 \cos\alpha_1]\cos\phi + [\Sigma a_1 \sin\alpha_1]\sin\phi,$$

and the momentary intensity is

$$[\Sigma a_1 \cos\alpha_1]^2 + [\Sigma a_1 \sin\alpha_1]^2,$$

or

$$a_1^2 + a_2^2 + \ldots + 2a_1a_2\cos(\alpha_1 - \alpha_2) + \ldots.$$

The phase-relations being unknown, this quantity is quite indeterminate. But, since each cosine varies from moment to moment, and on the whole is as much positive as negative, the *mean* intensity is

$$M(a_1^2) + M(a_2^2) + \ldots,$$

that is to say, is to be found by simple addition of the separate intensities.

Let us now dispense with the restriction to one direction of vibration, and consider in the first place the character of a *regular* vibration, of given frequency. The general expression will be

$$x = a\cos(\phi - \alpha), \qquad y = b\cos(\phi - \beta), \qquad (2)$$

where a, α, b, β are constants. If $\beta=\alpha$, the vibrations are executed entirely in the plane $x/y=a/b$, or the light is plane-polarized. Or if $\beta=\pi-\alpha$, the light is again plane-polarized, the plane of vibration being $x/y=-a/b$. In other cases the vibrations are not confined to one plane, so that the light is not plane-polarized, but, in conformity with the path denoted by (2), it is said to be ***elliptically***-polarized. If one of the constituents of elliptically-polarized light be suitably accelerated or retarded relatively to the other, it may be converted into plane-polarized light, and so identified by the usual tests. Or, conversely, plane-polarized light may be converted into elliptically-polarized by a similar operation. The relative acceleration in question is readily effected by a plate of doubly refracting crystal cut parallel to the axis.

If $\beta=\alpha\pm\frac{1}{2}\pi$, whether in the first instance or after the action of a crystalline plate,

$$x=a\cos(\phi-\alpha), \qquad y=\pm b\sin(\phi-\alpha). \quad\text{..............(3)}$$

The maxima and minima values of the one coordinate here occur synchronously with the evanescence of the other, and the coordinate axes are the *principal* axes of the elliptic path.

An important particular case arises when further $b=a$. The path is then a circle, and the light is said to be *circularly*-polarized. According to the sign adopted in the second equation (3), the circle is described in the one direction or in the other.

Circularly-polarized light can be resolved into plane-polarized components in *any* two rectangular directions, which are such that the intensities are equal and the phases different by a quarter period. If a crystalline plate be of such thickness that it retards one component by a quarter of a wave-length (or indeed by any odd multiple thereof) relatively to the other, it will convert plane-polarized light into circularly-polarized, and conversely,—in the latter case without regard to the azimuth in which it is held.

The property of circularly-polarized light whereby it is capable of resolution into oppositely plane-polarized components of equal intensities is possessed also by natural unpolarized light; but the discrimination may be effected experimentally with the aid of the quarter-wave plate. By this agency the circularly-polarized ray is converted into plane-polarized, while the natural light remains apparently unaltered. The difficulty which remains is rather to explain the physical character of natural light. To this we shall presently return; but in the meantime it is obvious that the constitution of natural light is essentially irregular, for we have seen that absolutely regular, *i.e.*, absolutely homogeneous, light ***is necessarily (elliptically) polarized.***

In discussing the vibration represented by (2), we have considered the amplitudes and phases to be constant; but in nature this is no more attainable than in the case of plane-polarized light. In order that the elliptic polarization may be of a definite character, it is only necessary that the *ratio* of amplitudes and the *difference* of phases should be absolute constants, and this of course is consistent with the same degree of irregularity as was admitted for plane vibrations.

The intensity of elliptically-polarized light is the sum of the intensities of its rectangular components. This we may consider to be an experimental fact, as well as a consequence of the theory of transverse vibrations. In whatever form such a theory may be adopted, the energy propagated will certainly conform to this law. When the constants in (2) are regarded as subject to variation, the apparent intensity is represented by

$$M(a^2)+M(b^2). \quad \text{.........................(4)}$$

We are now in a position to examine the constitution which must be ascribed to natural light. The conditions to be satisfied are that when resolved in any plane the mean intensity of the vibrations shall be independent of the orientation of the plane, and, further, that this property shall be unaffected by any previous relative retardation of the rectangular components into which it may have been resolved. The original vibration being represented by

$$x=a\cos(\phi-\alpha), \qquad y=b\cos(\phi-\beta),$$

or, as we may write it, since we are concerned only with phase *differences*,

$$x=a\cos\phi, \qquad y=b\cos(\phi-\delta), \quad \text{.................(5)}$$

let us suppose that the second component is subjected to a retardation ϵ. Thus

$$x=a\cos\phi, \qquad y=b\cos(\phi-\delta-\epsilon), \quad \text{..............(6)}$$

in which a, b, δ will be regarded as subject to rapid variation, while ϵ remains constant. If the vibration represented by (6) be now resolved in a direction x', making an angle ω with x, we have

$$\begin{aligned} x' &= a\cos\phi\cos\omega+b\cos(\phi-\delta-\epsilon)\sin\omega \\ &= [a\cos\omega+b\sin\omega\cos(\delta+\epsilon)]\cos\phi+b\sin\omega\sin(\delta+\epsilon)\sin\phi; \end{aligned}$$

and the intensity is

$$a^2\cos^2\omega+b^2\sin^2\omega+2ab\cos\omega\sin\omega\cos(\delta+\epsilon). \quad \text{...........(7)}$$

Of this expression we take the mean, ω and ϵ remaining constant. Thus the apparent intensity may be written

$$M(x'^2)=M(a^2)\cos^2\omega+M(b^2)\sin^2\omega+2M[ab\cos(\delta+\epsilon)]\cos\omega\sin\omega. \quad \text{...(8)}$$

In order now that the stream may satisfy the conditions laid down as necessary for natural light, (8) must be independent of ω and ϵ; so that

$$M(a^2)=M(b^2), \quad \text{...............................(9)}$$

$$M(ab\cos\delta)=M(ab\sin\delta)=0. \quad \text{....................(10)*}$$

* Verdet, *Leçons d'Optique Physique*, Vol. II. p. 83.

In these equations a^2 and b^2 represent simply the intensities, or squares of amplitudes, of the x and y vibrations; and the other two quantities admit also of a simple interpretation. The value of y may be written

$$y = b \cos \delta \cos \phi + b \sin \delta \sin \phi; \quad \text{..................(11)}$$

from which we see that $b \cos \delta$ is the coefficient of that part of the y vibration which has the same phase as the x vibration. Thus $ab \cos \delta$ may be interpreted as the product of the coefficients of the parts of the x and y vibrations which have the same phase. Next suppose the phase of y accelerated by writing $\frac{1}{2}\pi + \phi$ in place of ϕ. We should thus have

$$y = -b \cos \delta \sin \phi + b \sin \delta \cos \phi,$$

and $ab \sin \delta$ represents the product of the coefficients of the parts which are now in the same phase, or (which is the same) the product of the coefficients of the x vibration and of that part of the y vibration which was 90° behind in phase. In general, if

$$x = h \cos \phi + h' \sin \phi, \qquad y = k \cos \phi + k' \sin \phi, \quad \text{.........(12)}$$

the first product is $hk + h'k'$ and the second is $hk' - h'k$.

Let us next examine how the quantities which we have been considering are affected by a transformation of coordinates in accordance with the formulæ

$$x' = x \cos \omega + y \sin \omega, \qquad y' = -x \sin \omega + y \cos \omega. \quad \text{.........(13)}$$

We find

$$x' = \cos \phi \{a \cos \omega + b \sin \omega \cos \delta\} + \sin \phi \,.\, b \sin \delta \sin \omega, \quad \text{......(14)}$$

$$y' = \cos \phi \{-a \sin \omega + b \cos \omega \cos \delta\} + \sin \phi \,.\, b \sin \delta \cos \omega; \quad \text{...(15)}$$

whence

$$\text{amp.}^2 \text{ of } x' = a^2 \cos^2 \omega + b^2 \sin^2 \omega + 2ab \cos \delta \sin \omega \cos \omega, \quad \text{...(16)}$$

$$\text{amp.}^2 \text{ of } y' = a^2 \sin^2 \omega + b^2 \cos^2 \omega - 2ab \cos \delta \sin \omega \cos \omega. \quad \text{...(17)}$$

In like manner

$$\text{First product} = (b^2 - a^2) \sin \omega \cos \omega + ab \cos \delta (\cos^2 \omega - \sin^2 \omega), \quad \text{......(18)}$$

$$\text{Second product} = ab \sin \delta. \quad \text{..(19)}$$

The second product, representing the circulating part of the motion, is thus unaltered by the transformation.

Let us pass on to the consideration of the mean quantities which occur in (9), (10), writing for brevity

$$M(a^2) = A, \qquad M(b^2) = B, \qquad M(ab \cos \delta) = C, \qquad M(ab \sin \delta) = D.$$

From (16), (17), (18), (19), if A', B', C', D' denote the corresponding quantities after transformation,

$$A' = A\cos^2\omega + B\sin^2\omega + 2C\cos\omega\sin\omega, \qquad (20)$$

$$B' = A\sin^2\omega + B\cos^2\omega - 2C\cos\omega\sin\omega, \qquad (21)$$

$$C' = C(\cos^2\omega - \sin^2\omega) + (B - A)\cos\omega\sin\omega, \qquad (22)$$

$$D' = D. \qquad (23)$$

These formulæ prove that, if the conditions (9), (10), shown to be necessary in order that the light may behave as natural light, be satisfied for one set of axes, they are equally satisfied with any other. It is thus a matter of indifference with respect to what axes the retardation ϵ is supposed to be introduced, and the conditions (9), (10) are sufficient, as well as necessary, to characterize natural light.

Reverting to (8), we see that, whether the light be natural or not, its character, so far as experimental tests can show, is determined by the values of A, B, C, D. The effect of a change of axes is given by (20), &c., and it is evident that the new axes may always be so chosen that $C' = 0$. For this purpose it is only necessary to take ω such that

$$\tan 2\omega = 2C/(A - B).$$

If we choose these new axes as fundamental axes, the values of the constants for any others inclined to them at angle ω will be of the form

$$\left.\begin{aligned} A &= A_1\cos^2\omega + B_1\sin^2\omega \\ B &= A_1\sin^2\omega + B_1\cos^2\omega \\ C &= (B_1 - A_1)\cos\omega\sin\omega \end{aligned}\right\}. \qquad (24)$$

If A_1 and B_1 are here equal, then $C = 0$, $A = B$ for all values of ω. In this case, the light cannot be distinguished from natural light by mere resolution; but if D be finite, the difference may be made apparent with the aid of a retarding plate.

If A_1 and B_1 are unequal, they represent the maximum and minimum values of A and B. The intensity is then a function of the plane of resolution, and the light may be recognized as partially polarized by the usual tests. If either A_1 or B_1 vanishes, the light is plane-polarized*.

When several independent streams of light are combined, the values, not only of A and B, but also of C and D, for the mixture, are to be found by simple addition. It must here be distinctly understood that there are no permanent phase-relations between one component and another. Suppose, for example, that there are two streams of light, each of which satisfies the relations $A = B$, $C = 0$, but makes the value of D finite. If the two values of D are equal and opposite, and the streams are independent, the mixture

* In this case D_1 necessarily vanishes.

constitutes natural light. A particular case arises when each component is circularly-polarized ($D = \pm A = \pm B$), one in the right-handed and the other in the left-handed direction. The intensities being equal, the mixture is equivalent to natural light, but only under the restriction that the streams are without phase-relation. If, on the contrary, the second stream be similar to the first, affected merely with a constant retardation, the resultant is not natural, but completely (plane) polarized light.

We will now prove that the most general mixture of light may be regarded as compounded of one stream of light elliptically-polarized in a definite manner, and of an independent stream of natural light. The theorem is due to Stokes*, but the method that we shall follow is that of Verdet†.

In the first place, it is necessary to observe that the values of the fundamental quantities A, B, C, D are not free from restriction. It will be shown that in no case can $C^2 + D^2$ exceed AB.

In equations (2), expressing the vibration at any moment, let $a_1, b_1, \alpha_1, \beta_1$, be the values of a, b, α, β during an interval of time proportional to m_1, and in like manner let the suffixes 2, 3, ... correspond to times proportional to $m_2, m_3, \ldots$. Then

$$AB = m_1^2a_1^2b_1^2 + m_2^2a_2^2b_2^2 + \ldots + m_1m_2(a_1^2b_2^2 + a_2^2b_1^2) + \ldots.$$

Again, by (12),

$$C = m_1a_1b_1(\cos\alpha_1\cos\beta_1 + \sin\alpha_1\sin\beta_1) + \ldots$$

$$= m_1a_1b_1\cos\delta_1 + m_2a_2b_2\cos\delta_2 + \ldots,$$

$$D = m_1a_1b_1\sin\delta_1 + m_2a_2b_2\sin\delta_2 + \ldots;$$

where, as before,

$$\delta_1 = \beta_1 - \alpha_1, \qquad \delta_2 = \beta_2 - \alpha_2, \ldots.$$

Thus,

$$C^2 + D^2 = m_1^2a_1^2b_1^2 + m_2^2a_2^2b_2^2 + \ldots + m_1m_2a_1b_1a_2b_2\cos(\delta_2 - \delta_1) + \ldots.$$

From these equations we see that $AB - C^2 - D^2$ reduces itself to a sum of terms of the form

$$m_1m_2[a_1^2b_2^2 + a_2^2b_1^2 - 2a_1b_1a_2b_2\cos(\delta_2 - \delta_1)],$$

each of which is essentially positive.

The only case in which the sum can vanish is when

$$\delta_1 = \delta_2 = \delta_3 = \ldots,$$

and further
$$b_1 : a_1 = b_2 : a_2 = b_3 : a_3 = \ldots.$$

Under these conditions the light is reduced to be of a definite elliptic

* "On the Composition and Resolution of Streams of Light from Different Sources," *Camb. Phil. Trans.* 1852.

† *Loc. cit.* p. 94.

character, although the amplitude and phase of the system *as a whole* may be subject to rapid variation. The elliptic constants are given by

$$b^2/a^2 = B/A, \qquad \tan\delta = D/C. \qquad (25)$$

In general AB exceeds $(C^2 + D^2)$; but it will always be possible to find a positive quantity H, which when subtracted from A and B (themselves necessarily positive) shall reduce the product to equality with $C^2 + D^2$, in accordance with

$$(A - H)(B - H) = C^2 + D^2. \qquad (26)$$

The original light may thus be resolved into two groups. For the first group the constants are $H, H, 0, 0$; and for the second $A - H, B - H, C, D$. Each of these is of a simple character; for the first represents natural light, and the second light elliptically-polarized. It is thus proved that in general a stream of light may be regarded as composed of one stream of natural light and of another elliptically-polarized. The intensity of the natural light is $2H$, where from (26)

$$H = \tfrac{1}{2}(A + B) - \tfrac{1}{2}\sqrt{\{(A - B)^2 + 4(C^2 + D^2)\}}. \qquad (27)$$

The elliptic constants of the second component are given by

$$b^2/a^2 = (B - H)/(A - H), \qquad \tan\delta = D/C, \qquad (28)$$

and

$$M(a^2) = A - H. \qquad (29)$$

If $D = 0$, and therefore by (28) $\delta = 0$, the second component is plane-polarized. This is regarded as a particular case of elliptic polarization. Again, if $A = B$, $C = 0$, the polarization is circular.

The laws of interference of polarized light, discovered by Fresnel and Arago, are exactly what the theory of transverse vibrations would lead us to expect, when once we have cleared up the idea of unpolarized light. Ordinary sources, such as the sun, emit unpolarized light. If this be resolved in two opposite directions, the polarized components are not only each irregular, but there is no permanent phase-relation between them. No light derived from one can therefore ever interfere regularly with light derived from the other. If, however, we commence with plane-polarized light, we have only one series of irregularities to deal with. When resolved in two rectangular directions, the components cannot then interfere, but only on account of the perpendicularity. If brought back by resolution to the same plane of polarization, interference becomes possible, because the same series of irregularities are to be found in both components.

§ 21. *Double Refraction.*

The construction by which Huygens explained the ordinary and extraordinary refraction of Iceland spar has already been given (Light, *Enc. Brit.* Vol. XIV. p. 610). The wave-surface is in two sheets, composed of a sphere and of an ellipsoid of revolution, in contact with one another at the extremities of the polar axis. In biaxal crystals the wave-surface is of a more complicated character, including that of Huygens as a particular case.

It is not unimportant to remark that the essential problem of double refraction is to determine the two velocities with which plane waves are propagated, when the direction of the normal to the wave-front is assigned. When this problem has been solved, the determination of the wave-surface is a mere matter of geometry, not absolutely necessary for the explanation of the leading phenomena, but convenient as affording a concise summary of the principal laws. In all cases the wave-surface is to be regarded as the envelope at any subsequent time of all the plane wave-fronts which at a given instant may be supposed to be passing through a particular point.

In singly refracting media, where the velocity of a wave is the same in all directions, the wave-normal coincides with the *ray.* In doubly refracting crystals this law no longer holds good. The principles by which the conception of a ray is justified (§ 10), when applied to this case, show that the centre of the zone system is not in general to be found at the foot of the perpendicular upon the primary wave-front. The surface whose contact with the primary wave-front determines the element from which the secondary disturbance arrives with least retardation is now not a sphere, but whatever wave-surface is appropriate to the medium. The direction of the ray, corresponding to any tangent plane of the wave-surface, is thus not the normal, but the radius vector drawn from the centre to the point of contact.

The velocity of propagation (reckoned always perpendicularly to the wave-front) may be conceived to depend upon the direction of the wave-front, or wave-normal, and upon what we may call (at any rate figuratively) the direction of vibration. If the velocity depended exclusively upon the wave-normal, there could be no *double,* though there might be *extraordinary,* refraction, *i.e.,* refraction deviating from the law of Snell; but of this nothing is known in nature. The fact that there are in general two velocities for one wave-front proves that the velocity depends upon the direction of vibration.

According to the Huygenian law, confirmed to a high degree of accuracy by the observations of Brewster and Swan*, a ray polarized in a principal plane (*i.e.,* a plane passing through the axis) of a uniaxal crystal suffers ordinary refraction only, that is, propagates itself with the same velocity in

* *Edin. Trans.* Vol. XVI. p. 375.

all directions. The interpretation which Fresnel put upon this is that the vibrations (understood now in a literal sense) are perpendicular to the plane of polarization, and that the velocity is constant because the direction of vibration is in all cases similarly related (perpendicular) to the axis. The development of this idea in the fertile brain of Fresnel led him to the remarkable discovery of the law of refraction in biaxal crystals.

The hypotheses upon which Fresnel based his attempt at a mechanical theory are thus summarized by Verdet:—

(1) The vibrations of polarized light are perpendicular to the plane of polarization;

(2) The elastic forces called into play during the propagation of a system of plane waves (of rectilinear transverse vibrations) differ from the elastic forces developed by the parallel displacement of a single molecule only by a constant factor, independent of the particular direction of the plane of the wave;

(3) When a plane wave propagates itself in any homogeneous medium, the components parallel to the wave-front of the elastic forces called into play by the vibrations of the wave are alone operative;

(4) The velocity of a plane wave which propagates itself with type unchanged in any homogeneous medium is proportional to the square root of the effective component of the elastic force developed by the vibrations.

Fresnel himself was perfectly aware that his theory was deficient in rigour, and indeed there is little to be said in defence of his second hypothesis. Nevertheless, the great historical interest of this theory, and the support that experiment gives to Fresnel's conclusion as to the actual form of the wave-surface in biaxal crystals, render some account of his work in this field imperative.

The potential energy of displacement of a single molecule from its position of equilibrium is ultimately a quadratic function of the three components reckoned parallel to any set of rectangular axes. These axes may be so chosen as to reduce the quadratic function to a sum of squares, so that the energy may be expressed,

$$V = \tfrac{1}{2}a^2\xi^2 + \tfrac{1}{2}b^2\eta^2 + \tfrac{1}{2}c^2\zeta^2, \quad \text{.......................(1)}$$

where ξ, η, ζ are the three component displacements. The corresponding forces of restitution, obtained at once by differentiation, are

$$X = a^2\xi, \qquad Y = b^2\eta, \qquad Z = c^2\zeta. \quad \text{.............(2)}$$

The force of restitution is thus in general inclined to the direction of displacement. The relation between the two directions X, Y, Z and ξ, η, ζ is the same as that between the normal to a tangent plane and the radius vector ρ to the point of contact in the ellipsoid

$$a^2\xi^2 + b^2\eta^2 + c^2\zeta^2 = 1. \quad \text{.........................(3)}$$

If a^2, b^2, c^2 are unequal, the directions of the coordinate axes are the only ones in which a displacement calls into operation a parallel force of restitution. If two of the quantities a^2, b^2, c^2 are equal, the ellipsoid (3) is of revolution, and every direction in the plane of the equal axes possesses the property in question. This is the case of a uniaxal crystal. If the three quantities a^2, b^2, c^2 are all equal, the medium is isotropic.

If we resolve the force of restitution in the direction of displacement, we obtain a quantity dependent upon this direction in a manner readily expressible by means of the ellipsoid of elasticity (3). For, when the total displacement is given, this quantity is proportional to

$$\frac{a^2\xi^2 + b^2\eta^2 + c^2\zeta^2}{\xi^2 + \eta^2 + \zeta^2},$$

that is to say, to the inverse square of the radius vector ρ in (3).

We have now to inquire in what directions, limited to a particular plane, a displacement may be so made that the *projection* of the force of restitution upon the plane may be parallel to the displacement. The answer follows at once from the property of the ellipsoid of elasticity. For, if in any section of the ellipsoid we have a radius vector such that the plane containing it and the normal to the corresponding tangent plane is perpendicular to the plane of the section, the tangent line to the section must be perpendicular to the radius vector, that is, the radius vector must be a principal axis of the section. There are therefore two, and in general only two, directions in any plane satisfying the proposed condition, and these are perpendicular to one another. If, however, the plane be one of those of circular section, every line of displacement is such that the component of the force, resolved parallel to the plane, coincides with it.

According to the principles laid down by Fresnel, we have now complete data for the solution of the problem of double refraction. If the direction of the wave-front be given, there are (in general) only two directions of vibration such that a single wave is propagated. If the actual displacements do not conform to this condition, they will be resolved into two of the required character, and the components will in general be propagated with different velocities. The two directions are the principal axes of the section of (3) made by the wave-front, and the velocities of propagation are inversely proportional to the lengths of these axes.

The law connecting the lengths of the axes with the direction $(l, m. n)$ of the plane is a question of geometry*; and indeed the whole investigation of the wave-surface may be elegantly carried through geometrically with the aid of certain theorems of MacCullagh respecting *apsidal* surfaces (Salmon,

* See Salmon's *Analytical Geometry of Three Dimensions*, Dublin 1882, § 102.

ch. XIV.). For this, however, we have not space, and must content ourselves with a sketch of the analytical method of treatment.

If v be the velocity of propagation in direction l, m, n, the wave-surface is the envelope of planes

$$lx + my + nz = v, \qquad \text{.............................(4)}$$

where v is a function of l, m, n, whose form is to be determined. If (λ, μ, ν) be the corresponding direction of vibration, then

$$l\lambda + m\mu + n\nu = 0. \qquad \text{.............................(5)}$$

According to the principles laid down by Fresnel, we see at once that the force of restitution $(a^2\lambda, b^2\mu, c^2\nu)$, corresponding to a displacement unity, is equivalent to a force v^2 along (λ, μ, ν), together with some force (P) along (l, m, n). Resolving parallel to the coordinate axes, we get

$$lP = a^2\lambda - v^2\lambda, \qquad mP = b^2\mu - v^2\mu, \qquad nP = c^2\nu - v^2\nu,$$

or

$$\lambda = \frac{lP}{a^2 - v^2}, \qquad \mu = \frac{mP}{b^2 - v^2}, \qquad \nu = \frac{nP}{c^2 - v^2}. \qquad \text{.........(6)}$$

Multiplying these by l, m, n respectively, and taking account of (5), we see that

$$\frac{l^2}{a^2 - v^2} + \frac{m^2}{b^2 - v^2} + \frac{n^2}{c^2 - v^2} = 0 \qquad \text{.......................(7)}$$

is the relation sought for between v and (l, m, n). In this equation b, c are the velocities when the direction of propagation is along x, the former being applicable when the vibration is parallel to y, and the latter when it is parallel to x.

The directions of vibration are determined by (5) and by the consideration that (l, m, n), (λ, μ, ν), and $(a^2\lambda, b^2\mu, c^2\nu)$ lie in a plane, or (as we may put it) are all perpendicular to one direction (f, g, h). Thus

$$\left.\begin{aligned} lf + mg + nh &= 0 \\ \lambda f + \mu g + \nu h &= 0 \\ a^2\lambda f + b^2\mu g + c^2\nu h &= 0 \end{aligned}\right\}. \qquad \text{.......................(8)}$$

The determinant expressing the result of the elimination of $f : g : h$ may be put into the form

$$\frac{l}{\lambda}(b^2 - c^2) + \frac{m}{\mu}(c^2 - a^2) + \frac{n}{\nu}(a^2 - b^2) = 0, \qquad \text{.................(9)}$$

which with (5) suffices to determine (λ, μ, ν) as a function of (l, m, n).

The fact that the system of equations (5), (8) is symmetrical as between (λ, μ, ν) and (f, g, h) proves that the two directions of vibration corresponding to a given (l, m, n) are perpendicular to one another.

The direct investigation of the wave-surface from (4) and (7) was first effected by Ampère, but his analytical process was very laborious. Fresnel had indeed been forced to content himself with an indirect method of verification. But in the following investigation of A. Smith* the eliminations are effected with comparatively little trouble.

In addition to (4) and (7), we know that

$$l^2 + m^2 + n^2 = 1. \quad\text{.............................(10)}$$

To find the equation to the envelope, we have to differentiate these equations, making l, m, n, v vary. Eliminating the differentials by the method of multipliers, we obtain the following:—

$$x = Al + \frac{Bl}{v^2 - a^2}, \qquad y = Am + \frac{Bm}{v^2 - b^2}, \qquad z = An + \frac{Bn}{v^2 - c^2}, \quad\text{...(11, 12, 13)}$$

and

$$1 = Bv\left\{\frac{l^2}{(v^2 - a^2)^2} + \frac{m^2}{(v^2 - b^2)^2} + \frac{n^2}{(v^2 - c^2)^2}\right\}. \quad\text{...........(14)}$$

The equations (11), (12), (13) multiplied by l, m, n respectively, and added, give

$$v = A. \quad\text{...................................(15)}$$

The same equations, squared and added, give

$$x^2 + y^2 + z^2 = A^2 + B/v.$$

If we put r^2 for $x^2 + y^2 + z^2$, and for A the value just found, we obtain

$$B = v(r^2 - v^2). \quad\text{.............................(16)}$$

If these values of A and B be substituted in (11),

$$x = lv\left\{1 + \frac{r^2 - v^2}{v^2 - a^2}\right\} = lv\,\frac{r^2 - a^2}{v^2 - a^2},$$

or

$$l = \frac{v^2 - a^2}{r^2 - a^2}\,\frac{x}{v}. \quad\text{.............................(17)}$$

If we substitute this value of l, and the corresponding values of m, n in (4), we get

$$\frac{(v^2 - a^2)\,x^2}{r^2 - a^2} + \frac{(v^2 - b^2)\,y^2}{r^2 - b^2} + \frac{(v^2 - c^2)\,z^2}{r^2 - c^2} = v^2 = \frac{v^2x^2}{r^2} + \frac{v^2y^2}{r^2} + \frac{v^2z^2}{r^2},$$

whence

$$\frac{x^2a^2}{r^2 - a^2} + \frac{y^2b^2}{r^2 - b^2} + \frac{z^2c^2}{r^2 - c^2} = 0, \quad\text{....................(18)}$$

as the equation of the wave-surface.

By (6) equation (11) may be written

$$x = Al + BP^{-1}\lambda,$$

from which and the corresponding equations we see that the direction (x, y, z) lies in the same plane as (l, m, n) and (λ, μ, ν). Hence in any tangent plane

* *Camb. Trans.* VI. 1835.

of the wave-surface the direction of vibration is that of the line joining the foot of the perpendicular and the point of contact (x, y, z).

The equation (18) leads to another geometrical definition of Fresnel's wave-surface. If through the centre of the ellipsoid reciprocal to the ellipsoid of elasticity (3), viz.,

$$x^2/a^2 + y^2/b^2 + z^2/c^2 = 1, \qquad\qquad (19)$$

a plane be drawn, and on the normal to this plane two lengths be marked off proportional to the axes of the elliptic section determined by the plane, the locus of the points thus obtained, the apsidal surface of (19), is the wave-surface (18).

Fully developed in integral powers of the coordinates, (18) takes the form

$$(x^2 + y^2 + z^2)(a^2x^2 + b^2y^2 + c^2z^2) - a^2(b^2 + c^2)x^2 - b^2(c^2 + a^2)y^2 - c^2(a^2 + b^2)z^2 + a^2b^2c^2 = 0. \qquad (20)$$

The section of (20) by the coordinate plane $y = 0$ is

$$(x^2 + z^2 - b^2)(a^2x^2 + c^2z^2 - a^2c^2) = 0, \qquad\qquad (21)$$

representing a circle and an ellipse (Fig. 24). That the sections by each of the principal planes would be a circle and an ellipse might have been foreseen independently of a general solution of the envelope problem. The forms of the sections prescribed in (21) and the two similar equations are sufficient to determine the character of the wave-surface, if we assume that it is of the *fourth degree*, and involves only the even powers of the coordinates. It was somewhat in this way that the equation was first obtained by Fresnel.

Fig. 24.

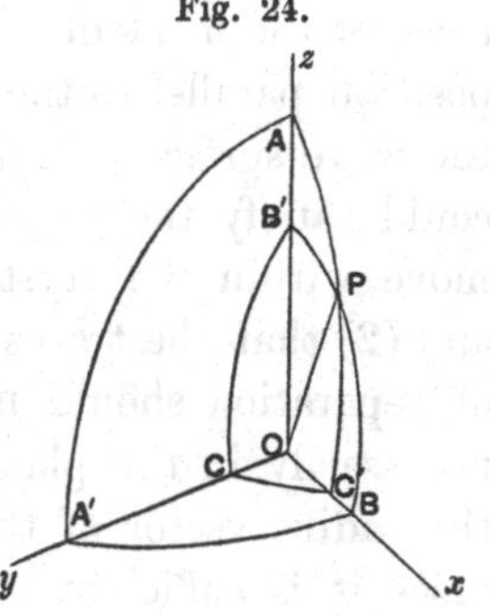

If two of the principal velocities, *e.g.*, a and b, are equal, (20) becomes

$$(x^2 + y^2 + z^2 - a^2)(a^2x^2 + a^2y^2 + c^2z^2 - a^2c^2) = 0, \qquad (22)$$

so that the wave-surface degenerates into the Huygenian sphere and ellipsoid of revolution appropriate to a uniaxal crystal. The two sheets touch one another at the points $x = 0$, $y = 0$, $z = \pm a$. If $c > a$, as in Iceland spar, the ellipsoid is external to the sphere. On the other hand, if $c < a$, as in quartz, the ellipsoid is internal.

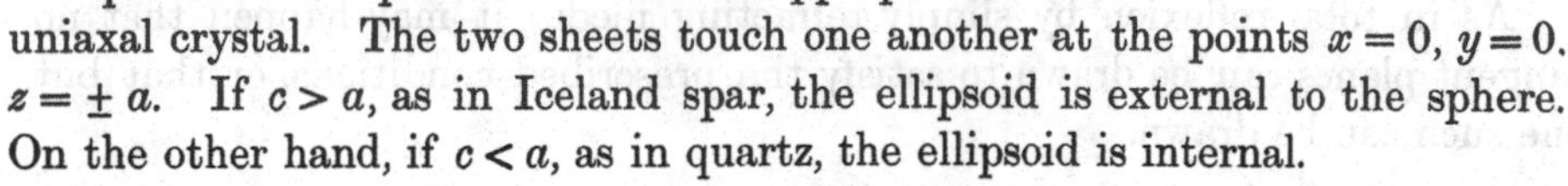

We have seen that when the wave-front is parallel to the circular sections of (3), the two wave-velocities coincide. Thus in (7), if a^2, b^2, c^2 be in descending order of magnitude, we have $m = 0$, $v = b$; so that

$$\frac{l^2}{a^2 - b^2} = \frac{n^2}{b^2 - c^2} = \frac{1}{a^2 - c^2}. \qquad\qquad (23)$$

In general, if θ, θ' be the angles which the normal to the actual wave-front makes with the optic axes, it may be proved that the difference of the squares of the two roots of (7) is given by

$$v_2^2 - v_1^2 = (a^2 - c^2) \sin \theta \sin \theta'. \qquad (24)$$

In a uniaxal crystal the optic axes coincide with the axis of symmetry, and there is no distinction between θ' and θ.

Since waves in a biaxal crystal propagated along either optic axis have but one velocity, it follows that tangent planes to the wave-surface, perpendicular to these directions, touch both sheets of the surface. It may be proved further that each plane touches the surface not merely at two, but at an infinite number of points which lie upon a circle.

The directions of the optic axes, and the angle included between them, are found frequently to vary with the colour of the light. Such a variation is to be expected, in view of dispersion, which renders a^2, b^2, c^2 functions of the wave-length.

A knowledge of the form of the wave-surface determines in all cases the law of refraction according to the construction of Huygens. We will suppose for simplicity that the first medium is air, and that the surface of separation between the media is plane. The incident wave-front at any moment of time cuts the surface of separation in a straight line. On this line take any point, and with it as centre construct the wave-surface in the second medium corresponding to a certain interval of time. At the end of this interval the trace of the incident wave-front upon the surface will have advanced to a new position, parallel to the former. Planes drawn through this line so as to touch the wave-surface give the positions of the refracted wave-fronts. None other could satisfy the two conditions—(1) that the refracted wave-front should move within the crystal with the normal velocity suitable to its direction, and (2) that the traces of the incident and refracted waves upon the surface of separation should move together. The normal to a refracted wave lies necessarily in the plane of incidence, but the refracted *ray*, coinciding with the radius vector of the wave-surface, in general deviates from it. In most cases it is sufficient to attend to the wave-normal.

As in total reflexion by simply refracting media, it may happen that no tangent planes can be drawn to satisfy the prescribed conditions, or that but one such can be drawn.

When the crystal is uniaxal, one wave is refracted according to the ordinary law of Snell. The accuracy of both the sphere and the ellipsoid of the Huygenian construction has been fully verified by modern observations *.

* Stokes, *Proc. Roy. Soc.* Vol. xx. p. 443, 1872; Glazebrook, *Phil. Trans.* 1880, p. 421; Hastings, *Amer. Jour.* Jan. 1888.

The simplest case of uniaxal refraction is when the axis of the crystal is perpendicular to the plane of incidence, with respect to which every thing then becomes symmetrical. The section of the wave-surface with which we have to deal reduces to two concentric circles; so that *both* waves are refracted according to the ordinary law, though of course with different indices.

In biaxal crystals one wave follows the ordinary law of refraction, if the plane of incidence coincide with a principal plane of the crystal. This consequence of his theory was verified by Fresnel himself, and subsequently by Rudberg and others. But the most remarkable phenomena of biaxal refraction are undoubtedly those discovered by Hamilton and Lloyd, generally known as conical refraction.

In general there are two refracted rays, corresponding to two distinct waves. But the refracted waves coalesce when they are perpendicular to either optic axis, and (as we have seen) this wave touches the wave-surface along a circle. Thus corresponding to one wave direction there are an infinite number of rays, lying upon a cone. The division of a single incident ray into a cone of refracted rays is called internal conical refraction. If the second face of the crystal is parallel to the first, each refracted ray resumes on emergence its original direction, so that the emergent bundle forms a hollow cylinder.

External conical refraction depends upon the singular points in the principal plane of zx, where the two sheets of the surface cross one another (Fig. 24). At such a point (P) an infinite number of tangent planes may be drawn to the surface, and each of the perpendiculars from O represents a wave direction, corresponding to the single *ray* OP. On emergence these waves will be differently refracted; and thus corresponding to a single internal ray there are an infinite number of external rays, lying upon a cone.

It has already been admitted that the dynamical foundations of Fresnel's theory are unsound; and it must be added that the rigorous theory of crystalline solids investigated by Cauchy and Green does not readily lend itself to the explanation of Fresnel's laws of double refraction. On this subject the reader should consult Prof. Stokes's Report. Sir W. Thomson has recently shown* that an originally isotropic medium, pressed unequally in different directions, may be so constituted as to vibrate in accordance with Fresnel's laws.

It may perhaps be worth while to remark that the equations, analogous to (2) § 24, which lead to these laws are

$$\frac{d^2\xi}{dt^2} = \frac{dp}{dx} + a^2\nabla^2\xi, \qquad \frac{d^2\eta}{dt^2} = \frac{dp}{dy} + b^2\nabla^2\eta, \text{ \&c.,} \qquad \text{...........(25)}$$

* "On Cauchy's and Green's Doctrine of Extraneous Force to explain dynamically Fresnel's Kinematics of Double Refraction," *Phil. Mag.* Feb. 1888.

where a, b, c are the principal wave-velocities. If we here assume

$$\xi = \lambda\theta, \qquad \eta = \mu\theta, \qquad \zeta = \nu\theta,$$
$$\theta/\theta_0 = p/p_0 = e^{ik(lx+my+nz-Vt)},$$

and substitute in (25), the condition of transversality leads at once to the desired results. But the equations (25) are not applicable to the vibrations of a crystalline solid.

In the electromagnetic theory double refraction is attributed to æolotropic inductive capacity, and appears to offer no particular difficulty.

If the present position of the theory of double refraction is still somewhat unsatisfactory, it must be remembered that the uncertainty does not affect the general principle. Almost any form of wave-theory involving transverse vibrations will explain the leading phenomenon, viz., the bifurcation of the ray. It is safe to predict that when ordinary refraction is well understood there will be little further trouble over double refraction.

The wave-velocity is not the only property of light rendered unsymmetrical by crystalline structure. In many cases the two polarized rays are subject to a different rate of *absorption*. Tourmalines and other crystals may be prepared in plates of such thickness that one ray is sensibly stopped and the other sensibly transmitted, and will then serve as polarizing (or analysing) apparatus. Although for practical purposes Nicol's prisms (Light, *Enc. Brit.* Vol. XIV. p. 612) are usually to be preferred, the phenomenon of double absorption is of great theoretical interest. The explanation is doubtless closely connected with that of double refraction.

§ 22. *Colours of Crystalline Plates.*

When polarized light is transmitted through a moderately thin plate of doubly refracting crystal, and is then analysed, *e.g.*, with a Nicol, brilliant colours are often exhibited, analogous in their character to the tints of Newton's scale. With his usual acuteness, Young at once attributed these colours to interference between the ordinary and extraordinary waves, and showed that the thickness of crystal required to develop a given tint, inversely proportional to the doubly refracting power, was in agreement with this view. But the complete explanation, demanding a fuller knowledge of the laws of interference of polarized light, was reserved for Fresnel and Arago. The subject is one which admits of great development*; but the interest turns principally upon the beauty of the effects, and upon the facility with which many of them may be obtained in experiment. We must limit ourselves to a brief treatment of one or two of the simpler cases.

* See Verdet's *Leçons*, Vol. II.

The incident vibration being plane-polarized, we will suppose that its plane makes an angle α with the principal plane of the crystal. On entering the crystal it is accordingly resolved into the two components represented by

$$\cos\alpha\cos\phi, \qquad \sin\alpha\cos\phi, \qquad \text{where } \phi = 2\pi t/\tau.$$

In traversing the crystal both waves are retarded, but we are concerned only with the difference of the retardations. Denoting the difference by ρ, we may take as the expressions of the waves on emergence

$$\cos\alpha\cos\phi, \qquad \sin\alpha\cos(\phi-\rho).$$

It may be remarked that, in the absence of *dispersion*, ρ would be inversely proportional to λ; but in fact there are many cases where it deviates greatly from this law.

Now let the plane of analysation be inclined at the angle β to that of primitive polarization (Fig. 25). Then for the sum of the two resolved components we have

$$\cos\alpha\cos(\alpha-\beta)\cos\phi + \sin\alpha\sin(\alpha-\beta)\cos(\phi-\rho),$$

of which the intensity is

$$\{\cos\alpha\cos(\alpha-\beta) + \sin\alpha\sin(\alpha-\beta)\cos\rho\}^2 + \sin^2\alpha\sin^2(\alpha-\beta)\sin^2\beta$$
$$= \cos^2\beta - \sin 2\alpha\sin 2(\alpha-\beta)\sin^2\tfrac{1}{2}\rho. \quad \ldots(1)$$

If in (1) we write $\beta+\frac{1}{2}\pi$ in place of β, we get

$$\sin^2\beta + \sin 2\alpha\sin 2(\alpha-\beta)\sin^2\tfrac{1}{2}\rho; \quad \ldots\ldots\ldots\ldots\ldots\ldots(2)$$

and we notice that the sum of (1) and (2) is unity under all circumstances. The effect of rotating the analyser through 90° is thus always to transform the tint into its complementary. The two complementary tints may be seen at the same time if we employ a double-image prism. In the absence of an analyser we may regard the two images as superposed, and there is no colour.

Fig. 25.

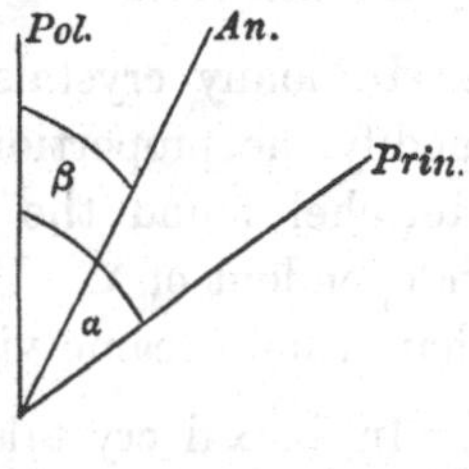

These expressions may be applied at once to the explanation of the colours of thin plates of mica or selenite. In this case the retardation ρ is proportional to the thickness, and approximately independent of the precise direction of the light, supposed to be nearly perpendicular to the plate, viz., nearly parallel to a principal axis of the crystal.

The most important cases are when $\beta=0$, $\beta=\frac{1}{2}\pi$. In the latter the field would be dark were the plate removed; and the actual intensity is

$$\sin^2 2\alpha\sin^2\tfrac{1}{2}\rho. \quad \ldots\ldots\ldots\ldots\ldots\ldots\ldots\ldots\ldots\ldots(3)$$

The composition of the light is thus independent of the azimuth of the plate (α); but the *intensity* varies greatly, vanishing four times during the complete revolution. The greatest brightness occurs when the principal plane bisects the angle between the planes of polarization and analysis. If $\beta = 0$, the light is complementary to that represented by (3).

If two plates be superposed, the retardations are added if the azimuths correspond; but they are subtracted if one plate be rotated relatively to the other through 90°. It is thus possible to obtain colour by the superposition of two nearly similar plates, although they may be too thick to answer the purpose separately.

If dispersion be neglected, the law of the colours in (3) is the same as that of the reflected tints of Newton's scale. The thicknesses of the plates of mica (acting by double refraction) and of air required to give the same colour are as 400 : 1. When a plate is too thick to show colour, its action may be analysed with the aid of a spectroscope.

Still thicker plates may be caused to exhibit colour, if the direction of the light within them makes but a small angle with an optic axis. Let us suppose that a plate of Iceland spar, or other uniaxal crystal (except quartz), cut perpendicularly to the axis, is interposed between the polarizing and analysing apparatus, and that the latter is so turned that the field is originally dark. The ray which passes perpendicularly is not doubly refracted, so that the centre of the field remains dark. At small angles to the optic axis the relative retardation is evidently proportional to the square of the inclination, so that the colours are disposed in concentric rings. But the intensity is not the same at the various parts of the circumference. In the plane of polarization and in the perpendicular plane there is no double refraction, or rather one of the refracted rays vanishes. Along the corresponding lines in the field of view there is no revival of light, and the ring system is seen to be traversed by a black cross.

In many crystals the influence of dispersion is sufficient to sensibly modify the proportionality of ρ to λ. In one variety of uniaxal apophyllite Herschel found the rings nearly achromatic, indicating that ρ was almost independent of λ. Under these circumstances a much larger number of rings than usual became visible.

In biaxal crystals, cut so that the surfaces are equally inclined to the optic axes, the rings take the form of lemniscates.

A medium originally isotropic may acquire the doubly refracting property under the influence of strain; and, if the strain be *homogeneous*, the conditions are optically identical with those found in a natural crystal. The principal axes of the wave-surface coincide with those of strain. If the strain be symmetrical, the medium is optically uniaxal. In general, if P, Q, R be the principal stresses, the difference of velocities for waves propagated parallel to R is evidently proportional to $(P - Q)$, and so on.

More often it happens that the strain is not homogeneous. Even then the small parts may be compared to crystals, but the optical constants vary from point to point. The comparatively feeble doubly refracting power thus developed in glass may best be made evident by the production of the colours of polarized light. Thus, in an experiment due to Brewster, a somewhat stout slab of glass, polished on the edges, is interposed between crossed Nicols. When the slab is bent in a plane perpendicular to that of vision, a revival of light takes place along the edges, where the elongation and contraction is greatest. If the width (in the direction of vision) be sufficient, the effect may be increased until the various colours of Newton's scale are seen. These colours vary from point to point of the thickness in the plane of bending, the "neutral axis" remaining dark. The optic axis, being everywhere coincident with the direction of elongation (or contraction), is parallel to the length of the slab. To this direction the plane of polarization should be inclined at about 45°.

The condition of internal strain is not necessarily due to forces applied from without. Thus, if glass originally free from strain be unequally heated, the accompanying expansions give rise to internal strains which manifest themselves in polarized light. If the heating be moderate, so as not to approach the softening point, the state of ease is recovered upon cooling, and the double refraction disappears. But if the local temperature be raised further, the hot parts may relieve themselves of the temporary strain, and then upon cooling they and other parts may be left in a condition of permanent strain. Sudden cooling of glass heated to the softening point leads to a similar result. The outer parts harden while the interior is still at a higher temperature, so that, when the whole is cooled down, the outside, being as it were too large for the inside, is in a condition of radial tension and circumferential compression. An examination in polarized light shows that the strains thus occasioned are often very severe. If any small part be relieved by fracture from the constraint exercised upon it by the remainder, the doubly refracting property almost or wholly disappears. In this respect unannealed glass differs essentially from a crystal, all parts of which are similar and independent. It may be remarked that it is difficult to find large pieces of glass so free from internal strain as to show no revival of light when examined between crossed Nicols.

§ 23. *Rotatory Polarization.*

In general a polarized ray travelling along the axis of a uniaxal crystal undergoes no change; but it was observed by Arago that, if quartz be used in this experiment, the plane of polarization is found to be rotated through an angle proportional to the thickness of crystal traversed. The subject was further studied by Biot, who ascertained that the rotation due to a given thickness is inversely as the square of the wave-length of the light, thus varying very rapidly with the colour. In some specimens of quartz (called in consequence right-handed) the rotation is to the right, while in others it is to the left. Equal thicknesses of right- and left-handed quartz may thus compensate one another.

Fresnel has shown that the rotation of the plane may be interpreted as indicating a different velocity of propagation of the two circularly-polarized components into which plane-polarized light may always be resolved. In ordinary media the right- and left-handed circularly-polarized rays travel at the same speed, and at any stage of their progress recompound a ray rectilinearly-polarized in a fixed direction. But it is otherwise if the velocities of propagation of the circular components be even slightly different.

The first circularly-polarized wave may be expressed by

$$\xi_1 = r\cos(nt - k_1 z), \qquad \eta_1 = r\sin(nt - k_1 z)\,; \quad \text{............(1)}$$

and the second (of equal amplitude) by

$$\xi_2 = r\cos(nt - k_2 z), \qquad \eta_2 = -r\sin(nt - k_2 z). \quad \text{.........(2)}$$

The resultant of (1) and (2) is

$$\xi = \xi_1 + \xi_2 = 2r\cos\tfrac{1}{2}(k_2 - k_1)\,z\,.\cos\{nt - \tfrac{1}{2}(k_1 + k_2)\,z\},$$

$$\eta = \eta_1 + \eta_2 = 2r\sin\tfrac{1}{2}(k_2 - k_1)\,z\,.\cos\{nt - \tfrac{1}{2}(k_1 + k_2)\,z\}\,;$$

so that

$$\eta/\xi = \tan\tfrac{1}{2}(k_2 - k_1)\,z, \quad \text{..........................(3)}$$

which shows that for any fixed value of z the light is plane-polarized. The direction of this plane, however, varies with z. Thus, if $\eta/\xi = \tan\theta$, so that θ gives the angular position of the plane in reference to ξ, we have

$$\theta = \tfrac{1}{2}(k_2 - k_1)\,z, \quad \text{........................ .....(4)}$$

indicating a rotation proportional to z. The quantities k_1, k_2 are inversely as the wave-lengths of the two circular components for the same periodic time. When the relative retardation amounts to an entire period, $(k_2 - k_1)\,z = 2\pi$, and then, by (4), $\theta = \pi$. The revolution of the plane through two right angles restores the original state of polarization. In quartz the rotation is very rapid, amounting in the case of yellow light to about 24° for each millimetre traversed.

It is interesting to observe with what a high degree of accuracy the comparison of the velocities of the two waves can be effected. If the plane of polarization be determined to one minute of angle, a relative retardation of $\lambda/10800$ is made manifest. If l be the thickness traversed, v and $v + \delta v$ the two velocities, the relative retardation is $l\delta v/v$. To take an example, suppose that $l = 20$ inches, $\lambda = \frac{1}{40000}$ inch; so that if $\delta v/v$ exceed 10^{-8}, the fact might be detected. [inch = 2·54 cm.]

In quartz the rotation of the plane depends upon the crystalline structure, but there are many liquids, *e.g.*, oil of turpentine and common syrup, which exhibit a like effect. In such cases the rotation is of course independent of the direction of the light; it must be due to some peculiarity in the constitution of the molecules.

A remarkable connexion has been observed between the rotatory property and the crystalline form. Thus Herschel found that in many specimens the right-handed and left-handed varieties of quartz could be distinguished by the disposition of certain subordinate faces. The crystals of opposite kinds are symmetrical in a certain sense, but are yet not *superposable.* The difference is like that between otherwise similar right- and left-handed screws. The researches of Pasteur upon the rotatory properties of tartaric acid have opened up a new and most interesting field of chemistry. At that time two isomeric varieties were known,—ordinary tartaric acid, which rotates to the right, and racemic acid, which is optically inactive, properties of the acids shared also by the salts. Pasteur found that the crystals of tartaric acid and of the tartrates possessed a right-handed structure, and endeavoured to discover corresponding bodies with a left-handed structure. After many trials crystallizations of the double racemate of soda and ammonia were obtained, including crystals of opposite kinds. A selection of the right-handed specimens yielded ordinary dextro-tartaric acid, while a similar selection of the left-handed crystals gave a new variety—lævo-tartaric acid, rotating the plane of polarization to the left in the same degree as ordinary tartaric acid rotates it to the right. A mixture in equal proportions of the two kinds of tartaric acid, which differ scarcely at all in their chemical properties*, reconstitutes racemic acid.

The possibility of inducing the rotatory property in bodies otherwise free from it was one of the finest of Faraday's discoveries. He found that, if heavy glass, bisulphide of carbon, &c., are placed in a magnetic field, a ray of polarized light, propagated along the lines of magnetic force, suffers rotation. The laws of the phenomenon were carefully studied by Verdet, whose conclusions may be summed up by saying that in a given medium the rotation of the plane for a ray proceeding in any direction is proportional to the difference of magnetic potential at the initial and final points. In bisulphide of carbon, at 18° and for a difference of potential equal to unity C.G.S., the rotation of the plane of polarization of a ray of soda light is ·04202 minute of angle†.

A very important distinction should be noted between the magnetic rotation and that natural to quartz, syrup, &c. In the latter the rotation is

* It would seem that the two varieties could be chemically distinguished only by their relations with bodies themselves right-handed or left-handed.

† *Phil. Trans.* 1885, p. 343. [Vol. II. p. 377.]

always right-handed or always left-handed with respect to the direction of the ray. Hence when the ray is reversed the absolute direction of rotation is reversed also. A ray which traverses a plate of quartz in one direction, and then after reflexion traverses the same thickness again in the opposite direction, recovers its original plane of polarization. It is quite otherwise with the rotation under magnetic force. In this case the rotation is in the same absolute direction even though the ray be reversed. Hence, if a ray be reflected backwards and forwards any number of times along a line of magnetic force, the rotations due to the several passages are all accumulated. The non-reversibility of light in a magnetized medium proves the case to be of a very exceptional character, and (as was argued by Thomson) indicates that the magnetized medium is itself in rotatory motion independently of the propagation of light through it*.

The importance of polarimetric determinations has led to the contrivance of various forms of apparatus adapted to the special requirements of the case. If the light be bright enough, fairly accurate measurements may be made by merely rotating a Nicol until the field appears dark. Probably the best form of analyser, when white light is used and the plane is the same for all the coloured components, is the Jellet†, formed by the combination of two portions of Iceland spar. By this instrument the field of view is duplicated, and the setting is effected by turning it until the two portions of the field, much reduced in brightness, appear *equally* dark. A similar result is attained in the Laurent, which, however, is only applicable to homogeneous light. In this apparatus, advantage is taken of the action of a half-wave plate. In passing such a plate the plane of polarization is as it were *reflected* by the principal section, that is, rotated until it makes the same angle with the principal section as at first, but upon the further side. The plate covers only half of the field of view, and the eye is focused upon the dividing edge. The planes of polarization of the two halves of the field are different, unless the original plane be parallel (or perpendicular) to the principal section. In the Laurent analyser the half-wave plate is rigidly combined with a Nicol in such a position that the principal section of the latter makes a small but finite angle with that of the plate. The consequence is that the two halves of the field of view cannot be blackened simultaneously, but are rendered equally dark when the instrument is so turned that the principal section of the plate is parallel to the plane of original polarization, which is also that of the uncovered half of the field. A slight rotation in either direction darkens one half of the field and brightens the other half.

In another form of "half-shade" polarimeter, invented by Poynting, the half-wave plate of the Laurent is dispensed with, a small rotation of one half

* Maxwell's *Electricity and Magnetism*, Vol. II. chap. XXI.

† A description is given in Glazebrook's *Physical Optics*, London 1883.

of the field with respect to the other half being obtained by quartz (cut perpendicularly to the axis) or by syrup. In the simplest construction the syrup is contained in a small cell with parallel glass sides, and the division into two parts is effected by the insertion of a small piece of plate glass about $\frac{3}{16}$ inch thick, a straight edge of which forms the dividing line. If the syrup be strong, the difference of thickness of $\frac{3}{16}$ inch gives a relative rotation of about 2°. In this arrangement the sugar cell is a fixture, and only the Nicol rotates. The reading of the divided circle corresponds to the mean of the planes for the two halves of the field, and this of course differs from the original position of the plane before entering the sugar. This circumstance is usually of no importance, the object being to determine the *rotation* of the plane of polarization when some of the conditions are altered.

A discussion of the accuracy obtainable in polarimetry will be found in a recent paper by Lippich*.

In Soleil's apparatus, designed for practical use in the estimation of the strength of sugar solutions, the rotation due to the sugar is compensated by a wedge of quartz. Two wedges, one of right-handed and the other of left-handed quartz, may be fitted together, so that a movement of the combination in either direction increases the thickness of one variety traversed and diminishes that of the other. The linear movement required to compensate the introduction of a tube of syrup measures the quantity of sugar present.

§ 24. *Dynamical Theory of Diffraction.*

The explanation of diffraction phenomena given by Fresnel and his followers is independent of special views as to the nature of the ether, at least in its main features; but in the absence of a more complete foundation it is impossible to treat rigorously the mode of action of a solid obstacle such as a screen. The full solution of problems of this kind is scarcely to be expected. Even in the much simpler case of sound, where we know what we have to deal with, the mathematical difficulties are formidable; and we are not able to solve even such an apparently elementary question as the transmission of sound past a rigid infinitely thin plane screen, bounded by a straight edge†, or perforated with a circular aperture. But, without entering upon matters of this kind, we may inquire in what manner a primary wave may be resolved into elementary secondary waves, and in particular as to the law of intensity and polarization in a secondary wave as dependent upon its direction of propagation, and upon the character as regards polarization of the primary wave. This question is treated by Stokes in his "Dynamical Theory of Diffraction"‡ on the basis of the elastic solid theory.

* *Wien. Ber.* LXXXV. 9th Feb. 1882. See also *Phil. Trans.* 1885, p. 360. [Vol. II. p. 378.]

† [1901. We owe to Sommerfeld some advance in this direction.]

‡ *Camb. Phil. Trans.* Vol. IX. p. 1; Stokes' *Collected Papers*, Vol. II. p. 243.

Let x, y, z be the coordinates of any particle of the medium in its natural state, and ξ, η, ζ the displacements of the same particle at the end of time t, measured in the directions of the three axes respectively. Then the first of the equations of motion may be put under the form

$$\frac{d^2\xi}{dt^2} = b^2\left(\frac{d^2\xi}{dx^2} + \frac{d^2\xi}{dy^2} + \frac{d^2\xi}{dz^2}\right) + (a^2 - b^2)\frac{d}{dx}\left(\frac{d\xi}{dx} + \frac{d\eta}{dy} + \frac{d\zeta}{dz}\right),$$

where a^2 and b^2 denote the two arbitrary constants. Put for shortness

$$\frac{d\xi}{dx} + \frac{d\eta}{dy} + \frac{d\zeta}{dz} = \delta, \quad\text{.............................(1)}$$

and represent by $\nabla^2\xi$ the quantity multiplied by b^2. According to this notation, the three equations of motion are

$$\left.\begin{aligned} \frac{d^2\xi}{dt^2} &= b^2\nabla^2\xi + (a^2 - b^2)\frac{d\delta}{dx} \\ \frac{d^2\eta}{dt^2} &= b^2\nabla^2\eta + (a^2 - b^2)\frac{d\delta}{dy} \\ \frac{d^2\zeta}{dt^2} &= b^2\nabla^2\zeta + (a^2 - b^2)\frac{d\delta}{dz} \end{aligned}\right\} \quad\text{.....................(2)}$$

It is to be observed that δ denotes the dilatation of volume of the element situated at (x, y, z). In the limiting case in which the medium is regarded as absolutely incompressible δ vanishes; but, in order that equations (2) may preserve their generality, we must suppose a at the same time to become infinite, and replace $a^2\delta$ by a new function of the coordinates.

These equations simplify very much in their application to plane waves. If the ray be parallel to OX, and the direction of vibration parallel to OZ, we have $\xi = 0$, $\eta = 0$, while ζ is a function of x and t only. Equation (1) and the first pair of equations (2) are thus satisfied identically. The third equation gives

$$\frac{d^2\zeta}{dt^2} = b^2\frac{d^2\zeta}{dx^2}, \quad\text{..............................(3)}$$

of which the solution is

$$\zeta = f(bt - x), \quad\text{..............................(4)}$$

where f is an arbitrary function.

The question as to the law of the secondary waves is thus answered by Stokes. "Let $\xi = 0$, $\eta = 0$, $\zeta = f(bt - x)$ be the displacements corresponding to the incident light; let O_1 be any point in the plane P (of the wave-front), dS an element of that plane adjacent to O_1; and consider the disturbance due to that portion only of the incident disturbance which passes continually across dS. Let O be any point in the medium situated at a distance from the point O_1 which is large in comparison with the length of a wave; let $O_1O = r$, and let this line make an angle θ with the direction of propagation of the incident light, or the axis of x, and ϕ with the direction of vibration, or axis of z. Then the displacement at O will take place in a direction perpendicular to O_1O, and lying in the plane ZO_1O; and, if ζ' be the displacement at O, reckoned positive in the direction nearest to that in which the incident vibrations are reckoned positive,

$$\zeta' = \frac{dS}{4\pi r}(1 + \cos\theta)\sin\phi\, f'(bt - r).$$

In particular, if

$$f(bt - x) = c\sin\frac{2\pi}{\lambda}(bt - x), \quad \text{.....................(5)}$$

we shall have

$$\zeta' = \frac{c\,dS}{2\lambda r}(1 + \cos\theta)\sin\phi\cos\frac{2\pi}{\lambda}(bt - r). \quad \text{...........(6)''}$$

It is then verified that, after integration with respect to dS, (6) gives the same disturbance as if the primary wave had been supposed to pass on unbroken.

The occurrence of $\sin\phi$ as a factor in (6) shows that the relative intensities of the primary light and of that diffracted in the direction θ depend upon the condition of the former as regards polarization. If the direction of primary vibration be perpendicular to the plane of diffraction (containing both primary and secondary rays), $\sin\phi = 1$; but, if the primary vibration be in the plane of diffraction, $\sin\phi = \cos\theta$. This result was employed by Stokes as a criterion of the direction of vibration; and his experiments, conducted with gratings, led him to the conclusion that the vibrations of polarized light are executed in a direction *perpendicular* to the plane of polarization.

The factor $(1 + \cos\theta)$ shows in what manner the secondary disturbance depends upon the direction in which it is propagated with respect to the front of the primary wave.

If, as suffices for all practical purposes, we limit the application of the formulæ to points in advance of the plane at which the wave is supposed to be broken up, we may use simpler methods of resolution than that above considered. It appears indeed that the purely mathematical question has no definite answer. In illustration of this the analogous problem for sound may be referred to. Imagine a flexible lamina to be introduced so as to coincide with the plane at which resolution is to be effected. The introduction of the lamina (supposed to be devoid of inertia) will make no difference to the propagation of plane parallel sonorous waves through the position which it occupies. At every point the motion of the lamina will be the same as would have occurred in its absence, the pressure of the waves impinging from behind being just what is required to generate the waves in front. Now it is evident that the aerial motion in front of the lamina is determined by what happens at the lamina without regard to the cause of the motion there existing. Whether the necessary forces are due to aerial pressures acting on the rear, or to forces directly impressed from without, is a matter of indifference. The conception of the lamina leads immediately to two schemes, according to which a primary wave may be supposed to be broken up. In the first of these the element dS, the effect of which is to be estimated, is supposed to execute its actual motion, while every other element of the plane lamina is maintained at rest. The resulting aerial motion in front is readily calculated*; it is symmetrical with respect to the origin, *i.e.*, independent of θ. When the secondary disturbance thus obtained is integrated with respect to dS over the entire plane of the lamina, the result is necessarily the same as would have been obtained had the primary wave been supposed to pass on without resolution, for this is precisely the motion generated when every element of the lamina vibrates with a common motion, equal to that attributed to dS. The only assumption here involved is the evidently legitimate one that, when two systems of variously distributed motion at the lamina are superposed, the corresponding motions in front are superposed also.

The method of resolution just described is the simplest, but it is only one of an indefinite number that might be proposed, and which are all equally legitimate, so long as the question is regarded as a merely mathematical one, without reference to the physical properties of actual screens. If, instead of supposing the *motion* at dS to be that of the primary wave, and to be zero elsewhere, we suppose the *force* operative over the element dS of the lamina to be that corresponding to the primary wave, and to vanish elsewhere, we obtain a secondary wave following quite a different law†. In this case the motion in different directions varies as $\cos\theta$, vanishing at right angles to the direction of propagation of the primary wave. Here again, on integration over the entire lamina, the aggregate effect of the secondary waves is necessarily the same as that of the primary.

In order to apply these ideas to the investigation of the secondary wave of light, we require the solution of a problem, first treated by Stokes‡, viz., the determination of the motion in an infinitely extended elastic solid due to a locally applied periodic force. If we suppose that the force impressed upon the element of mass $D\,dx\,dy\,dz$ is

$$DZ\,dx\,dy\,dz;$$

being everywhere parallel to the axis of Z, the only change required in our equations (1), (2) is the addition of the term Z to the second member of the third equation (2). In the forced vibration, now under consideration, Z, and

* *Theory of Sound*, § 278. † *Loc. cit.* equation (10).

‡ *Loc. cit.* §§ 27—30.

the quantities ξ, η, ζ, δ expressing the resulting motion, are to be supposed proportional to e^{int}, where $i=\sqrt{(-1)}$, and $n=2\pi/\tau$, τ being the periodic time. Under these circumstances the double differentiation with respect to t of any quantity is equivalent to multiplication by the factor $-n^2$, and thus our equations take the form

$$\left.\begin{aligned}(b^2\nabla^2+n^2)\,\xi+(a^2-b^2)\frac{d\delta}{dx}&=0\\(b^2\nabla^2+n^2)\,\eta+(a^2-b^2)\frac{d\delta}{dy}&=0\\(b^2\nabla^2+n^2)\,\zeta+(a^2-b^2)\frac{d\delta}{dz}&=-Z\end{aligned}\right\}. \quad \text{............(7)}$$

It will now be convenient to introduce the quantities $\varpi_1, \varpi_2, \varpi_3$, which express the *rotations* of the elements of the medium round axes parallel to those of coordinates, in accordance with the equations

$$\varpi_3=\frac{d\xi}{dy}-\frac{d\eta}{dx},\qquad \varpi_1=\frac{d\eta}{dz}-\frac{d\zeta}{dy},\qquad \varpi_2=\frac{d\zeta}{dx}-\frac{d\xi}{dz}.\text{......(8)}$$

In terms of these we obtain from (7), by differentiation and subtraction,

$$\left.\begin{aligned}(b^2\nabla^2+n^2)\,\varpi_3&=0\\(b^2\nabla^2+n^2)\,\varpi_1&=dZ/dy\\(b^2\nabla^2+n^2)\,\varpi_2&=-dZ/dx\end{aligned}\right\}. \quad \text{....................(9)}$$

The first of equations (9) gives

$$\varpi_3=0. \text{..................................(10)}$$

For ϖ_1 we have

$$\varpi_1=-\frac{1}{4\pi b^2}\iiint\frac{dZ}{dy}\frac{e^{-ikr}}{r}\,dxdydz, \quad \text{....................(11)*}$$

where r is the distance between the element $dxdydz$ and the point where ϖ_1 is estimated, and

$$k=n/b=2\pi/\lambda, \quad \text{..............................(12)}$$

λ being the wave-length.

We will now introduce the supposition that the force Z acts only within a small space of volume T, situated at (x, y, z), and for simplicity suppose that it is at the origin of coordinates that the rotations are to be estimated. Integrating by parts in (11), we get

$$\int\frac{e^{-ikr}}{r}\frac{dZ}{dy}\,dy=\left[Z\frac{e^{-ikr}}{r}\right]-\int Z\frac{d}{dy}\left(\frac{e^{-ikr}}{r}\right)dy,$$

in which the integrated terms at the limits vanish, Z being finite only within the region T. Thus

$$\varpi_1=\frac{1}{4\pi b^2}\iiint Z\,\frac{d}{dy}\left(\frac{e^{-ikr}}{r}\right)dxdydz.$$

* This solution may be verified in the same manner as Poisson's theorem, in which $k=0$.

Since the dimensions of T are supposed to be very small in comparison with λ, the factor $\frac{d}{dy}\left(\frac{e^{-ikr}}{r}\right)$ is sensibly constant; so that, if Z stand for the mean value of Z over the volume T, we may write

$$\varpi_1 = \frac{TZ}{4\pi b^2}\cdot\frac{y}{r}\cdot\frac{d}{dr}\left(\frac{e^{-ikr}}{r}\right). \quad\text{.......................(13)}$$

In like manner we find

$$\varpi_2 = -\frac{TZ}{4\pi b^2}\cdot\frac{x}{r}\cdot\frac{d}{dr}\left(\frac{e^{-ikr}}{r}\right).\quad\text{.......................(14)}$$

From (10), (13), (14) we see that, as might have been expected, the rotation at any point is about an axis perpendicular both to the direction of the force and to the line joining the point to the source of disturbance. If the resultant rotation be ϖ, we have

$$\varpi = \frac{TZ}{4\pi b^2}\cdot\frac{\sqrt{(x^2+y^2)}}{r}\cdot\frac{d}{dr}\left(\frac{e^{-ikr}}{r}\right) = \frac{TZ\sin\phi}{4\pi b^2}\,\frac{d}{dr}\left(\frac{e^{-ikr}}{r}\right),$$

ϕ denoting the angle between r and z. In differentiating e^{-ikr}/r with respect to r, we may neglect the term divided by r^2 as altogether insensible, kr being an exceedingly great quantity at any moderate distance from the origin of disturbance. Thus

$$\varpi = -\frac{ik\,.\,TZ\sin\phi}{4\pi b^2}\cdot\frac{e^{-ikr}}{r}, \quad\text{.....................(15)}$$

which completely determines the rotation at any point. For a disturbing force of given integral magnitude it is seen to be everywhere about an axis perpendicular to r and to the direction of the force, and in magnitude dependent only upon the angle (ϕ) between these two directions and upon the distance (r).

The intensity of light is, however, more usually expressed in terms of the actual displacement in the plane of the wave. This displacement, which we may denote by ζ', is in the plane containing z and r, and perpendicular to the latter. Its connexion with ϖ is expressed by $\varpi = d\zeta'/dr$; so that

$$\zeta' = \frac{TZ\sin\phi}{4\pi b^2}\cdot\frac{e^{i(nt-kr)}}{r}, \quad\text{.......................(16)}$$

where the factor e^{int} is restored.

Retaining only the real part of (16), we find, as the result of a local application of force equal to

$$DTZ\cos nt, \quad\text{............................(17)}$$

the disturbance expressed by

$$\zeta' = \frac{TZ\sin\phi}{4\pi b^2}\cdot\frac{\cos(nt-kr)}{r}. \quad\text{....................(18)}$$

The occurrence of $\sin\phi$ shows that there is no disturbance radiated in the direction of the force, a feature which might have been anticipated from considerations of symmetry.

We will now apply (18) to the investigation of a law of secondary disturbance, when a primary wave

$$\zeta = \sin(nt - kx) \quad \text{.........................(19)}$$

is supposed to be broken up in passing the plane $x = 0$. The first step is to calculate the force which represents the reaction between the parts of the medium separated by $x = 0$. The force operative upon the positive half is parallel to OZ, and of amount per unit of area equal to

$$-b^2 D d\zeta/dx = b^2 kD \cos nt;$$

and to this force acting over the whole of the plane the actual motion on the positive side may be conceived to be due. The secondary disturbance corresponding to the element dS of the plane may be supposed to be that caused by a force of the above magnitude acting over dS and vanishing elsewhere; and it only remains to examine what the result of such a force would be.

Now it is evident that the force in question, supposed to act upon the positive half only of the medium, produces just double of the effect that would be caused by the same force if the medium were undivided, and on the latter supposition (being also localized at a point) it comes under the head already considered. According to (18), the effect of the force acting at dS parallel to OZ, and of amount equal to

$$2b^2 kD\, dS \cos nt,$$

will be a disturbance

$$\zeta' = \frac{dS \sin\phi}{\lambda r} \cos(nt - kr), \quad \text{....................(20)}$$

regard being had to (12). This therefore expresses the secondary disturbance at a distance r and in a direction making an angle ϕ with OZ (the direction of primary vibration) due to the element dS of the wave-front.

The proportionality of the secondary disturbance to $\sin\phi$ is common to the present law and to that given by Stokes, but here there is no dependence upon the angle θ between the primary and secondary rays. The occurrence of the factor $(\lambda r)^{-1}$, and the necessity of supposing the phase of the secondary wave accelerated by a quarter of an undulation, were first established by Archibald Smith, as the result of a comparison between the primary wave, supposed to pass on without resolution, and the integrated effect of all the secondary waves (§ 10). The occurrence of factors such as $\sin\phi$, or $\frac{1}{2}(1 + \cos\theta)$, in the expression of the secondary wave has no influence upon the result of the integration, the effects of all the elements for which the factors differ appreciably from unity being destroyed by mutual interference.

The choice between various methods of resolution, all mathematically admissible, would be guided by physical considerations respecting the mode of action of obstacles. Thus, to refer again to the acoustical analogue in which plane waves are incident upon a perforated rigid screen, the circumstances of the case are best represented by the first method of resolution, leading to symmetrical secondary waves, in which the normal motion is supposed to be zero over the unperforated parts. Indeed, if the aperture is very small, this method gives the correct result, save as to a constant factor. In like manner our present law (20) would apply to the kind of obstruction that would be caused by an actual physical division of the elastic medium, extending over the whole of the area supposed to be occupied by the intercepting screen, but of course not extending to the parts supposed to be perforated. In the present state of our ignorance this law seems to be at least as plausible as any other.

§ 25. *The Diffraction of Light by Small Particles.*

The theory of the diffraction, dispersion, or scattering of light by small particles, as it has variously been called, is of importance, not only from its bearings upon fundamental optical hypotheses, but on account of its application to explain the origin and nature of the light from the sky. The view, suggested by Newton and advocated in more recent times by such authorities as Herschel* and Clausius†, that the light of the sky is a blue of the first order reflected from aqueous particles, was connected with the then prevalent notion that the suspended moisture of clouds and mists was in the form of vesicles or bubbles. Experiments such as those of Brücke‡ pointed to a different conclusion. When a weak alcoholic solution of mastic is agitated with water, the precipitated gum scatters a blue light, obviously similar in character to that from the sky. Not only would it be unreasonable to attribute a vesicular structure to the mastic, but (as Brücke remarked) the dispersed light is much richer in quality than the blue of the first order. Another point of great importance is well brought out in the experiments of Tyndall§ upon clouds precipitated by the chemical action of light. Whenever the particles are sufficiently fine, the light emitted laterally is blue in colour, and, in a direction perpendicular to the incident beam, is *completely polarized.*

About the colour there can be no *primâ facie* difficulty; for, as soon as the question is raised, it is seen that the standard of linear dimension, with

* Article "Light," *Enc. Metrop.* 1830, § 1143.

† *Pogg. Ann.* Vols. LXXII. LXXVI. LXXXVIII.; *Crelle*, Vols. XXXIV. XXXVI.

‡ *Pogg. Ann.* Vol. LXXXIII.

§ *Phil. Mag.* [4], Vol. CXXXVII. p. 388.

reference to which the particles are called small, is the wave-length of light, and that a given set of particles would (on any conceivable view as to their mode of action) produce a continually increasing disturbance as we pass along the spectrum towards the more refrangible end.

On the other hand, that the direction of complete polarization should be independent of the refracting power of the matter composing the cloud has been considered mysterious. Of course, on the theory of thin plates, this direction would be determined by Brewster's law; but, if the particles of foreign matter are small in all their dimensions, the circumstances are materially different from those under which Brewster's law is applicable.

The investigation of this question upon the elastic solid theory will depend upon how we suppose the solid to vary from one optical medium to another. The slower propagation of light in glass or water than in air or vacuum may be attributed to a greater density, or to a less rigidity, in the former case; or we may adopt the more complicated supposition that both these quantities vary, subject only to the condition which restricts the ratio of velocities to equality with the known refractive index. It will presently appear that the original hypothesis of Fresnel, that the rigidity remains the same in both media, is the only one that can be reconciled with the facts; and we will therefore investigate upon this basis the nature of the secondary waves dispersed by small particles.

Conceive a beam of plane-polarized light to move among a number of particles, all small compared with any of the wave-lengths. According to our hypothesis, the foreign matter may be supposed to *load* the ether, so as to increase its *inertia* without altering its resistance to distortion. If the particles were away, the wave would pass on unbroken and no light would be emitted laterally. Even with the particles retarding the motion of the ether, the same will be true if, to counterbalance the increased inertia, suitable forces are caused to act on the ether at all points where the inertia is altered. These forces have the same period and direction as the undisturbed luminous vibrations themselves. The light actually emitted laterally is thus the same as would be caused by forces exactly the opposite of these acting on the medium otherwise free from disturbance, and it only remains to see what the effect of such force would be.

On account of the smallness of the particles, the forces acting throughout the volume of any individual particle are all of the same intensity and direction, and may be considered as a whole. The determination of the motion in the ether, due to the action of a periodic force at a given point, is a problem with which we have recently been occupied (§ 24). But, before applying the solution to a mathematical investigation of the present question, it may be well to consider the matter for a few moments from a more general point of view.

In the first place. there is necessarily a complete symmetry round the direction of the force. The disturbance, consisting of transverse vibrations, is propagated outwards in all directions from the centre; and, in consequence of the symmetry, the direction of vibration in any ray lies in the plane containing the ray and the axis of symmetry; that is to say, the direction of vibration in the scattered or diffracted ray makes with the direction of vibration in the incident or primary ray the least possible angle. The symmetry also requires that the intensity of the scattered light should vanish for the ray which would be propagated along the axis; for there is nothing to distinguish one direction transverse to the ray from another. The application of this is obvious. Suppose, for distinctness of statement, that the primary ray is vertical, and that the plane of vibration is that of the meridian. The intensity of the light scattered by a small particle is constant, and a maximum, for rays which lie in the vertical plane running east and west, while there is *no scattered ray along the north and south line.* If the primary ray is unpolarized, the light scattered north and south is entirely due to that component which vibrates east and west, and is therefore *perfectly polarized,* the direction of its vibration being also east and west. Similarly any other ray scattered horizontally is perfectly polarized, and the vibration is performed in the horizontal plane. In other directions the polarization becomes less and less complete as we approach the vertical.

The observed facts as to polarization are thus readily explained, and the general law connecting the intensity of the scattered light with the wave-length follows almost as easily from considerations of *dimensions.*

The object is to compare the intensities of the incident and scattered light, for these will clearly be proportional. The number (i) expressing the ratio of the two amplitudes is a function of the following quantities:—(T) the volume of the disturbing particle; (r) the distance of the point under consideration from it; (λ) the wave-length; (b) the velocity of propagation of light; (D) and (D') the original and altered densities: of which the first three depend only upon space, the fourth on space and time, while the fifth and sixth introduce the consideration of mass. Other elements of the problem there are none, except mere numbers and angles, which do not depend upon the fundamental measurements of space, time, and mass. Since the ratio (i), whose expression we seek, is of no dimensions in mass, it follows at once that D and D' occur only under the form $D:D'$, which is a simple number and may therefore be disregarded. It remains to find how i varies with T, r, λ, b.

Now, of these quantities, b is the only one depending on time; and therefore, as i is of no dimensions in time, b cannot occur in its expression.

Moreover, since the same amount of energy is propagated across all spheres concentric with the particle, we recognize that i varies as r. It is equally evident that i varies as T, and therefore that it must be proportional to $T/\lambda^2 r$, T being of three dimensions in space. In passing from one part of the spectrum to another, λ is the only quantity which varies, and we have the important law:—

When light is scattered by particles which are very small compared with any of the wave-lengths, the ratio of the amplitudes of the vibrations of the scattered and incident lights varies inversely as the square of the wave-length, and the ratio of *intensities* as the inverse fourth power.

The light scattered from small particles is of a much richer blue than the blue of the first order as reflected from a very thin plate. From the general theory (§ 8), or by the method of dimensions, it is easy to prove that in the latter case the intensity varies as λ^{-2}, instead of λ^{-4}.

The principle of energy makes it clear that the light emitted laterally is not a new creation, but only diverted from the main stream. If I represent the intensity of the primary light after traversing a thickness x of the turbid medium, we have

$$dI = -hI\lambda^{-4}\,dx,$$

where h is a constant independent of λ. On integration,

$$\log(I/I_0) = -h\lambda^{-4}x, \quad \text{..........................(1)}$$

if I_0 correspond to $x = 0$,—a law altogether similar to that of absorption, and showing how the light tends to become yellow and finally red as the thickness of the medium increases*.

Captain Abney has found that the above law agrees remarkably well with his observations on the transmission of light through water in which particles of mastic are suspended†.

We may now investigate the mathematical expression for the disturbance propagated in any direction from a small particle upon which a beam of light strikes. Let the particle be at the origin of coordinates, and let the expression for the primary vibration be

$$\zeta = \sin(nt - kx). \quad \text{..............................(2)}$$

The acceleration of the element at the origin is $-n^2 \sin nt$; so that the force which would have to be applied to the parts where the density is D' (instead of D), in order that the waves might pass on undisturbed, is per unit of volume

$$-(D' - D)\,n^2 \sin nt.$$

To obtain the total force which must be supposed to act, the factor T (representing the volume of the particle) must be introduced. The opposite

* "On the Light from the Sky, its Polarization and Colour," *Phil. Mag.* Feb. 1871.

† *Proc. Roy. Soc.* May 1886.

of this, conceived to act at O, would give the same disturbance as is actually caused by the presence of the particle. Thus by (18) (§ 24) the secondary disturbance is expressed by

$$\zeta' = \frac{D'-D}{D}\frac{n^2 T \sin\phi}{4\pi b^2}\frac{\sin(nt-kr)}{r}$$

$$= \frac{D'-D}{D}\frac{\pi T\sin\phi}{\lambda^2 r}\sin(nt-kr). \quad \text{.................(3)*}$$

The preceding investigation is based upon the assumption that in passing from one medium to another the rigidity of the ether does not change. If we forego this assumption, the question is necessarily more complicated; but, on the supposition that the changes of rigidity (ΔN) and of density (ΔD) are relatively small, the results are fairly simple. If the primary wave be represented by

$$\zeta = e^{-ikx}, \quad \text{....................................(4)}$$

the component rotations in the secondary wave are

$$\left.\begin{aligned} \varpi_3 &= P\left(-\frac{\Delta N}{N}\frac{yz}{r^2}\right) \\ \varpi_1 &= P\left(\frac{\Delta D}{D}\frac{y}{r}+\frac{\Delta N}{N}\frac{xy}{r^2}\right) \\ \varpi_2 &= P\left(-\frac{\Delta D}{D}\frac{x}{r}+\frac{\Delta N}{N}\frac{z^2-x^2}{r^2}\right) \end{aligned}\right\}, \quad \text{.................(5)}$$

where

$$P = \frac{ik^3 T}{4\pi}\frac{e^{-ikr}}{r}. \quad \text{.............................(6)}$$

The expression for the resultant rotation in the general case would be rather complicated, and is not needed for our purpose. It is easily seen to be about an axis perpendicular to the scattered ray (x, y, z), inasmuch as

$$x\varpi_1 + y\varpi_2 + z\varpi_3 = 0.$$

Let us consider the more special case of a ray scattered normally to the incident ray, so that $x = 0$. We have

$$\varpi^2 = \varpi_1{}^2 + \varpi_2{}^2 + \varpi_3{}^2 = P^2\left(\frac{\Delta N}{N}\right)^2\frac{z^2}{r^2} + P^2\left(\frac{\Delta D}{D}\right)^2\frac{y^2}{r^2}. \quad \text{.........(7)}$$

If ΔN, ΔD be both finite, we learn from (7) that there is no direction perpendicular to the primary (polarized) ray in which the secondary light vanishes. Now experiment tells us plainly that there is such a direction, and therefore we are driven to the conclusion that either ΔN or ΔD must vanish.

* In strictness the force must be supposed to act upon the medium in its actual condition, whereas in (18) the medium is supposed to be absolutely uniform. It is not difficult to prove that (3) remains unaltered, when this circumstance is taken into account; and it is evident in any case that a correction would depend upon the square of $(D'-D)$.

The consequences of supposing ΔN to be zero have already been traced. They agree very well with experiment, and require us to suppose that the vibrations are perpendicular to the plane of polarization. So far as (7) is concerned, the alternative supposition that ΔD vanishes would answer equally well, if we suppose the vibrations to be executed in the plane of polarization; but let us now revert to (5), which gives

$$\varpi_3 = -\frac{P\Delta N}{N}\frac{yz}{r^2}, \qquad \varpi_1 = +\frac{P\Delta N}{N}\frac{xy}{r^2}, \qquad \varpi_2 = +\frac{P\Delta N}{N}\frac{z^2-x^2}{r^2}. \;\ldots\ldots(8)$$

According to these equations there would be, in all, six directions from O along which there is no scattered light,—two along the axis of y normal to the original ray, and four ($y=0$, $z=\pm x$) at angles of 45° with that ray. So long as the particles are small no such vanishing of light in oblique directions is observed, and we are thus led to the conclusion that the hypothesis of a finite ΔN and of vibrations in the plane of polarization cannot be reconciled with the facts. No form of the elastic solid theory is admissible except that in which the vibrations are supposed to be perpendicular to the plane of polarization, and the difference between one medium and another to be a difference of density only*.

Fig. 26.

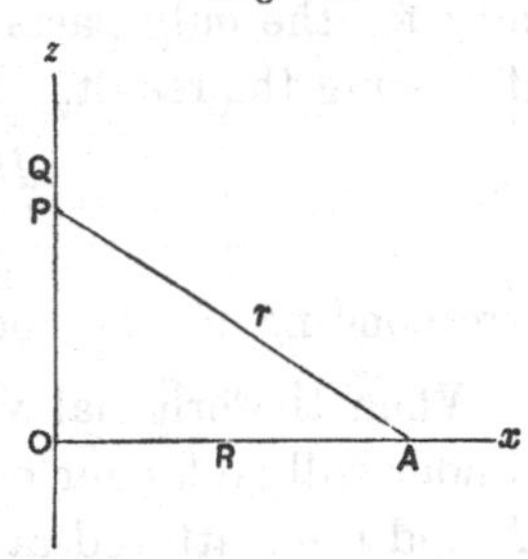

Before leaving this subject it may be instructive to show the application of a method, similar to that used for small particles, to the case of an obstructing *cylinder*, whose axis is parallel to the fronts of the primary waves. We will suppose (1) that the variation of optical properties depends upon a difference of density ($D'-D$), and is small in amount; (2) that the diameter of the cylinder is very small in comparison with the wave-length of light.

Let the axis of the cylinder be the axis of z (Fig. 26), and (as before) let the incident light be parallel to x. The original vibration is thus, in the principal cases, parallel to either z or y. We will take first the former case, where the disturbance due to the cylinder must evidently be symmetrical round OZ and parallel to it. The element of the disturbance at A, due to PQ (dz), will be proportional to dz in amplitude, and will be retarded in phase by an amount corresponding to the distance r. In calculating the effect of the whole bar we have to consider the integral

$$\int_0^\infty \frac{dz}{r}\sin(nt-kr) = \int_R^\infty \frac{dr\sin(nt-kr)}{\sqrt{(r^2-R^2)}}.$$

* See a paper, "On the Scattering of Light by Small Particles," *Phil. Mag.* June 1871. [Vol. I. p. 104.]

The integral on the left may be treated as in § 15, and we find

$$\int_{-\infty}^{\infty} r^{-1} \sin(nt - kr)\, dz = \sqrt{(\lambda/R)} \sin(nt - kR - \tfrac{1}{4}\pi),$$

showing that the total effect is retarded $\frac{1}{8}\lambda$ behind that due to the central element at O. We have seen (3) that, if σ be the sectional area, the effect of the element PQ is

$$\frac{D' - D}{D} \frac{\pi\sigma\, dz \sin\phi}{\lambda^2 r} \sin(nt - kr),$$

where ϕ is the angle OPA. In strictness this should be reckoned perpendicular to PA, and therefore, considered as a contribution to the resultant at A, should be multiplied by $\sin\phi$. But the factor $\sin^2\phi$, being sensibly equal to unity for the only parts which are really operative, may be omitted without influencing the result. In this way we find, for the disturbance at A,

$$\frac{D' - D}{D} \frac{\pi\sigma}{\lambda^{\frac{3}{2}} R^{\frac{1}{2}}} \sin(nt - kR - \tfrac{1}{4}\pi), \quad \text{....................(9)}$$

corresponding to the incident wave $\sin(nt - kx)$.

When the original vibration is parallel to y, the disturbance due to the cylinder will no longer be symmetrical about OZ. If α be the angle between OX and the scattered ray, which is of course always perpendicular to OZ, it is only necessary to introduce the factor $\cos\alpha$ in order to make the previous expression (9) applicable.

The investigation shows that the light diffracted by an ideal wire-grating would, according to the principles of Fresnel, follow the law of polarization enunciated by Stokes. On the other hand, this law would be departed from, were we to suppose that there is any difference of rigidity between the cylinder and the surrounding medium.

§ 26. *Reflexion and Refraction.*

So far as the directions of the rays are concerned, the laws of reflexion and refraction were satisfactorily explained by Huygens on the principles of the wave-theory. The question of the *amount* of light reflected, as dependent upon the characters of the media and upon the angle of incidence, is a much more difficult one, and cannot be dealt with *à priori* without special hypotheses as to the nature of the luminous vibrations, and as to the cause of the difference between various media. By a train of reasoning, not strictly dynamical, but of great ingenuity, Fresnel was led to certain formulæ, since known by his name, expressing the ratio of the reflected to the incident vibration in terms of one constant (μ). If θ be the angle of incidence and θ_1 the angle of refraction, Fresnel's expression for light polarized in the plane of incidence is

$$\frac{\sin(\theta - \theta_1)}{\sin(\theta + \theta_1)}, \quad \text{................................(1)}$$

where the relation between the angles θ, θ_1, and μ (the relative refractive index) is, as usual,

$$\sin\theta = \mu \sin\theta_1. \qquad (2)$$

In like manner, for light polarized perpendicularly to the plane of incidence, Fresnel found

$$\frac{\tan(\theta-\theta_1)}{\tan(\theta+\theta_1)}. \qquad (3)$$

In the particular case of perpendicular incidence, both formulæ coincide with one previously given by Young, viz.,

$$(\mu-1)/(\mu+1). \qquad (4)$$

Since these formulæ agree fairly well with observation, and are at any rate the simplest that can at all represent the facts, it may be advisable to consider their significance a little in detail. As θ increases from 0 to $\frac{1}{2}\pi$, the sine-formula increases from Young's value to unity. We may see this most easily with the aid of a slight transformation:—

$$\frac{\sin(\theta-\theta_1)}{\sin(\theta+\theta_1)} = \frac{1-\tan\theta_1/\tan\theta}{1+\tan\theta_1/\tan\theta} = \frac{\mu-\cos\theta/\cos\theta_1}{\mu+\cos\theta/\cos\theta_1}.$$

Now, writing $\cos\theta/\cos\theta_1$ in the form

$$\sqrt{\left\{\frac{1-\sin^2\theta}{1-\mu^{-2}\sin^2\theta}\right\}},$$

we recognize that, as θ increases from 0 to $\frac{1}{2}\pi$, $\cos\theta/\cos\theta_1$ diminishes continuously from 1 to 0, and therefore (1) increases from $(\mu-1)/(\mu+1)$ to unity.

It is quite otherwise with the tangent-formula. Commencing at Young's value, it diminishes, as θ increases, until it attains zero, when $\theta+\theta_1=\frac{1}{2}\pi$, or $\sin\theta_1=\cos\theta$; or by (2) $\tan\theta=\mu$. This is the polarizing angle defined by Brewster. It presents itself here as the angle of incidence for which there is no reflexion of the polarized light under consideration. As the angle of incidence passes through the polarizing angle, the reflected vibration changes sign, and increases in numerical value until it attains unity at a grazing incidence ($\theta=\frac{1}{2}\pi$).

We have hitherto supposed that the second medium (into which the light enters at the refracting surface) is the denser. In the contrary case, total reflexion sets in as soon as $\sin\theta=\mu^{-1}$, at which point θ_1 becomes imaginary. We shall be able to follow this better in connexion with a mechanical theory.

If light falls upon the first surface of a parallel plate at the polarizing angle, the refracted ray also meets the second surface of the plate at the appropriate polarizing angle. For if μ be the index of the second medium relatively to the first, the tangent of the angle of incidence, which is also the cotangent of the angle of refraction, is equal to μ. At the second surface (the third medium being the same as the first) the angles of incidence and refraction are interchanged, and therefore the condition for the polarizing angle is satisfied, since the index for the second refraction is μ^{-1}.

The principal formulæ apply to light polarized in, and perpendicular to, the plane of incidence. If the plane of polarization make an angle α with that of incidence, the original vibration may be resolved into two,—$\cos\alpha$ polarized in the plane of incidence, and $\sin\alpha$ polarized in the perpendicular plane. These components are reflected according to the laws already considered, and reconstitute plane-polarized light, of intensity

$$\cos^2\alpha \frac{\sin^2(\theta-\theta_1)}{\sin^2(\theta+\theta_1)} + \sin^2\alpha \frac{\tan^2(\theta-\theta_1)}{\tan^2(\theta+\theta_1)}. \quad \text{.................(5)}$$

If the incident light be polarized in a plane making 45° with the plane of incidence, or be circularly-polarized (§ 20), or be unpolarized, (5) applies to the reflected light, with substitution of $\frac{1}{2}$ for $\cos^2\alpha$ and $\sin^2\alpha$. If β denote in the general case the angle between the plane of incidence and that in which the reflected light is polarized,

$$\tan\beta = \tan\alpha \frac{\cos(\theta+\theta_1)}{\cos(\theta-\theta_1)}, \quad \text{........................(6)}$$

a result the approximate truth of which has been verified by Fresnel and Brewster.

The formulæ for the intensities of the refracted light follow immediately from the corresponding formulæ relative to the reflected light in virtue of the principle of energy. The simplest way to regard the matter is to suppose the refracted light to emerge from the second medium into a third medium similar to the first without undergoing loss from a second reflexion, a supposition which would be realized if the transition between the two media were very gradual instead of abrupt. The intensities of the different lights may then be measured in the same way; and the supposition that no loss of energy is incurred when the incident light gives rise to the reflected and refracted lights requires that the sum of the squares of the vibrations representing the latter shall be equal to the square of the vibration representing the former, viz., unity. We thus obtain, in the two cases corresponding to (1) and (3),

$$1 - \frac{\sin^2(\theta-\theta_1)}{\sin^2(\theta+\theta_1)} = \frac{\sin 2\theta \sin 2\theta_1}{\sin^2(\theta+\theta_1)}, \quad \text{.........................(7)}$$

$$1 - \frac{\tan^2(\theta-\theta_1)}{\tan^2(\theta+\theta_1)} = \frac{\sin 2\theta \sin 2\theta_1}{\sin^2(\theta+\theta_1)\cos^2(\theta-\theta_1)}. \quad \text{...........(8)}$$

A plate of glass, or a pile of parallel plates, is often convenient as a polarizer, when it is not necessary that the polarization be quite complete.

At the precise angle of incidence ($\tan^{-1}\mu$) there would be, according to Fresnel's formulæ, only one kind of polarized light reflected, even when the incident light is unpolarized. The polarization of the transmitted light, on the other hand, is imperfect; but it improves as the number of plates is increased.

If we suppose that there is no regular interference, the intensity (r) of the light reflected from a plate is readily calculated by a geometric series when the intensity (ρ) of the light reflected from a single surface is known. The light reflected from the first surface is ρ. That transmitted by the first surface, reflected at the second, and then transmitted at the first, is $\rho(1-\rho)^2$. The next component, reflected three times and transmitted twice, is $\rho^3(1-\rho)^2$, and so on. Hence

$$r = \rho + (1-\rho)^2\{\rho + \rho^3 + \rho^5 + \ldots\} = \frac{2\rho}{1+\rho}. \quad \text{.............(9)}$$

The intensity of the light reflected from a pile of plates has been investigated by Provostaye and Desains*. If $\phi(m)$ be the reflexion from m plates, we may find as above for the reflexion from $(m+1)$ plates,

$$\phi(m+1) = r + (1-r)^2\phi(m)\{1 + r\phi(m) + r^2[\phi(m)]^2 + r^3[\phi(m)]^3 + \ldots\}$$
$$= \frac{r + (1-2r)\phi(m)}{1 - r\phi(m)}.$$

By means of this expression we may obtain in succession the values of $\phi(2)$, $\phi(3)$, &c., in terms of $\phi(1)$, viz., r. The general value is

$$\phi(m) = \frac{mr}{1+(m-1)r}, \quad \text{.........................(10)}$$

as may easily be verified by substitution.

The corresponding expression for the light *transmitted* by a pile of m plates is

$$\psi(m) = 1 - \phi(m) = \frac{1-r}{1+(m-1)r}. \quad \text{..................(11)}$$

The investigation has been extended by Stokes so as to cover the case in which the plates exercise an absorbing influence†.

The verification of Fresnel's formulæ by direct photometric measurement is a matter of some difficulty. The proportion of perpendicularly incident light transmitted by a glass plate has been investigated by Rood‡; but the deficiency may have been partly due to absorption. If we attempt to deal directly with the reflected light, the experimental difficulties are much increased; but the evidence is in favour of the approximate correctness of

* *Ann. d. Chim.* xxx. p. 159, 1850.

† *Proc. Roy. Soc.* xi. p. 545, 1862.

‡ *Am. Jour.* Vol. L. July 1870.

Fresnel's formulæ when light is reflected nearly perpendicularly from a recently polished glass surface. When the surface is old, even though carefully cleaned, there may be a considerable falling off of reflecting power*.

We have seen that according to Fresnel's tangent-formula there would be absolutely no reflexion of light polarized perpendicularly to the plane of incidence, when the angle of incidence is $\tan^{-1}\mu$, or, which comes to the same thing, common light reflected at this angle could be perfectly extinguished with a Nicol's prism.

It was first observed by Airy that in the case of the diamond and other highly refracting media this law is only approximately in accordance with the facts. It is readily proved by experiment that, whatever be the angle of incidence, sunlight reflected from a plate of black glass is incapable of being quenched by a Nicol, and is therefore imperfectly plane-polarized. [1901. If however the glass has recently been repolished with putty powder, the reflexion is much reduced.]

This subject has been studied by Jamin. The character of the reflected vibration can be represented, as regards both amplitude and phase, by the situation in a plane of a point P relatively to the origin of coordinates O. The length of the line OP represents the amplitude, while the inclination of OP to the axis of x represents the phase. According to Fresnel's formula appropriate to light polarized perpendicularly to the plane of incidence, P is situated throughout on the axis of x, passing through O when the angle of incidence is $\tan^{-1}\mu$. Jamin found, however, that in general P does not pass through O, but above or below it. When P is on the axis of y, the amplitude is a minimum, and the phase is midway between the extreme phases. For one class of bodies the phase is in arrear of that corresponding to perpendicular incidence, and for another class of bodies in advance. In a few intermediate cases P passes sensibly through O; and then the change of phase is sudden, and the minimum amplitude is zero.

In the case of metals the polarization produced by reflexion is still more incomplete. Light polarized perpendicularly to the plane of incidence is reflected at all angles, the amount, however, decreasing as the angle of incidence increases from 0° to about 75°, and then again increasing up to a grazing incidence. The most marked effect is the relative retardation of one polarized component with respect to the other. At an angle of about 75° this retardation amounts to a quarter-period.

The intensity of reflexion from metals is often very high. From silver, even at perpendicular incidence, as much as 95 per cent. of the incident light

* "On the Intensity of Light reflected from Certain Surfaces at nearly Perpendicular Incidence," *Proc. Roy. Soc.* 1886. [Vol. II. p. 522.]

is reflected. There is reason for regarding the high reflecting power of metals as connected with the intense absorption which they exercise. Many aniline dyes reflect in abnormal proportion from their surfaces those rays of the spectrum to which they are most opaque. The peculiar absorption spectrum of permanganate of potash is reproduced [with reversal] in the light reflected from a surface of a crystal*.

§ 27. *Reflexion on the Elastic Solid Theory.*

On the theory which assimilates the æther to an elastic solid, the investigation of reflexion and refraction presents no very serious difficulties, but the results do not harmonize very well with optical observation. It is, however, of some importance to understand that reflexion and refraction can be explained, at least in their principal features, on a perfectly definite and intelligible theory, which, if not strictly applicable to the æther, has at any rate a distinct mechanical significance. The refracting surface and the wave-fronts may for this purpose be supposed to be plane.

When the vibrations are perpendicular to the plane of incidence ($z=0$), the solution of the problem is very simple. We suppose that the refracting surface is $x=0$, the rigidity and density in the first medium being N, D, and in the second N_1, D_1. The displacements in the two media are in general denoted by ξ, η, ζ; ξ_1, η_1, ζ_1; but in the present case ξ, η, ξ_1, η_1 all vanish. Moreover ζ, ζ_1 are independent of z. The equations to be satisfied in the interior of the media are accordingly (§ 24)

$$\frac{d^2\zeta}{dt^2}=\frac{N}{D}\left(\frac{d^2\zeta}{dx^2}+\frac{d^2\zeta}{dy^2}\right), \quad\text{.........................(1)}$$

$$\frac{d^2\zeta_1}{dt^2}=\frac{N_1}{D_1}\left(\frac{d^2\zeta_1}{dx^2}+\frac{d^2\zeta_1}{dy^2}\right). \quad\text{.......................(2)}$$

At the boundary the conditions to be satisfied are the continuity of displacement and of stress; so that, when $x=0$,

$$\zeta=\zeta_1, \qquad N\frac{d\zeta}{dx}=N_1\frac{d\zeta_1}{dx}. \quad\text{.......................(3)}$$

The incident waves may be represented by

$$\zeta=e^{i(ax+by+ct)},$$

where

$$Dc^2=N(a^2+b^2); \quad\text{.............................(4)}$$

* Stokes, "On the Metallic Reflection exhibited by Certain Non-Metallic Substances," *Phil. Mag.* Dec. 1853.

and $ax + by =$ const. gives the equation of the wave-fronts. The reflected and refracted waves may be represented by

$$\zeta = \zeta' e^{i(-ax+by+ct)}, \qquad (5)$$

$$\zeta_1 = \zeta_1' e^{i(a_1x+by+ct)}. \qquad (6)$$

The coefficient of t is necessarily the same in all three waves on account of the periodicity, and the coefficient of y must be the same since the traces of all the waves upon the plane of separation must move together. With regard to the coefficient of x, it appears by substitution in the differential equations that its sign is changed in passing from the incident to the reflected wave; in fact

$$c^2 = V^2 \{(\pm a)^2 + b^2\} = V_1^2 \{a_1^2 + b^2\}, \qquad (7)$$

where V, V_1 are the velocities of propagation in the two media given by

$$V^2 = N/D, \qquad V_1^2 = N_1/D_1. \qquad (8)$$

Now $b/\surd(a^2 + b^2)$ is the sine of the angle included between the axis of x and the normal to the plane of waves--in optical language, the sine of the angle of incidence, and $b/\surd(a_1^2 + b^2)$ is in like manner the sine of the angle of refraction. If these angles be denoted (as before) by θ, θ_1, (7) asserts that $\sin\theta : \sin\theta_1$ is equal to the constant ratio $V : V_1$, the well-known law of sines. The laws of reflexion and refraction follow simply from the fact that the velocity of propagation normal to the wave-fronts is constant in each medium, that is to say, independent of the *direction* of the wave-front, taken in connexion with the equal velocities of the traces of all the waves on the plane of separation ($V/\sin\theta = V_1/\sin\theta_1$).

The boundary conditions (3) now give

$$1 + \zeta' = \zeta_1', \qquad Na(1 - \zeta') = N_1 a_1 \zeta_1', \qquad (9)$$

whence

$$\zeta' = \frac{Na - N_1 a_1}{Na + N_1 a_1}, \qquad (10)$$

a formula giving the reflected wave in terms of the incident wave (supposed to be unity). This completes the symbolical solution. If a_1 (and θ_1) be real, we see that, if the incident wave be

$$\zeta = \cos(ax + by + ct),$$

or in terms of V, λ, and θ,

$$\zeta = \cos\frac{2\pi}{\lambda}(x\cos\theta + y\sin\theta + Vt), \qquad (11)$$

the reflected wave is

$$\zeta = \frac{N\cot\theta - N_1\cot\theta_1}{N\cot\theta + N_1\cot\theta_1}\cos\frac{2\pi}{\lambda}(-x\cos\theta + y\sin\theta + Vt). \qquad (12)$$

The formula for intensity of the reflected wave is here obtained on the supposition that the waves are of harmonic type; but, since it does not involve λ and there is no change of phase, it may be extended by Fourier's theorem to waves of any type whatever. It may be remarked that when the first and second media are interchanged the coefficient in (12) simply changes sign, retaining its numerical value.

The amplitude of the reflected wave, given in general by (12), assumes special forms when we introduce more particular suppositions as to the nature of the difference between media of diverse refracting power. According to Fresnel and Green the rigidity does not vary, or $N = N_1$. In this case

$$\frac{N \cot\theta - N_1 \cot\theta_1}{N \cot\theta + N_1 \cot\theta_1} = \frac{\cot\theta - \cot\theta_1}{\cot\theta + \cot\theta_1} = \frac{\sin(\theta_1 - \theta)}{\sin(\theta_1 + \theta)}.$$

If, on the other hand, the density is the same in various media,

$$N_1 : N = V_1^2 : V^2 = \sin^2\theta_1 : \sin^2\theta,$$

and then

$$\frac{N \cot\theta - N_1 \cot\theta_1}{N \cot\theta + N_1 \cot\theta_1} = \frac{\tan(\theta_1 - \theta)}{\tan(\theta_1 + \theta)}.$$

If we assume the complete accuracy of Fresnel's expressions, either alternative agrees with observation; only, if $N = N_1$, light must be supposed to vibrate normally to the plane of polarization; while, if $D = D_1$, the vibrations are parallel to that plane.

An intermediate supposition, according to which the refraction is regarded as due partly to a difference of density and partly to a difference of rigidity, could scarcely be reconciled with observation, unless one variation were very subordinate to the other. But the most satisfactory argument against the joint variation is that derived from the theory of the diffraction of light by small particles (§ 25).

We will now, limiting ourselves for simplicity to Fresnel's supposition ($N_1 = N$), inquire into the character of the solution when total reflexion sets in. The symbolical expressions for the reflected and refracted waves are

$$\zeta = \frac{a - a_1}{a + a_1} e^{i(-ax+by+ct)}, \quad \text{.........................(13)}$$

$$\zeta_1 = \frac{2a}{a + a_1} e^{i(a_1x+by+ct)}, \quad \text{.........................(14)}$$

and so long as a_1 is real they may be interpreted to indicate

$$\zeta = \frac{a - a_1}{a + a_1} \cos(-ax + by + ct), \quad \text{....................(15)}$$

$$\zeta_1 = \frac{2a}{a + a_1} \cos(a_1x + by + ct), \quad \text{....................(16)}$$

corresponding to the incident wave

$$\zeta = \cos(ax + by + ct). \quad \text{..........................(17)}$$

In this case there is a refracted wave of the ordinary kind, conveying away a part of the original energy. When, however, the second medium is the rarer ($V_1 > V$), and the angle of incidence exceeds the so-called critical angle $\{\sin^{-1}(V/V_1)\}$, there can be no refracted wave of the ordinary kind. In whatever direction it may be supposed to lie, its trace must necessarily outrun the trace of the incident wave upon the separating surface. The quantity a_1, as defined by our equations, is then imaginary, so that (13) and (14) no longer express the real parts of the symbolical expressions (5) and (6).

If $-ia_1'$ be written in place of a_1, the symbolical equations are

$$\zeta = \frac{a + ia_1'}{a - ia_1'} e^{i(-ax+by+ct)}, \qquad \zeta_1 = \frac{2a}{a - ia_1'} e^{i(-ia_1'x+by+ct)};$$

from which, by discarding the imaginary parts, we obtain

$$\zeta = \cos(-ax + by + ct + 2\epsilon), \quad\quad\quad (18)$$

$$\zeta_1 = \frac{2a}{\sqrt{(a^2 + a_1'^2)}} e^{a_1'x} \cos(by + ct + \epsilon), \quad\quad\quad (19)$$

where

$$\tan \epsilon = a_1'/a. \quad\quad\quad (20)$$

Since x is supposed to be negative in the second medium, we see that the disturbance is there confined to a small distance (a few wave-lengths) from the surface, and no energy is propagated into the interior. The whole of the energy of the incident waves is to be found in the reflected waves, or the reflexion is *total*. There is, however, a change of phase of 2ϵ, given by (20), or in terms of V, V_1, and θ,

$$\tan \epsilon = \sqrt{\{\tan^2\theta - \sec^2\theta\,(V^2/V_1^2)\}}. \quad\quad\quad (21)$$

The principal application of the formulæ being to reflexions when the second medium is air, it will be convenient to denote by μ the index of the *first* medium *relatively to the second*, so that $\mu = V_1/V$. Thus

$$\tan \epsilon = \sqrt{\{\tan^2\theta - \sec^2\theta/\mu^2\}}. \quad\quad\quad (22)$$

The above interpretation of his formula $\sin(\theta_1 - \theta)/\sin(\theta_1 + \theta)$, in the case where θ_1 becomes imaginary, is due to the sagacity of Fresnel. His argument was perhaps not set forth with full rigour, but of its substantial validity there can be no question. By a similar process Fresnel deduced from his tangent-formula for the change of phase ($2\epsilon'$) accompanying total reflexion when the vibrations are executed in the plane of incidence,

$$\tan \epsilon' = \mu\sqrt{\{\mu^2\tan^2\theta - \sec^2\theta\}}. \quad\quad\quad (23)$$

The phase-differences represented by 2ϵ and $2\epsilon'$ cannot be investigated experimentally, but the *difference* $(2\epsilon' - 2\epsilon)$ is rendered evident when the incident light is polarized obliquely so as to contribute components in both the principal planes. If in the act of reflexion one component is retarded more or less than the other, the resultant light is no longer plane but elliptically polarized.

From (22) and (23) we have

$$\tan(\epsilon' - \epsilon) = \cos\theta \sqrt{\{1 - \mu^{-2}\operatorname{cosec}^2\theta\}},$$

whence

$$\cos(2\epsilon' - 2\epsilon) = \frac{2\mu^2 \sin^4\theta - (1+\mu^2)\sin^2\theta + 1}{(1+\mu^2)\sin^2\theta - 1}. \quad \text{.........(24)}$$

The most interesting case occurs when the difference of phase amounts to a quarter of a period, corresponding to light circularly polarized. If, however, we put $\cos(2\epsilon' - 2\epsilon) = 0$, we find

$$4\mu^2 \sin^2\theta = 1 + \mu^2 \pm \sqrt{\{(1+\mu^2)^2 - 8\mu^2\}},$$

from which it appears that, in order that $\sin\theta$ may be real, μ^2 must exceed $3 + \sqrt{8}$. So large a value of μ^2 not being available, the conversion of plane-polarized into circularly-polarized light by one reflexion is impracticable.

The desired object may, however, be attained by two successive reflexions. The angle of incidence may be so accommodated to the index that the alteration of phase amounts to $\frac{1}{8}$ period, in which case a second reflexion under the same conditions will give rise to light circularly polarized. Putting $(2\epsilon - 2\epsilon') = \frac{1}{4}\pi$, we get

$$2\mu^2 \sin^4\theta = (1 + \sqrt{\tfrac{1}{2}})\{(1+\mu^2)\sin^2\theta + 1\}, \quad \text{..............(25)}$$

an equation by which θ is determined when μ is given. It appears that, when $\mu = 1{\cdot}51$, $\theta = 48° \, 37'$ or $54° \, 37'$. These results were verified by Fresnel by means of the rhomb shown in Fig. 27.

The problem of reflexion upon the elastic solid theory, when the vibrations are executed in the plane of incidence, is more complicated, on account of the tendency to form waves of dilatation. In order to get rid of these, to which no optical phenomena correspond, it is necessary to follow Green in supposing that the velocity of such waves is infinite, or that the media are incompressible*. Even then we have to introduce in the neighbourhood of the interface waves variously called longitudinal, pressural, or surface waves; otherwise it is impossible to satisfy the conditions of continuity of strain and stress. These waves, analogous in this respect to those occurring in the second medium when total reflexion is in progress (19), extend to a depth of a few wave-lengths only, and they are so

Fig. 27.

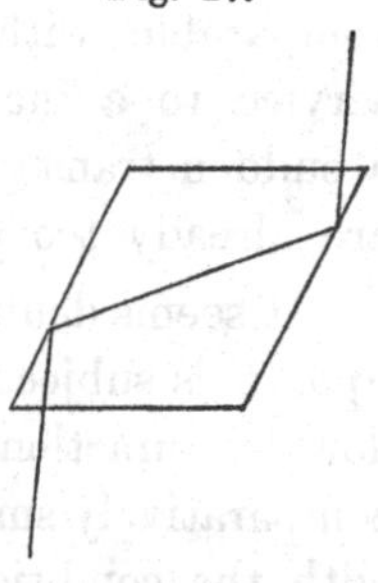

* The supposition that the velocity is zero, favoured by some writers, is inadmissible. Even dilatational waves involve a shearing of the medium, and must therefore be propagated at a finite rate, unless the resistance to compression were *negative*. But in that case the equilibrium would be unstable. [1901. Lord Kelvin has since (*Phil. Mag.* XXVI. p. 415, 1888) shown that, if the medium be held fast at the boundary, negative resistance to compression need not involve instability.]

constituted that there is neither dilatation nor rotation. On account of them the final formulæ are less simple than those of Fresnel. If we suppose the densities to be the same in the two media, there is no correspondence whatever between theory and observation. In this case, as we have seen, vibrations perpendicular to the plane of incidence are reflected according to Fresnel's tangent-formula; and thus vibrations in the plane of incidence should follow the sine-formula. The actual result of theory is, however, quite different. In the case where the relative index does not differ greatly from unity, polarizing angles of $22\frac{1}{2}°$ and $67\frac{1}{2}°$ are indicated, a result totally at variance with observation. As in the case of diffraction by small particles, an elastic solid theory, in which the densities in various media are supposed to be equal, is inadmissible. If, on the other hand, following Green, we regard the rigidities as equal, we get results in better agreement with observation. To a first approximation indeed (when the refraction is small) Green's formula coincides with Fresnel's tangent-formula; so that light vibrating in the plane of incidence is reflected according to this law, and light vibrating in the perpendicular plane according to the sine-formula. The vibrations are accordingly perpendicular to the plane of polarization.

The deviations from the tangent-formula, indicated by theory when the refraction is not very small, are of the same general character as those observed by Jamin, but of much larger amount. The minimum reflexion at the surface of glass ($\mu = \frac{3}{2}$) would be $\frac{1}{49}$*, nearly the half of that which takes place at perpendicular incidence, and very much in excess of the truth. This theory cannot therefore be considered satisfactory as it stands, and various suggestions have been made for its improvement. The only variations from Green's suppositions admissible in strict harmony with an elastic solid theory is to suppose that the transition from one medium to the other is gradual instead of abrupt, that is, that the transitional layer is of thickness comparable with the wave-length. This modification would be of more service to a theory which gave Fresnel's tangent-formula as the result of a sudden transition than to one in which the deviations from that formula are already too great.

It seems doubtful whether there is much to be gained by further discussion upon this subject, in view of the failure of the elastic solid theory to deal with double refraction. The deviations from Fresnel's formulæ for reflexion are comparatively small; and the whole problem of reflexion is so much concerned with the condition of things at the interface of two media, about which we know little, that valuable guidance can hardly be expected from this quarter. It is desirable to bear constantly in mind that reflexion depends entirely upon an approach to discontinuity in the properties of the medium. If the thickness of the transitional layer amounted to a few wave-lengths, there would be no sensible reflexion at all.

* Green's *Papers*, by Ferrers, p. 333.

Another point may here be mentioned. Our theories of reflexion take no account of the fact that one at least of the media is dispersive. The example of a stretched string, executing transverse vibrations, and composed of two parts, one of which in virtue of stiffness possesses in some degree the dispersive property, shows that the boundary conditions upon which reflexion depends are thereby modified. We may thus expect a finite reflexion at the interface of two media, if the dispersive powers are different, even though the indices be absolutely the same for the waves under consideration, in which case there is no refraction. But a knowledge of the dispersive properties of the media is not sufficient to determine the reflexion without recourse to hypothesis*.

§ 28. *The Velocity of Light.*

According to the principles of the wave-theory, the dispersion of refraction can only be explained as due to a variation of velocity with wave-length or period. In aerial vibrations, and in those propagated through an elastic solid, there is no such variation; and so the existence of dispersion was at one time considered to be a serious objection to the wave-theory. Dispersion *in vacuo* would indeed present some difficulty, or at least force upon us views which at present seem unlikely as to the constitution of free æther. The weight of the evidence is, however, against the existence of dispersion *in vacuo*. "Were there a difference of one hour in the times of the blue and red rays reaching us from Algol, this star would show a well-marked coloration in its phases of increase or decrease. No trace of coloration having been noticed, the difference of times cannot exceed a fraction of an hour. It is not at all probable that the parallax of this star amounts to one-tenth of a second, so that its distance, probably, exceeds two million radii of the earth's orbit, and the time which is required for its light to reach us probably exceeds thirty years, or a quarter of a million hours. It is therefore difficult to see how there can be a difference as great as four parts in a million between the

* The reader who desires to pursue this subject may consult Green, "On the Laws of Reflexion and Refraction of Light at the Common Surface of Two Non-Crystallized Media," *Camb. Trans.* 1838 (Green's *Works*, London 1871, pp. 242, 283); Lorenz, "Ueber die Reflexion des Lichts an der Gränzfläche zweier isotropen, durchsichtigen Mittel," *Pogg. Ann.* CXI. p. 460 (1860), and "Bestimmung der Schwingungsrichtung des Lichtæthers durch die Reflexion und Brechung des Lichtes," *ibid.* CXIV. p. 238 (1861); Strutt (Rayleigh), "On the Reflexion of Light from Transparent Matter," *Phil. Mag.* [4] XIII. (1871); Von der Mühll, "Ueber die Reflexion und Brechung des Lichtes an der Grenze unkrystallinischen Medien," *Math. Ann.* V. 470 (1872), and "Ueber Greens Theorie der Reflexion und Brechung des Lichtes," *Math. Ann.* XXVII. 506 (1886); Thomson, *Baltimore Lectures*; Glazebrook, "Report on Optical Theories," *Brit. Ass. Rep.* 1886; Rayleigh, "On Reflexion of Vibrations at the Confines of Two Media between which the Transition is gradual," *Proc. Math. Soc.* XI.; and Walker, "An Account of Cauchy's Theory of Reflexion and Refraction of Light," *Phil. Mag.* XXIII. p. 151 (1887). References to recent German writers, Ketteler, Lommel, Voigt, &c., will be found in Glazebrook's Report.

velocities of light coming from near the two ends of the bright part of the spectrum*."

For the velocity of light *in vacuo*, as determined in kilometres per second by terrestrial methods (Light, *Enc. Brit.* Vol. XIV. p. 585), Newcomb gives the following tabular statement:—

Michelson, at Naval Academy, in 1879	299,910
Michelson, at Cleveland, 1882	299,853
Newcomb, at Washington, 1882, using only results supposed to be nearly free from constant errors	299,860
Newcomb, including all determinations	299,810

To these may be added, for reference—

Foucault, at Paris, in 1862	298,000
Cornu, at Paris, in 1874	298,500
Cornu, at Paris, in 1878	300,400
This last result, as discussed by Listing	299,990
Young and Forbes, 1880—1881	301,382

Newcomb concludes, as the most probable result—

Velocity of light *in vacuo* = 299,860 ± 30 kilometres [per second].

It should be mentioned that Young and Forbes inferred from their observations a difference of velocities of blue and red light amounting to about 2 per cent., but that neither Michelson nor Newcomb, using Foucault's method, could detect any trace of such a difference.

When we come to consider the propagation of light through ponderable media, there seems to be little reason for expecting to find the velocity independent of wave-length. The interaction of matter and æther may well give rise to such a degree of complication that the differential equation expressing the vibrations shall contain more than one constant. The law of constant velocity is a special property of certain very simple media. Even in the case of a stretched string, vibrating transversely, the velocity becomes a function of wave-length as soon as we admit the existence of finite stiffness.

As regards the law of dispersion, a formula, derived by Cauchy from theoretical considerations, was at one time generally accepted. According to this,

$$\mu = A + B\lambda^{-2} + C\lambda^{-4} + \dots; \qquad (1)$$

and there is no doubt that even the first two terms give a good representation of the truth in media not very dispersive, and over the more luminous portion of the spectrum. A formula of this kind treats dispersion as due to the smallness of wave-lengths, giving a definite limit to refraction (A) when the wave-length is very large. Recent investigations by Langley on the law of dispersion for rock-salt in the ultra-red region of the spectrum are not

* Newcomb, *Astron. Papers*, Vol. II. parts III. and IV., Washington 1885.

very favourable to this idea. The phenomena of abnormal dispersion indicate a close connexion between refraction and absorption, and Helmholtz has formulated a general theory of dispersion based upon the hypothesis that it may be connected with an absorbing influence operative upon invisible portions of the spectrum. Upon this subject, which is as yet little understood, the reader may consult Glazebrook's "Report on Optical Theories*." [1901. Since this article was written, great advances have been made by the German physicists, of whom Rubens may specially be named.]

The limits of this article do not permit the consideration of the more speculative parts of our subject. We will conclude by calling attention to two recent experimental researches by Michelson, the results of which cannot fail to give valuable guidance to optical theorists. The first of these† was a repetition under improved conditions of a remarkable experiment of Fizeau, by which it is proved that when light is propagated through water, itself in rapid movement in the direction of the ray, the velocity is indeed influenced, but not to the full extent of the velocity of the water (v). Within the limits of experimental error, the velocity agrees with a formula suggested by Fizeau on the basis of certain views of Fresnel, viz.,

$$V = V_0 \pm \frac{\mu^2 - 1}{\mu^2} v, \qquad (2)$$

V_0 being the velocity when the medium is stationary. In the case of water, $(\mu^2 - 1)/\mu^2 = \cdot 437$. Conformably with (2), a similar experiment upon air, moving at a velocity of 25 metres per second, gave no distinct effect.

From the result of the experiments upon water we should be tempted to infer that at the surface of the earth, moving through space, the æther still retains what must be coarsely called relative motion. Nevertheless, the second research above alluded to‡ appears to negative this conclusion, and to prove that, at any rate within the walls of a building, the æther must be regarded as fully partaking in the motion of material bodies.

* *Brit. Assoc. Rep.* 1886. In this matter, as in most others, the advantage lies with the electro-magnetic theory. See J. W. Gibbs, *Amer. Journ.* XXIII. 1882.

† "Influence of Motion of the Medium on the Velocity of Light," by A. Michelson and E. W. Morley, *Amer. Journ.* XXXI. May, 1886.

‡ "On the Relative Motion of the Earth and the Luminiferous Æther," by Michelson and Morley, *Phil. Mag.* Dec. 1887.

149.

ON THE REFLEXION OF LIGHT AT A TWIN PLANE OF A CRYSTAL.

[*Phil. Mag.* Vol. XXVI. pp. 241—255, 1888.]

MY object in the present paper is to calculate *à priori* the reflexion of light at the surface between twin crystals, and to obtain formulæ analogous to those discovered by Fresnel for the case where both media are isotropic. It is evident that success can only be attained upon the basis of a theory capable of explaining at once Fresnel's laws of double refraction in crystals and those just referred to, governing the intensity of reflexion when light passes from one isotropic medium to another. So far as I am aware the electric theory of Maxwell is the only one satisfying these conditions*; and I have accordingly employed the equations of this theory. It will be remembered that the electric theory of double refraction was worked out by Maxwell himself, and that the application to the problem of reflexion was successfully effected by von Helmholtz and Lorentz†. The present investigation starts, however, independently from the fundamental equations, as given in Maxwell's *Electricity and Magnetism.*

Equations of a Dialectric Medium, of which the Magnetic Permeability is Unity throughout.

In Maxwell's notation the various components are represented as follows:—

Electric Displacement..f, g, h;
Current...u, v, w;
Magnetic Force (or Induction)...........................a, b, c;
Electromagnetic Momentum ....F, G, H;
Electromotive Force ....P, Q, R;

* See Prof. Willard Gibbs's excellent "Comparison of the Elastic and the Electrical Theories of Light with respect to the Law of Double Refraction and the Dispersion of Colours" (*Am. Journ. Sci.* June, 1888), which reaches me while revising the present investigation for the press.

† References to the works of previous writers will be found in Glazebrook's "Report on Optical Theories," *Brit. Assoc. Rep.* 1886.

and the equations connecting them may be written

$$u=\frac{df}{dt},\quad v=\frac{dg}{dt},\quad w=\frac{dh}{dt},\ \ldots\ldots\ldots\ldots\ldots\ldots\ldots(1)$$

$$\frac{df}{dx}+\frac{dg}{dy}+\frac{dh}{dz}=0,\ \ldots\ldots\ldots\ldots\ldots\ldots\ldots\ldots\ldots(2)$$

$$4\pi u=\frac{dc}{dy}-\frac{db}{dz},\quad 4\pi v=\frac{da}{dz}-\frac{dc}{dx},\quad 4\pi w=\frac{db}{dx}-\frac{da}{dy},\ \ldots\ldots(3)$$

$$a=\frac{dH}{dy}-\frac{dG}{dz},\quad b=\frac{dF}{dz}-\frac{dH}{dx},\quad c=\frac{dG}{dx}-\frac{dF}{dy},\ldots\ldots(4)$$

$$P=-\frac{dF}{dt}-\frac{d\Phi}{dx},\quad Q=-\frac{dG}{dt}-\frac{d\Phi}{dy},\quad R=-\frac{dH}{dt}-\frac{d\Phi}{dz}\ldots(5)$$

In (1) it is assumed that the medium is a perfect insulator. Equations (4) and (5) may be replaced by

$$\frac{da}{dt}=\frac{dQ}{dz}-\frac{dR}{dy},\quad \frac{db}{dt}=\frac{dR}{dx}-\frac{dP}{dz},\quad \frac{dc}{dt}=\frac{dP}{dy}-\frac{dQ}{dx},\ \ldots\ldots(6)$$

from which Φ disappears. Thus

$$4\pi\frac{d^2f}{dt^2}=4\pi\frac{du}{dt}=\frac{d}{dy}\frac{dc}{dt}-\frac{d}{dz}\frac{db}{dt}$$

$$=\frac{d^2P}{dy^2}+\frac{d^2P}{dz^2}-\frac{d}{dx}\left\{\frac{dQ}{dy}+\frac{dR}{dz}\right\}$$

$$=\nabla^2P-\frac{d}{dx}\left\{\frac{dP}{dx}+\frac{dQ}{dy}+\frac{dR}{dz}\right\},\ \ldots\ldots\ldots\ldots\ldots\ldots\ldots(7)$$

where as usual

$$\nabla^2=d^2/dx^2+d^2/dy^2+d^2/dz^2.$$

In (7) and the similar equations in g and h there is involved no assumption as to the homogeneity or isotropy of the dielectric medium. If, however, these conditions are fulfilled,

$$\frac{dP}{dx}+\frac{dQ}{dy}+\frac{dR}{dz}=0,$$

P, Q, R being proportional to f, g, h; and the equations then assume a specially simple form.

The boundary conditions which must be satisfied at the transition from one homogeneous medium to another are obtained without difficulty from the differential equations. We will suppose that the surface of transition is the plane $x=0$. The first condition follows immediately from (2). It is that f must be continuous across the surface $x=0$. Equation (7) shows that $dQ/dy+dR/dz$ must be continuous. From the similar equation in g, viz.:—

$$4\pi\frac{d^2g}{dt^2}=\frac{d}{dz}\frac{da}{dt}-\frac{d}{dx}\frac{dc}{dt}=\nabla^2Q-\frac{d}{dy}\left\{\frac{dP}{dx}+\frac{dQ}{dy}+\frac{dR}{dz}\right\},\ \ldots\ldots(8)$$

we see not only that dc/dt, or c, must be continuous, but also that Q must be continuous. In like manner from the corresponding equation in h it follows that R and b must be continuous. The continuity of Q and R secures that of $dQ/dy + dR/dz$; so that it is sufficient to provide for the continuity of

$$f,\ Q,\ R,\ b,\ c. \qquad\qquad \text{(A)*}$$

Isotropic Reflexion.

If both media are isotropic, the problem of reflexion of plane waves is readily solved. When the electric displacements are perpendicular to the plane of incidence (xy), f and g vanish, while h and the other remaining functions are independent of z. The only boundary conditions requiring attention are that R and b should be continuous, or by (6) that R and dR/dx should be continuous. This leads, as is well known, to Fresnel's sine-formula as the expression for the reflected wave.

When the electric displacements are in the plane of incidence, $h = 0$, and (as before) all the remaining functions are independent of z. As an introduction to the more difficult investigation before us, it may be well to give a sketch of the solution for this case. In the upper medium we have as the relation between force and displacement,

$$P,\ Q,\ R = 4\pi V^2 (f,\ g,\ h), \qquad\qquad (9)$$

and in the lower,

$$P,\ Q,\ R = 4\pi V_1^2 (f,\ g,\ h), \qquad\qquad (10)$$

V, V_1 being the two wave-velocities, whose ratio gives the refractive index. Since $h = 0$, $R = 0$; and since $R = 0$, $dP/dz = 0$, it follows by (6) that $b = 0$. The only conditions (A) requiring further consideration are thus the continuity of f, Q or $V^2 g$, and c.

As the expression for the incident wave we take

$$f = q e^{i(px+qy+st)}, \qquad g = -p e^{i(px+qy+st)}; \qquad\qquad (11)$$

the ratio of the coefficients being determined by the consideration that the directions f, g, h and p, q, r are perpendicular†. In like manner for the reflected wave we have

$$f = q\theta' e^{i(-px+qy+st)}, \qquad g = p\theta' e^{i(-px+qy+st)}, \qquad\qquad (12)$$

and for the refracted wave

$$f = q\theta_1 e^{i(p_1x+qy+st)}, \qquad g = -p_1\theta_1 e^{i(p_1x+qy+st)}. \qquad\qquad (13)$$

* Of these conditions the first is really superfluous. If we differentiate (7) &c. with respect to x, y, z respectively and add, we see that the truth of (2) is involved. In some cases it would shorten the analytical expressions if we took P, Q, R as fundamental variables, in place of f, g, h.

† In the present case $r=0$.

The coefficient of y is the same for all the waves, since their traces on the plane $x=0$ must move together. The multipliers θ', θ_1 determine the amplitudes of the reflected and refracted waves, and may be regarded as the quantities whose expression is sought. The velocity of propagation in the first medium is $s/\sqrt{(p^2+q^2)}$, so that

$$V^2(p^2+q^2) = V_1^2(p_1^2+q^2). \quad \text{.....................(14)}$$

We have now to consider the boundary conditions. The continuity of f, when $x=0$, requires that

$$1+\theta' = \theta_1; \quad \text{..................................(15)}$$

and the continuity of V^2g requires that

$$V^2p(1-\theta') = V_1^2p_1\theta_1. \quad \text{...........................(16)}$$

These two equations suffice for the determination of θ', θ_1; and we may infer that the third boundary condition is superfluous. It is easily proved to be so; for in the upper medium,

$$\frac{dc}{dt} = \frac{dP}{dy} - \frac{dQ}{dx} = V^2\left\{\frac{df}{dy} - \frac{dg}{dx}\right\}$$

$$= V^2(1+\theta')(p^2+q^2)\,e^{i(qy+st)}$$

when $x=0$. In the lower medium, when $x=0$,

$$\frac{dc}{dt} = V_1^2\theta_1(p_1^2+q^2)\,e^{i(qy+st)};$$

so that by (14) the continuity of dc/dt leads to the same condition as the continuity of f.

The usual formula for the reflected wave is readily obtained from (15), (16). If ϕ, ϕ_1 be the angles of incidence and refraction,

$$V_1^2/V^2 = \sin^2\phi_1/\sin^2\phi,$$

$$p_1/p = (p_1/q) \div (p/q) = \cot\phi_1/\cot\phi;$$

so that

$$\frac{1-\theta'}{1+\theta'} = \frac{\sin^2\phi_1\cot\phi_1}{\sin^2\phi\cot\phi} = \frac{\sin 2\phi_1}{\sin 2\phi}.$$

Accordingly,

$$\theta' = \frac{\sin 2\phi - \sin 2\phi_1}{\sin 2\phi + \sin 2\phi_1} = \frac{\tan(\phi-\phi_1)}{\tan(\phi+\phi_1)}. \quad \text{..............(17)}$$

The insertion of this value of θ' in (12) gives the expression for the reflected wave corresponding to the incident wave (11). The ratio of amplitudes in the two cases, being proportional to $\sqrt{(f^2+g^2)}$, is represented by θ', and (17) is the well-known tangent-formula of Fresnel.

Propagation in a Crystal.

In a homogeneous crystalline medium, the relation of force to strain may be expressed

$$P,\ Q,\ R = 4\pi\,(a_1^2 f,\ b_1^2 g,\ c_1^2 h) \quad\text{.....................(18)}$$

where a_1, b_1, c_1 are the principal wave-velocities. We here suppose that the axes of coordinates are chosen so as to be parallel to the principal axes of the crystal. The introduction of these relations into (7), &c., gives

$$\frac{d^2 f}{dt^2} = a_1^2 \nabla^2 f - \frac{d\Pi}{dx},\quad \frac{d^2 g}{dt^2} = b_1^2 \nabla^2 g - \frac{d\Pi}{dy},\quad \frac{d^2 h}{dt^2} = c_1^2 \nabla^2 h - \frac{d\Pi}{dz},\ \text{...(19)}$$

where

$$\Pi = a_1^2 df/dx + b_1^2 dg/dy + c_1^2 dh/dz. \quad\text{..................(20)}$$

The principal problem of double refraction is the investigation of the form of the wave-surface. By means of (19) we can readily determine the law of velocity (V) for various directions of wave-front (l, m, n). For this purpose we assume

$$f,\ g,\ h = (\lambda,\ \mu,\ \nu)\, e^{ik\omega}, \quad\text{..........................(21)}$$

where

$$\omega = lx + my + nz - Vt, \quad\text{..........................(22)}$$

and $k = 2\pi \div$ wave-length. In accordance with (2) we must have

$$l\lambda + m\mu + n\nu = 0, \quad\text{.............................(23)}$$

signifying that the electric displacement is in the plane of the wave-front. If we now write

$$\Pi = \Pi_0 e^{ik\omega},$$

and substitute the values of f, g, h from (21) in (19) we find

$$\lambda\,(V^2 - a_1^2) = ik^{-1}\,\Pi_0 \,.\, l, \quad \&c.,$$

so that by (23)

$$\frac{l^2}{V^2 - a_1^2} + \frac{m^2}{V^2 - b_1^2} + \frac{n^2}{V^2 - c_1^2} = 0, \quad\text{.....................(24)}$$

which is Fresnel's law of velocities, leading to the wave-surface discovered by him.

Reflexion at a Twin Plane.

We are now prepared for the consideration of our special problem, viz., the reflexion of plane waves at a twin surface of a crystal. We suppose that the plane of separation is $x = 0$, and we assume that there is a plane perpendicular to this ($z = 0$), with respect to which each twin is symmetrical. The only difference between the two media is that which corresponds to a rotation through 180° about the axis of x, perpendicular to the twin plane.

In consequence of the symmetry the axis of z is a principal axis in both media; but the axes of x and y are not principal axes. For the relation between force and strain in the first medium we may take

$$P = 4\pi (Af + Bg), \qquad Q = 4\pi (Bf + Cg), \qquad R = 4\pi Dh. \quad \text{......(25)}$$

In the second medium we may in the first instance assume similar expressions with accented letters; but the peculiar relation between the two media demands that $A' = A$, $C' = C$, $D' = D$, $B' = -B$. Thus for the second twin medium,

$$P = 4\pi (Af - Bg), \qquad Q = 4\pi (-Bf + Cg), \qquad R = 4\pi Dh, \quad \text{......(26)}$$

the only difference being the change in the sign of B. If B vanish, all optical distinction between the twins disappears, and there can be no reflexion. The magnitude of B depends upon the intensity of the double refraction in the twins, and also upon the angles between the principal axes and the twin plane. If one of these angles were to vanish, B would disappear, in spite of a powerful double refraction.

For a general solution of the problem of reflexion from a twin plane, we should have to suppose the plane of incidence to be inclined at an arbitrary angle to the plane of symmetry (x, y); but we may limit ourselves without much loss of interest to the two principal cases, when the plane of incidence (1) coincides with the plane of symmetry, (2) is perpendicular to it.

Incidence in the Plane of Symmetry.

Under the first head there are two problems which may be considered separately. The simplest is that which arises when the vibrations are perpendicular to the plane of incidence, that is, are parallel to z. It is not difficult to see that in this case the difference between the twins never comes into operation, and that accordingly the reflexion vanishes; but it may be well to apply the general method.

Since f, g, and therefore {by (25), (26)} P and Q, vanish throughout, while h and R are independent of z, the two first of equations (7) are satisfied identically, and the third becomes

$$4\pi \frac{d^2h}{dt^2} = \frac{d^2R}{dx^2} + \frac{d^2R}{dy^2},$$

or by (25)

$$\frac{d^2h}{dt^2} = D\left(\frac{d^2h}{dx^2} + \frac{d^2h}{dy^2}\right). \quad \text{......................(27)}$$

This equation applies to both media, since there is no change in the value of D. Thus, so far as the equations to be satisfied in the interior are concerned, the incident wave may be supposed to continue its course without alteration.

It is equally evident that the general boundary conditions are also satisfied. For f, Q, c vanish throughout, and by (6) the continuity of R and b merely requires the continuity of h and dh/dx. Since all the conditions are satisfied by supposing the incident wave to pass on without alteration, it is clear that there can be no reflected wave.

We have next to consider the case when the vibrations are executed in the plane of incidence, so that h vanishes, while (as before) all the remaining functions are independent of z. On account of the symmetry there can be but one reflected and but one refracted wave, and in each h must vanish. We may, therefore, take the following expressions as applicable to the various waves:—

Incident wave:

$$f = q\, e^{i(px+qy+st)}, \qquad g = -p\, e^{i(px+qy+st)} \quad \text{..................(28)}$$

satisfying

$$pf + qg = 0;$$

Reflected wave:

$$f = q\,\theta'\, e^{i(p'x+qy+st)}, \quad g = -p'\,\theta'\, e^{i(p'x+qy+st)}; \quad \text{...........(29)}$$

Refracted wave:

$$f = q\,\theta_1\, e^{i(p_1x+qy+st)}, \quad g = -p_1\theta_1\, e^{i(p_1x+qy+st)}. \quad \text{...........(30)}$$

The coefficient of the time (s) is necessarily the same throughout on account of the periodicity; and the coefficient of y is the same, since the traces of all three waves upon the plane of separation $x = 0$ must move together. The relations between p, q, s; p', q, s; p_1, q, s are to be obtained by substitution in the differential equations. Of these the equation in h is satisfied identically, since $R = 0$. The other equations for the upper medium are by (7), (8), (25),

$$\frac{d^2f}{dt^2} = \frac{d^2}{dy^2}(Af + Bg) - \frac{d^2}{dx\,dy}(Bf + Cg),$$

$$\frac{d^2g}{dt^2} = \frac{d^2}{dx^2}(Bf + Cg) - \frac{d^2}{dx\,dy}(Af + Bg).$$

These must be satisfied by the incident and reflected waves. On substitution we find that both equations lead to the same conditions, viz.:—

$$s^2 = Aq^2 - 2Bpq + Cp^2, \quad \text{......................(31)}$$

a quadratic equation of which the two roots give p and p' in terms of q and s.

In the second medium we get in like manner for the refracted wave

$$s^2 = Aq^2 + 2Bp_1q + Cp_1^2, \quad \text{.......................(32)}$$

the sign of B being changed. Equating the two values of s^2, we find

$$C(p^2 - p_1^2) = 2Bq(p + p_1),$$

or

$$C(p - p_1) = 2Bq. \quad \text{................................(33)}$$

We have now to consider the boundary conditions (A). The functions R and b vanish throughout; but it remains to provide for the continuity of f, Q, and c, when $x = 0$. The first of these conditions gives at once

$$1 + \theta' = \theta_1. \quad \text{................................(34)}$$

Again, the continuity of Q, equal to $Bf+Cg$ in the first medium, and to $-Bf+Cg$ in the second, gives

$$Bq-Cp+\theta'(Bq-Cp')=-\theta_1(Bq+Cp_1). \quad \text{..........(35)}$$

The continuity of c leads, when regard is paid to (31), (32), merely to the repetition of the condition (34).

If we eliminate θ_1 between (34), (35), we find

$$\theta'\{2Bq-Cp'-Cp_1\}=C(p-p_1)-2Bq=0 \text{ by (33).}$$

Hence θ' vanishes. Neither in this case, nor when the vibrations are perpendicular to the plane of incidence, is there any reflexion of light incident in the plane of symmetry. And this conclusion may of course be extended to natural light, and to light plane or elliptically polarized in any way whatever.

Plane of Incidence perpendicular to that of Symmetry.

We have now to consider the case when the plane of incidence is the plane $y=0$, perpendicular to that of symmetry. Here f, g, h are all finite, but they (as well as P, Q, R, &c.) are independent of the coordinate y. The problem is more complicated than when the plane of incidence coincides with that of symmetry, because an incident wave is here attended by *two* reflected waves, and *two* refracted waves.

The equation of the incident wave in the upper medium may be expressed

$$f, g, h=(\lambda, \mu, \nu)\,e^{i(px+rz+st)};$$

or, since by (2) $\lambda p+\nu r=0$,

$$f, g, h=(r, \mu, -p)\,e^{i(px+rz+st)}. \quad \text{..................(36)}$$

The differential equations to be satisfied in the upper medium assume the form

$$\frac{d^2f}{dt^2}=\frac{d^2}{dz^2}(Af+Bg)-\frac{d^2}{dx\,dz}(Dh),$$

$$\frac{d^2g}{dt^2}=\left(\frac{d^2}{dx^2}+\frac{d^2}{dz^2}\right)(Bf+Cg),$$

$$\frac{d^2h}{dt^2}=\frac{d^2}{dx^2}(Dh)-\frac{d^2}{dx\,dz}(Af+Bg).$$

If we substitute for f, g, h from (36), the first and third equations give

$$s^2=r(Ar+B\mu)+p^2D, \quad \text{...............(37)}$$

and the second equation gives

$$\mu s^2=(p^2+r^2)(Br+C\mu). \quad \text{.....................(38)}$$

These two equations determine p and μ, when r, s are given. Since the elimination of μ leads to a quadratic in p^2, it is evident that there are four admissible values $\pm p_1$, $\pm p_2$, corresponding to waves of given periodicity, whose trace on the plane of separation moves with a given velocity. Of these two (say with the + sign) are waves approaching the surface, and two are waves receding from it. If we limit ourselves to a single incident wave $(+p_1)$ we shall have still to take into account two reflected waves corresponding to $-p_1$, $-p_2$. The equations show that the value of μ is the same whether p be positive or negative; we shall suppose that μ_1 corresponds to $\pm p_1$, μ_2 to $\pm p_2$.

In applying the equations to the second medium we have to change the sign of B; and it is evident that they are satisfied by the same values of p as before, and that the preceding values of μ are to be taken negatively. Hence in the second medium $-\mu_1$ corresponds to $\pm p_1$, $-\mu_2$ to $\pm p_2$. For the purposes of our present problem, where there is no incident wave in the second medium, we are concerned only with $+p_1$ and $+p_2$.

The complete specification of the system of waves corresponding to a single incident wave (p_1) in the first medium is thus:—

Incident wave:

$$f, g, h = (r, \mu_1, -p_1)\,\Theta_1\, e^{i(p_1x+rz+st)}\,; \quad\text{.................(39)}$$

Two reflected waves:

$$f, g, h = (r, \mu_1, p_1)\,\theta'\, e^{i(-p_1x+rz+st)} + (r, \mu_2, p_2)\,\theta''\, e^{i(-p_2x+rz+st)}\,; \quad\text{.................(40)}$$

Two refracted waves:*

$$f, g, h = (r, -\mu_1, -p_1)\,\theta_1\, e^{i(p_1x+rz+st)} + (r, -\mu_2, -p_2)\,\theta_2\, e^{i(p_2x+rz+st)}. \quad\text{...............(41)}$$

The next step is the introduction of the boundary conditions (A). The continuity of f requires that

$$\Theta_1 + \theta' + \theta'' = \theta_1 + \theta_2. \quad\text{..........................(42)}$$

The continuity of R, or Dh, or h, gives with equal facility

$$p_1\Theta_1 - p_1\theta' - p_2\theta'' = p_1\theta_1 + p_2\theta_2. \quad\text{.................(43)}$$

Again, the continuity of Q, equal to $Bf + Cg$ in the first medium and to $-Bf + Cg$ in the second, gives

$$(Br + C\mu_1)\,\Theta_1 + (Br + C\mu_1)\,\theta' + (Br + C\mu_2)\,\theta'' = -(Br + C\mu_1)\,\theta_1 - (Br + C\mu_2)\,\theta_2. \quad\text{...........(44)}$$

The continuity of b, or db/dt, or by (6) $dR/dx - dP/dz$, is found, when regard is paid to (37), to be already secured by (42); and we have only further to

* It should be noticed that one of the refracted waves is not refracted in the literal sense, being parallel to the incident wave.

consider the continuity of dc/dt, or by (6) of dQ/dx, since P is here independent of y. Thus

$$p_1(Br + C\mu_1)\Theta_1 - p_1(Br + C\mu_1)\theta' - p_2(Br + C\mu_2)\theta'' = -p_1(Br + C\mu_1)\theta_1 - p_2(Br + C\mu_2)\theta_2. \quad \text{......(45)}$$

The coefficients which occur in (44), (45) may be expressed more briefly in terms of the *velocities* of the various waves. For

$$V^2 = s^2/(p^2 + r^2), \quad \text{.........................(46)}$$

and thus by (38),

$$Br + C\mu_1 = \mu_1 V_1^2, \qquad Br + C\mu_2 = \mu_2 V_2^2. \quad \text{...........(47)}$$

Setting now

$$p_2/p_1 = \varpi, \qquad \mu_2 V_2^2/\mu_1 V_1^2 = \sigma, \quad \text{.................(47')}$$

the four equations of condition take the form

$$\left.\begin{aligned} \Theta_1 + \theta' + \theta'' &= \theta_1 + \theta_2, \\ \Theta_1 - \theta' - \varpi\theta'' &= \theta_1 + \varpi\theta_2, \\ \Theta_1 + \theta' + \sigma\theta'' &= -\theta_1 - \sigma\theta_2, \\ \Theta_1 - \theta' - \varpi\sigma\theta'' &= -\theta_1 - \varpi\sigma\theta_2. \end{aligned}\right\}. \quad \text{..............(48)}$$

If we equate the values of θ_1, θ_2 obtained from the first and second pairs of equations (48), we find

$$\left.\begin{aligned} (\sigma + 1)\theta' + (\varpi + 1)\sigma\theta'' &= 0, \\ (\varpi - 1)\Theta_1 + (\varpi + 1)\theta' + \varpi(\sigma + 1)\theta'' &= 0, \end{aligned}\right\}, \quad \text{........(49)}$$

and from these again

$$\theta' = -\frac{\sigma(\varpi^2 - 1)\Theta_1}{(\varpi - \sigma)(\varpi\sigma - 1)}, \qquad \theta'' = \frac{(\sigma + 1)(\varpi - 1)\Theta_1}{(\varpi - \sigma)(\varpi\sigma - 1)}, \quad \text{...(50, 51)}$$

by which the two reflected waves are determined.

These reflected waves correspond to the incident wave (Θ_1, p_1, μ_1), and it is the wave θ' which is reflected according to the ordinary law. If there be a second incident wave (Θ_2, p_2, μ_2), the corresponding reflected waves are to be found from (50), (51) by interchanging θ', θ'', and by writing for ϖ, σ the reciprocals of these ratios. If both incident waves coexist,

$$\theta' = \frac{(1 - \varpi)\sigma}{(\varpi - \sigma)(\varpi\sigma - 1)}\{(1 + \varpi)\Theta_1 + \varpi(1 + \sigma)\Theta_2\}, \quad \text{........(52)}$$

$$\theta'' = \frac{\varpi - 1}{(\varpi - \sigma)(\varpi\sigma - 1)}\{(1 + \sigma)\Theta_1 + (1 + \varpi)\sigma\Theta_2\}. \quad \text{........(53)}$$

It will be observed that although the fronts of the two incident waves Θ_1, Θ_2 are not parallel, they are the waves that would be generated by the double refraction of a single wave incident from an isotropic medium upon a face of the crystal parallel to the twin plane.

Doubly Refracting Power Small.

Thus far our equations are general. But the interpretation will be very much facilitated if we introduce a supposition, which does not deviate far from the reality of nature, viz. that the doubly refracting energy is comparatively small. There is no new limitation upon the direction of the principal axes relatively to those of coordinates, but we assume that A, C, D are nearly equal, and that B is small. We may imagine the two twin crystals to be bounded by faces parallel to the twin face, and to be embedded in an isotropic medium of nearly similar optical power. Under these circumstances p_1, p_2; V_1, V_2 are nearly equal, so that approximately $\varpi = 1$, $\sigma = \mu_2/\mu_1$; and we may write (52), (53) in the form

$$\theta' = \frac{p_2 - p_1}{p(\mu_2 - \mu_1)^2}\{2\mu_1\mu_2\Theta_1 + \mu_2(\mu_1 + \mu_2)\Theta_2\}, \quad \text{...........(54)}$$

$$\theta'' = \frac{p_1 - p_2}{p(\mu_2 - \mu_1)^2}\{\mu_1(\mu_1 + \mu_2)\Theta_1 + 2\mu_1\mu_2\Theta_2\}. \quad \text{...........(55)}$$

It should be remarked that the intensities of the waves represented by Θ_1, &c. are not simply proportional to Θ_1^2, &c. Referring to (39), (40), we see that the intensity of Θ_1, θ' is measured by $(r^2 + p^2 + \mu_1^2)(\Theta_1^2, \theta'^2)$; and that of Θ_2, θ'' by $(r^2 + p^2 + \mu_2^2)(\Theta_2^2, \theta''^2)$.

Plate bounded by Surfaces parallel to Twin Plane.

Let us now regard the waves Θ_1, Θ_2 as due to the passage into the crystal of waves from an isotropic medium, under such conditions (of gradual transition, if necessary) that there is no loss by reflexion. The interface is supposed to be parallel to the twin reflecting plane, and the optical power to be so nearly equal to that of the crystal that the refraction is negligible. Then, if the vibration parallel to y (perpendicular to the plane of incidence) be M, and that in plane of incidence be N, we have

$$M = \mu_1\Theta_1 + \mu_2\Theta_2, \qquad N = \sqrt{(p^2 + r^2)}\{\Theta_1 + \Theta_2\}. \quad \text{......(56, 57)}$$

In like manner, if the vibrations of the emergent reflected wave perpendicular and parallel to the plane of incidence be M', N',

$$M' = \mu_1\theta' + \mu_2\theta'', \qquad N' = \sqrt{(p^2 + r^2)}\{\theta' + \theta''\}. \quad \text{......(58, 59)}$$

If we are prepared to push to an extreme our supposition as to the smallness of the doubly refracting power, Θ, θ in these equations may be identified with the corresponding quantities in (54), (55); for a retardation of phase in crossing and recrossing the stratum *alike for all the waves* might be disregarded. We shall presently return to this question; but we will in the meantime trace out the consequences which ensue when the double refraction, if not extremely small in itself, is at least so small in relation to the distances through which it acts (the thickness of the stratum), that the relative changes of phase may be neglected. Then

$$M' = \frac{(p_2 - p_1)\,\mu_1\mu_2}{p\,(\mu_1 - \mu_2)}\{\Theta_2 + \Theta_1\} = \frac{(p_2 - p_1)\,\mu_1\mu_2}{p\,(\mu_1 - \mu_2)}\,\frac{N}{\sqrt{(p^2 + r^2)}}, \quad \text{......(60)}$$

$$N' = \frac{(p_2 - p_1)\sqrt{(p^2 + r^2)}\,.\,M}{p\,(\mu_2 - \mu_1)}. \quad \text{......................(61)}$$

We have now to introduce certain relations derived from (37), (38). By elimination of s, we get

$$Br\,.\,\mu^2 + \mu\,\{(A - C)\,r^2 + (D - C)\,p^2\} - Br\,(p^2 + r^2) = 0. \quad \text{......(62)}$$

If we here disregard the difference between p_1 and p_2, we may treat it as a quadratic, by which the two values of μ are determined; and it follows that

$$-\,\mu_1\mu_2 = p^2 + r^2. \quad \text{..............................(63)}$$

We might have arrived at this conclusion more quickly from the consideration that in the limit the two directions of displacement $(r;\ \mu_1, p_1)$, (r, μ_2, p_2) in the reflected waves must be perpendicular to one another.

Again, from the general equation (37) we see that

$$Br\,(\mu_1 - \mu_2) + (p_1^2 - p_2^2)\,D = 0,$$

whence approximately,

$$\frac{p_2 - p_1}{\mu_1 - \mu_2} = \frac{rB}{2pD}. \quad \text{..............................(64)}$$

Introducing these relations into (60), (61), we find

$$M' = -\frac{r\sqrt{(p^2 + r^2)}\,.\,B\,.\,N}{2p^2D}, \quad \text{....................(65)}$$

$$N' = -\frac{r\sqrt{(p^2 + r^2)}\,.\,B\,.\,M}{2p^2D}. \quad \text{....................(66)}$$

These equations indicate that the intensity of the reflected light $(M'^2 + N'^2)$ is proportional to that of the incident, without regard to the polarization of the latter. Again, if the incident light be unpolarized (M and N equal, and without permanent phase relation), so also is the reflected light. But what is more surprising is, that if the incident light be polarized in or perpendicular to the plane of incidence, the reflected light is polarized *in the opposite manner*.

The intensity of reflexion may be expressed in terms of the angle of incidence ϕ, for

$$p/\sqrt{(p^2 + r^2)} = \cos\phi, \qquad r/\sqrt{(p^2 + r^2)} = \sin\phi,$$

so that

$$M'^2 + N'^2 = \frac{B^2 \sin^2 \phi}{4D^2 \cos^4 \phi}(M^2 + N^2). \quad \text{..................(67)}$$

When the angle of incidence is small, the intensity is proportional to its square. And, as was to be expected, the reflexion is proportional to B^2.

The laws here arrived at are liable to modification when, as must usually happen in practice, the thickness of the plate cannot be neglected. The incident light, on its way to the twin surface, and the reflected light on its way back, is subject to a depolarizing influence, which in most cases complicates the relation between the polarizations of the light before entering and after leaving the crystal. One law, however, remains unaffected. If the light impinging upon the crystal be unpolarized, it retains this character upon arrival at the twin face. We have shown that it does not lose it in the act of reflexion, neither can it lose it in the return passage after reflexion. Hence, if the light originally incident upon the layer of crystal be unpolarized, so is the reflected light ultimately emergent from it.

If, on the other hand, the incident light be polarized, whether plane or elliptically, the character of the emergent light must depend upon the precise thickness of the crystalline layer, and will vary rapidly from one part of the spectrum to another. The simplest case that we can consider is when the polarization of the incident rays is such that one or other of Θ_1, Θ_2 vanish. We will suppose that it is Θ_2; so that after reflexion,

$$\frac{\text{Intensity of } \theta'}{\text{Intensity of } \theta''} = \frac{p^2 + r^2 + \mu_1^2}{p^2 + r^2 + \mu_2^2} \cdot \frac{\theta'^2}{\theta''^2}$$

$$= \frac{\mu_1^2 - \mu_1\mu_2}{\mu_2^2 - \mu_1\mu_2} \frac{4\mu_1^2\mu_2^2}{\mu_1^2(\mu_1 + \mu_2)^2} = \frac{-4\mu_1\mu_2}{(\mu_1 + \mu_2)^2} \quad \text{.........(68)}$$

by (54), (55). This is the ratio of intensities that would be observed with an analyzing nicol held so as to retain in succession θ' and θ''. If the crystalline layer be moderately thick, and the light be of mixed wave-lengths, there will be no interference observable between θ' and θ'', and thus the ratio just found is the extreme ratio of intensities. By means of (62) we may express it in terms of the angle of incidence (ϕ), and of the fundamental optical constants of the crystal. Thus

$$\sqrt{\left\{\frac{-\mu_1\mu_2}{(\mu_1 + \mu_2)^2}\right\}} = \frac{\sqrt{\{p^2 + r^2\}} \,.\, Br}{(A - C) r^2 + (D - C) p^2}$$

$$= \frac{B \sin \phi}{(A - C) \sin^2 \phi + (D - C) \cos^2 \phi}. \quad \text{..............(69)}$$

This expression shows that in general the emergent light will be fully polarized only when ϕ is very small. In this case we virtually fall back upon our original investigation where the thickness of the layer was neglected. Since only Θ_1 is present, there is no depolarization in the first passage; and when $\phi = 0$ there is no depolarization upon the return passage in consequence of the disappearance of θ'. The polarizations corresponding in this case to Θ_1, Θ_2 are obviously those in and perpendicular to the plane of incidence; and we learn that, *when the angle of incidence is small*, polarizations in and perpendicular to the plane of incidence are reversed in the reflected ray. If the incident ray be polarized in other directions than these, the reflected ray is in general not fully polarized, even though the angle of incidence be small.

150.

ON THE REMARKABLE PHENOMENON OF CRYSTALLINE REFLEXION DESCRIBED BY PROF. STOKES.

[*Phil. Mag.* XXVI. pp. 256—265, 1888.]

THE phenomenon in question is that exhibited by certain crystals of chlorate of potash, consisting of a peculiar internal coloured reflexion. The following, stated very briefly, are its leading features as described by Stokes*:—

(1) If one of the crystalline plates be turned round in its own plane, without alteration of the angle of incidence, the peculiar reflexion vanishes twice in a revolution, viz. when the plane of incidence coincides with the plane of symmetry of the crystal.

(2) As the angle of incidence is increased, the reflected light becomes brighter and rises in refrangibility.

(3) The colours are not due to absorption, the transmitted light being strictly complementary to the reflected.

(4) The coloured light is not polarized. It is produced indifferently whether the incident light be common light or light polarized in any plane, and is seen whether the reflected light be viewed directly or through a Nicol's prism turned in any way.

(5) The spectrum of the reflected light is frequently found to consist almost entirely of a comparatively narrow band. When the angle of incidence is increased, the band moves in the direction of increasing refrangibility, and at the same time increases rapidly in width. In many cases the reflexion appears to be almost total.

Prof. Stokes has proved that the seat of the colour is a narrow layer, about a thousandth of an inch in thickness, in the interior of the crystal; and

* *Proc. Roy. Soc.* Feb. 1885.

he gives reasons for regarding this layer as a twin stratum. But the phenomenon remains a mystery. "It is certainly very extraordinary and paradoxical that light should suffer total or all but total reflexion at a transparent stratum of the very same substance, merely differing in orientation, in which the light had been travelling, and that, independently of its polarization."

From the first reading of Prof. Stokes's paper, I have been much impressed with the difficulty so clearly set forth. It seemed impossible that a combination of two surfaces merely could determine either so copious or so highly selected a reflexion. If light of a particular wave-length is almost totally reflected, what hinders the reflexion when the wave-length is altered, say, by one twentieth part? Such a result may arise from the interference of two streams under a relative retardation of many periods; but in that case there are necessarily a whole series of wave-lengths all equally effective. The prism should reveal a number of bright bands and not merely a single band. The selection of a particular wave-length reminds one rather of what takes place in gratings; and I was from the first inclined to attribute the colours to a periodic structure, in which the twins alternate a large number of times. Such a view explains not only the high degree of selection, but also the copiousness of the reflexion.

Partly with a view to this question, I have discussed in a recent paper* the propagation of waves in an infinite laminated medium (where, however, the properties are supposed to vary continuously according to the harmonic law), and have shown that, however slight the variation, reflexion is ultimately total, provided the agreement be sufficiently close between the wave-length of the structure and the half wave-length of the vibration. The number of alternations of structure necessary in order to secure a practically perfect reflexion will evidently depend upon the other circumstances of the case. If the variation be slight, so that a single reflexion is but feeble, a large number of alternations are necessary for the full effect, and a correspondingly accurate adjustment of wave-lengths is then required. If the variation be greater, or act to better advantage, so that a single reflexion is more powerful, there is no need to multiply so greatly the number of alternations; and at the same time the demand for precision of adjustment becomes less exacting. The application of this principle to the case of an actual crystal, supposed to include a given number of alternations, presents no difficulty. At perpendicular incidence symmetry requires (and observation verifies) that the reflexion vanish; but, as the angle of incidence increases, a transition from one twin to the other becomes more and more capable of causing reflexion.

* "On the Maintenance of Vibrations by Forces of Double Frequency, and on the Propagation of Waves through a Medium endowed with a Periodic Structure," *Phil. Mag.* Aug. 1887. [Vol. III. p. 1.]

Hence if the number of alternations be large, the spectrum of the reflected light is at first limited to a narrow band (whose width determines in fact the number of alternations). As the angle of incidence increases, the reflexion at the centre soon becomes sensibly total, and at the same time the band begins to widen*, in consequence of the less precise adjustment of wave-lengths now necessary. At higher angles the reflexion may be sensibly total over a band of considerable width. All this agrees precisely with Prof. Stokes's description of the case considered by him to be typical. The movement of the band towards the blue end of the spectrum is to be attributed to the increasing obliquity within the crystal, as in the ordinary theory of thin plates.

It thus appears that if we allow ourselves to invent a suitable crystalline structure, there need be no difficulty in explaining the vigour and purity of the reflexion; but such an exercise of ingenuity is of little avail unless we can at the same time render an account of the equally remarkable circumstances stated in (1) and (4). When the incidence is in the plane of symmetry, no reflexion takes place. As Prof. Stokes remarks, this might be expected as regards light polarized in the plane of symmetry; but that there should be no reflexion of the other polarized component is curious, to say the least. Not less remarkable is it that when the incidence is in the perpendicular plane, the reflected light should show no signs of polarization. The phenomenon being certainly connected with the doubly refracting property, we should naturally have expected the contrary.

The investigation of the reflexion from a twin-plane, contained in the preceding paper [Vol. III. p. 194], shows, however, that the actually observed results are in conformity with theory. In the plane of symmetry there should be no reflexion of either polarized component, at least to the same degree of approximation as is attained in Fresnel's well-known formulæ for isotropic reflexion. As regards light reflected in the perpendicular plane, theory indicates that if the incident light be unpolarized, so also will be the reflected light. Again, the intensity of the (unanalyzed) reflected light should be independent of the polarization of the incident. So far there is complete agreement with the observations of Prof. Stokes. But there is a further peculiarity to be noticed. Theory shows that in the act of reflexion at a twin plane, *the polarization is reversed.* If the incident light be polarized in the plane of incidence, the reflected light is polarized in the perpendicular plane, and *vice versâ.* When I first obtained this result, I

* It should be observed that if the spectrum be a prismatic one, there is a cause of widening which must be regarded as purely instrumental. According to Cauchy's law ($\mu = A + B\lambda^{-2}$),

$$\delta\mu = -2B\lambda^{-3}\delta\lambda;$$

so that if the band correspond in every position to a given relative range of λ, its apparent width (reckoned as proportional to $\delta\mu$) will vary as λ^{-2}. In a diffraction-spectrum this cause of widening with diminishing λ would be non-existent.

thought it applicable without reservation in the actual experiment, and on trial was disappointed to find that the reflected light was nearly unpolarized, even when the incident light was fully polarized, whether in the plane of incidence or in the perpendicular plane. When, however, the *angle of incidence was diminished*, the expected phenomenon was observed, provided that the original polarization were in, or perpendicular to, the plane of incidence. If the original polarization were oblique, the reflected light was not fully polarized, even though the angle of incidence were small*.

Further consideration appeared to show that the loss of polarization usually observed could be explained by the depolarizing action of the layer of crystal through which the light passes, both on its way to the reflecting plane and on its return therefrom. As is shown in the preceding paper, this depolarizing action does not occur when the angle of incidence is small, and the polarization in, or perpendicular to, the plane of incidence. It seems scarcely too much to say that the theory not only explains the laws laid down by Stokes, but also predicts a very peculiar law not before suspected†.

The theory, as so far developed, is indeed limited to incidences in the two principal planes. It could probably be treated more generally without serious difficulty; but there seems no reason to suppose that anything very distinctive would emerge. It is not unlikely that the intensity would prove to be proportional to the square of the sine of the angle between the planes of incidence and of symmetry. If this theory be accepted—and I see no reason for distrusting it—the brilliant reflexion cannot be explained as due to a single twin stratum. The simplest case which we can consider is when the angle of incidence is small and the polarization in or perpendicular to the plane of incidence. There is then sensibly but one wave reflected at the first twin plane. On the arrival of the transmitted wave at the hinder surface of the twin stratum, a second reflexion ensues, similar to the first, except for the reversal of phase due to the altered circumstances. The relation to one another of the two reflected waves is exactly the same as in the ordinary theory of thin plates, and does not appear to admit of the production of anything unusual. I think we may even go further, and conclude that in conformity with our theory it is impossible to find an

* Whatever the angle of incidence, the arrangement of crossed nicols may sometimes be conveniently applied in order to isolate the light under investigation from that reflected at the front surface of the crystalline plate. In the observations described in the text the crystal was mounted with Canada balsam between thick plates of glass, so that there was no difficulty in observing separately the various reflexions. At small angles of incidence the coloured image is at its brightest when the analyzing nicol is so turned that the white image (reflected from the glass) vanishes, and *vice versâ*, the incident light being polarized in, or perpendicularly to, the plane of incidence.

† The wording of Prof. Stokes's description is perhaps a little ambiguous, but I gather that he did not examine the result of a *simultaneous* operation of polarizer and analyzer.

explanation of the brilliant and highly selected reflexion, unless upon the supposition that there is a repeated alternation of structure.

The optical evidence in favour of the view that there is a large number of twin planes thus appears to be very strong; the difficulty is rather to understand how such a structure can originate. And yet if we admit, as we must, the possibility of the formation of one twin plane, and of two twin planes at a very small distance asunder*, there seems nothing to forbid a structure regularly periodic, which may perhaps be due to causes vibratory in their nature.

It would undoubtedly be far more satisfactory to be able to speak of the periodic structure as a matter of direct observation, and it is to be desired that some practised microscopist should turn his attention to the subject. *Ex hypothesi*, we could not expect to see the ruled pattern upon a section cut perpendicularly to the twin planes, as it would lie upon, or beyond, the microscopic limit. I have tried to detect it upon a surface inclined to the planes at a very small angle, but hitherto without success†.

In the absence of complete evidence it is proper to treat the views here put forward with a certain reserve; but it is perhaps not premature to consider a little further what may be expected to result from a structure more or less regular. If the periodicity be nearly perfect, the bright central band in the spectrum would be accompanied by subordinate bands of inferior and decreasing brilliancy. If the angle of incidence be small, so that the aggregate reflexion is but feeble, each stratum may be considered to act independently, and the various reflected waves to be simply superposed. The resultant intensity will depend of course upon the phase relations. At the centre of the band the partial reflexions agree in phase, and the intensity is a maximum. As we leave this point in either direction, the phases begin to separate. When the alteration of wave-length is such that the phases of the reflected waves range over a complete cycle, the resultant vanishes, and a dark band appears in the spectrum. The same thing occurs whenever the relative retardation of the extreme components amounts to a complete number of periods. At points approximately midway between these, the resultant is a maximum, but the values of the successive maxima diminish‡. Near the central band, where (when the number of alternations is great)

* This is the simplest supposition open to us, when, as in most of the coloured crystals, the parts on either side of a very thin lamina are similarly oriented.

† [1901. In *Manchester Proceedings* for 1889, Vol. III. p. 117, it is reported that "Dr Hodgkinson exhibited a specimen of iridescent chlorate of potash mounted in a special way in order to demonstrate that the colour is produced, not by the interference of one thin plate, but by numerous thin plates. The thin plates were readily seen in the specimen by means of a hand magnifier, and the exhibit confirmed a prediction made several months since by Lord Rayleigh."]

‡ The case is similar to that of the distribution of brightness in the neighbourhood of a "principal maximum," when light of given wave-length is diffracted by a grating.

a considerable fraction of the incident light is reflected by the system of layers, this way of regarding the matter may cease to be applicable, for then the anterior and the posterior layers act under sensibly different conditions.

Apart from the magnitude of the complete linear period, something will depend upon the manner in which it is divided between the twins. The most favourable, as it is also perhaps the most probable, arrangement is that in which the thicknesses are equal. In that case every partial reflexion may agree in phase. If the thicknesses, though regular, are unequal, we may first form the resultant for contiguous pairs, and then consider the manner in which the partial resultants aggregate.

It will be seen that even if the thicknesses of the twins are equal, there are still *two* ways in which a regularly laminated crystal may vary, as compared with the single kind of variation open to a simple twin stratum. These are the magnitude of the linear period, and the number of periods. Comparison of a number of coloured crystals* seems to favour the view that there are important differences of constitution, even when the colour is the same at a given incidence.

In many cases the appearances are such as to suggest that the periodicity is imperfect. A little irregularity might alter or obliterate the subordinate bands, while leaving the central band practically unaffected. Sometimes there is evidence of two or more distinct periods, each sustained through a number of alternations. If the period were subject to a gradual change, the central band in the spectrum of the reflected light would be diffused, even at small angles of incidence. The mere broadening of the band might be due to fewness of alternations; but this case would be distinguished from the other by the accompanying feebleness of illumination.

On the whole, the character of the reflected light appears to me to harmonize generally with the periodical theory. One objection, however, should be mentioned. It might be supposed that the total number of twin planes was as likely to be odd as to be even. In the former case the layers of crystal on either side of the thin lamina (which is the seat of the colour) would be of opposite orientations. In many crystals the character of the twinning is difficult of observation, but I have not noticed any instance of *brilliant coloration* answering to this description. So far as it goes this argument is in favour of the simple stratum theory; but, in view of our ignorance as to how the twin planes originate, it can hardly be considered decisive.

I have also examined a number of what appeared to be simply twinned crystals, kindly sent me by Mr Stanford, of the North British Chemical Works. The light reflected from the twin plane is not easily observed on

* For a rich collection of such crystals I am indebted to Mr Muspratt. He informs me that, though the result of a second crystallization from comparatively pure liquids, the coloured crystals are but rarely found when the chlorate is produced by the magnesium process.

account of its feeble character, at least when, as in the experiments now referred to, the incidence is limited by the requirement that the light must enter the crystal at a face parallel to the twin plane. Using, however, the method described by Prof. Stokes (§ 13), I was enabled to separate the reflexions at the twin plane from those at the external surfaces of the crystal. A narrow slit admitted sunlight into the dark room, and was focused upon the crystal by a good achromatic object-glass*. When the obliquely reflected light was examined with a hand magnifier, a ghost-like image corresponding to the twin plane could usually be detected. As the crystal was rotated in its own plane, this image *vanished twice during the revolution.*

It is worthy of notice that there is an evident difference both in the brightness and quality of the reflected light obtained from different crystals, even though apparently simply twinned. This suggests that, instead of a single twin plane, there may sometimes be in reality 3, 5, or a higher odd number of such in close juxtaposition. In other specimens, affording similar reflexions, the principal thicknesses on either side of a very thin layer are undoubtedly of the same kind, so that the number of twin planes must be even. Here, again, the reflected light exhibited marked differences, when various crystals were examined. In none of those now referred to could the light reflected from the thin layer be observed without very special arrangements.

In these experiments the light entered and left the crystal by a face parallel to the twin planes. In one specially well-formed and apparently simply twinned crystal I was able to observe a much more oblique reflexion from the internal surface or surfaces. The light here entered and left the crystal by cleavage faces making a large angle with the reflecting planes, and thus under conditions widely different from those considered hitherto, and in the latter part of the preceding theoretical discussion. Three reflected images were seen, all completely polarized (the original light being unpolarized), two in one direction and the third in the opposite direction. These images are coloured, and present tolerably discontinuous spectra, giving rise to a suspicion that the twin plane is not really single. These observations were made without special arrangements by merely examining the reflected images of a candle-flame, when the crystal was held close to the eye.

I have made many experiments on the crystallization of chlorate of potash in the hope of tracing the genesis of the coloured crystals, but without decisive results. Besides the usually small but highly coloured crystals, found by Stokes, I have obtained many larger ones in which the reflexion is feebler and less pure. These appear to be distinct from the exceedingly thin plates which at the early stage of crystallization swim about in the solution. Mounted in Canada balsam the crystals in question show colours of varying degrees of brightness and purity; and under these circumstances the effect

* I did not succeed in my first trials when I employed a common lens.

can hardly be due to the action of the external surfaces (in contact with the balsam). The light disappears twice during the revolution of the plates in azimuth, just as in the case of the more highly coloured specimens. It seems natural to suppose that the reflexion takes place from twin surfaces relatively few in number, and perhaps less regular in disposition. Altogether the existence of these crystals favours the view that fully formed colour is due to a large number of regular alternations.

Some interesting observations bearing upon our present subject have been recorded by Mr Madan*. Transparent crystals, free from twinning, were heated on an iron plate to the neighbourhood of the fusion-point. During the heating no change was observable, but "when the temperature had sunk a few degrees a remarkable change spread quickly and quietly over the crystal-plate causing it to reflect light almost as brilliantly as if a film of silver had been deposited on it." Subsequently examined, the altered crystals are found to "reflect little light at small angles of incidence, but at all angles greater than about 10° they reflect light with a brilliancy which shows that the reflexion must be almost total....When the plate is turned round in its own plane, two positions are found, differing in azimuth by 180°, in which the crystal reflects no more light than an ordinary crystal under the same conditions. In these cases the plane of incidence coincides with the plane of crystallographic symmetry."

Mr Madan worked with comparatively thick (1 millim.) plates, from which the associated twin had been removed by grinding. In repeating his experiments I found it more convenient to use thin plates, such as may be obtained without difficulty from crystallizations upon a moderate scale, and which appear to be free from twinning†. There seems to be little doubt that the altered crystals are composed of twinned layers. Except in respect of colour, there is no difference between the behaviour of these crystals and that of the brilliantly iridescent ones described by Stokes. If light be incident at a small angle, and be polarized in or perpendicularly to the plane of incidence, the polarization of the reflected light is the *opposite* to that of the incident.

The only difference that I should suppose to exist between the constitution of these crystals and that of the iridescent ones is, that in the former case the alternations are irregular, and also probably more numerous. Mr Madan conceives that there are actual cavities between the layers in the heated crystals, comparing them to films of decomposed glass‡. It is,

* "On the Effect of Heat in changing the Structure of Crystals of Potassium Chlorate," *Nature*, May 20, 1886.

† It is not clear why composite crystals free from included mother-liquor should suffer disruption upon heating. A line drawn on the twin plane would tend to expand equally, to whichever crystal it be considered to belong.

‡ "Although a large amount of light must escape reflexion at a single cavity, yet if the transmitted rays encountered a large number of precisely similar and similarly situated cavities

however, certain that no closeness of contact could obviate the optical discontinuity at a twin plane; and there is besides a marked experimental distinction between the cases in question. It is easy to observe, and was, I think, observed by Brewster, that the application of water to a film of decomposed glass destroys the effect. The water insinuates itself into the cavities, and greatly attenuates the reflecting power. If a corresponding experiment be tried, by wetting the edge of one of Mr Madan's crystals with saturated solution of the salt, no change is observed to ensue.

Whether there are cavities or no, the fact that during the preparation the silvery reflexion does not set in until the crystal has sensibly cooled is of great interest. I have found that if a crystal in which the silvery lustre has already been produced be reheated, the lustre disappears, to return again upon a fall of temperature. The operation may be repeated any number of times.

The existence of twin strata in Iceland spar was observed by Brewster*, and Reusch† has shown that such strata can be induced artificially by suitably applied pressure (Fig. 1) in rhombs originally homogeneous. The planes of these strata truncate the polar edges, *i.e.* the edges which meet symmetrically at the obtuse trihedral angle (O). Being desirous of examining whether the reflexion from these strata would conform to the law deduced from theory, I submitted a rhomb to the treatment prescribed by Reusch with the effect of developing several exceedingly thin twin laminæ (four or five at least) in close juxtaposition. When light is reflected from these strata in a plane perpendicular to the edge (OD) which they truncate, the brilliancy is considerable. But the observation which I wished to make required that the plane of incidence should be perpendicular to this, so as to include the truncated edge and the optic axis. Without much difficulty it was proved that in this plane *the reflexion vanished,* reviving on either side as the plane of incidence deviated a little from the plane of symmetry. The observation was facilitated by immersing the crystal in a small cell containing water or bisulphide of carbon, the twin strata being horizontal, and the plane of symmetry parallel to two of the sides of the cell.

Fig. 1.

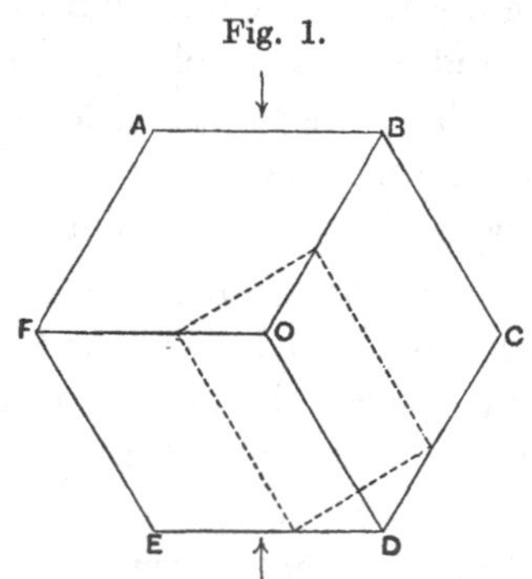

at slightly lower levels in the crystal, the sum of the partial reflexions would produce an effect almost equivalent to a total reflexion of the original incident ray, and a corresponding deficiency in the amount of light transmitted through the whole plate. The brilliancy of the colours in the light reflected from the well-known films of decomposed glass is accounted for in precisely the same way, and the successive separate films of glass can be easily seen under a microscope at the edges of the compound film, where they only partially overlap."

* *Treatise on Optics*, 1853, p. 349.

† *Pogg. Ann.* t. XII. p. 448 (1867).

151.

IS THE VELOCITY OF LIGHT IN AN ELECTROLYTIC LIQUID INFLUENCED BY AN ELECTRIC CURRENT IN THE DIRECTION OF PROPAGATION?

[*Brit. Ass. Report*, 1888, pp. 341—343.]

THE question here proposed has been considered by Roiti* and by Zecher†. My experiments were made in ignorance of the work of these observers, and the results would scarcely be worth recording were it not that the examination seems to have been pushed further than hitherto. It may be well to say at once that the result is negative.

The interference fringes were produced by the method of Michelson as used in his important investigation respecting "The Influence of Motion of the Medium upon the Velocity of Light‡." The incident ray *ab* meets a half-silvered surface at *b*, by which part of the light is reflected and part is transmitted. The reflected ray follows the course *abcdefbg*, being in all twice reflected in *b*. The transmitted ray takes the course *abfedcbg*, being twice transmitted at *b*. These rays having pursued identical paths are in a condition to form the centre of a system of fringes, however long and far apart may be the courses *cd*, *ef*.

There is here nothing to distinguish the ray *ab* from a neighbouring parallel ray. The incident plane *wave-front* perpendicular to *ab* gives rise eventually to two coincident wave-fronts perpendicular to *bg*. With a wave incident in another direction the case is different. The two emergent wave-fronts remain, indeed, necessarily parallel, both having experienced an even number of reflexions (four and six). But there will exist in general a relative retardation, of amount (for wave-fronts perpendicular to the plane of the diagram) proportional to the deviation from the principal wave-front.

* *Pogg. Ann.* CL. p. 164, 1873. † *Rep. de Phys.* XX. p. 151, 1884.
‡ *Am. Journal*, XXXI. p. 377, 1886.

Hence, if the incident light comes in all directions, a telescope at *g*, focused for indefinitely distant objects, reveals a system of interference bands, whose direction should be vertical, if the adjustments could be perfectly carried out in the manner intended.

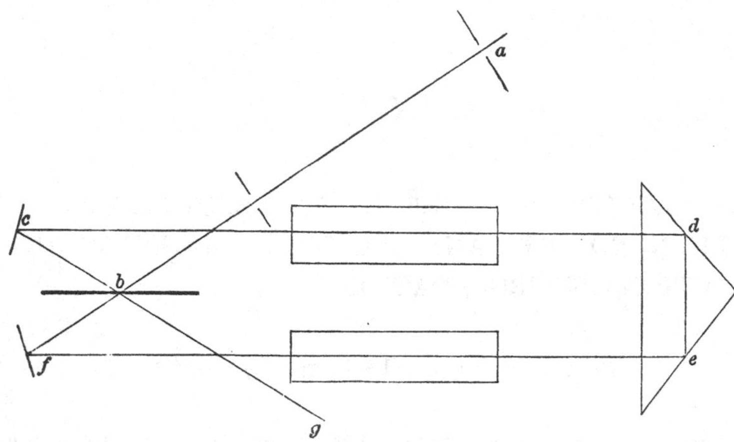

The success of the method does not require the complete symmetry of the diagram. If the reflexions at *d*, *e* are effected by a right-angled prism, it is necessary that *cd*, *ef* be parallel to one another but not that they be parallel to the surface *b*. Supposing all the surfaces to remain vertical in any case, the positions of *b*, *f*, and the incident ray *ab*, may be chosen arbitrarily. If the distance *de* between the parallel courses is not closely prescribed, *one* adjustment by rotation of the mirror *c* will suffice. In my experiments the optical parts were mounted upon a large iron plate, so that the movable pieces *c*, *de* could be shifted without loss of level. The incident ray *ab* was defined by a small hole near the paraffin lamp which served as a source of light, and by the centre of a moderately large circular aperture perforated in a screen and illuminated when necessary with a candle. The mirror *c* was then rotated until the rays *cd*, *fe* were parallel. This was tested by observing the equality of their mutual distances near the extremities of their course.

If the distance between the parallel rays is prescribed, the adjustment is more troublesome. The line *fe* being fixed, sights are laid down defining the desired position of *cd*. These sights, as well as those before referred to defining the incident ray, have now to be brought to apparent superposition as seen by an eye looking along *dc*. For this purpose *two* conditions have to be satisfied by, and two motions must be provided for, the mirror *c*. One of these should be a movement of rotation, and the other of translation in a direction nearly perpendicular to the plane of the mirror. Thus the mounting may consist of a circular turntable resting upon an iron plate, the curved edge of which is guided by the sides of a **V**, cut out of a flat piece of metal and clamped to the plate. In each position of the V the angular motions are easily swept over, and the double adjustment is effected without much difficulty. When the parallelism of the rays is secured, the insertion of the reflecting prism is all that remains. The adjustment of this is best effected with the eye at the observing telescope, which at this stage should be focused upon the small aperture in the neighbourhood of the flame. By a motion of the prism parallel to its hypothenuse the two images are brought to coincidence*, and then the bands appear, if not at once, when the telescope is accommodated for infinitely distant objects.

The half-silvered central plate would be at its best if it reflected light of the same intensity as it transmits. I have generally found the reflexion on the side next the air more powerful than upon the side next the glass; so that the ideal would require the geometric mean of the two reflexions to be equal to that of the two transmissions. A very slight silvering is all that is wanted, such as from its want of coherence and brilliancy would be useless for other purposes; and the bands appear tolerable black, even though the interfering lights are of decidedly unequal intensities. There is, of course, a reflexion from the unsilvered surface of the plate. Owing to want of parallelism in my apparatus, this image was distinctly separated from the other. The two back reflectors were of flat glass, silvered by the milk sugar process and used as specula.

The imperfections of the surfaces disturbed the formation of the bands from full accordance with theory. The definition was usually better when the pencils were limited, as by the screens employed to define the incident ray, than when all obstruction was removed. The final adjustments for the distinctness and desired width of bands were made with the eye at the telescope by shifting the reflecting prism and occasionally by slight displacements of one or other of the reflectors.

The tubes enclosing parts of *cd*, *ef*, and containing the electrolyte (diluted sulphuric acid of nearly maximum conductivity), were closed at the ends by plates of parallel glass. The current entered by lateral attachments, so arranged that liquid (or gas) rising or falling from the platinum electrodes would not at first enter the operative part of the tubes. The diameter of the tubes was about $\frac{3}{4}$ inch, and the effective length about 11 inches. [inch = 2·54 cms.]

It will be sufficient to give details of one experiment. The two tubes were connected in multiple arc, and of course in such a manner that the

* It should be noticed that if the object were at infinity, or if with the aid of a collimating lens an image of it were thrown to infinity, the two images as seen focused through the telescope would overlap in any case; for it may be proved that, whatever may be the positions of the five reflecting surfaces, the two emergent rays, corresponding to any incident ray, are necessarily parallel.

current travelled in opposite directions. The magnitude of the whole current (say from eight Grove cells) was 1·5 ampere; so that the current *density*, in amperes per sq. cm., was

$$\frac{\cdot 75}{\pi \times \cdot 38^2 \times 2\cdot 54^2} = \cdot 26.$$

Now one of the interfering rays travelled 22 inches, or 56 centimetres, with the current, and the other ray the same distance against the current. On *reversal* of the current no shift of the bands could be perceived under conditions where a shift of $\frac{1}{10}$ of a band* must have been evident. Hence we may conclude that a current of the above-mentioned density does not accelerate or retard the propagation of light in the ratio of $\frac{1}{10}\lambda$ to 224 cms. In the liquid we may take $\lambda = 4 \times 10^{-5}$ cms., and if we reduce the result so as to correspond to density unity, we may say that in dilute sulphuric acid a current of one ampere per square centimetre does not alter the velocity of light by 1 part in 13 millions, or by 15 metres per second.

It would probably be possible to carry the test ten or fifteen times further by the use of much larger tubes and a more powerful battery, but there seems to be no sufficient encouragement at present to make the attempt. The case would, of course, be very different were anyone to show by *à priori* argument a reason for expecting an effect of this order of magnitude.

* Probably I might say $\frac{1}{20}$, but it is best to be upon the safe side. When the contact was maintained, a slight shift was observed, but in a direction independent of that of the current.

152.

ON THE BENDING AND VIBRATION OF THIN ELASTIC SHELLS, ESPECIALLY OF CYLINDRICAL FORM.

[*Proceedings of the Royal Society*, XLV. pp. 105—123, 1888.]

IN a former publication* "On the Infinitesimal Bending of Surfaces of Revolution," I have applied the theory of bending to explain the deformation and vibration of thin elastic shells which are symmetrical about an axis, and have worked out in detail the case where the shell is a portion of a sphere. The validity of this application depends entirely upon the principle that when the shell is thin enough and is vibrating in one of the graver possible modes, the middle surface behaves as if it were inextensible. "When a thin sheet of matter is subjected to stress, the force which it opposes to extension is great in comparison with that which it opposes to bending. Under ordinary circumstances, the deformation takes place approximately as if the sheet were inextensible as a whole, a condition which, in a remarkable degree, facilitates calculation, though (it need scarcely be said) even bending implies an extension of all but the central layers." If we fix our attention upon one of the terms involving sines or cosines of multiples of the longitude, into which, according to Fourier's theorem, the whole deformation may be resolved, the condition of inextensibility is almost enough to define the type. If there are two edges, *e.g.*, parallel to circles of latitude, the solution contains two arbitrary constants; but if a pole be included, as when the shell is in the form of a hemisphere, one of the constants vanishes, and the type of deformation is wholly determined, without regard to any other mechanical condition, to be satisfied at the edge or elsewhere. It will be convenient to restate, analytically, the type of deformation arrived at {equation (5)}. If the point upon the middle surface, whose coordinates were originally a, θ, ϕ, moves to $a+\delta r$, $\theta+\delta\theta$, $\phi+\delta\phi$, the solution is

$$\left.\begin{aligned} \delta\phi &= A\tan^s\tfrac{1}{2}\theta\cos s\phi \\ \delta\theta &= -A\sin\theta\tan^s\tfrac{1}{2}\theta\sin s\phi \\ \delta r &= Aa(s+\cos\theta)\tan^s\tfrac{1}{2}\theta\sin s\phi \end{aligned}\right\}, \quad \ldots\ldots\ldots\ldots\ldots(1)$$

* *London Math. Soc. Proc.* Vol. XIII. p. 4, November 1881. [Vol. I. Art. 78.]

θ being the colatitude measured from the pole through which the shell is complete. Any integral value higher than unity is admissible for s. The values 0 and 1 correspond to displacements not involving strain.

In a recent paper* Mr Love dissents from the general principle involved in the theory above briefly sketched, and rejects the special solutions founded upon it as inapplicable to the vibration of thin shells. The argument upon which I proceeded in my former paper, and which still seems to me valid, may be put thus: It is a general mechanical principle† that, if given displacements (not sufficient by themselves to determine the configuration) be produced in a system originally in equilibrium by forces of corresponding types, the resulting deformation is determined by the condition that the potential energy of deformation shall be as small as possible. Apply this to an elastic shell, the given displacements being such as not of themselves to involve a stretching of the middle surface‡. The resulting deformation will, in general, include both stretching and bending, and any expression for the energy will contain corresponding terms proportional to the first and third powers respectively of the thickness. This energy is to be as small as possible. Hence, when the thickness is diminished without limit, the actual displacement will be one of pure bending, if such there be, consistent with the given conditions. Otherwise the energy would be of the first order (in thickness) instead of, as it might be, of the third order, in violation of the principle.

It will be seen that this argument takes no account of special conditions to be satisfied at the edge of the shell. This is the point at which Mr Love concentrates his objections. He considers that the general condition necessary to be satisfied at a free edge is in fact violated by such a deformation as (1). But the condition in question§ contains terms proportional to the first and to the third powers respectively of the thickness, the coefficients of the former involving as factors the extensions and shear of the middle surface. It appears to me that when the thickness is diminished without limit, the fulfilment of the boundary condition requires only that the middle surface be unstretched, precisely the requirement satisfied by solutions such as (1).

Of course, so long as the thickness is finite, the forces in operation will entail some stretching of the middle surface, and the amount of this stretching will depend on circumstances. A good example is afforded by a circular cylinder with plane edges perpendicular to the axis. Let normal forces locally applied at the extremities of one diameter of the central section cause

* "On the small free Vibrations and Deformation of a thin elastic Shell," *Phil. Trans.* A, 1888.

† *Phil. Mag.* March 1875, [Vol. I. p. 236]; *Theory of Sound*, § 74.

‡ There are cases where no displacement (involving strain at all) is possible without stretching of the middle surface, *e.g.*, the complete sphere.

§ See his equation (33).

a given shortening of that diameter. That the potential energy may be a minimum, the deformation must assume more and more the character of mere bending as the thickness is reduced. The only kind of bending that can occur in this case is the purely cylindrical one in which every normal section is similarly deformed, and then the potential energy is proportional to the total length of the cylinder. We see, therefore, that if the cylinder be very long, the energy of bending corresponding to the given local contraction of the central diameter may become very great, and a heavy strain is thrown upon the principle that the deformation of minimum energy is one of pure bending.

If the small thickness of the shell be regarded as given, a point will at last be attained when the energy can be made least by a sensible local stretching of the middle surface such as will dispense with the uniform bending otherwise necessary over so great a length. But even in this extreme case it seems correct to say that, when the thickness is *sufficiently* reduced, the deformation tends to become one of pure bending.

At first sight it may appear strange that of two terms in an expression of the potential energy, the one proportional to the cube of the thickness is to be retained, while that proportional to the first power may be omitted. The fact, however, is that the large potential energy which would accompany any stretching of the middle surface is the very reason why such stretching will not occur. The comparative largeness of the coefficient (proportional to the first power of the thickness) is more than neutralised by the smallness of the stretching itself, to the *square* of which the energy is proportional.

In general, if ψ_1 be the coordinate measuring the violation of the tie which is supposed to be more and more insisted upon by increasing stiffness, and if the other coordinates be suitably chosen, the potential energy of the system may be expressed

$$V = \tfrac{1}{2}c_1\psi_1^2 + \tfrac{1}{2}c_2\psi_2^2 + \tfrac{1}{2}c_3\psi_3^2 + \dots.$$

This follows from the general theorem that V and T may always be reduced to sums of squares simply, if we suppose that $T = \tfrac{1}{2}a_1\dot{\psi}_1^2$.

The equations of equilibrium under the action of external forces $\Psi_1, \Psi_2, \dots$ are thus

$$\Psi_1 = c_1\psi_1, \qquad \Psi_2 = c_2\psi_2, \qquad \&c.;$$

hence if the forces are regarded as given, the effect of increasing c_1 without limit is not merely to annul ψ_1, but also the term in V which depends upon it.

An example might be taken from the case of a rod clamped at one end A, and deflected by a lateral force, whose stiffness from the end A up to a neighbouring place B, is conceived to increase indefinitely. In the limit we may regard the rod as clamped at B, and neglect the energy of the part AB, in spite of, or rather in consequence of, its infinite stiffness.

If it be admitted that the deformations to be considered are pure bendings, the next step is the calculation of the potential energy corresponding thereto. In my former paper, the only case for which this part of the problem was attempted was that of the sphere. After bending, "the principal curvatures differ from the original curvature of the sphere in opposite directions, and to an equal amount*, and the potential energy of bending corresponding to any element of the surface is proportional to the square of this excess or defect of curvature, without regard to the direction of the principal planes." Though he agrees with my conclusions, Mr Love appears to regard the argument as insufficient. But clearly in the case of a given spherical shell, there are no other elements upon which the energy of bending could depend. "Thus the energy corresponding to the element of surface $a^2 \sin\theta\, d\theta\, d\phi$ may be denoted by

$$a^2 H (\delta\rho^{-1})^2 \sin\theta\, d\theta\, d\phi, \qquad\qquad (2)$$

where H depends upon the material and upon the thickness."

By the nature of the case H is proportional to the elastic constants and to the cube of the thickness, from which it follows by the method of dimensions that it is independent of a, the radius of the sphere. I did not, at the time, attempt the further determination of H, not needing it for my immediate purpose. Mr Love has shown that

$$H = \tfrac{4}{3} n h^3, \qquad\qquad (3)$$

where $2h$ represents the thickness, and n is the constant of rigidity. Why n alone should occur, to the exclusion of the constant of compressibility, will presently appear more clearly.

The application of (2) to the displacements expressed in (1) gave {equation (18)}

$$V = 2\pi \Sigma (s^3 - s) A_s^2 \int_0^\theta H \sin^{-3}\theta \tan^{2s} \tfrac{1}{2}\theta\, d\theta, \qquad\qquad (4)$$

θ being the colatitude of the (circular) edge. In the case of the hemisphere of uniform thickness

$$V = \tfrac{1}{2}\pi H \Sigma (s^3 - s)(2s^2 - 1) A_s^2. \qquad\qquad (5)$$

The calculation of the pitch of free vibration then presented no difficulty. If σ denote the superficial density, and $\cos pt$ represent the type of vibration, p_2 corresponding to $s = 2$, p_3 to $s = 3$, and so on, it appeared that

$$p_2 = \frac{\sqrt{H}}{a^2\sigma} \times 5{\cdot}2400, \qquad p_3 = \frac{\sqrt{H}}{a^2\sigma} \times 14{\cdot}726, \qquad p_4 = \frac{\sqrt{H}}{a^2\sigma} \times 28{\cdot}462;$$

so that

$$p_3/p_2 = 2{\cdot}8102, \qquad p_4/p_3 = 5{\cdot}4316,$$

determining the *intervals* between the graver notes.

* This is in virtue of Gauss's theorem that the product of the principal curvatures is unaffected by bending.

If the form of the shell be other than spherical, the middle surface is no longer symmetrical with respect to the normal at any point, and the expression of the potential energy is more complicated. The question is now not merely one of the curvature of the deformed surface; account must also be taken of the correspondence of normal sections before and after deformation*. A complete investigation has been given by Love; but the treatment of the question now to be explained, even if less rigorous, may help to throw light upon this somewhat difficult subject.

In the actual deformation of a material sheet of finite extent there will usually be at any point not merely a displacement of the point itself, but a rotation of the neighbouring parts of the sheet, such as a rigid body may undergo. All this contributes nothing to the energy. In order to take the question in its simplest form, let us refer the original surface to the normal and principal tangents at the point in question as axes of coordinates, and let us suppose that after deformation, the lines in the sheet originally coincident with the principal tangents are brought back (if necessary) to occupy the same positions as at first. The possibility of this will be apparent when it is remembered that in virtue of the inextensibility of the sheet, the angles of intersection of all lines traced upon it remain unaltered. The equation of the original surface in the neighbourhood of the point being

$$z = \tfrac{1}{2}\left(\frac{x^2}{\rho_1} + \frac{y^2}{\rho_2}\right), \quad \text{..............................(6)}$$

that of the deformed surface may be written

$$z = \tfrac{1}{2}\left\{\frac{x^2}{\rho_1 + \delta\rho_1} + \frac{y^2}{\rho_2 + \delta\rho_2} + 2\tau xy\right\}. \quad \text{.....................(7)}$$

In strictness $(\rho_1 + \delta\rho_1)^{-1}$, $(\rho_2 + \delta\rho_2)^{-1}$ are the curvatures of the sections made by the planes $x = 0$, $y = 0$; but since principal curvatures are a maximum or a minimum, they represent with sufficient accuracy the new principal curvatures, although these are to be found in slightly different planes. The condition of inextensibility shows that points which have the same x and y in (6) and (7) are *corresponding* points, and by Gauss's theorem it is further necessary that

$$\frac{\delta\rho_1}{\rho_1} + \frac{\delta\rho_2}{\rho_2} = 0. \quad \text{..............................(8)}$$

It thus appears that the energy of bending will depend upon two quantities, one giving the alterations of principal curvature, and the other τ depending upon the shift (in the material) of the principal planes.

* An extreme case may serve as an illustration. Suppose that the bending is such that the principal planes retain their positions relatively to the material surface, but that the principal curvatures are exchanged. The nature of the curvature at the point in question is the same after deformation as before, and by a rotation through 90° round the normal the surfaces may be made to fit; nevertheless the energy of bending is finite.

In calculating the energy we may regard it as due to the stretchings and contractions under tangential forces of the various infinitely thin laminæ into which the shell may be divided. The middle lamina, being unstretched, makes no contribution. Of the other laminæ, the stretching is in proportion to the distance from the middle surface, and the energy of stretching is therefore as the square of this distance. When the integration over the whole thickness of the shell is carried out, the result is accordingly proportional to the cube of the thickness.

The next step is to estimate more precisely the energy corresponding to a small element of area of a lamina. The general equations in three dimensions, as given in Thomson and Tait's *Natural Philosophy*, § 694, are

$$na = S, \qquad nb = T, \qquad nc = U, \qquad \text{.....................(9)}$$

$$Me = P - \sigma(Q + R), \qquad Mf = Q - \sigma(R + P), \qquad Mg = R - \sigma(P + Q), \quad \text{...(10)}$$

where

$$\sigma = \frac{m - n}{2m} \qquad \text{................................(11)*}$$

The energy w, corresponding to the unit of volume, is given by

$$2w = (m + n)(e^2 + f^2 + g^2) + 2(m - n)(fg + ge + ef) + n(a^2 + b^2 + c^2). \qquad (12)$$

In the application to a lamina, supposed parallel to xy, we are to take $R = 0$, $S = 0$, $T = 0$; so that

$$g = -\sigma\frac{e + f}{1 - \sigma}, \qquad a = 0, \qquad b = 0.$$

Thus in terms of the elongations e, f, parallel to x, y, and of the shear c, we get

$$w = n\left\{e^2 + f^2 + \frac{m - n}{m + n}(e + f)^2 + \tfrac{1}{2}c^2\right\}. \qquad \text{.................(13)}$$

We have now to express the elongations of the various laminæ of a shell when bent, and we will begin with the case where $\tau = 0$, that is, when the principal planes of curvature remain unchanged. It is evident that in this case the shear c vanishes, and we have to deal only with the elongations e and f parallel to the axes. In the section by the plane of zx, let s, s' denote corresponding infinitely small arcs of the middle surface and of a lamina distant h from it. If ψ be the angle between the terminal normals, $s = \rho_1\psi$, $s' = (\rho_1 + h)\psi$, $s' - s = h\psi$. In the bending, which leaves s unchanged,

$$\delta s' = h\delta\psi = hs\delta(1/\rho_1).$$

Hence

$$e = \delta s'/s' = h\delta(1/\rho_1),$$

* M is Young's modulus, σ is Poisson's ratio, n is the constant of rigidity, and $(m - \frac{1}{3}n)$ that of cubic compressibility. In terms of Lamé's constants (λ, μ), $m = \lambda + \mu$, $n = \mu$.

and in like manner $f = h\delta(1/\rho_2)$. Thus for the energy U per unit of *area* we have

$$dU = nh^2dh\left\{\left(\delta\frac{1}{\rho_1}\right)^2 + \left(\delta\frac{1}{\rho_2}\right)^2 + \frac{m-n}{m+n}\left(\delta\frac{1}{\rho_1} + \delta\frac{1}{\rho_2}\right)^2\right\},$$

and on integration over the whole thickness of the shell $(2h)$*

$$U = \frac{2nh^3}{3}\left\{\left(\delta\frac{1}{\rho_1}\right)^2 + \left(\delta\frac{1}{\rho_2}\right)^2 + \frac{m-n}{m+n}\left(\delta\frac{1}{\rho_1} + \delta\frac{1}{\rho_2}\right)^2\right\}. \quad \text{.........(14)}$$

This conclusion may be applied at once, so as to give the result applicable to a spherical shell; for, since the original principal planes are arbitrary, they can be taken so as to coincide with the principal planes after bending. Thus $\tau = 0$; and by Gauss's theorem

$$\delta\frac{1}{\rho_1} + \delta\frac{1}{\rho_2} = 0,$$

so that

$$U = \frac{4nh^3}{3}\left(\delta\frac{1}{\rho_1}\right)^2, \quad \text{..............................(15)}$$

where $\delta\rho^{-1}$ denotes the change of principal curvature. Since $e = -f$, $g = 0$, the various laminæ are simply sheared, and that in proportion to their distance from the middle surface. The energy is thus a function of the constant of rigidity only.

The result (14) is applicable directly to the plane plate; but this case is peculiar in that, on account of the infinitude of ρ_1, ρ_2, (8) is satisfied without any relation between $\delta\rho_1$ and $\delta\rho_2$. Thus for a plane plate

$$U = \frac{2nh^3}{3}\left\{\frac{1}{\rho_1^2} + \frac{1}{\rho_2^2} + \frac{m-n}{m+n}\left(\frac{1}{\rho_1} + \frac{1}{\rho_2}\right)^2\right\}, \quad \text{..............(16)}$$

where ρ_1^{-1}, ρ_2^{-1}, are the two independent principal curvatures after bending.

We have thus far considered τ to vanish; and it remains to investigate the effect of the deformations expressed by

$$\delta z = \tau xy = \tfrac{1}{2}\tau(\xi^2 - \eta^2), \quad \text{.......................(17)}$$

where ξ, η relate to new axes inclined at 45° to those of x, y. The curvatures defined by (17) are in the planes of ξ, η, equal in numerical value and opposite in sign. The elongations in these directions for any lamina within the thickness of the shell are $h\tau$, $-h\tau$, and the corresponding energy (as in the case of the sphere just considered) takes the form

$$U' = \tfrac{4}{3}nh^3\tau^2. \quad \text{............................(18)}$$

* It is here assumed that m and n are independent of h, that is, that the material is homogeneous. If we discard this restriction, we may form the conception of a shell of given thickness, whose middle surface is physically inextensible, while yet the resistance to bending is moderate. In this way we may realise the types of deformation discussed in the present paper, *without supposing the thickness to be infinitely small;* and the independence of such types upon conditions to be satisfied at a free edge is perhaps rendered more apparent.

This energy is to be added* to that already found in (14); and we get finally

$$U=\frac{2nh^3}{3}\left\{\left(\delta\frac{1}{\rho_1}\right)^2+\left(\delta\frac{1}{\rho_2}\right)^2+\frac{m-n}{m+n}\left(\delta\frac{1}{\rho_1}+\delta\frac{1}{\rho_2}\right)^2+2\tau^2\right\}, \quad\text{......(19)}$$

as the complete expression of the energy, when the deformation is such that the middle surface is unextended. We may interpret τ by means of the angle χ, through which the principal planes are shifted; thus

$$\tau=2\chi\left(\frac{1}{\rho_2}-\frac{1}{\rho_1}\right). \quad\text{..........................(20)}$$

It will now be in our power to treat more completely a problem of great interest, viz., the deformation and vibration of a cylindrical shell. In my former paper [Art. 78] I investigated the types of bending, but without a calculation of the corresponding energy. The results were as follows†. If the cylinder be referred to columnar coordinates z, r, ϕ, so that the displacements of a point whose equilibrium coordinates are z, a, ϕ are denoted by δz, δr, $a\,\delta\phi$, the equations expressing inextensibility take the form

$$\frac{d\delta z}{dz}=0, \qquad \delta r+a\frac{d\delta\phi}{d\phi}=0, \qquad \frac{d\delta z}{d\phi}+a^2\frac{d\delta\phi}{dz}=0, \quad\text{.........(21)}$$

from which we may deduce

$$d^2\delta\phi/dz^2=0. \quad\text{.............................(22)}$$

By (22), if $\delta\phi \propto \cos s\phi$, we may take

$$a\delta\phi=(A_s a+B_s z)\cos s\phi, \quad\text{.......................(23)}$$

and then, by (21)

$$\delta r=s\,(A_s a+B_s z)\sin s\phi, \qquad \delta z=-\,s^{-1}B_s a\sin s\phi. \quad\text{...(24, 25)}$$

If the cylinder be complete, s is integral; A_s and B_s are independent constants, either of which may vanish. In the latter case the displacement is in two dimensions only‡. It is unnecessary to stop to consider the demonstrations of (21), inasmuch as these equations will present themselves independently in the course of the investigation which follows.

It will be convenient to replace δz, δr, $a\,\delta\phi$ by single letters, which, however, it is difficult to choose so as not to violate some of the usual conventions. In conformity with Mr Love's general notation, I will write

$$\delta z=u, \qquad a\,\delta\phi=v, \qquad \delta r=w. \quad\text{..............(26)}$$

* There are clearly no terms involving the products of τ with the changes of principal curvature $\delta(\rho_1^{-1})$, $\delta(\rho_2^{-1})$; for a change in the sign of τ can have no influence upon the energy of the deformation defined by (7).

† The method of investigation is similar to that employed by Jellet in his memoir ("On the Properties of Inextensible Surfaces," *Irish Acad. Trans.* Vol. XXII. p. 179, 1855), to which reference should have been made.

‡ See *Theory of Sound*, § 233.

The problem before us is the expression of the changes of principal curvature and shifts of principal planes at any point $P(z, \phi)$ of the cylinder in terms of the displacements u, v, w. As in (6), take as fixed coordinate axes the principal tangents and normal to the undisturbed cylinder at the point P, the axis of x being parallel to that of the cylinder, that of y tangential to the circular section, and that of ζ normal, measured inwards. If, as it will be convenient to do, we measure z and ϕ from the point P, we may express the undisturbed coordinates of a material point Q in the neighbourhood of P, by

$$x = z, \qquad y = a\phi, \qquad \zeta = \tfrac{1}{2}a\phi^2. \quad \text{.................(27)}$$

During the displacement the coordinates of Q will receive the increments

$$u, \quad w \sin\phi + v\cos\phi, \quad -w\cos\phi + v\sin\phi;$$

so that after displacement

$$x = z + u, \qquad y = a\phi + w\phi + v(1 - \tfrac{1}{2}\phi^2),$$
$$\zeta = \tfrac{1}{2}a\phi^2 - w(1 - \tfrac{1}{2}\phi^2) + v\phi;$$

or if u, v, w be expanded in powers of the small quantities z, ϕ,

$$x = z + u_0 + \frac{du}{dz_0}z + \frac{du}{d\phi_0}\phi + \dots \quad \text{.......................(28)}$$

$$y = a\phi + w_0\phi + v_0 + \frac{dv}{dz_0}z + \frac{dv}{d\phi_0}\phi + \dots \quad \text{..............(29)}$$

$$\zeta = \tfrac{1}{2}a\phi^2 - w_0 - \frac{dw}{dz_0}z - \frac{dw}{d\phi_0}\phi + v_0\phi$$
$$+ \tfrac{1}{2}w_0\phi^2 - \tfrac{1}{2}\frac{d^2w}{dz_0^2}z^2 - \frac{d^2w}{dz_0\,d\phi_0}z\phi - \tfrac{1}{2}\frac{d^2w}{d\phi_0^2}\phi^2$$
$$+ \frac{dv}{dz_0}z\phi + \frac{dv}{d\phi_0}\phi^2, \quad \text{.......................................(30)}$$

u_0, v_0, ... being the values of u, v at the point P.

These equations give the coordinates of the various points of the deformed sheet. We have now to suppose the sheet moved as a rigid body so as to restore the position (as far as the first power of small quantities is concerned) of points infinitely near P. A purely translatory motion by which the displaced P is brought back to its original position will be expressed by the simple omission in (28), (29), (30) of the terms u_0, v_0, w_0 respectively, which are independent of z, ϕ. The effect of an arbitrary rotation is represented by the additions to x, y, ζ respectively of $y\theta_3 - \zeta\theta_2$, $\zeta\theta_1 - x\theta_3$, $x\theta_2 - y\theta_1$; where for the present purpose θ_1, θ_2, θ_3 are small quantities of the order of the deformation, the square of which is to be neglected throughout. If we make these additions to (28), &c., substituting for x, y, ζ in the terms containing θ their approximate values, we find so far as the first powers of z, ϕ

$$x = z + \frac{du}{dz_0} z + \frac{du}{d\phi_0} \phi + a\phi\theta_3,$$

$$y = a\phi + w_0\phi + \frac{dv}{dz_0} z + \frac{dv}{d\phi_0} \phi - z\theta_3,$$

$$\zeta = -\frac{dw}{dz_0} z - \frac{dw}{d\phi_0} \phi + v_0\phi + z\theta_2 - a\phi\theta_1.$$

Now, since the sheet is assumed to be inextensible, it must be possible so to determine θ_1, θ_2, θ_3 that to this order $x = z$, $y = a\phi$; $\zeta = 0$.

Hence
$$\frac{du}{dz_0} = 0, \qquad \frac{du}{d\phi_0} + a\theta_3 = 0,$$

$$\frac{dv}{dz_0} - \theta_3 = 0, \qquad w_0 + \frac{dv}{d\phi_0} = 0,$$

$$-\frac{dw}{dz_0} + \theta_2 = 0, \qquad \frac{dw}{d\phi_0} - v_0 + a\theta_1 = 0.$$

The conditions of inextensibility are thus (if we drop the suffixes as no longer required)

$$\frac{du}{dz} = 0, \qquad w + \frac{dv}{d\phi} = 0, \qquad \frac{du}{d\phi} + a\frac{dv}{dz} = 0, \qquad \text{......(31)}$$

which agree with (21).

Returning to (28), &c., as modified by the addition of the translatory and rotatory terms, we get

$$x = z + \text{terms of 2nd order in } z, \phi,$$

$$y = a\phi + \quad " \quad "$$

$$\zeta = \tfrac{1}{2}a\phi^2 + \tfrac{1}{2}w_0\phi^2 - \tfrac{1}{2}\frac{d^2w}{dz_0^2}z^2 - \frac{d^2w}{dz_0 d\phi_0} z\phi$$

$$- \tfrac{1}{2}\frac{d^2w}{d\phi_0^2}\phi^2 + \frac{dv}{dz_0} z\phi + \frac{dv}{d\phi_0}\phi^2;$$

or since by (31) $d^2w/dz^2 = 0$, and $dv/d\phi = -w$,

$$\zeta = \tfrac{1}{2}a\phi^2 - \tfrac{1}{2}w_0\phi^2 - \frac{d^2w}{dz_0 d\phi_0} z\phi - \tfrac{1}{2}\frac{d^2w}{d\phi_0^2}\phi^2 + \frac{dv}{dz_0} z\phi.$$

The equation of the deformed surface after transference is thus

$$\zeta = xy\left\{\frac{1}{a}\frac{dv}{dz_0} - \frac{1}{a}\frac{d^2w}{dz_0 d\phi_0}\right\} + y^2\left\{\frac{1}{2a} - \frac{1}{2a^2}w_0 - \frac{1}{2a^2}\frac{d^2w}{d\phi_0^2}\right\}. \qquad \text{......(32)}$$

Comparing with (7) we see that

$$\delta\frac{1}{\rho_1} = 0, \qquad \delta\frac{1}{\rho_2} = -\frac{1}{a^2}\left(w + \frac{d^2w}{d\phi^2}\right), \qquad \tau = \frac{1}{a}\left(\frac{dv}{dz} - \frac{d^2w}{dz\,d\phi}\right), \qquad \text{......(33)}$$

so that by (19)

$$U=\frac{4nh^3}{3a^2}\left\{\frac{m}{m+n}\frac{1}{a^2}\left(w+\frac{d^2w}{d\phi^2}\right)^2+\left(\frac{dv}{dz}-\frac{d^2w}{dz\,d\phi}\right)^2\right\} \quad\text{..........(34)}$$

This is the potential energy of bending reckoned per unit of area. It can if desired be expressed by (31) entirely in terms of v*.

We will now apply (34) to calculate the whole potential energy of a complete cylinder, bounded by plane edges $z=\pm l$, and of thickness which, if variable at all, is a function of z only. Since u, v, w are periodic when ϕ increases by 2π, their most general expression in accordance with (31) is {compare (23), &c.}

$$v=\Sigma\left[(A_s a+B_s z)\cos s\phi-(A_s'a+B_s'z)\sin s\phi\right], \quad\text{...........(35)}$$

$$w=\Sigma\left[s(A_s a+B_s z)\sin s\phi+s(A_s'a+B_s'z)\cos s\phi\right], \quad\text{........(36)}$$

$$u=\Sigma\left[-s^{-1}B_s a\sin s\phi-s^{-1}B_s'a\cos s\phi\right], \quad\text{....................(37)}$$

in which the summation extends to all integral values of s from 0 to ∞. But the displacements corresponding to $s=0$, $s=1$ are such as a rigid body might undergo, and involve no absorption of energy. When the values of u, v, w are substituted in (34) all the terms containing products of sines or cosines with different values of s vanish in the integration with respect to ϕ, as do also those which contain $\cos s\phi \sin s\phi$. Accordingly

$$\int_0^{2\pi}Ua\,d\phi=\frac{4\pi nh^3}{3a}\left[\frac{m}{m+n}\frac{1}{a^2}\Sigma(s^3-s)^2\right.$$

$$\left.\{(A_s a+B_s z)^2+(A_s'a+B_s'z)^2\}+\Sigma(s^2-1)^2(B_s^2+B_s'^2)\right]. \quad\text{...(38)}$$

Thus far we might consider h to be a function of z; but we will now treat it as a constant. In the integration with respect to z the odd powers of z will disappear, and we get as the energy of the whole cylinder of radius a, length $2l$, and thickness $2h$,

$$V=\int_{-l}^{+l}\int_0^{2\pi}Ua\,d\phi\,dz$$

$$=\frac{8\pi nh^3l}{3a}\Sigma(s^2-1)^2\left[\frac{m\,.\,s^2}{m+n}\left\{A_s^2+A_s'^2+\frac{l^2}{3a^2}(B_s^2+B_s'^2)\right\}+B_s^2+B_s'^2\right], \quad\text{.....(39)}$$

in which $s=2, 3, 4, \ldots$.

* From Mr Love's general equations (12), (13), (18) a concordant result may be obtained by introduction of the special conditions—

$$h_1=0, \qquad h_2=1/a, \qquad 1/\rho_1=0, \qquad 1/\rho_2=1/a,$$

limiting the problem to the case of the cylinder, and of those

$$\sigma_1=\sigma_2=\varpi=0,$$

which express the inextensibility of the middle surface.

The expression (39) for the potential energy suffices for the solution of statical problems. As an example we will suppose that the cylinder is compressed along a diameter by equal forces F, applied at the points $z = z_1$, $\phi = 0$, $\phi = \pi$, although it is true that so highly localised a force hardly comes within the scope of the investigation in consequence of the stretchings of the middle surface, which will occur in the immediate neighbourhood of the points of application*.

The work done upon the cylinder by the forces F during the hypothetical displacement indicated by δA_s, &c., will be by (36)

$$-F\Sigma s\,(a\delta A_s' + z_1\delta B_s')\,(1 + \cos s\pi),$$

so that the equations of equilibrium are

$$dv/dA_s = 0, \qquad dv/dB_s = 0.$$

$$dv/dA_s' = -(1 + \cos s\pi)\, saF, \qquad dv/dB_s' = -(1 + \cos s\pi)\, sz_1F.$$

Thus for all values of s, $A_s = B_s = 0$; and for odd values of s, $A_s' = B_s' = 0$. But when s is even,

$$\frac{ms^2}{m+n} A_s' = -\frac{3sa^2F}{8\pi nh^3 l\,(s^2-1)^2}, \quad \text{.....................(40)}$$

$$\left\{\frac{ms^2}{m+n}\frac{l^2}{3a^2} + 1\right\} B_s' = -\frac{3saz_1F}{8\pi nh^3 l\,(s^2-1)^2}; \quad \text{.....................(41)}$$

and the displacement w at any point (z, ϕ) is given by

$$w = 2\,(A_2'a + B_2'z)\cos 2\phi + 4\,(A_4'a + B_4'z)\cos 4\phi + \ldots, \quad \text{...(42)}$$

where A_2', B_2', A_4', ... are determined by (40), (41).

If the cylinder be moderately long in proportion to its diameter, the second term in the left-hand member of (41) may be neglected, so that

$$\frac{l^2}{3a^2}\frac{B_s'}{z_1} = \frac{A_s'}{a}.$$

In this case (42) may be written

$$w = \left(1 + \frac{3z_1z}{l^2}\right)\{2A_2'a\cos 2\phi + 4A_4'a\cos 4\phi + \ldots\}, \quad \text{......(43)}$$

showing that, except as to magnitude and sign, the curve of deformation is the same for all values of z_1 and z†.

If $z = \pm z_1$, the amplitudes are in the ratio $1 \pm 3z_1^2/l^2$; and if, further, $z_1 = l$, *i.e.*, if the force be applied at one of the ends of the cylinder, the

* Whatever the curvature of the surface, an area upon it may be taken so small as to behave like a plane, and therefore bend, in violation of Gauss's condition, when subjected to a force which is so nearly discontinuous that it varies sensibly within the area.

† That w is unaltered when z and z_1 are interchanged is an example of the general law of reciprocity.

amplitudes are as $2:-1$. The section where the deformation (as represented by w) is zero, is given by $3zz_1 + l^2 = 0$, in which if $z_1 = l$, $z = -\frac{1}{3}l$.

When the condition as to the length of the cylinder is not imposed, the ratio $B_s' : A_s'$ is dependent upon s, and therefore the curves of deformation vary with z, apart from mere magnitude and sign. If, however, we limit ourselves to the more important term $s = 2$, we have

$$\frac{4m}{m+n}\frac{A_2'}{a} = \left\{\frac{4m}{m+n}\frac{l^2}{3a^2} + 1\right\}\frac{B_2'}{z_1},$$

and

$$w = 2B_2'\left\{\frac{a^2}{z_1}\left(\frac{l^2}{3a^2} + \frac{m+n}{4m}\right) + z\right\}\cos 2\phi;$$

so that w vanishes when

$$\frac{zz_1}{a^2} + \frac{l^2}{3a^2} + \frac{m+n}{4m} = 0. \quad\text{.......................(44)}$$

This equation may be applied to find what is the length of the cylinder when the deformation just vanishes at one end if the force is applied at the other. If $z_1 = -z = l$,

$$\frac{l}{a} = \sqrt{\left\{\frac{3(m+n)}{8m}\right\}}.$$

For many materials σ {equation (11)} is about $\frac{1}{4}$, or $m = 2n$. In such cases the condition is

$$l = \tfrac{3}{4}a.$$

It should not be overlooked that although w may vanish, u remains finite.

Reverting to (23), (24), (25) we see that, if the cylinder is open at both ends, there are two types of deformation possible for each value of s. If we suppose the cylinder to be closed at $z = 0$ by a flat disk attached to it round the circumference, the inextensibility of the disk imposes the conditions, $w = \delta r = 0$, $v = a\,\delta\phi = 0$, when $z = 0$*. Hence $A_s = 0$, and the only deformation now possible is

$$v = a\,\delta\phi = B_s z \cos s\phi, \qquad w = \delta r = sB_s z \sin s\phi. \quad\text{.........(45)}$$

Another disk, attached where z has a finite value, would render the cylinder rigid.

Instead of a plane disk let us next suppose that the cylinder is closed at $z = 0$ by a hemisphere attached to it round the circumference. By (1) the three component displacements at the edge of the hemisphere ($\theta = \frac{1}{2}\pi$) are of the form

$$v = a\,\delta\phi = a\cos s\phi, \qquad u = a\,\delta\theta = -a\sin s\phi, \qquad w = \delta r = sa\sin s\phi.$$

Equating these to the corresponding values for the cylinder, as given by (23), (24), (25), we get $A_s = 1$, $B_s = s$; so that the deformation of the cylinder is now limited to the type

$$v = (a + sz)\cos s\phi, \qquad w = s(a + sz)\sin s\phi, \qquad u = -a\sin s\phi, \quad\text{...(46)}$$

* s being greater than 1.

in which we may, of course, introduce an arbitrary multiplier and an arbitrary addition to ϕ. If the convexity of the hemisphere be turned outwards, z is to be considered positive.

In like manner any other convex additions at one end of the cylinder might be treated. There are apparently three conditions to be satisfied by only two constants, but one condition is really redundant, being already secured by the inextensibility of the edges provided for in the types of deformations determined separately for the two shells. Convex additions, closing both ends of the cylinder, render it rigid, in accordance with Jellet's theorem that a closed oval shell cannot be bent.

It is of importance to notice how a cylinder, or a portion of a cylinder, can *not* be bent. Take, for example, an elongated strip, bounded by two generating lines subtending at the axis a small angle. Equations (31) $\{$giving $d^2w/dr^2 = 0\}$ show that the strip cannot be bent in the plane containing the axis and the middle generating line*. The only bending symmetrical with respect to this plane is a purely cylindrical one which leaves the middle generating line straight. There are two ways in which we may conceive the strip altered so as to render it susceptible of the desired kind of bending. The first is to take out the original cylindrical curvature, which reduces it to a plane strip. The second is to replace it by one in which the middle line is curved from the beginning, like the equator of a sphere or ellipsoid of revolution. In this case the total curvature being finite, the Gaussian condition can be satisfied by a change of meridional curvature compensating the supposed change of equatorial curvature. It is easy to calculate the actual stiffness from (8) and (14), for here $\tau = 0$. We have

$$U = \frac{2nh^3}{3}\left(\delta\frac{1}{\rho_1}\right)^2\left\{1 + \frac{\rho_1^2}{\rho_2^2} + \frac{m-n}{m+n}\left(1 - \frac{\rho_1}{\rho_2}\right)^2\right\}, \quad \text{.........(47)}$$

which expresses the work per unit of area corresponding to a given bending $\delta\rho_1^{-1}$ along the equator. If $\rho_1 = \infty$, the cylindrical strip is infinitely stiff. If the curvature be spherical, $\rho_2 = \rho_1$, and

$$U = \frac{4nh^3}{3}\left(\delta\frac{1}{\rho_1}\right)^2; \quad \text{.........................(48)}$$

and if $\rho_2 = \infty$,

$$U = \frac{4nh^3}{3}\cdot\frac{m}{m+n}\left(\delta\frac{1}{\rho_1}\right)^2. \quad \text{....................(49)}$$

Whatever the equatorial curvature may be, the ratio of stiffnesses in the two cases is equal to $m : m+n$, or about $2:3$, the spherically curved strip being the stiffer.

The same principle applies to the explanation of Bourdon's gauge. In this instrument there is a tube whose axis lies along an arc of a circle and

* This is the principle upon which metal is corrugated.

whose section is elliptical, the longer axis of the ellipse being perpendicular to the general plane of the tube. If we now consider the curvature at points which lie upon the axial section, we learn from Gauss's theorem that a diminished curvature along the axis will be accompanied by a nearer approach to a circular section, and reciprocally. Since a circular form has the largest area for a given perimeter, internal pressure tends to diminish the eccentricity of the elliptic section and with it the general curvature of the tube. Thus, if one end be fixed, a pointer connected with the free end may be made to indicate the internal pressure*.

We will now proceed with the calculation for the frequencies of vibration of the complete cylindrical shell of length $2l$. If the volume-density be ρ†, we have as the expression of the kinetic energy by means of (35), (36), (37)

$$T = \tfrac{1}{2} . 2h\rho . \iint (\dot{u}^2 + \dot{v}^2 + \dot{w}^2)\, a\, d\phi\, dz$$

$$= 2\pi\rho h l a\, \Sigma \{a^2 (1+s^2)(\dot{A}_s^2 + \dot{A}_s'^2) + [\tfrac{1}{3} l^2 (1+s^2) + s^{-2} a^2] (\dot{B}_s^2 + \dot{B}_s'^2)\}. \; \ldots (50)$$

From these expressions for V and T in (39), (50) the types and frequencies of vibration can be at once deduced. The fact that the squares, and not the products, of A_s, B_s, are involved, shows that these quantities are really the principal coordinates of the vibrating system. If A_s, or A_s', vary as $\cos p_s t$ we have

$$p_s^2 = \tfrac{4}{3} \frac{mn}{m+n} \frac{h^2}{\rho a^4} \frac{(s^3 - s)^2}{s^2 + 1}. \qquad \ldots\ldots\ldots\ldots (51)$$

This is the equation for the frequencies of vibration in two dimensions‡. For a given material, the frequency is proportional to the thickness and inversely as the square on the diameter of the cylinder§.

* Dec. 19.—It appears, however, that the bending of a curved tube of elliptical section cannot be pure, since the parts of the walls which lie furthest from the [plane of the] circular axis are necessarily stretched. The difficulty thus arising may be obviated by replacing the two halves of the ellipse, which lie on either side of the major axis, by two symmetrical curves which meet on the major axis at a *finite angle*. [See Art. 171 below.]

According to the equations (in columnar coordinates) of my former paper, the conditions that δr, δz shall be independent of ϕ lead to—

$$\delta r = Cr, \qquad \frac{d\delta z}{dz} + C\left(\frac{dr}{dz}\right)^2 = 0,$$

where C is an absolute constant.

The case where the section is a rhombus ($dr/dz = \pm \tan \alpha$) may be mentioned.

The difficulty referred to above arises when $dr/dz = \infty$.

† This can scarcely be confused with the notation for the curvature in the preceding parts of the investigation.

‡ See *Theory of Sound*, § 233.

§ There is nothing in these laws special to the cylinder. In the case of similar shells of any form, vibrating by pure bending, the frequency will be as the thicknesses and inversely as corresponding areas. If the similarity extend also to the thickness, then the frequency is inversely as the linear dimension, in accordance with the general law of Cauchy.

In like manner if B_s, or B_s', vary as $\cos p_s't$, we find

$$p_s'^2 = \tfrac{4}{3}\frac{mn}{m+n}\frac{h^2}{\rho a^4}\frac{(s^3-s)^2}{s^2+1}\frac{1+\dfrac{3a^2}{s^2l^2}\dfrac{m+n}{m}}{1+\dfrac{3a^2}{(s^4+s^2)\,l^2}}. \quad\text{............(52)}$$

If the cylinder be at all long in proportion to its diameter, the difference between p_s' and p_s becomes very small. Approximately in this case

$$p_s'/p_s = 1 + \frac{3a^2}{2s^2l^2}\left(\frac{m+n}{m} - \frac{1}{s^2+1}\right);$$

or if we take $m = 2n$, $s = 2$,

$$p_2'/p_2 = 1 + \frac{39\,a^2}{80\,l^2}.$$

In my former paper I gave the types of vibration for a circular cone, of which the cylinder may be regarded as a particular case. In terms of columnar coordinates (z, r, ϕ) we have

$$\delta\phi = (A_s + B_s z^{-1})\cos s\phi, \quad\text{..........................(53)}$$

$$\delta r = s\tan\gamma\,(A_s z + B_s)\sin s\phi, \quad\text{.....................(54)}$$

$$\delta z = \tan^2\gamma\,[s^{-1}B_s - s\,(A_s z + B_s)]\sin s\phi, \quad\text{.........(55)}$$

γ being the semi-vertical angle of the cone. For the calculation of the energy of bending it would be simpler to use polar coordinates (r, θ, ϕ), r being measured from the vertex instead of from the axis.

If the cone be complete up to the vertex, we must suppose, in (53) &c., $B_s = 0$. And if we proceed to calculate the potential energy, we shall find it infinite, at least when the thickness is uniform. For since A_s is of no dimensions in length, the square of the change of curvature must be proportional to $A_s^2 z^{-2}$. When this is multiplied by $z\,dz$, and integrated, a logarithm is introduced, which assumes an infinite value when $z = 0$. The complete cone must therefore be regarded as infinitely stiff, just as the cylinder would be if one rim were held fast.

If two similar cones (bounded by circular rims) are attached so that the common rim is a plane of symmetry, the bending may be such that the common rim remains plane. If the distance of this plane from the vertex be z_1, the condition to be satisfied in (53) &c., is that $\delta z = 0$ where $z = z_1$. Hence

$$\delta\phi = A_s\left\{1 - \frac{s^2}{s^2-1}\frac{z_1}{z}\right\}\cos s\phi, \quad\text{.....................(56)}$$

$$\delta r = s\tan\gamma\; A_s\left\{z - \frac{s^2 z_1}{s^2-1}\right\}\sin s\phi, \quad\text{...............(57)}$$

$$\delta z = s\tan^2\gamma\; A_s\{z_1 - z\}\sin s\phi. \quad\text{.....................(58)}$$

153.

ON THE COMPOSITION OF WATER.

[*Proceedings of the Royal Society*, XLV. pp. 425—430, 1889.]

DURING the past year I have continued the work described in a former communication on the relative densities of hydrogen and oxygen*, in the hope of being able to prepare lighter hydrogen than was then found possible. To this end various modifications have been made in the generating apparatus. Hydrogen has been prepared from potash in place of acid. In one set of experiments the gas was liberated by aluminium. In this case the generator consisted of a large closed tube sealed to the remainder of the apparatus; and the aluminium was attached to an iron armature so arranged that by means of an external electro-magnet it was possible to lower it into the potash, or to remove it therefrom. The liberated gas passed through tubes containing liquid potash†, corrosive sublimate, finely powdered solid potash, and, lastly, a long length of phosphoric anhydride. But the result was disappointing; for the hydrogen proved to be no lighter than that formerly obtained from sulphuric acid.

I have also tried to purify hydrogen yet further by absorption in palladium. In his recent important memoir‡, "On the Combustion of weighed Quantities of Hydrogen and the Atomic Weight of Oxygen," Mr Keiser describes experiments from which it appears that palladium will not occlude nitrogen—a very probable impurity in even the most carefully prepared gas. My palladium was placed in a tube sealed, as a lateral attachment, to the middle of that containing the phosphoric anhydride; so that the hydrogen was submitted in a thorough manner to this reagent both before and after absorption by the palladium. Any impurity that

* *Roy. Soc. Proc.* February, 1888 (Vol. XLIII. p. 356). [Vol. III. Art. 146.]

† Of course this tube was superfluous in the present case, but it was more convenient to retain it.

‡ *Amer. Chem. Journ.* Vol. x. No. 4.

might be rejected by the palladium was washed out of the tube by a current of hydrogen before the gas was collected for weighing. But as the result of even this treatment I have no improvement to report, the density of the gas being almost exactly as before.

Hitherto the observations have related merely to the densities of hydrogen and oxygen, giving the ratio 15·884, as formerly explained. To infer the composition of water by weight, this number had to be combined with that found by Mr Scott as representing the ratio of volumes*. The result was

$$\frac{2 \times 15{\cdot}884}{1{\cdot}9965} = 15{\cdot}914.$$

The experiments now to be described are an attempt at an entirely independent determination of the relative weights by actual combustion of weighed quantities of the two gases. It will be remembered that in Dumas's investigation the composition of water is inferred from the weights of the oxygen and of the water, the hydrogen being unweighed. In order to avoid the very unfavourable conditions of this method, recent workers have made it a point to weigh the hydrogen, whether in the gaseous state as in the experiments of Professor Cooke and my own, or occluded in palladium as in Mr Keiser's practice. So long as the hydrogen is weighed, it is not very material whether the second weighing relate to the water or to the oxygen. The former is the case in the work of Cooke and Keiser, the latter in the preliminary experiments now to be reported.

Nothing could be simpler in principle than the method adopted. Globes of the same size as those employed for the density determinations are filled to atmospheric pressure with the two gases, and are then carefully weighed. By means of Sprengel pumps the gases are exhausted into a mixing chamber, sealed below with mercury, and thence by means of a third Sprengel are conducted into a eudiometer, also sealed below with mercury, where they are fired by electric sparks in the usual way. After sufficient quantities of the gases have been withdrawn, the taps of the globes are turned, the leading tubes and mixing chamber are cleared of all remaining gas, and, after a final explosion in the eudiometer, the nature and amount of the residual gas are determined. The quantities taken from the globes can be found from the weights before and after operations. From the quantity of that gas which proved to be in excess, the calculated weight of the residue is subtracted. This gives the weight of the two gases which actually took part in the combustion.

In practice, the operation is more difficult than might be supposed from the above description. The efficient capacity of the eudiometer being

* [1901. Dr Scott's final number (*Proc. Roy. Soc.* Vol. LIII. p. 133, 1893) was 2·00245.]

necessarily somewhat limited, the gases must be fed in throughout in very nearly the equivalent proportions; otherwise there would soon be such an accumulation of residue that no further progress could be made. For this reason nothing could be done until the intermediate mixing chamber was provided. In starting a combustion, this vessel, originally full of mercury, was charged with equivalent quantities of the two gases. The oxygen was first admitted until the level of the mercury had dropped to a certain mark, and subsequently the hydrogen down to a second mark, whose position relatively to the first was determined by preliminary measurements of volume. The mixed gases might then be drawn off into the eudiometer until exhausted, after which the chamber might be recharged as before. But a good deal of time may be saved by replenishing the chamber from the globes simultaneously with the exhaustion into the eudiometer. In order to do this without losing the proper proportion, simple mercury manometers were provided for indicating the pressures of the gases at any time remaining in the globes. But even with this assistance close attention was necessary to obviate an accumulation of residual gas in the eudiometer, such as would endanger the success of the experiment, or, at least, entail tedious delay. To obtain a reasonable control, two sparking places were provided, of which the upper was situate nearly at the top of the eudiometer. This was employed at the close, and whenever in the course of the combustion the residual gas chanced to be much reduced in quantity; but, as a rule, the explosions were made from the lower sparking point. The most convenient state of things was attained when the tube contained excess of oxygen down to a point somewhat below the lower sparking wires. Under these circumstances, each bubble of explosive gas readily found its way to the sparks, and there was no tendency to a dangerous accumulation of mixed gas before an explosion took place. When the gas in excess was hydrogen, the manipulation was more difficult, on account of the greater density of the explosive gas retarding its travel to the necessary height.

In spite of all precautions several attempted determinations have failed from various causes, such as fracture of the eudiometer and others which it is not necessary here to particularise, leading to the loss of much labour. Five results only can at present be reported, and are as follows:—

December	24, 1888	15·93
January	3, 1889	15·98
„	21, „	15·98
February	2, „	15·93
„	13, „	15·92
	Mean	15·95

This number represents the atomic ratio of oxygen and hydrogen as deduced immediately from the weighings with allowance for the unburnt residue. It is subject to the correction for buoyancy rendered necessary by the shrinkage of the external volume of the globes when internally exhausted, as explained in my former communication*. In these experiments, the globe which contained the hydrogen was the same (14) as that employed for the density determinations. The necessary correction is thus four parts in a thousand, reducing the final number for the atomic weight of oxygen to

$$15{\cdot}89,$$

somewhat lower than that which I formerly obtained (15·91) by the use of Mr Scott's value of the volume ratio. It may be convenient to recall that the corresponding number obtained by Cooke and Richards (corrected for shrinkage) is 15·87, while that of Keiser is 15·95.

In the present incomplete state of the investigation, I do not wish to lay much stress upon the above number, more especially as the agreement of the several results is not so good as it should be. The principal source of error, of a non-chemical character, is in the estimation of the weight of the hydrogen. Although this part of the work cannot be conducted under quite such favourable conditions as in the case of a density determination, the error in the difference of the two weighings should not exceed 0·0002 gram. The whole weight of the hydrogen used is about 0·1 gram†; so that the error should not exceed three in the last figure of the final number. It is thus scarcely possible to explain the variations among the five numbers as due merely to errors of the weighings.

The following are the details of the determination of February 2, chosen at random:—

Before combustion ... $G_{14} + H + 0{\cdot}2906 = G_{11}$ ... pointer 20·05
After „ ... $G_{14} + H + 0{\cdot}4006 = G_{11}$ ... pointer 20·31

$$\text{Hydrogen taken} = 0{\cdot}1100 - 0{\cdot}00005 = 0{\cdot}10995 \text{ gram.}$$

Before combustion ... $G_{13} + O = G_{11} + 2{\cdot}237$ ... pointer 20·00
After „ ... $G_{13} + O = G_{11} + 1{\cdot}357$ ... pointer 19·3

$$\text{Oxygen taken} = 0{\cdot}8800 + 0{\cdot}0001 = 0{\cdot}8801 \text{ gram.}$$

At the close of operations the residue in the eudiometer was oxygen, occupying 7·8 c.c. This was at a total pressure of $29{\cdot}6 - 16{\cdot}2 = 13{\cdot}4$ inches

* The necessity of this correction was recognised at an early stage, and, if I remember rightly, was one of the reasons which led me to think that a redetermination of the density of hydrogen was desirable. In the meantime, however, the question was discussed by Agamennone (*Atti (Rendiconti) d. R. Accad. dei Lincei*, 1885), and some notice of his work reached me. When writing my paper last year I could not recall the circumstances; but since the matter has attracted attention I have made inquiry, and take this opportunity of pointing out that the credit of first publication is due to Agamennone.

† It was usual to take for combustion from two-thirds to three-fourths of the contents of the globe.

of mercury. Subtracting 0·4 inch for the pressure of the water vapour, we get 13·0 as representing the oxygen pressure. The temperature was about 12° C. Thus, taking the weight of a cub. cm. of oxygen at 0° C. and under a pressure of 76·0 cm. of mercury to be 0·00143 gram, we get as the weight of the residual oxygen

$$0{\cdot}00143 \frac{7{\cdot}8}{1+12\times 0{\cdot}00367} \frac{13{\cdot}0\times 2{\cdot}54}{76{\cdot}0} = 0{\cdot}0046 \text{ gram.}$$

The weight of oxygen burnt was, therefore, 0·8801 − 0·0046 = 0·8755 gram.

Finally, for the ratio of atomic weights,

$$\frac{\text{Oxygen}}{\frac{1}{2}\text{ Hydrogen}} = 15{\cdot}926.$$

In several cases the residual gas was subjected to analysis. Thus, after the determination of February 2, the volume was reduced by additions of hydrogen to 1·2 c.c. On introduction of potash there was shrinkage to about 0·9, and, on addition of pyrogallic acid, to 0·1 or 0·2. These volumes of gas are here measured at a pressure of $\frac{1}{3}$ atmosphere, and are, therefore, to be divided by 3 if we wish to estimate the quantities of gas under standard conditions. The final residue of (say) 0·05 c.c. should be nitrogen, and, even if originally mixed with the hydrogen—the most unfavourable case—would involve an error of only $\frac{1}{2000}$ in the final result. The 0·1 c.c. of carbonic anhydride, if originally contained in the hydrogen, would be more important; but this is very improbable. If originally mixed with the oxygen, or due to leakage through india-rubber into the combustion apparatus, it would lead to no appreciable error.

The aggregate impurity of 0·15, here indicated, is tolerably satisfactory in comparison with the total quantity of gas dealt with—2000 c.c. It is possible, however, that nitrogen might be oxidised, and thus not manifest itself under the above tests. In another experiment the water of combustion was examined for acidity, but without definite indications of nitric acid. The slight reddening observed appeared to be rather that due to carbonic acid, some of which, it must be remembered, would be dissolved in the water. These and other matters demand further attention.

The somewhat complicated glass blowing required for the combustion apparatus has all been done at home by my assistant, Mr Gordon, on whom has also fallen most of the rather tedious work connected with the evacuation of globes and other apparatus, and with the preparation of the gases.

[1901. Further work upon this subject is recorded in *Proc. Roy. Soc.* Vol. L. p. 449, 1892. *Vide infra.*]

154.

THE HISTORY OF THE DOCTRINE OF RADIANT ENERGY.

[*Philosophical Magazine*, XXVII. pp. 265—270, 1889.]

IN his interesting Address* to the American Association for the Advancement of Science, Prof. Langley sketches the development of the modern doctrine of Radiant Energy, and deduces important lessons to be laid to heart by all concerned in physical investigation. This is a most useful undertaking; but in the course of it there occur one or two statements which, in the interest of scientific history, ought not to be allowed to pass without a protest.

After quoting Melloni's very unequivocal conclusion of 1843, that "Light is merely a series of calorific indications sensible to the organs of sight; or, *vice versâ*, the radiations of obscure heat are veritable invisible radiations of light," Prof. Langley goes on to say, "So far as I know, no physicist of eminence reasserted Melloni's principle with equal emphasis till J. W. Draper, in 1872. Only sixteen years ago, or in 1872, it was almost universally believed that there were three different entities in the spectrum, represented by actinic, luminous, and thermal rays."

These words struck me strangely as I first read them. My own scientific ideas were formed between 1860 and 1866, and I certainly never believed in the three entities. Having on a former occasion referred to this question† as an illustration of the difference of opinion which is sometimes to be found between the theoretical and experimental schools of workers, I was sufficiently interested in the matter to look up a few references, with results which are, I think, difficult to reconcile with Prof. Langley's view.

In Young's *Lectures*‡ we read:—"Dr Herschel's experiments have shown that radiant heat consists of various parts which are differently refrangible,

* *Amer. Journ. Sci.* Jan. 1889.

† Address to Section A, *Brit. Assoc. Report*, 1882. [Vol. II. p. 122.]

‡ Vol. I. p. 638 (1807).

and that, in general, invisible heat is less refrangible than light. This discovery must be allowed to be one of the greatest that have been made since the days of Newton....

"It was first observed in Germany by Ritter, and soon afterwards in England by Dr Wollaston, that the muriate of silver is blackened by invisible rays, which extend beyond the prismatic spectrum, on the violet side. It is therefore probable that these black or invisible rays, the violet, blue, green, perhaps the yellow, and the red rays of light, and the rays of invisible heat, constitute seven different degrees of the same scale, distinguished from each other into this limited number, not by natural divisions, but by their effects on our senses: and we may also conclude that there is some similar relation between heated and luminous bodies of different kinds."

And, again, on p. 654: "If heat is not a substance, it must be a quality; and this quality can only be motion. It was Newton's opinion that heat consists in a minute vibratory motion of the particles of bodies, and that this motion is communicated through an apparent vacuum by the undulations of an elastic medium, which is also concerned in the phenomena of light. If the arguments which have been lately advanced in favour of the undulatory theory of light be deemed valid, there will be still stronger reasons for admitting this doctrine respecting heat; and it will only be necessary to suppose the vibrations and undulations principally constituting it to be larger and stronger than those of light, while at the same time the smaller vibrations of light, and even the blackening rays, derived from still more minute vibrations, may perhaps, when sufficiently condensed, concur in producing the effects of heat. These effects, beginning from the blackening rays, which are invisible, are a little more perceptible in the violet, which still possess but a faint power of illumination; the yellow-green afford the most light; the red give less light, but much more heat; while the still larger and less frequent vibrations, which have no effect upon the sense of sight, may be supposed to give rise to the least refrangible rays, and to constitute invisible heat."

It is doubtless true that Young's views did not at the time of the publication of these lectures* command the authority which now attaches to them. But when the undulatory theory gained acceptance, there was no room left for the distinct entities.

J. B. Reade, one of the pioneers of photography, in a letter to R. Hunt †,

* I may remark, in passing, that Brougham knew a little of experimenting, as of everything else, except law! [1901. The reference is to a contemporary gibe at Lord Chancellor Brougham that "had he known a little law, he would have known a little of everything." Young's views were violently attacked by Brougham in the *Edinburgh Review*.]

† Hunt's "Researches on Light," *Longmans*, 1854, p. 374. Hunt himself, not being an undulationist, was upon the other side.

of date Feb. 1854, thus speaks of Young:—"Dr Young's propositions are, that radiant light consists in undulations of the luminiferous æther, that light differs from heat only in the frequency of its undulations, that undulations less frequent than those of light produce heat, and that undulations more frequent than those of light produce chemical and photographic action,—all proved by experiments."

Sir John Herschel's presentation of the matter* is not very explicit. "The solar rays, then, possess at least three distinct powers: those of heating, illuminating, and effecting chemical combinations or decompositions; and these powers are distributed among the differently refrangible rays in such a manner as to show their complete independence on each other. Later experiments have gone a certain way to add another power to the list—that of exciting magnetism." Although the marginal index runs "Calorific, luminous, and chemical rays," the choice of words in the text, as well as the reference to magnetism (for surely no one believed in a special magnetizing entity), points to the conclusion that Herschel held the modern view.

For the decade between 1850 and 1860, the citation upon which I most rely as indicative of the view held by the highest authorities, and by those capable of judging where the highest authority was to be found, is from Prof. Stokes's celebrated memoir upon Fluorescence †. On p. 465 we read:—"Now according to the Undulatory Theory, the nature of light is defined by two things, its period of vibration, and its state of polarization. To the former corresponds its refrangibility, and, so far as the eye is a judge of colour, its colour." And in a footnote here appended:—

"It has been maintained by some philosophers of the first eminence that light of definite refrangibility might still be compound; and though no longer decomposable by prismatic refraction may still be so by other means. I am not now speaking of compositions and resolutions dependent upon polarization. It has been suggested by advocates of the undulatory theory, that possibly a difference of properties in lights of the same refrangibility might correspond to a difference in the law of vibration, and that lights of given refrangibility may differ in tint, just as musical notes of given pitch differ in quality. Were it not for the strong conviction I felt that light of definite refrangibility is in the strict sense of the word homogeneous, I should probably have been led to look in this direction for an explanation of the remarkable phenomena presented by a solution of sulphate of quinine. It would lead me too far from the subject of the present paper to explain the grounds of this conviction. I will only observe that I have not overlooked the remarkable effect of absorbing media in causing apparent changes of colour in a pure spectrum; but this I believe to be a subjective phenomenon depending upon contrast."

It can scarcely be necessary to insist that "light" is used here in the wider sense, a large part of the memoir dealing with the transformation of invisible into visible light.

* Art. Light, *Enc. Met.* 1830, § 1147.

† "On a Change of Refrangibility of Light." *Phil. Trans.* 1852.

The allusion in the note is, of course, to Brewster. This distinguished discoverer never accepted the wave-theory, and was thus insensible to the repugnance with which his doctrine of three different kinds of luminous radiation was regarded by every undulationist. The matter was not finally set at rest until Helmholtz showed that Brewster's effects depended upon errors of experiment not previously recognized.

The following, from W. Thomson*, is almost equally significant:—

"It is assumed in this communication that the undulatory theory of radiant heat and light, according to which light is merely radiant heat, of which the vibrations are performed in periods between certain limits of duration, is true. 'The chemical rays' beyond the violet end of the spectrum consist of undulations of which the full vibrations are executed in periods shorter than those of the extreme visible violet light, or than about the eight hundred million millionth of a second. The periods of the vibrations of visible light lie between this point and another, about double as great, corresponding to the extreme visible red light. The vibrations of the obscure radiant heat beyond the red end are executed in longer periods than this; the longest which has yet been experimentally tested being about the eighty million millionth of a second."

Again, in Lloyd's "Wave Theory of Light"†, we find the following passage:—"It appears, then, that sensibility of the eye is confined within much narrower limits than that of the ear; the ratio of the times of the extreme vibrations which affect the eye being only that of 1·58 to 1, which is less than the ratio of the times of vibration of a fundamental note and its octave. There is no reason for supposing, however, that the vibrations themselves are confined within these limits. In fact, we know that there are *invisible* rays *beyond* the two extremities of the spectrum, whose periods of vibration (and lengths of wave) must fall without the limits now stated to belong to the visible rays."

I believe that it would be not too much to say that during the decade 1850—1860 nearly all the leading workers in physics, with the exception of Brewster, held the modern view of radiation. It would be quite consistent with this that many chemists, photographers, and workers in other branches of science, who trusted to more or less antiquated text-books for their information, should have clung to a belief in the three entities. After 1860, and the discussions respecting the discoveries of Stewart and Kirchhoff, I should have supposed that there were scarcely two opinions. Stewart's *Elementary Treatise on Heat* was published in 1866, and was widely used

* "On the Mechanical Action of Radiant Heat or Light"; &c. *Proc. Roy. Soc. Edinb.* Feb. 1852.

† *Longmans*, 1857, p. 16.

in schools and colleges. In book II. ch. II. he elaborately discusses the whole question, summing up in favour of the view that "radiant light and heat are only varieties of the same physical agent, and that when once the spectrum of a luminous object has been obtained, the separation of the different rays from one another is physically complete; so that, if we take any region of the visible spectrum, its illuminating and heating effects are caused by precisely the same rays." What there was further for Draper or any one else to say in 1872 I am at a loss to comprehend*.

To pass on to another point. I have followed the excellent advice to read W. Herschel's original memoirs; but I must confess that the impression produced upon my mind is different in some respects from that expressed by Prof. Langley. It seems to me that Herschel fully established the diversity of radiant heat. In the first memoir† a paragraph is headed "*Radiant Heat is of different Refrangibility*," the question being fully discussed; and from the following memoir (p. 291) it is evident that this proposition extends to invisible radiation. "The four last experiments prove that the maximum of the heating power is vested among the invisible rays; and is probably not less than half an inch beyond the last visible ones, when projected in the manner before mentioned. The same experiments also show that the sun's invisible rays, in their less refrangible state, and considerably beyond the maximum, still exert a heating power fully equal to that of red-coloured light...." Can it then be said of De la Roche that he, in 1811, before anyone else, "derives the just and most important, as well as the then most novel conception, that radiant heat is of different *kinds*"? It was doubtless a most important step when De la Roche and Melloni exhibited the diversity of radiant heat *by means of selective absorption*; but I do not see how we can regard them as the discoverers of the fact.

It would take too long to establish by quotations, but it is pretty evident that in his two earlier papers‡ Herschel leaned to the view that light was not "essentially different from radiant heat." Why then, after laying hands upon the truth, did he let it go, and decide that light and heat are not occasioned by the same rays?

"The question§, which we are discussing at present, may therefore at once be reduced to this single point. Is the heat which has the refrangibility of the red rays occasioned by the light of these rays? For, should that be the case, as there will be then only one set of rays, one fate only can attend them, in being either transmitted or stopped, according to the power of the glass applied to them. We are now to appeal to our prismatic experiment

* I have limited myself to citations from English writers, but I have no reason to think that the course of opinion was different in France and Germany.

† *Phil. Trans.* 1800, p. 255.

‡ See pp. 272, 291, 292.

§ Third Memoir, p. 520.

upon the subject, which is to decide the question." The issue could not be more plainly stated. The experiment is discussed, and this is the conclusion:—"Here then we have a direct and simple proof, in the case of the red glass, that the rays of light are transmitted, while those of heat are stopped, and that thus they have nothing in common but a certain equal degree of refrangibility...."

I am disposed to think that it was this erroneous conclusion from experiment*, more, perhaps, than preconceived views about caloric, that retarded progress in radiant heat for so many years. We are reminded of Darwin's saying that a bad observation is more mischievous than unsound theory. It would be interesting to inquire upon what grounds we now reject the plain answer which Herschel thought himself to have received from experiment. I do not recall a modern investigation in which the heat and light absorptions are proved to be equal for the various parts of the visible spectrum. Can it be that after all we have nothing but theory to oppose to Herschel's facts?

I hope it will be understood that these criticisms, even if they are sound, do not touch the substance of Prof. Langley's address, which is doubly interesting as coming from one who has done so much himself to enlarge our knowledge of this branch of science.

* See Whewell's *History of the Inductive Sciences*, Vol. II. p. 548 (1847).

155.

NOTE ON THE FREE VIBRATIONS OF AN INFINITELY LONG CYLINDRICAL SHELL.

[*Proceedings of the Royal Society*, XLV. pp. 443—448, 1889.]

IN a recent memoir* Mr Love has considered this question among others; but he has not discussed his result {equation (95)}, except in its application to a rather special case involving the existence of a free edge. When the cylinder is regarded as infinitely long, the problem is naturally of a simpler character; and I have thought that it might be worth while to express more fully the frequency equation, as applicable to all vibrations, independent of the thickness of the shell, which are periodic with respect both to the length and the circumference of the cylinder.

In order to prevent misunderstanding, it may be well to premise that the vibrations, whose frequency is to be determined, do not include the gravest of which a thin shell is capable. If the middle surface be simply bent, the potential energy of deformation is of a higher order of magnitude than in the contrary case, and according to the present method of treatment the frequency of vibration will appear to be zero. It is known, however, that the only possible modes of bending of a cylindrical shell are such as are not periodic along the length, or rather have the wave-length in this direction infinitely long†. When the middle surface is stretched, as well as bent, the potential energy of bending may be neglected, except in certain very special cases.

* "On the small Free Vibrations and Deformation of a thin Elastic Shell," *Phil. Trans.* A, Vol. CLXXIX. (1888), p. 491.

† "On the Bending and Vibration of thin Elastic Shells, especially of Cylindrical Form," *Roy. Soc. Proc. supra*, p. 105. [Vol. III. p. 217.]

Taking cylindrical coordinates (r, ϕ, z), and denoting the displacements parallel to z, ϕ, r by u, v, w respectively, we have for the principal elongations and shear at any point (a, ϕ, z)*—

$$\sigma_1 = \frac{du}{dz}, \qquad \sigma_2 = \frac{w}{a} + \frac{1}{a}\frac{dv}{d\phi}, \qquad \varpi = \frac{1}{a}\frac{du}{d\phi} + \frac{dv}{dz}; \quad\text{.........(1)}$$

and the energy per unit of area is expressed by

$$2nh\left\{\sigma_1^2 + \sigma_2^2 + \tfrac{1}{2}\varpi^2 + \frac{m-n}{m+n}(\sigma_1+\sigma_2)^2\right\}, \quad\text{..............(2)}$$

where $2h$ denotes the thickness of the shell, and m, n are the elastic constants of Thomson and Tait's notation.

The functions u, v, w are to be assumed proportional to the sines, or cosines, of μz and $s\phi$. These may be combined in various ways, but a sufficient example is

$$u = U\cos s\phi \cos \mu z, \qquad v = V \sin s\phi \sin \mu z, \qquad w = W\cos s\phi \sin \mu z; \text{...(3)}$$

so that

$$\sigma_1 = -\mu U \cos s\phi \sin \mu z, \quad\text{.............................(4)}$$

$$\sigma_2 = (W + sV)\cos s\phi \sin\mu z, \quad\text{.......................(5)}$$

$$\varpi = (-sU + \mu V)\sin s\phi \cos \mu z, \quad\text{....................(6)}$$

unity being written for convenience in place of a. The energy per unit area is thus

$$2nh\left[\cos^2 s\phi \sin^2\mu z\left\{\mu^2U^2 + (W+sV)^2 + \frac{m-n}{m+n}(W+sV-\mu U)^2\right\} + \tfrac{1}{2}\sin^2 s\phi\cos^2\mu z\,(-sU+\mu V)^2\right]. \quad\text{......(7)}$$

Again, the kinetic energy per unit area is, if ρ be the volume density,

$$\rho h\left[\left(\frac{dU}{dt}\right)^2\cos^2 s\phi\cos^2\mu z + \left(\frac{dV}{dt}\right)^2\sin^2 s\phi \sin^2\mu z + \left(\frac{dW}{dt}\right)^2\cos^2 s\phi\sin^2\mu z\right]. \text{ (8)}$$

In the integration of these expressions with respect to ϕ and z, the mean value of each $\sin^2$ or $\cos^2$ is $\frac{1}{2}$†. We may then apply Lagrange's method. If the type of vibration be $\cos pt$, and $p^2\rho/n = k^2$, the resulting equations may be written

$$\{2(M+1)\mu^2 + s^2 - k^2\}U - (2M+1)\mu s V - 2M\mu W = 0, \text{ ...(9)}$$

$$-(2M+1)\mu s U + \{\mu^2 + 2(M+1)s^2 - k^2\}V + 2(M+1)sW = 0, \text{...(10)}$$

$$-2M\mu U + 2(M+1)sV + \{2(M+1) - k^2\}W = 0, \text{...(11)}$$

where

$$M = \frac{m-n}{m+n} \quad\text{................................(12)}$$

* See a paper on the "Infinitesimal Bending of Surfaces of Revolution" (*London Math. Soc. Proc.* Vol. XIII. p. 4, Nov. 1881), and those already cited. [Vol. I. p. 551.]

† In the physical problem the range of integration for ϕ is from 0 to 2π; but mathematically we are not confined to one revolution. We may conceive the shell to consist of several superposed convolutions, and then s is not limited to be a whole number.

The frequency equation is that expressing the evanescence of the determinant of this triad of equations.

We will consider for a moment the simple case which arises when $\mu = 0$, that is, when the displacements are independent of z. The three equations reduce to

$$(s^2 - k^2)\, U = 0, \qquad\qquad (13)$$

$$\{2(M+1)s^2 - k^2\}\, V + 2(M+1)\, sW = 0, \qquad\qquad (14)$$

$$2(M+1)\, sV + \{2(M+1) - k^2\}\, W = 0; \qquad\qquad (15)$$

and they may be satisfied in two ways. First let $V = W = 0$; then U may be finite, provided

$$s^2 - k^2 = 0. \qquad\qquad (16)$$

The corresponding type for U is

$$U = \cos s\phi \cos pt, \qquad\qquad (17)$$

where

$$p^2 = \frac{ns^2}{\rho a^2}, \qquad\qquad (18)$$

a being restored, as can be done at any moment by consideration of dimensions. In this motion the material is sheared without extension, every generating line of the cylinder moving along its own length. The frequency depends upon the circumferential wave-length, and not upon the curvature of the cylinder.

The second kind of vibrations are those in which $U = 0$, so that the motion is strictly in two dimensions. The elimination of the ratio V/W from (14), (15) gives

$$k^2\{k^2 - 2(M+1)(1+s^2)\} = 0, \qquad\qquad (19)$$

as the frequency equation. The first root is $k^2 = 0$, indicating infinitely slow motion. These are the flexural vibrations already referred to, and the corresponding relation between V and W is by (14)

$$sV + W = 0, \qquad\qquad (20)$$

giving by (4), (5), (6),

$$\sigma_1 = \sigma_2 = \varpi = 0.$$

The other root of (19) gives, on restoration of a,

$$k^2a^2 = \frac{4m}{m+n}(1+s^2), \qquad\qquad (21)$$

or

$$p^2 = \frac{4mn}{m+n}\,\frac{1+s^2}{a^2\rho}; \qquad\qquad (22)$$

while the relation between V and W is

$$-V + sW = 0. \qquad\qquad (23)$$

It will be observed that when s is very large, the flexural vibrations tend to become exclusively normal, and the extensional vibrations to become exclusively tangential, as might have been expected from the theory of plane plates.

Returning now to the general case, the determinant of (9), (10), (11) gives on reduction

$$[k^2-\mu^2-s^2]\{k^2[k^2-2(M+1)(\mu^2+s^2+1)]+4(2M+1)\mu^2\} + 4(2M+1)\mu^2 s^2 = 0. \quad \text{......(24)}$$

If $\mu=0$, we have the three solutions already considered,

$$k^2=0, \qquad k^2=s^2, \qquad k^2=2(M+1)(s^2+1).$$

If $s=0$, that is, if the deformation be symmetrical about the axis, we have

$$k^2=\mu^2, \text{ or } k^2[k^2-2(M+1)(\mu^2+1)]+4(2M+1)\mu^2=0. \quad \text{...(25)}$$

Corresponding to the first root we have $U=0$, $W=0$, as is readily proved on reference to the original equations with $s=0$. The vibrations are the purely torsional ones represented by

$$v=\sin \mu z \cos pt, \quad \text{..........................(26)}$$

where

$$p^2=n\mu^2/\rho. \quad \text{..........................(27)}$$

The frequency depends upon the wave-length parallel to the axis, and not upon the radius of the cylinder.

The remaining roots of (25) correspond to motions for which $V=0$, or which take place in planes through the axis. The general character of these vibrations may be illustrated by the case where μ is small, or the wave-length a large multiple of the radius of the cylinder. We find approximately from the quadratic (on restoration of a)

$$\frac{k^2a^2}{M+1}=2+\frac{2M^2\mu^2a^2}{(M+1)^2}, \quad \text{......................(28)}$$

or

$$k^2=\frac{2(2M+1)\mu^2}{(M+1)}. \quad \text{......................(29)}$$

The vibrations of (28) are nearly purely radial. If we suppose that μ vanishes, we fall back upon

$$k^2a^2=2(M+1),$$

or

$$p^2=\frac{4mn}{m+n}\frac{1}{a^2\rho}, \quad \text{......................(30)*}$$

as may be seen from (22), by putting $s=0$.

On the other hand, the vibrations of (29) are nearly purely axial. In terms of m and n,

$$p^2=\frac{n\mu^2}{\rho}\cdot\frac{3m-n}{m}, \quad \text{......................(31)}$$

* This equation is given, in a slightly different form, by Love (*loc. cit.* p. 523).

Now, if q denote Young's modulus,

$$q = \frac{n(3m-n)}{m}; \qquad \text{.............................(32)}$$

so that

$$p^2 = \frac{q\mu^2}{\rho}. \qquad \text{.......................................(33)}$$

This is the ordinary formula for the longitudinal vibrations of a rod, the fact that the section is here a thin annulus not influencing the result to this order of approximation.

Another extreme case worthy of notice occurs when s is very great. Equation (24) then reduces to

$$k^2 [k^2 - \mu^2 - s^2][k^2 - 2(M+1)(\mu^2 + s^2)] = 0; \qquad \text{...........(34)}$$

so that k^2 becomes a function of μ and s only through $(\mu^2 + s^2)$, as might have been expected from the theory of plane plates. The first root relates to flexural vibrations; the second to vibrations of shearing, as in (18); the third to vibrations involving extension of the middle surface, analogous to those in (22).

It is scarcely necessary to add, in conclusion, that the most general deformation of the middle surface can be expressed by means of a series of such as are periodic with respect to z and ϕ, so that the problem considered is really the most general small motion of an infinite cylindrical shell.

Another particular case worth notice arises when $s = 1$, so that (24) assumes the form

$$k^2 (k^2 - \mu^2 - 1)[k^2 - 2(M+1)(\mu^2 + 2)] + 4\mu^2 (k^2 - \mu^2)(2M+1) = 0. \text{ ...(35)}$$

As we have already seen, if μ be zero, one of the values of k^2 vanishes. If μ be small, the corresponding value of k^2 is of the order μ^4. Equation (35) gives in this case

$$k^2 = \frac{2M+1}{M+1}\mu^4; \qquad \text{..........................(36)}$$

or in terms of p, q, and with restoration of a,

$$p^2 = \frac{q\mu^4 a^2}{2\rho}. \qquad \text{.................................(37)}$$

This agrees with the usual formula* for the transverse vibrations of rods.

* *Theory of Sound*, § 181.

156.

ON THE FREE VIBRATIONS OF AN INFINITE PLATE OF HOMOGENEOUS ISOTROPIC ELASTIC MATTER.

[*Proceedings of the London Mathematical Society*, XX. pp. 225—234, 1889.]

THE solid here contemplated is that bounded by two infinite planes parallel to xy; and the vibrations are supposed to be periodic, not only with respect to the time (e^{ipt}), but also with respect to x and y. The results, so far as thin plates are concerned, have long been known; but the method may not be without interest in view of the difficulties which beset the rigorous treatment of the theory of thin plates, and of the fact that it is not limited to the case of small thickness. A former investigation*, "On Waves propagated along the Plane Surface of an Elastic Solid," may be regarded as a particular case of that now before us.

In conformity with the suppositions as to periodicity, we might assume that all the functions concerned involve x and y only through the factors e^{ifx}, e^{igy}. But, by a rotation of the axes, $e^{i(fx+gy)}$ may be replaced by e^{ifx} without loss of generality, and it will considerably simplify our equations if we limit them to the latter form. Any function of x, y (*e.g.*, the dilatation) may be expanded in a series of such terms as $\cos fx \cos gy$, and this may be resolved into two of the form

$$\cos(fx+gy), \qquad \cos(fx-gy).$$

But between these forms there is no essential difference, for on account of the symmetry of the plane we shall have to deal in either case only with $\sqrt{(f^2+g^2)}$. The assumption of proportionality with e^{ifx} is not, however, equivalent to a limitation of the problem to two dimensions, as might at first be supposed; inasmuch as β, the displacement parallel to y, is allowed to remain finite.

* *Proc. Lond. Math. Soc.* Vol. XVII. Nov. 1885. [Vol. II. p. 441.]

If θ be the dilatation, the usual equations are

$$\rho \frac{d^2\alpha}{dt^2} = m \frac{d\theta}{dx} + n\nabla^2\alpha, \qquad \&c., \quad \ldots\ldots\ldots\ldots\ldots\ldots(1)$$

in which

$$\theta = \frac{d\alpha}{dx} + \frac{d\beta}{dy} + \frac{d\gamma}{dz}, \quad \ldots\ldots\ldots\ldots\ldots\ldots(2)$$

and m, n denote the elastic constants of the material according to Thomson and Tait's notation*.

If α, β, γ all vary as e^{ipt}, equations (1) become

$$m \frac{d\theta}{dx} + n\nabla^2\alpha + \rho p^2\alpha = 0, \qquad \&c. \quad \ldots\ldots\ldots\ldots\ldots\ldots(3)$$

Differentiating equations (3) in order with respect to x, y, z, and adding, we get

$$(\nabla^2 + h^2)\,\theta = 0, \quad \ldots\ldots\ldots\ldots\ldots\ldots(4)$$

in which

$$h^2 = \rho p^2/(m+n). \quad \ldots\ldots\ldots\ldots\ldots\ldots(5)$$

Again, if we put

$$k^2 = \rho p^2/n, \quad \ldots\ldots\ldots\ldots\ldots\ldots(6)$$

equations (3) take the form

$$(\nabla^2 + k^2)\,\alpha = \left(1 - \frac{k^2}{h^2}\right)\frac{d\theta}{dx}, \qquad \&c. \quad \ldots\ldots\ldots\ldots\ldots\ldots(7)$$

A particular solution of (7) is†

$$\alpha = -\frac{1}{h^2}\frac{d\theta}{dx}, \qquad \beta = -\frac{1}{h^2}\frac{d\theta}{dy}, \qquad \gamma = -\frac{1}{h^2}\frac{d\theta}{dz}; \ldots\ldots\ldots(8)$$

in order to complete which it is only necessary to add complementary terms u, v, w satisfying the equations

$$(\nabla^2 + k^2)\,u = 0, \qquad (\nabla^2 + k^2)\,v = 0, \qquad (\nabla^2 + k^2)\,w = 0, \ldots\ldots\ldots(9)$$

$$\frac{du}{dx} + \frac{dv}{dy} + \frac{dw}{dz} = 0. \quad \ldots\ldots\ldots\ldots\ldots\ldots(10)$$

According to our present suppositions, x and y are involved only through e^{ifx}, that is, y is not involved at all. Thus

$$d\theta/dy = 0, \qquad dv/dy = 0.$$

The displacement β is thus identical with v, and satisfies the differential equation

$$(\nabla^2 + k^2)\,\beta = 0. \quad \ldots\ldots\ldots\ldots\ldots\ldots(11)$$

Again, in virtue of (9) and (10), we may write

$$u = d\chi/dz, \qquad w = -\,d\chi/dx, \quad \ldots\ldots\ldots\ldots\ldots\ldots(12)$$

* Lamé's constants λ, μ are related to m, n according to $\lambda + \mu = m$, $\mu = n$.

† Lamb, "On the Vibrations of an Elastic Sphere," *Math. Soc. Proc.* May 1882.

where χ is a function of x and z, which satisfies

$$(\nabla^2 + k^2)\chi = 0; \quad\quad (13)$$

and

$$\alpha = -\frac{1}{h^2}\frac{d\theta}{dx} + \frac{d\chi}{dz}, \quad \gamma = -\frac{1}{h^2}\frac{d\theta}{dz} - \frac{d\chi}{dx}. \quad\quad (14)^*$$

We have not yet made use of the supposition that x occurs only in the factor e^{ifx}. Under this condition we get from (4)

$$\theta = P\cosh rz + Q\sinh rz, \quad\quad (15)$$

where

$$r^2 = f^2 - h^2; \quad\quad (16)$$

and from (13), (11),

$$\chi = A\sinh sz + B\cosh sz, \quad\quad (17)$$

$$\beta = C\cosh sz + D\sinh sz, \quad\quad (18)$$

where

$$s^2 = f^2 - k^2. \quad\quad (19)$$

The arbitrary quantities P, Q, A, B, C, D may be supposed to include the factors e^{ipt}, e^{ifx}, but are otherwise constants.

The evanescence of the three component stresses at the two bounding surfaces gives in all six equations. The components of tangential stress are, in general, proportional to

$$\frac{d\beta}{dz} + \frac{d\gamma}{dy}, \quad \frac{d\gamma}{dx} + \frac{d\alpha}{dz}.$$

As regards the first of these, we have at present $d\gamma/dy = 0$; so that the condition to be satisfied at each surface is simply

$$d\beta/dz = 0. \quad\quad (20)$$

The evanescence of the second tangential stress gives

$$-\frac{2}{h^2}\frac{d^2\theta}{dx\,dz} - \frac{d^2\chi}{dx^2} + \frac{d^2\chi}{dz^2} = 0. \quad\quad (21)$$

These equations are to hold good at both surfaces. If we take the origin at the middle of the thickness, the bounding surfaces may be represented by $z = \pm z_1$; and equations (20), (21) must be satisfied by the odd and even functions separately. Thus, from (18), (20),

$$C\sinh sz_1 = 0, \quad D\cosh sz_1 = 0, \quad\quad (22)$$

a pair of equations which may be satisfied in two ways. We may suppose $D = 0$, so that

$$\beta = C\cosh sz, \quad\quad (23)$$

in conjunction with

$$\sinh sz_1 = 0; \quad\quad (24)$$

or, on the other hand,

$$\beta = D\sinh sz, \quad\quad (25)$$

under the condition

$$\cosh sz_1 = 0. \quad\quad (26)$$

* Green, *Camb. Trans.* 1837; Reprint of Green's Works, p. 261.

During these vibrations the solid is simply sheared. In the vibrations of the first class represented by (23), β is an even function of z, α and γ vanishing. In the vibrations of the second class, β is an odd function of z, and therefore vanishes at the middle surface. The roots of (24) are

$$sz_1 = iq\pi,$$

where q is an integer; so that, by (19),

$$k^2 = f^2 + q^2\pi^2/z_1^2, \quad \text{.........................(27)}$$

and the stationary vibrations are of the type

$$\beta = \cos pt \cos fx \cos \frac{q\pi z}{z_1}, \quad \text{....................(28)}$$

p being given by (6) and (27).

In like manner, for the vibrations of the second class,

$$\beta = \cos pt \cos fx \sin \frac{(q+\frac{1}{2})\pi z}{z_1}, \quad \text{.................(29)}$$

where
$$k^2 = f^2 + \frac{(q+\frac{1}{2})^2\pi^2}{z_1^2}. \quad \text{..........................(30)}$$

In (28), (29), we may of course replace $\cos pt$, or $\cos fx$, by $\sin pt$, or $\sin fx$, respectively*.

The kind of vibrations just considered are those for which β is finite, while α and γ vanish. In the second kind of vibrations, β vanishes, so that the motion is strictly in two dimensions. There are four boundary conditions to be satisfied, two derived from (21), and two expressive of the evanescence of the normal stress. The latter condition is that

$$(m - n)\,\theta + 2n\,d\gamma/dz = 0,$$

when $z = \pm z_1$; or, in terms of k^2 and h^2,

$$(k^2 - 2h^2)\,\theta + 2h^2 d\gamma/dz = 0. \quad \text{.......................(31)}$$

Substituting from (14), (15), (17), in (21), (31), we obtain, with use of (16), (19),

$$2ifrh^{-2}(P \sinh rz + Q \cosh rz) + (k^2 - 2f^2)(A \sinh sz + B \cosh sz) = 0, \text{...(32)}$$

$$(k^2 - 2f^2)(P \cosh rz + Q \sinh rz) - 2h^2 ifs\,(A \cosh sz + B \sinh sz) = 0. \text{...(33)}$$

* In the present investigation the section of the solid perpendicular to y is an infinitely elongated rectangle. It may be worth notice that the corresponding solutions (in which every linear element parallel to the axis moves as a rigid body along its own length) may readily be obtained for cylinders of other sections, *e.g.*, the finite rectangle and the circle. There is complete mathematical analogy with the vibrations of a stretched membrane having the form of the section of the cylinder, under the condition that the boundary is free to move perpendicularly to the plane of the membrane. (*Theory of Sound*, § 227.)

These equations are to hold when $z = \pm z_1$, and must therefore be true for the odd and even parts separately. Thus

$$2ifrh^{-2} P \sinh rz_1 + (k^2 - 2f^2) A \sinh sz_1 = 0, \qquad (34)$$

$$(k^2 - 2f^2) P \cosh rz_1 - 2h^2 ifs\, A \cosh sz_1 = 0; \qquad (35)$$

$$2ifrh^{-2} Q \cosh rz_1 + (k^2 - 2f^2) B \cosh sz_1 = 0, \qquad (36)$$

$$(k^2 - 2f^2) Q \sinh rz_1 - 2h^2 ifs\, B \sinh sz_1 = 0. \qquad (37)$$

It will be seen that in these equations the constants P, A are separated from Q, B. The system can therefore be satisfied in two distinct ways. For the first class of vibrations $Q = 0$, $B = 0$. Equations (36), (37) are thus disposed of; while the first pair serve to determine the ratio $P : A$, and in addition impose a relation between the other quantities. Equations (14) show that θ and α are even functions of z, but that γ is an odd function. In this case of vibrations, therefore, the middle surface remains plane, but undergoes extension.

The frequency equation is found by elimination of $P : A$ between (34), (35):—

$$4f^2 rs \sinh rz_1 \cosh sz_1 = (k^2 - 2f^2)^2 \cosh rz_1 \sinh sz_1;$$

or, as it may be written,

$$4f^2 rs \tanh rz_1 = (k^2 - 2f^2)^2 \tanh sz_1. \qquad (38)$$

Again, from (35),
$$\frac{P}{2h^2 ifs \cosh sz_1} = \frac{A}{(k^2 - 2f^2) \cosh rz_1};$$

so that the type of vibration is, by (14),

$$\alpha = e^{ipt} e^{ifx} \{2sf^2 \cosh sz_1 \cosh rz + s(k^2 - 2f^2) \cosh rz_1 \cosh sz\}, \qquad (39)$$

$$\gamma = - e^{ipt} e^{ifx} \{2ifrs \cosh sz_1 \sinh rz + if(k^2 - 2f^2) \cosh rz_1 \sinh sz\}. \qquad (40)$$

We may apply these results to the case where the plate is *thin*, so that fz_1 is small. If rz_1, sz_1, in (38), be small, we find

$$(k^2 - 2f^2)^2 = 4f^2 r^2 = 4f^2 (f^2 - h^2),$$

or
$$k^4 = 4f^2 (k^2 - h^2). \qquad (41)$$

This equation determines k^2, since the ratio h^2/k^2 depends only upon the elastic quality of the material. In terms of m and n, from (5) and (6),

$$k^2 = \frac{4mf^2}{m+n}, \qquad (42)$$

or
$$p^2 = \frac{4f^2}{\rho} \frac{mn}{m+n}. \qquad (43)$$

At the same time, (39), (40) give approximately

$$\alpha = k^2 s\, e^{ipt} e^{ifx}, \qquad \gamma = -ifsz\,(k^2 - 2h^2)\, e^{ipt} e^{ifx},$$

or, if we throw out the common factor $k^2 s$,

$$\alpha = e^{ipt} e^{ifx}, \qquad \gamma = -\frac{m-n}{m+n}\, ifz\, e^{ipt} e^{ifx}. \quad\text{..................(44)}$$

This gives the same relation between the principal strains as is obtained in the ordinary theory of thin plates *, viz.,

$$\frac{d\gamma}{dz} = -\frac{m-n}{m+n}\left(\frac{d\alpha}{dx} + \frac{d\beta}{dy}\right).$$

A complete discussion of (38) would lead rather far, but we may easily find a second approximation in which the square of z_1 is included. Thus, since

$$\tanh rz_1 = rz_1\,(1 - \tfrac{1}{3} r^2 z_1^2 + \ldots),$$

$$4f^2 r^2 \frac{1 - \frac{1}{3} r^2 z_1^2}{1 - \frac{1}{3} s^2 z_1^2} = (k^2 - 2f^2)^2,$$

or

$$4f^2 r^2 \{1 - \tfrac{1}{3} z_1^2 (r^2 - s^2)\} = (k^2 - 2f^2)^2;$$

whence, on substitution of the values of r^2 and s^2 from (16), (19),

$$k^4 = 4f^2 (k^2 - h^2)\, \{1 - \tfrac{1}{3} z_1^2 (f^2 - h^2)\}. \quad\text{..................(45)}$$

From the first approximation we know that r^2, or $f^2 - h^2$, is positive. Hence k^2 diminishes with z_1^2, or the pitch falls as the thickness increases. An exception occurs when $r^2 = 0$; but this can happen only when $k^2 = 2f^2 = 2h^2$, or the material is such that $m = n$. If the character of the material be of this description, $k^2 = 2f^2$ satisfies (38), whatever may be the value of z_1. Each lamina parallel to xy vibrates unconstrained by its neighbours, and $\gamma = 0$ throughout.

If the material be incompressible, $h^2 = 0$, and (45) assumes the simplified form

$$k^2 = 4f^2 \{1 - \tfrac{1}{3} f^2 z_1^2\}. \quad\text{..........................(46)}$$

In any of these equations, if we suppose that the functions vary as e^{igy}, as well as e^{ifx}, the generalized result is obtained by merely writing $(f^2 + g^2)$ for f^2.

We now pass on to consider the second class of vibrations, for which, in (34), &c., $P = 0$, $A = 0$. Here θ and α are odd functions of z_1, while γ is an even function, so that the middle surface is bent without extension. As regards the equations (36), (37), which involve Q and B, it will be seen that

* See, for example, *Proc. Roy. Soc.* Dec. 1888. [Vol. III. p. 222.]

they differ from the first pair of equations involving P and A merely by the interchange everywhere of cosh and sinh. We have, therefore, in place of (38),

$$4f^2 rs \coth rz_1 = (k^2 - 2f^2)^2 \coth sz_1; \quad \text{.................(47)}$$

and in place of (39), (40),

$$\alpha = e^{ipt} e^{ifx} \{2sf^2 \sinh sz_1 \sinh rz + s(k^2 - 2f^2) \sinh rz_1 \sinh sz\}, \quad \text{......(48)}$$

$$\gamma = -e^{ipt} e^{ifx} \{2ifrs \sinh sz_1 \cosh rz + if(k^2 - 2f^2) \sinh rz_1 \cosh sz\}. \quad \text{...(49)}$$

If we now introduce the assumption that the plate is thin, we find, by expanding the hyperbolic functions in (47),

$$4f^2(f^2 - k^2)\{1 + \tfrac{1}{3}z_1^2(k^2 - h^2)\} = (k^2 - 2f^2)^2.$$

The first approximation gives $k^2 = 0$, signifying that the notes are infinitely grave. The second approximation is

$$k^4 = \tfrac{4}{3}z_1^2 f^4 (k^2 - h^2), \quad \text{..........................(50)}$$

or, in terms of p, m, n, ρ,

$$p^2 = \frac{mn}{m+n}\frac{4f^4 z_1^2}{3\rho}. \quad \text{..........................(51)}$$

Again, if we drop out a common factor $(k^2 r z_1)$, (48), (49) take the forms

$$\alpha = f^2 z e^{ipt} e^{ifx}, \qquad \gamma = if e^{ipt} e^{ifx}. \quad \text{...................(52)}$$

Hence $\alpha = -z\,d\gamma/dx$, signifying that to this order of approximation every line originally perpendicular to the middle surface retains its straightness and perpendicularity during the vibrations.

The third approximation to the value of k^2 from (47) gives

$$p^2 = \frac{mn}{m+n}\frac{4f^4 z^2}{3\rho}\left\{1 - f^2 z_1^2\left[\frac{4m}{3(m+n)} + \frac{7}{15}\right]\right\}; \quad \text{........(53)}$$

so that, when the thickness is increased beyond a certain point, the rise of pitch begins to be less rapid than according to the second approximation (51).

When z_1 is infinitely great, we get, from (38) or (47),

$$4f^2 rs = (k^2 - 2f^2)^2, \quad \text{..........................(54)*}$$

the equation considered in the paper, already referred to, upon surface-waves.

From (43), (53) we learn that p^2 is positive, or the equilibrium is stable, so long as m is positive. On the other hand, it was proved by Green many years ago that a solid body would be unstable if m were less than $\tfrac{1}{3}n$, $m - \tfrac{1}{3}n$ being in fact the dilatation modulus. The reconciliation of these apparently contradictory results depends upon principles similar to those recently applied

* This is upon the supposition that r and s are real. In the contrary case the equation would have no definite limit.

by Sir W. Thomson*, to show that a solid, every part of the boundary of which is held fixed, is stable, so long as m is greater than $-n$, and this in spite of the fact that, if the boundary were freed, the solid would at once collapse or expand indefinitely. In the present case of an infinite slab, the assumption that the displacements are periodic with respect to x and y is tantamount to the imposition of a constraint at infinity, rendering stability possible under circumstances which would otherwise lead to indefinite collapse or expansion of the medium.

The general expression for the energy of a strained isotropic solid is†

$$2w = (m+n)(\mathrm{e}^2+\mathrm{f}^2+\mathrm{g}^2) + 2(m-n)(\mathrm{fg}+\mathrm{ge}+\mathrm{ef}) + n(a^2+b^2+c^2), \quad \ldots(55)$$

e, f, g being the principal extensions; a, b, c the shears, relatively to the coordinate axes. Since e, f, g may vanish, it is clear that the stability of the medium requires that n be positive; and again, since a, b, c may all vanish, the terms in e, f, g must of themselves be positive in all cases that may arise.

Thus, leaving out a, b, c, we write

$$2w = (3m-n)(\mathrm{e}^2+\mathrm{f}^2+\mathrm{g}^2) + (n-m)\{(\mathrm{e}-\mathrm{f})^2+(\mathrm{f}-\mathrm{g})^2+(\mathrm{g}-\mathrm{e})^2\}, \quad \ldots(56)$$

from which it follows that, if $n > m > \frac{1}{3}n$, the equilibrium is stable. If, however, $m < \frac{1}{3}n$, it will be possible to make w negative by taking $\mathrm{e}=\mathrm{f}=\mathrm{g}$. If $m > n$, the equilibrium is stable, as may be seen by writing $2w$ in the form

$$2w = (m-n)(\mathrm{e}+\mathrm{f}+\mathrm{g})^2 + 2n(\mathrm{e}^2+\mathrm{f}^2+\mathrm{g}^2). \quad \ldots\ldots\ldots\ldots(57)$$

Hence, if there be no limitation on the strains, the necessary and sufficient conditions of stability are that n should be positive and m greater than $\frac{1}{3}n$.

But now suppose that the strains are limited to be in two dimensions, so that (for example) $\mathrm{g}=0$. The supposition $\mathrm{e}=\mathrm{f}=\mathrm{g}$ is then not admissible, and the criterion of stability is altered. We have

$$\begin{aligned} 2w &= (m+n)(\mathrm{e}^2+\mathrm{f}^2) + 2(m-n)\,\mathrm{ef} \\ &= (n-m)(\mathrm{e}-\mathrm{f})^2 + 2m(\mathrm{e}^2+\mathrm{f}^2). \quad \ldots\ldots\ldots\ldots(58) \end{aligned}$$

This shows that there is stability if m be positive and less than n, and instability if m be negative. That the equilibrium is stable if m be greater than n is shown, as in (57), by putting $2w$ into the form

$$2w = (m-n)(\mathrm{e}+\mathrm{f})^2 + 2n(\mathrm{e}^2+\mathrm{f}^2). \quad \ldots\ldots\ldots\ldots(59)$$

Hence, under the limitation $\mathrm{g}=0$, the necessary and sufficient conditions of stability are that n and m be positive.

Comparing the results, we see that, as m diminishes, instability sets in when $m=\frac{1}{3}n$, if the boundary be free; when $m=0$, if (as virtually in our

* *Phil. Mag.* Nov. 1888.

† Thomson and Tait's *Natural Philosophy*, § 695.

present problem) the strains be limited to two dimensions; when $m = -n$, if the boundary be everywhere held fast.

I have endeavoured to investigate the two-dimensional free vibrations of an infinitely long cylindrical shell directly from the fundamental equations, as in the foregoing theory of the plane plate. The preliminary analysis is simple, and there is no difficulty in obtaining the solutions analogous to (42). If a be the radius of the cylinder, and the wave-length measured round the circumference be $2\pi/f$, we have

$$k^2a^2 = \frac{4m}{m+n}(f^2a^2+1), \qquad \text{.........................(60)}$$

and

$$p^2 = \frac{4(f^2a^2+1)}{\rho a^2}\,\frac{mn}{m+n}. \qquad \text{.........................(61)}$$

But this solution is much more readily obtained by the special methods applicable to thin plates, as to the legitimacy of which for this purpose there can be no question. And if, in order to investigate the flexural vibrations of the shell, we retain the lower powers of the thickness, the reduction of the resulting determinant becomes a very complicated affair. I have not succeeded in verifying by a rigorous application of this method the equation analogous to (51), viz.:

$$p^2 = \frac{mn}{m+n}\,\frac{4f^2z_1^2}{3\rho a^2}\,\frac{(f^2a^2-1)^2}{f^2a^2+1}, \qquad \text{.....................(62)}$$

$2z_1$ being the thickness, and as before fa the number of wave-lengths in the circumference. Putting $a = \infty$, we fall back, of course, upon the formulæ for the plane plate.

157.

ON THE LIMIT TO INTERFERENCE WHEN LIGHT IS RADIATED FROM MOVING MOLECULES.

[*Philosophical Magazine*, XXVII. pp. 298—304, 1889.]

IN a recent number of *Wiedemann's Annalen*, Ebert* discusses the application of Doppler's principle to the radiation from the moving molecules of an incandescent gas†, and arrives at the conclusion that the widths of the spectral lines, as calculated upon the basis of the principle, are much greater than is consistent with experiments upon interference with a large relative retardation. This is a matter of no small importance. Unless the discrepancy can be explained, the dynamical theory of gases would, it appears to me, have received a heavy blow from which it could with difficulty recover. If it be true that a gas consists of molecules in irregular motion, and that for the most part each molecule radiates independently, there seems no escape from the conclusion that the character of the aggregate radiation must be governed by Doppler's principle.

If v be the velocity of a molecule, θ the inclination of its motion to the line of sight, the natural wave-frequency N is changed by the motion into n, where

$$n = N\frac{V + v\cos\theta}{V}, \qquad \text{.........................(1)}$$

and V is the velocity of light. If Λ, λ be the original and altered wave-lengths, so that

$$\Lambda = V/N, \qquad \lambda = V/n; \qquad \text{.........................(2)}$$

then

$$\lambda = \Lambda\frac{V}{V + v\cos\theta} = \Lambda\left(1 - \frac{v}{V}\cos\theta\right) \text{ approximately}, \qquad \text{......(3)}$$

when v/V is small.

* *Wied. Ann.* XXXVI. p. 466 (1889).

† Lippich, *Pogg. Ann.* CXXXIX. p. 465 (1870). Rayleigh, *Nature*, VIII. p. 474 (1873) [Vol. I. p. 183].

As a first approximation, Ebert supposes that the velocity v of every molecule is the same. In this case the spectral band, into which what would otherwise be a mathematical line is dilated, has the limiting wave-frequencies

$$N(1+v/V), \qquad N(1-v/V), \quad \text{.....................(4)}$$

and between these limits is of uniform brightness. For the number of molecules whose lines of motion lie between θ and $\theta + d\theta$ is proportional to $\sin\theta\, d\theta$, and this again by (1) is proportional to dn. It is here assumed that the spectrum is formed upon a scale of wave-frequencies; but for the present purpose the range concerned is so small that it becomes a matter of indifference upon what principle the spectrum is disposed.

The typical case of interference arises when two streams of homogeneous light are superposed, which differ in nothing but phase. If δ denote this difference of phase, the vibrations may be represented by $\cos\psi + \cos(\psi+\delta)$, or by

$$2\cos\tfrac{1}{2}\delta \,.\, \cos(\psi + \tfrac{1}{2}\delta); \quad \text{.......................(5)}$$

and the intensity is

$$I = 4\cos^2\tfrac{1}{2}\delta. \quad \text{................................(6)}$$

If the two streams are obtained by reflexion at the opposite faces of a parallel plate, the circumstances are somewhat more complicated. But the simple theory is applicable even here as a first approximation, which becomes more and more rigorous as the difference of optical quality between the plate and the medium in contact with it is supposed to diminish. If μ be the index of the plate, Δ its thickness,

$$\delta = \pi + \frac{4\pi\mu\Delta}{\lambda} = \pi + \frac{4\pi n\mu\Delta}{V}. \quad \text{...................(7)}$$

If the plate be of air, $\mu = 1$. In any case the variation of μ is small compared to that of n; so that if Δ denote the equivalent thickness of air, we may take

$$I = 4\sin^2\{2\pi n\Delta/V\}, \quad \text{........................(8)}$$

a function of n—the frequency, as well as of Δ and V.

If now the light be heterogeneous, we have nothing further to do than to integrate (8) with respect to n, after introduction of a factor i such that $i\,dn$ represents the illumination corresponding to dn*. In the present case, where the intensity is supposed to be *uniform* within limits n_1 and n_2, and to vanish outside them, we have

$$\int I\,dn = 4i\int_{n_1}^{n_2}\sin^2(2\pi n\Delta/V)\,dn$$

$$= 2\int i\,dn \,.\left[1 - \frac{\sin\{2\pi\Delta(n_2-n_1)/V\}}{2\pi\Delta(n_2-n_1)/V}\,.\,\cos\{2\pi\Delta(n_2+n_1)/V\}\right]. \quad \text{......(9)}$$

* It is here assumed that the range included is too small to give rise to sensible chromatic variation.

From this we fall back on (8), if we suppose that $(n_2 - n_1)$ is infinitely small, so that

$$\int I\,dn = 2\int i\,dn\,.\,[1 - \cos(4\pi n\Delta/V)].$$

The difference between (8) and (9) thus depends upon the factor

$$\frac{\sin\{2\pi\Delta(n_2 - n_1)/V\}}{2\pi\Delta(n_2 - n_1)/V}, \quad \text{......................(10)}$$

which multiplies the second term of (9). If we introduce the special values of n_1, n_2 from (4), and denote the angle in (10) by α,

$$\alpha = 2\pi\Delta(n_2 - n_1)/V = \frac{4\pi\Delta}{\Lambda}\cdot\frac{v}{V}. \quad \text{.................(11)}$$

So long as α is small, the mode of interference is nearly the same as if $v = 0$. This will be the case when Δ is sufficiently small, so that at first the bands are absolutely black. As Δ increases, the distinctness of the bands will depend mainly upon the relative brightnesses of the least and most illuminated parts. If we call this ratio h, and denote by a the *numerical* value of (10), we have

$$h = (1 - a)/(1 + a), \quad \text{.........................(12)}$$

or

$$a = (1 - h)/(1 + h). \quad \text{.........................(13)}$$

Now from (10) it appears that when α is equal to π, or to any multiple of π, $a = 0$, and the field is absolutely uniform. Between values of α equal to π and 2π, 2π and 3π, and so on, there are revivals of distinctness, the maxima of which occur at values not far removed from $\frac{3}{2}\pi$, $\frac{5}{2}\pi$, &c. Thus, between π and 2π there is to be found a value of a at least equal to $2/3\pi$, corresponding to $h = \frac{2}{3}$ nearly. At this stage the bands should certainly be visible.

In order to estimate at what point the interference-bands would first disappear as Δ increases, we must make some supposition as to the largest value of h indistinguishable in experiment from unity. Under favourable circumstances in other respects we may perhaps assume for this purpose $h = \cdot 95$, so that $a = \cdot 025$. Since a is small, α is nearly equal to π. We may take approximately $\sin\alpha = \cdot 025\pi$, or $\alpha = \cdot 975\pi$. In fact, so long as we take h nearly equal to unity, the precise value makes very little difference to the corresponding value of α, and for the purposes of such a discussion as the present we may suppose with sufficient accuracy $\alpha = \pi$. In this case, by (11),

$$\frac{2\Delta}{\Lambda} = \frac{V}{2v}, \quad \text{...............................(14)}$$

which gives the retardation (2Δ) measured in wave-lengths in the neighbourhood of which the bands would first disappear. This estimate differs widely from that put forward by Ebert. The latter is equivalent to

$$\frac{2\Delta}{\Lambda} = \cdot 15\,\frac{V}{v}. \quad \text{...............................(15)}$$

According to my calculation the value of α corresponding to (15) would be 54°, a would be ·86, and h would be ·075; so that the bands should be hardly distinguishable from those which occur when $\Delta = 0$.

For the grounds of his estimate Ebert refers to an earlier paper*, in which, however, the calculation seems to relate to a problem materially different from the present, that, namely, in which the refrangibility of the light is limited to two distinct values (as approximately in the case of the soda lines), instead of being distributed equally over the same range. In this case (9) is replaced by

$$4\left\{1 - \cos\frac{2\pi\Delta(n_2 - n_1)}{V} \cdot \cos\frac{2\pi\Delta(n_1 + n_2)}{V}\right\}; \quad \text{..........(16)}$$

so that, if α have the same form as in (11), and a' denote the numerical value of $\cos\alpha$,

$$h = (1 - a')/(1 + a'), \quad \text{..........................(17)}$$

as before.

According to (16) the field is first uniform when $\alpha = \frac{1}{2}\pi$, instead of π, as from (9). When $\alpha = \pi$, the bands are again black, and as Δ further increases there is a strictly periodic alternation between blackness and absolute disappearance of the bands.

The substitution for a spectral band of uniform brightness of one in which the illumination is all condensed at the edges explains a large part of the discrepancy between (14) and (15); but even in the latter problem (15) seems to be a very small estimate of Δ. According to (15), $\alpha = 54°$, $\cos\alpha = ·59$; so that from (17) $h = ·26$. Bands of which the darkest parts are of only one quarter of the illumination of the brightest parts could hardly be invisible.

The more nearly correct formula (14) is itself, however, based upon the assumption that all the vibrating molecules move with the same velocity. This is the origin of the law expressed in (9), according to which the bands should reappear at a retardation greater than that of first disappearance. But the real law of the distribution of velocity is that discovered by Maxwell, if there is any truth in the molecular theory†. That such is the case is recognized by Ebert; and he argues that the broadening of the spectral band due to velocities higher than the mean, will entail a further diminution in the maximum retardation consistent with visible interference‡. I proceed

* *Wied. Ann.* xxxiv. p. 39 (1888).

† It is here assumed that we are dealing with a gas in approximate temperature equilibrium. The case of luminosity under electric discharge may require further consideration.

‡ In the earlier memoir (*Wied. Ann.* xxxiv.) Ebert appears to regard the capability of interference (*Interferenz-fähigkeit*) of a spectral line as dependent upon other causes than the width of the line and the distribution of brightness over it. In this view I cannot agree. "The narrowness of the bright line of light seen in the spectroscope, and the possibility of a large

to the actual calculation of the maximum retardation on the basis of Maxwell's law.

If ξ, η, ζ be the rectangular components of v, the number of molecules whose component velocities lie at any time between ξ and $\xi + d\xi$, η and $\eta + d\eta$, ζ and $\zeta + d\zeta$, will be proportional to

$$e^{-\beta(\xi^2+\eta^2+\zeta^2)}\, d\xi\, d\eta\, d\zeta.$$

If ξ be the direction of the line of sight, the component velocities η, ζ are without influence in the present problem. All that we require to know is that the number of molecules for which the component ξ lies between ξ and $\xi + d\xi$ is proportional to

$$e^{-\beta\xi^2} d\xi. \qquad \text{.................................(18)}$$

The relation of β to the mean (resultant) velocity v* is

$$v = \frac{2}{\sqrt{(\pi\beta)}}. \qquad \text{..............................(19)}$$

If the natural frequency of the waves emitted by the molecules be N, the actual frequency of the waves from a molecule travelling with component velocity ξ is by Doppler's principle

$$n = N(1 + \xi/V). \qquad \text{...........................(20)}$$

Hence by (8) the expression to be investigated, and corresponding to (9), is

$$4\int_{-\infty}^{+\infty} \sin^2 \frac{2\pi\Delta}{\Lambda}\left(1 + \frac{\xi}{V}\right) . \, e^{-\beta\xi^2}\, d\xi. \qquad \text{..................(21)}$$

In (21) we have

$$2\sin^2 \frac{2\pi\Delta}{\Lambda}\left(1 + \frac{\xi}{V}\right) = 1 - \cos\frac{4\pi\Delta}{\Lambda}\left(1 + \frac{\xi}{V}\right)$$

$$= 1 - \cos\frac{4\pi\Delta}{\Lambda}\cos\frac{4\pi\Delta\xi}{\Lambda V} + \sin\frac{4\pi\Delta}{\Lambda}\sin\frac{4\pi\Delta\xi}{\Lambda V}.$$

The last of the three terms, being of uneven order in ξ, vanishes when integrated. The first and second are included under the well-known formula

$$\int_0^\infty e^{-a^2x^2}\cos 2rx\, dx = \frac{\sqrt{\pi}}{2a} e^{-r^2/a^2};$$

and we obtain

$$2\sqrt{\left(\frac{\pi}{\beta}\right)} . \left[1 - \cos\frac{4\pi\Delta}{\Lambda} . \exp\left(-\frac{4\pi^2\Delta^2}{\beta\Lambda^2 V^2}\right)\right]. \qquad \text{...........(22)}$$

number of (interference) bands, depend upon precisely the same conditions; the one is in truth as much an interference phenomenon as the other" (*Enc. Brit.* "Wave Theory," Vol. XXIV. p. 425 [Vol. III. p. 60]). It is obvious that nothing could give rise in the spectroscope to a mathematical line of light, but an infinite train of waves of harmonic type and of absolute regularity.

* This must be distinguished from the velocity of mean square, with which the pressure is most directly connected.

In conformity with previous notation we may write

$$a'' = \exp\left(-\frac{4\pi^2\Delta^2}{\beta\Lambda^2 V^2}\right);$$

or, if we introduce the value of β from (19),

$$a'' = \exp\left\{-\pi\left(\frac{\pi\Delta}{\Lambda}\frac{v}{V}\right)^2\right\}. \qquad \text{.........................(23)}$$

The ratio of the least and greatest brightnesses is then, as before,

$$h = (1 - a'')/(1 + a''). \qquad \text{...........................(24)}$$

If we now assume as determining the limit of visibility $h = \cdot 95$, we find $a'' = \cdot 025$, and from (23)

$$\frac{2\Delta}{\Lambda} = \cdot 690\,\frac{V}{v}. \qquad \text{...............................(25)}$$

It appears therefore that the maximum admissible retardation is sensibly *greater* than that calculated (14) upon the supposition that all the molecules move with the mean velocity v, and as much as $4\frac{1}{2}$ times greater than that (15) taken by Ebert as the basis of his comparison with observation.

Under these circumstances it would seem that there is no discrepancy remaining to be explained. It is true that the width of spectral lines is not wholly due to movement of the molecules; but it is possible that this is the principal cause of dilatation when the flames are coloured by the spray of very dilute solutions, as in Ebert's use of the method of Gouy*. Again, it is true that interference-bands are often observed under conditions less favourable than is supposed in the above estimate of h. In Michelson's method, however, the bands may be very black at small retardations; and it seems very probable that at higher retardations bands involving even less than 5 per cent. of the brightness might be visible†. The question is one of very great interest, and I hope that Herr Ebert will pursue his investigations until it is thoroughly cleared up.

* *Ann. de Chim.* XVIII. p. 1 (1879).

† [See Art. 161 below.]

158.

IRIDESCENT CRYSTALS.

[*Proc. Roy. Inst.* XII. pp. 447—449, 1889; *Nature*, XL. pp. 227, 228, 1889.]

THE principal subject of the lecture is the peculiar coloured reflection observed in certain specimens of chlorate of potash. Reflection implies a high degree of discontinuity. In some cases, as in decomposed glass, and probably in opals, the discontinuity is due to the interposition of layers of air; but, as was proved by Stokes, in the case of chlorate crystals the discontinuity is that known as twinning. The seat of the colour is a very thin layer situated in the interior of the crystal and parallel to its faces.

The following laws were discovered by Stokes:—

(1) If one of the crystalline plates be turned round in its own plane, without alteration of the angle of incidence, the peculiar reflection vanishes twice in a revolution, viz. when the plane of incidence coincides with the plane of symmetry of the crystal. [Shown.]

(2) As the angle of incidence is increased the reflected light becomes brighter and rises in refrangibility. [Shown.]

(3) The colours are not due to absorption, the transmitted light being strictly complementary to the reflected.

(4) The coloured light is not polarised. It is produced indifferently, whether the incident light be common light or light polarised in any plane, and is seen whether the reflected light be viewed directly or through a Nicol's prism turned in any way. [Shown.]

(5) The spectrum of the reflected light is frequently found to consist almost entirely of a comparatively narrow band. When the angle of incidence is increased, the band moves in the direction of increasing refrangibility, and at the same time increases rapidly in width. In many cases the reflection appears to be almost total.

In order to project these phenomena a crystal is prepared by cementing a smooth face to a strip of glass, whose sides are not quite parallel. The white reflection from the anterior face of the glass can then be separated from the real subject of the experiment.

A very remarkable feature in the reflected light remains to be noticed. If the angle of incidence be small, and if the incident light be polarised in or perpendicularly to the plane of incidence, the reflected light is polarised in the *opposite* manner. [Shown.]

Similar phenomena, except that the reflection is white, are exhibited by crystals prepared in a manner described by Madan. If the crystal be heated beyond a certain point the peculiar reflection disappears, but returns upon cooling. [Shown.]

In all these cases there can be little doubt that the reflection takes place at twin surfaces, the theory of such reflection* reproducing with remarkable exactness most of the features above described. In order to explain the vigour and purity of the colour reflected in certain crystals, it is necessary to suppose that there are a considerable number of twin surfaces disposed at approximate equal intervals. At each angle of incidence there would be a particular wave-length for which the phases of the several reflections are in agreement. The selection of light of a particular wave-length would thus take place upon the same principle as in diffraction spectra, and might reach a high degree of perfection.

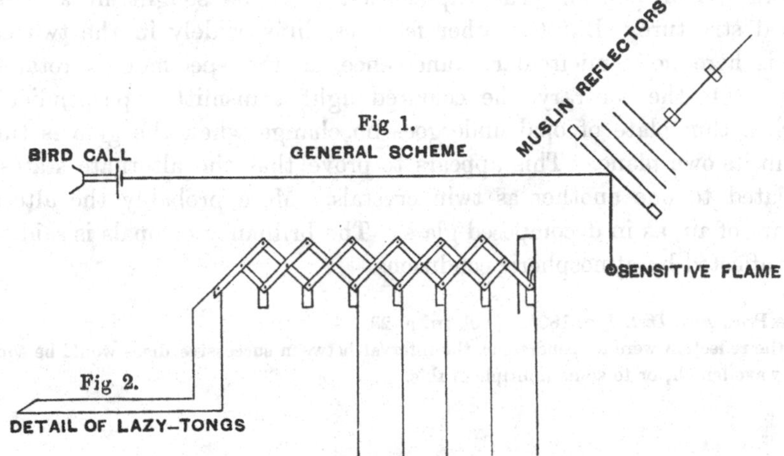

Fig 1. GENERAL SCHEME

Fig 2.

In illustration of this explanation an acoustical analogue is exhibited. The successive twin planes are imitated by parallel and equidistant discs of muslin (Figs. 1 and 2) stretched upon brass rings and mounted (with the aid of three lazy-tongs arrangements), so that there is but one degree of freedom

* *Phil. Mag.* Sept. 1888. [Vol. III. Art. 149.]

to move, and that of such a character as to vary the interval between the discs without disturbing their equidistance and parallelism.

The source of sound is a bird-call, giving a pure tone of high pitch (inaudible), and the percipient is a high pressure flame issuing from a burner so oriented that the direct waves are without influence upon the flame*. But the waves reflected from the muslin arrive in the effective direction, and if of sufficient intensity induce flaring. The experiment consists in showing that the action depends upon the distance between the discs. If the distance be such that the waves reflected from the several discs co-operate †, the flame flares, but for intermediate adjustments recovers its equilibrium. For full success it is necessary that the reflective power of a single disc be neither too great nor too small. A somewhat open fabric appears suitable.

It was shown by Brewster that certain natural specimens of Iceland spar are traversed by thin twin strata. A convergent beam, reflected at a nearly grazing incidence from the twin planes, depicts upon the screen an arc of light, which is interrupted by a dark spot corresponding to the plane of symmetry. [Shown.] A similar experiment may be made with small rhombs in which twin layers have been developed by mechanical force after the manner of Reusch.

The light reflected from fiery opals has been shown by Crookes to possess in many cases a high degree of purity, rivalling in this respect the reflection from chlorate of potash. The explanation is to be sought in a periodic stratified structure. But the other features differ widely in the two cases. There is here no semicircular evanescence, as the specimen is rotated in azimuth. On the contrary, the coloured light transmitted perpendicularly through a thin plate of opal undergoes no change when the gem is turned round in its own plane. This appears to prove that the alternate states are not related to one another as twin crystals. More probably the alternate strata are of air, as in decomposed glass. The brilliancy of opals is said to be readily affected by atmospheric conditions.

* See *Proc. Roy. Inst.* Jan. 1888. [Vol. III. p. 28.]

† If the reflection were perpendicular, the interval between successive discs would be equal to the half wave-length, or to some multiple of this.

159.

THE SAILING FLIGHT OF THE ALBATROSS.

[*Nature*, XL. p. 34, 1889.]

I HAVE been much interested by the letter of Mr A. C. Baines* upon this subject. In the year 1883 ("The Soaring of Birds," *Nature*, Vol. XXVII. p. 534 [Vol. II. p. 194]) I suggested that the explanation of these puzzling performances might be found in the increase of wind with height. To take advantage of this, the bird must rise against the wind and fall with it; but at the time referred to, I had before me only the observations of Mr Peal, in Assam, on the flight of pelicans, in which this feature is not alluded to. In Mr Baines's observations the omission is supplied, and there seems little reason to doubt that the true explanation of the flight of the albatross has been arrived at. In the case of the pelican soaring to a great elevation, it is less easy to understand how the differences of horizontal velocity can be sufficient.

Reference may be made to a paper by Mr H. Airy (*Nature*, Vol. XXVII. p. 590), in which the matter is further discussed. Similar views have also been put forward more recently by an American Author, whose name I have unfortunately forgotten. [1901. See further the Wilde Lecture on the Mechanical Principles of Flight (*Manchester Proceedings*, 1900).]

* *Nature*, May 2, p. 9. ["The birds go through a series of movements which are related to the direction of the wind. Starting from near the surface, they rise in a slanting direction against the wind, to a height which varies with the direction and strength of the wind. The average seemed to me about 20 feet. Then comes immediately a turn half round in a rather large circle, followed at once by a rapid descent down the wind. They then take a longer or shorter flight in various directions, almost touching the water. After that another ascent in the same manner, and so on, repeating this series of movements *ad libitum*."]

160.

ON THE CHARACTER OF THE COMPLETE RADIATION AT A GIVEN TEMPERATURE.

[*Philosophical Magazine*, XXVII. pp. 460—469, 1889.]

By complete radiation is here meant the radiation which would ultimately establish itself in an enclosure, whose walls are impervious, and are maintained at a uniform temperature*. It was proved by Stewart and Kirchhoff that this radiation is definite, not only in the aggregate, but also in its various parts; so that the energy radiated with wave-frequencies between n and $n + dn$ may be expressed by

$$F(n)\, dn, \qquad (1)$$

where, for a given temperature, $F(n)$ is a definite function of n. The reservation implied in the word *ultimately* is necessary in order to exclude radiation due to phosphorescence or to chemical action within the enclosure. The radiation commonly characterised, so far at any rate as its visible elements are concerned, by the term *white*, is supposed to be approximately similar to the complete radiation at a certain very high temperature.

As remarked by Kirchhoff, the function F, being independent of the properties of any particular kind of matter, is likely to be of a simple form; and speculations have naturally not been wanting. Within the last two years the subject has been considered by W. Michelson† and by H. F. Weber‡. The former, on the basis of an *à priori* argument of a not very convincing character, arrives at the conclusion that at temperature θ the radiation between the limits of wave-length λ and $\lambda + d\lambda$ may be expressed

$$I_\lambda d\lambda = B\theta^{-\frac{3}{2}} f(\theta)\, e^{-\frac{c}{\theta\lambda}}\, \lambda^{-2p-4}\, d\lambda. \qquad (2)$$

* [1901. The radiation, here characterised as *complete*, is sometimes described as *black*. To speak of a red-hot poker, or of the radiation from it, as black, does not seem happy.]

† *Journal de Physique*, t. VI. Oct. 1887; *Phil. Mag.* XXV. p. 425.

‡ *Berlin. Sitz.-Ber.* 1888.

According to Stephan the total radiation is proportional to θ^4. In conformity with this Michelson supposes that

$$p=1, \quad f(\theta)=K\theta^3;$$

so that (2) assumes the more special form

$$I_\lambda = B_1 \theta^{\frac{3}{2}} e^{-\frac{c}{\theta\lambda^2}} \lambda^{-6}. \qquad \text{(3)}$$

If, as appears to be preferable, we take n as independent variable, $F(n)\,dn$ is of the form

$$Ae^{-a^2n^2}n^4dn, \qquad \text{(4)}$$

A, a being functions of θ, but independent of n.

Weber's formula, so far as it here concerns us, is of a still simpler character. Expressed in terms of n, it differs from (4) merely by the omission of the factor n^4, thus corresponding to $p=-1$ in (2); so that

$$F(n)\,dn = Ae^{-a^2n^2}dn. \qquad \text{(5)}$$

The agreement between (5) and the measurements by Langley of the radiation at 178° C. is considered by Weber to be sufficiently good.

In contemplating such a formula as (5), it is impossible to refrain from asking in what sense we must interpret it in accordance with the principles of the Undulatory Theory, and whether we can form any distinct conception of the character of the vibration indicated by it. My object in the present paper is to offer some tentative suggestions towards the elucidation of these questions.

The first remark that I would make is that the formula must not be taken too literally. If there is one thing more certain than another, it is that a definite wave-frequency implies an infinite and unbroken succession of waves*. A good illustration is afforded by intermittent vibrations, as when a sound, itself constituting a pure tone, is heard through a channel which is periodically opened and closed. Such an intermittent vibration may be represented by†

$$2(1+\cos 2\pi mt)\cos 2\pi nt, \qquad \text{(6)}$$

where n is the frequency of the original vibration, and m the frequency of intermittence. By ordinary trigonometrical transformation (6) may be written

$$2\cos 2\pi nt + \cos 2\pi(n+m)t + \cos 2\pi(n-m)t; \qquad \text{(7)}$$

which shows that in this case the intermittent vibration is equivalent to *three* simple vibrations of frequencies n, $n+m$, $n-m$.

* "The pitch of a sonorous body vibrating freely cannot be defined with any greater closeness than corresponds with the total number of vibrations which it is capable of executing." (*Proc. Mus. Assoc.* Dec. 1878, p. 25.)

† "Acoustical Observations, III." *Phil. Mag.* April 1880. [Vol. I. p. 468.]

In order to distinguish wave-frequencies, whose difference is small, a correspondingly long series of waves is necessary; and of no finite train of irregular vibrations can it be said that waves of a certain frequency are present, and waves of a frequency infinitely little different therefrom absent. Neither can the proportions in which the two are present be assigned. In professing to assign these proportions, (5) and similar formulæ make assertions not directly supported by experiment. In a sense all the formulæ of mathematical physics are in this predicament; but here the assertion is of such a nature that it could not be tested otherwise than by experiments prolonged over all time.

In practice it is not time that brings the limitation, but the resolving power of our instruments. In gratings the resolving power is measured by the product of the total number of lines and the order of the spectrum under examination*. It will be allowing a good deal for the progress of experiment if we suppose that in measurements of energy it may be possible to discriminate wave-lengths (or frequencies) which differ by a millionth part. But a million wave-lengths of yellow light would occupy only 60 cm., and the waves would pass in 2×10^{-9} seconds! Waves whose frequencies differ by less than this are inextricably blended, even though we are at liberty to prolong our observations to all eternity.

At any point in the spectrum of a hot body there are, therefore, mingled waves of various frequencies lying within narrow limits. The resultant for any *very short interval of time* may be identified with a simple train, whose amplitude and phase, depending as they do upon the relative phases of the components, must be regarded as matters of chance. The probability of various amplitudes depends upon the principles explained in a former communication, "On the Resultant of a large number of Vibrations of the same Pitch and of Arbitrary Phase†." After an interval of time comparable with 10^{-9} second the amplitude is again practically a matter of chance; so that during the smallest interval of time of which our senses or our instruments could take cognizance, there are an immense number of independent combinations. But, under these circumstances, as was shown in the place referred to, we have to do merely with the sum of the individual intensities.

In his excellent memoir, *Sur le mouvement lumineux*‡, M. Gouy suggests that the nature of white light may be best understood by assimilating it to a

* *Phil. Mag.* Vol. XLVII. p. 200 (1874). [Vol. I. p. 216.]

† *Phil. Mag.* Aug. 1880. [Vol. I. p. 491.]

‡ *Journ. de Physique*, 1886, p. 354. I observe that M. Gouy had anticipated me (*Enc. Brit.* XXIV. p. 425 [Vol. III. p. 60]) in the remark that the production of a large number of interference-bands from originally white light is a proof of the resolving power of the spectroscope, and not of the regularity of the white light. It would be instructive if some one of the contrary opinion would explain what he means by regular white light The phrase certainly appears to me to be without meaning—what Clifford would have called *nonsense*.

sequence of entirely irregular impulses. It was by means of this idea that Young* explained the action of gratings; and although J. Herschel† took exception, there is no doubt that the method is perfectly sound. The question that I wish to raise is whether it is possible to define the kind of impulse of which an irregular sequence would represent the complete radiation of any temperature.

The first thing to be observed is that it will not do to suppose the impulses themselves to be arbitrary. In proof of this it may be sufficient to point out that in that case there would be no room for distinguishing the radiations of various temperatures. If the velocity [of disturbance] at every point [along the line of propagation] were arbitrary, that is independent of the velocity at neighbouring points however close, the radiation could have no special relation to any finite wave-length or frequency. In order to avoid this discontinuity we must suppose that the velocities at neighbouring points are determined by the same causes, so that it is only when the interval exceeds a certain amount that the velocities become independent of one another. This independence enters gradually. When the interval is very small, the velocities are the same. As the interval increases, the arbitrary element begins to assert itself. At a moderate distance the velocity at the second point is determined in part by agreement with the first, and in part independently. With augmenting distance the arbitrary part gains in importance until at last the common element is sensibly excluded‡.

Now this is precisely the condition of things that would result from the arbitrary distribution of a large number of impulses, in each of which the medium is disturbed according to a defined law. A simple case would be to suppose that each impulse is confined to a narrow region of given width, and within that region communicates a constant velocity§. An arbitrary distribution of such impulses over the whole length would produce a disturbance having, in many respects, the character we wish. But it is easy to see that this particular kind of impulse will not answer all requirements. For in the result of each impulse, and therefore in the aggregate of all the

* *Phil. Trans.* 1801.

† *Enc. Metrop.*, "Light," § 703 (1830).

‡ The following may serve as an illustration. Out of a very large number of men (say an army) let a regiment of 1000 be chosen by lot, and let the deviation of the mean height of the regiment from that of the army be exhibited as the ordinate of a curve. If a second set of 1000 be chosen by lot, the new ordinate will bear no relation to the old. But if at each step but one man of the regiment be eliminated by lot, and one successor be chosen in the same way, the new ordinate will be almost the same as the old one, and not until after a large number of steps (of the order of 1000) will the new ordinate become sensibly independent. If the abscissa be taken proportional to the number of steps (each finally treated as infinitesimal), the resulting curve will have the required property, and would exhibit a possible form for complete radiation. It seems not unlikely that the law is here the same as that obtained below on the basis of (8).

§ The reader may fix his ideas upon a stretched string vibrating transversely.

impulses, those wave-lengths would be excluded, which are submultiples of the length of the impulse. The objection could be met by combining impulses of different lengths; but then the whole question would be again open, turning upon the proportions in which the various impulses were introduced. What I propose here to inquire is whether any definite type can be suggested such that an arbitrary aggregation of them will represent complete radiation. It will be evident that in the definition of the type a constant factor may be left arbitrary. In other words, the impulses need only to be *similar*, and not necessarily to be *equal*.

Probably the simplest type of impulse, $\phi(x)$, that could at all meet the requirements of the case is that with which we are familiar in the theory of errors, viz.

$$\phi(x) = e^{-c^2x^2}. \qquad \text{.................................(8)}$$

It is everywhere finite, vanishes at an infinite distance, and is free from discontinuities. A single impulse of this type may be supposed to be the resultant of a very large number of localized infinitesimal simultaneous impulses, all *aimed* at a single point ($x = 0$), but liable to deviate from it owing to accidental causes. I do not at present attempt any physical justification of this point of view, but merely note the mathematical fact. The next step is to resolve the disturbance (8) into its elements in accordance with Fourier's theorem. We have*

$$\phi(x) = \frac{1}{\pi}\int_0^\infty \int_{-\infty}^{+\infty} \cos k(v-x)\,\phi(v)\,dk\,dv$$

$$= \frac{1}{\pi}\int_0^\infty \int_{-\infty}^{+\infty} \cos kv \cos kx\, e^{-c^2v^2}\,dk\,dv. \qquad \text{..............(9)}$$

Now

$$\int_{-\infty}^{+\infty} e^{-c^2v^2} \cos kv\, dv = \frac{\sqrt{\pi}}{c}\, e^{-k^2/4c^2}; \qquad \text{.....................(10)}$$

so that

$$e^{-c^2x^2} = \frac{1}{c\sqrt{\pi}}\int_0^\infty e^{-k^2/4c^2} \cos kx\, dk. \qquad \text{.................(11)}$$

This equation exhibits the resolution of (8) into its harmonic components; but it is not at once obvious how much energy we are to ascribe to each value of k, or rather to each small range of values of k. As in the theory of transverse vibrations of strings, we know that the energy corresponding to the product of any two distinct harmonic elements must vanish; but the application of this, when the difference between two values of k is infinitesimal, requires further examination. The following is an adaptation of Stokes's investigation† of a problem in diffraction.

* [1901. A slight change of notation is introduced.]

† *Edinb. Trans.* XX. p. 317 (1853); see also *Enc. Brit.* t. XXIV. p. 431. [Vol. III. p. 86.]

By Fourier's theorem (9) we have

$$\pi . \phi(x) = \int_0^\infty f_1(k) \cos kx \, dk + \int_0^\infty f_2(k) \sin kx \, dk, \quad \text{.........(16)}$$

where

$$f_1(k) = \int_{-\infty}^{+\infty} \cos kv \, \phi(v) \, dv, \qquad f_2(k) = \int_{-\infty}^{+\infty} \sin kv \, \phi(v) \, dv. \quad \text{...(17, 18)}$$

In order to shorten the expressions, we will suppose that, as in (11),

$$f_2(k) = 0.$$

We have

$$\pi^2 . \{\phi(x)\}^2 = \int_0^\infty \int_0^\infty f_1(k) f_1(k') \cos kx \cos k'x \, dk \, dk'.$$

This equation is now to be integrated with respect to x from $-\infty$ to $+\infty$; but, in order to avoid ambiguity, we will introduce the factor $e^{\mp \alpha x}$, where α is a small positive quantity. The positive sign in the alternative is to be taken when x is negative, and the negative sign when x is positive. The order of integration is then to be changed, so as to take first the integration with respect to x; and finally α is to be supposed to vanish. Thus

$$2\pi^2 . \{\phi(x)\}^2$$

$$= \text{Lim.} \int_{-\infty}^{+\infty} \int_0^\infty \int_0^\infty e^{\mp \alpha x} f_1(k) f_1(k') \{\cos x(k'-k) + \cos x(k'+k)\} \, dk \, dk' \, dx.$$

Now

$$\int_{-\infty}^{+\infty} e^{\pm \alpha x} \cos hx \, dx = \frac{2\alpha}{\alpha^2 + h^2};$$

so that

$$2\pi^2 \{\phi(x)\}^2 = \text{Lim.} \int_0^\infty \int_0^\infty \left\{ \frac{2\alpha}{\alpha^2 + (k'-k)^2} + \frac{2\alpha}{\alpha^2 + (k'+k)^2} \right\} f_1(k) f_1(k') \, dk \, dk'. \quad \text{.....(19)}$$

Of the right-hand member of (19) the second integral vanishes in the limit, since k and k' are both positive quantities. But in the first integral the denominator vanishes whenever k' is equal to k. If we put

$$k' = k = \alpha z, \qquad dk' = \alpha dz,$$

then, in the limit

$$\int_0^\infty \frac{2\alpha f_1(k') \, dk'}{\alpha^2 + (k'-k)^2} = \int_{-\infty}^{+\infty} \frac{2 f_1(k) \, dz}{1 + z^2} = 2\pi f_1(k).$$

Thus

$$\int_{-\infty}^{+\infty} \{\phi(x)\}^2 \, dx = \frac{1}{\pi} \int_0^\infty \{f_1(k)\}^2 \, dk. \quad \text{.................(20)}$$

If $f_2(k)$ be finite, we have, in lieu of (20),

$$\int_{-\infty}^{+\infty} \{\phi(x)\}^2 \, dx = \frac{1}{\pi} \int_0^\infty [\{f_1(k)\}^2 + \{f_2(k)\}^2] \, dk. \quad \text{...........(21)}$$

In M. Gouy's treatment of this question, the function $\phi(x)$ is supposed to be ultimately periodic. In this case $f(k)$ vanishes whenever k differs from one or other of the terms of an arithmetical progression; and the whole kinetic energy of the motion is equal to the sum of those of its normal components, as in all cases of vibration. The comparison of this method with the one adopted above, in which all values of k occur, throws light upon the nature of the harmonic expansion.

It is scarcely necessary to point out that vibrations started impulsively from rest divide themselves into two groups, constituting progressive waves in the two directions, and that the whole energy of each of these waves is the half of that communicated initially to the system in the kinetic form*.

The application of (21) to (11), where

$$f_1(k) = \frac{\sqrt{\pi}}{c} e^{-k^2/4c^2},$$

gives

$$\int_{-\infty}^{+\infty} e^{-2c^2x^2} dx = \frac{1}{c^2}\int_0^\infty e^{-k^2/2c^2} dk, \quad \text{.....................(22)}$$

as may be easily shown independently. The intensity, corresponding to the limits k and $k + dk$, is therefore

$$c^{-2} e^{-k^2/2c^2} dk;$$

and this, since k and n are proportional, is of the form (5).

If an infinite number of impulses, similar (but not necessarily equal) to (8), and of arbitrary sign, be distributed at random over the whole range from $-\infty$ to $+\infty$, the intensity of the resultant for an absolutely definite value of n would be indeterminate. Only the *probabilities* of various resultants could be assigned. And if the value of n were changed, by however little, the resultant would again be indeterminate. Within the smallest assignable range of n there is room for an infinite number of independent combinations. We are thus concerned only with an average, and the intensity of each component may be taken to be proportional to the total number of impulses (if equal) without regard to their phase-relations. In the aggregate vibration, the law according to which the energy is distributed is still for all practical purposes that expressed by (5).

If we decompose each impulse (8) in the manner explained, we may regard the whole disturbance as arising from an infinite number of simultaneous elementary impulses. These elementary impulses are distributed not entirely at random; for they may be arranged in groups such that the members of each group are of the same sign, and are, as it were, aimed at the same point under a law of error; while the different groups are without

* *Theory of Sound*, Vol. II. § 245.

relation, except that the law of error is the same for all. It is obviously not essential that the different groups should deliver their blows simultaneously. Further, it would have come to the same thing had we supposed all the impulses to be delivered at the same point in *space*, but to be distributed in *time* according to a similar law. In comparing the radiations at various temperatures, we should have to suppose that, as the temperature rises, not only does the total number of elementary impulses (of given magnitude) increase, but also the *accuracy of aim* of each group.

We have thus determined a kind of impulse such that a [random] aggregation of them will represent complete radiation according to Weber's law (5). One feature of this law is that $F(n)$ approaches a finite limit as n decreases. In this respect W. Michelson's special law (4) differs widely; for, according to it, $F(n)$ vanishes with n. This evanescence of $F(n)$ implies that the integrated value of each of our component impulses is zero. If we wish to inquire further into the law of the impulse, we have to determine $\phi(x)$ so that

$$f_1(k) = Ck^2 e^{-k^2/4c^2}. \quad \text{..........................(23)}$$

By successive differentiations of (10) with respect to k, it may be shown that

$$\frac{\sqrt{\pi}}{c}\frac{k^2}{2c^2}e^{-k^2/4c^2} = \int_{-\infty}^{+\infty} e^{-c^2x^2}(1-2c^2x^2)\cos kx\,dx. \quad \text{...........(24)}$$

Thus, if we take

$$\phi(x) = e^{-c^2x^2}(1-2c^2x^2), \quad \text{.......................(25)}$$

$f_1(k)$ will be of the required form. The curve representative of (25), viz.

$$y = e^{-x^2}(1-2x^2), \quad \text{..............................(26)}$$

is symmetrical with respect to $x=0$, vanishes when $x = \pm\infty$ and also when $x = \pm 2^{-\frac{1}{2}}$. The positive area between the last-named limits is numerically equal to the negative area lying outside them.

Other proposed forms for $f(k)$, such as those included in (2), might be treated in a similar way; but the above examples may suffice. The simplicity of (8) compared, *e.g.*, with (25), may be regarded as an argument in its favour. But we do not know enough of the mechanism of radiation to draw any confident conclusion. What we most require at present is more complete data from experiment, such as have been promised by Prof. Langley. As regards the radiation of very low frequency, a question may arise as to whether it is included in our present measurements. Some authorities have favoured the view that, when the frequency is sufficiently diminished, all kinds of matter become transparent; but the electric theory seems to point in the opposite direction. In comparing any theoretical formula with experiment, we must not forget that what we learn directly from the latter is the *difference* of radiations at *two* temperatures.

One more remark in conclusion. If the complete radiation for a given temperature be represented by (5), it follows that temperature may be defined by the value of a. The contrary would imply that the law of distribution is the same at all temperatures, and would be inconsistent with ordinary observation respecting "red" and "white heats." Now the dimensions of a are those of a time; so that temperature may be defined by a *time*, or (through the velocity of propagation) by a *line*. Thus in Prof. Langley's curves, which represent the distribution of energy in a diffraction spectrum, the wave-length corresponding to the maximum ordinate may be regarded as a linear specification of the temperature to which the curve relates*.

* [1901. On some of the questions here discussed reference may be made to Schuster's paper on "Interference Pheuomena" (*Phil. Mag.* XXXVII. p. 509, 1894). The progress of knowledge with respect to the law of complete radiation is not favourable to the idea that such radiation can be represented as a random sequence of impulses of simple type.]

161.

ON THE VISIBILITY OF FAINT INTERFERENCE-BANDS.

[*Philosophical Magazine*, XXVII. pp. 484—486, 1889.]

In a recent paper on the limit to interference when light is radiated from moving molecules*, it was necessary to form an estimate of the ratio of illuminations (h) at the darkest and brightest parts of a system of bands corresponding to the moment when they just cease to be visible from lack of contrast. In the comparison of uniformly illuminated surfaces, brought well into juxtaposition, h might be as great as ·99†; but in the case of bands, where the transition is gradual, a higher degree of contrast between the brightest and darkest parts may be expected to be necessary. In order to allow for this, I supposed that h might be estimated at ·95, the intensity of the light and the angular magnitude of the bands being assumed to be suitable. But since widely different estimates have been put forward by others, I have thought it worth while to test the matter with bands that are well under control.

In the first experiments light polarized by a Nicol fell upon a slit, against which was held a somewhat stout selenite. Direct examination of the slit through an analysing Nicol revealed no colour on account of the thickness of the selenite; but when a dispersing-prism was added, the resulting spectrum was marked out into bands, whose brightness and contrast depended upon the relative orientations of the Nicols and of the selenite. The theory of these bands is well known‡. If the Nicols be parallel, and if the principal sections of the Nicols and the selenite be inclined at the angle α, the expression for the brightness is

$$1 - \sin^2 2\alpha \sin^2 \tfrac{1}{2}\rho,$$

* *Phil. Mag.* April 1889. [Vol. III. p. 258.]

† See Helmholtz' *Physiological Optics*, § 21.

‡ See, for example, *Enc. Brit.* "Wave Theory," § 22. [Vol. III. p. 156.]

where ρ denotes the difference of retardations of the two rays to whose interference the bands are due. At the brightest place $\rho = 0$, and at the darkest $\sin^2 \frac{1}{2}\rho = 1$, so that

$$h = 1 - \sin^2 2\alpha.$$

The bands are thus invisible when $\alpha = 0$, and increase gradually in distinctness with α. When $\alpha = 45°$, the darkest place is absolutely black*.

The selenite was mounted upon a divided circle, and the observation consisted in finding the two positions, on either side of $\alpha = 0$, at which the bands manifested themselves with the desired degree of distinctness. The angular interval between the two positions was then taken as representing the value of 2α. In order that the bands should be recognizable with certainty it was found that 2α must be at least 14°. For a distinct and continuous impression $2\alpha = 17°$. Corresponding to these, we have for $1 - h$,

$$\sin^2 14° = {\cdot}0585, \qquad \sin^2 17° = {\cdot}0855.$$

In these observations the earliest recognition of the bands was somewhat interfered with by a want of smoothness in the spectrum due to irregularities in the selenite. Any irregularity, whether of this kind or caused by dust upon the edges of the slit, gives rise to horizontal markings in the spectrum which distract the eye. In a second set of experiments this difficulty was obviated by the substitution for the selenite of an accurately worked plate of quartz, cut parallel to the axis.

The following were the readings by myself (R) and by my assistant (G), when the bands were but just recognizable with certainty.

82° 6′	72° 13′	81° 2′	71° 59′
81 0	71 40	80 43	72 16
82 2	72 40	81 7	72 9
81 41	72 0		
80 31			
Mean...81 28	72 8	80 57	72 8

Hence

$$(R)\quad 2\alpha = 9°\ 20', \qquad (G)\quad 2\alpha = 8°\ 49';$$

so that, since $\sin^2 9° = {\cdot}0245$, the bands are visible when $1 - h$ is less than half as great as before.

* This presupposes an infinitely narrow slit. In practice the width must be reduced until, in this position, the bands are sensibly black.

The following were the readings when the bands were considered to be still *distinct*:—

83° 15′	70° 33′	82° 58′	70° 52′
83 28	69 53	82 41	71 2
83 0	70 30	83 37	71 20
Mean...83 14	70 19	83 5	71 5

Hence

$$(R)\quad 2\alpha = 12^\circ\ 55', \qquad (G)\quad 2\alpha = 12^\circ\ 0'.$$

Here $\sin^2 12^\circ\ 30' = {\cdot}0372$; so that a difference of 4 per cent. between the darkest and brightest parts is sufficient to show the bands with distinctness.

It seems therefore that I was well within the mark in assuming that bands involving 5 per cent. of the brightness might still be visible.

162.

ON THE UNIFORM DEFORMATION IN TWO DIMENSIONS OF A CYLINDRICAL SHELL OF FINITE THICKNESS, WITH APPLICATION TO THE GENERAL THEORY OF DEFORMATION OF THIN SHELLS.

[*Proceedings of the London Mathematical Society*, XX. pp. 372—381, 1889.]

THE theory of a thin uniform shell of elastic isotropic material, slightly deformed from an original curved condition, does not seem to be yet upon an entirely satisfactory footing. If the middle surface be extended, it is clear* that, to a first approximation, the potential energy per unit of area is

$$2nh\left\{\sigma_1^2+\sigma_2^2+\tfrac{1}{2}\varpi^2+\frac{m-n}{m+n}(\sigma_1+\sigma_2)^2\right\}, \quad\text{...............(1)}$$

where $2h$ denotes the thickness of the shell; m, n the elastic constants of Thomson and Tait's notation; σ_1, σ_2, ϖ the elongations and shear of the middle surface at the place under consideration. Again, if the deformation be such that the middle surface remain unextended, so that (1) vanishes, it is tolerably clear that the potential energy takes the form

$$\tfrac{2}{3}nh^3\left\{\left(\delta\frac{1}{\rho_1}\right)^2+\left(\delta\frac{1}{\rho_2}\right)^2+\frac{m-n}{m+n}\left(\delta\frac{1}{\rho_1}+\delta\frac{1}{\rho_2}\right)^2+2\tau^2\right\}, \quad\text{.........(2)}$$

where $\delta\rho_1^{-1}$, $\delta\rho_2^{-1}$ are the changes of principal curvatures of the middle surface, and τ is determined by the angle (χ) through which the principal planes are shifted according to the equation

$$\tau=2\chi\left(\frac{1}{\rho_2}-\frac{1}{\rho_1}\right). \quad\text{.............................(3)†}$$

* See Lamb, *Proc. Math. Soc.* Dec. 1882. Also *Proc. Roy. Soc.* XLV. (1888), p. 111, equation (13). [Vol. III. p. 222.]

† See Love, *Phil. Trans.* CLXXIX. (1888), A, pp. 505, 512; Rayleigh, *loc. cit.* p. 113. [Vol. III. p. 224.]

But when the middle surface undergoes stretching, so that (1) is finite, while yet the circumstances of the problem forbid us to remain satisfied with terms involving the first power of h, it is a more difficult question to determine the expression for the potential energy complete to the order h^3. An investigation of this problem has, however, been given by Mr Love, and his result* is exhibited in terms of σ_1, σ_2, ϖ, and of quantities depending upon these, and upon the alterations of curvature of the middle surface.

It may, indeed, be an under-statement of the case to speak of the problem as difficult, for to all appearance it may well be impossible in the form proposed. When the middle surface is plane, or when, though originally curved, it remains unstretched, there is no difficulty in supposing that the faces are exempt from imposed force. But when the middle surface of a shell is originally curved, and undergoes extension, equilibrium cannot be maintained without the cooperation of forces normal to the shell, and acting either upon the interior or upon the faces. It is easy to understand that the precise seat of these forces may be a matter of indifference, so far as the term of the first order (1) is concerned; but is there any reason for anticipating that there would be no effect upon the term of the third order? Rather, it would appear probable that there is no expression for the potential energy complete to the order h^3, in the absence of more definite suppositions as to the manner of application of the normal forces necessary in the general case. These doubts led me to think an investigation desirable, which should be based upon the general equations of elasticity, and conducted without the aid of approximations of ill-defined significance. For this purpose I have chosen the simplest problem involving the questions at issue—that namely of the deformation in two dimensions of a shell originally cylindrical.

Taking polar coordinates, let u, v† be the displacements at the point (r, θ) parallel to r and θ respectively. The displacement w, parallel to the axis of the cylinder, vanishes by hypothesis. The strains relative to these directions are‡

$$e=\frac{du}{dr},\qquad f=\frac{d}{d\theta}\left(\frac{v}{r}\right)+\frac{u}{r},\qquad g=0, \quad\text{.................(4)}$$

$$a=0,\qquad b=0,\qquad c=r\frac{d}{dr}\left(\frac{v}{r}\right)+\frac{1}{r}\frac{du}{d\theta}. \quad\text{..............(5)}$$

The stresses P, Q, R, S, T, U corresponding to these strains are given by

$$P=(m+n)\,e+(m-n)\,f,\qquad Q=(m+n)\,f+(m-n)\,e, \quad\text{......(6)}$$

$$S=0,\qquad T=0,\qquad U=nc. \quad\text{.......................(7)}$$

* *Loc. cit.* p. 505.

† This notation differs from that employed in my former papers, where u denoted the displacement parallel to the axis.

‡ Ibbetson's *Elastic Solids*, 1887, p. 238.

If there be no internal impressed forces, the equations of equilibrium are

$$\frac{d}{dr}(Pr)+\frac{dU}{d\theta}-Q=0, \qquad \frac{d}{dr}(Ur^2)+r\frac{dQ}{d\theta}=0. \quad\text{........(8, 9)}$$

We will now limit the problem by the supposition that the strains and stresses are independent of θ. Thus

$$dU/d\theta=0, \qquad dQ/d\theta=0; \quad\text{....................(10)}$$

and (8), (9) reduce to

$$\frac{d}{dr}(Pr)-Q=0, \qquad \frac{d}{dr}(Ur^2)=0. \quad\text{..............(11, 12)}$$

From (12) it follows that Ur^2 is an absolute constant. Hence if, as we will now suppose, U vanishes over the cylindrical faces of the shell, it necessarily vanishes throughout the interior. Thus, by (7),

$$c=0 \quad\text{...................................(13)}$$

throughout. From (5) and (13),

$$\frac{d}{dr}\left\{r^2\frac{d}{dr}\left(\frac{v}{r}\right)\right\}=-\frac{d}{d\theta}\frac{du}{dr}=-\frac{de}{d\theta}=0,$$

by hypothesis. Hence

$$v=C_1+C_2r, \quad\text{............................(14)}$$

where C_1, C_2 are independent of r, but may be functions of θ. Again, from (5) and (14),

$$\frac{du}{d\theta}=-r^2\frac{d}{dr}\left(\frac{v}{r}\right)=C_1;$$

so that, by (4),

$$\frac{df}{d\theta}=\frac{1}{r}\frac{d^2C_1}{d\theta^2}+\frac{d^2C_2}{d\theta^2}+\frac{C_1}{r}.$$

But $df/d\theta=0$, by supposition. Accordingly,

$$d^2C_1/d\theta^2+C_1=0, \qquad d^2C_2/d\theta^2=0;$$

or

$$C_1=H\cos\theta+K\sin\theta, \qquad C_2=C+D\theta, \quad\text{..............(15)}$$

where H, K, C, D are absolute constants. Thus, by (14),

$$v=H\cos\theta+K\sin\theta+(C+D\theta)\,r, \quad\text{.................(16)}$$

and

$$u=H\sin\theta-K\cos\theta+\phi(r), \quad\text{.........................(17)}$$

where $\phi(r)$ is a function of r which is, so far, arbitrary. Again, by (4),

$$e=\phi'(r), \qquad f=D+r^{-1}\phi(r), \quad\text{...................(18)}$$

indicating that the strains are independent of the coefficients H, K, C. The terms in H, K represent merely a displacement of the cylinder without rotation or strain, and the term in C represents simple rotation of the cylinder about its axis as a rigid body. They may be omitted without loss of anything material to the present inquiry.

So far, we have made no use of the condition (11) that there is no internal force in the radial direction. It is by means of this that the form of ϕ must be determined. From (6), (18),

$$P = (m+n)\,\phi'(r) + (m-n)\{D + r^{-1}\phi(r)\}\,; \quad\text{.............(19)}$$

$$Q = (m+n)\{D + r^{-1}\phi(r)\} + (m-n)\,\phi'(r)\,; \quad\text{.............(20)}$$

so that, by (11),

$$r^2\frac{d^2\phi}{dr^2} + r\frac{d\phi}{dr} - \phi = \frac{2nDr}{m+n}, \quad\text{.....................(21)}$$

—the differential equation which must be satisfied by ϕ.

The solution of (21) is

$$\phi = Ar + Br^{-1} + \frac{nD}{m+n}\,r\log r, \quad\text{.....................(22)}$$

where A and B are arbitrary constants. Corresponding to (22),

$$e = A - Br^{-2} + \frac{nD}{m+n}(\log r + 1), \quad\text{..................(23)}$$

$$f = D + A + Br^{-2} + \frac{nD}{m+n}\log r\,; \quad\text{..................(24)}$$

and from (16), (17), if $H = K = C = 0$,

$$u = Ar + Br^{-1} + \frac{nD}{m+n}\,r\log r, \qquad v = Dr\theta. \quad\text{.........(25, 26)}$$

We have now to consider the potential energy of strain. The general expression for the energy per unit of volume in a strained solid is

$$W = \tfrac{1}{2}(m+n)(e^2+f^2+g^2) + (m-n)(fg+ge+ef) + \tfrac{1}{2}n(a^2+b^2+c^2). \quad (27)$$

By (4), (5), (13), we have

$$a = 0, \qquad b = 0, \qquad c = 0, \qquad g = 0\,;$$

so that (27) reduces to

$$W = \tfrac{1}{2}(m+n)(e^2+f^2) + (m-n)ef = \tfrac{1}{2}m(e+f)^2 + \tfrac{1}{2}n(e-f)^2. \quad\text{...(28)}$$

In the present problem

$$e + f = D + 2A + \frac{nD}{m+n}(2\log r + 1), \quad\text{.......... ...(29)}$$

$$e - f = -D - 2Br^{-2} + \frac{nD}{m+n}. \quad\text{........................(30)}$$

Before proceeding further, we will consider in detail the very simple case which arises when $D = 0$. We have

$$e = A - Br^{-2}, \qquad f = A + Br^{-2}\,; \quad\text{................(31)}$$

$$u = Ar + Br^{-1}, \qquad v = 0. \quad\text{.........................(32)}$$

These equations constitute the solution of the problem of the deformation of a complete cylindrical shell (of finite thickness) under the action of hydrostatic pressures (or tractions) upon its inner and outer faces*. For the radial stress at any point, we have

$$P = 2mA - 2nBr^{-2}. \quad \text{.........................(33)}$$

Thus, if the stress upon the inner face $r = r_1$ be Π_1, and upon the outer face $r = r_2$ be Π_2,

$$\Pi_1 = 2mA - 2nBr_1^{-2}, \qquad \Pi_2 = 2mA - 2nBr_2^{-2}, \quad \text{.........(34)}$$

by which A and B are determined.

The expression for the energy becomes, by (28), (29), (30),

$$W = 2mA^2 + 2nB^2r^{-4}. \quad \text{.........................(35)}$$

The whole potential energy per unit of length parallel to the axis is given by

$$2\pi \int_{r_1}^{r_2} Wrdr = 2\pi \{mA^2(r_2^2 - r_1^2) - nB^2(r_2^{-2} - r_1^{-2})\}. \quad \text{......(36)}$$

In order to apply this result to a thin shell, we will write

$$r_1 = a - h, \qquad r_2 = a + h,$$

where $2h$ denotes the thickness of the shell, and a the radius of the middle surface. Thus

$$\int Wrdr = 4ah\left\{mA^2 + n\frac{B^2}{r_1^2r_2^2}\right\} = 4ah\{mA^2 + nB^2a^{-4} + 2nB^2a^{-6}h^2\}, \quad \text{...(37)}$$

approximately.

The extension of the middle surface is here, by (31),

$$\sigma = A + Ba^{-2}. \quad \text{.............................(38)}$$

Since there are two independent variables A, B, or Π_1, Π_2, in (37), it is clear that the potential energy cannot, in strictness, be determined by σ only. Let us, however, inquire to what order of approximation the energy is a function of σ, when h is regarded as small.

If ϖ denote the ratio of surface forces by which the deformation is maintained, we have, from (34),

$$mA(1 - \varpi) = nB(r_2^{-2} - \varpi r_1^{-2});$$

from which, and (38),

$$\sigma = A\left\{1 + \frac{m}{n}\frac{(1-\varpi)a^{-2}}{r_2^{-2} - \varpi r_1^{-2}}\right\} = Ba^{-2}\left\{1 + \frac{n}{m}\frac{r_2^{-2} - \varpi r_1^{-2}}{(1-\varpi)a^{-2}}\right\}, \quad \text{...(39)}$$

equations giving A and B in terms of σ and ϖ. Using these, we find, on reduction,

$$mA^2 + \frac{nB^2}{r_1^2r_2^2} = \frac{mn\sigma^2}{m+n}\left\{1 + \frac{2m}{m+n}\frac{h^2}{a^2} + \frac{4mn}{(m+n)^2}\frac{h^2}{a^2}\left(\frac{1+\varpi}{1-\varpi}\right)^2\right\},$$

* Ibbetson, *loc. cit.* pp. 313, 314.

the term containing the first power of h disappearing. Thus, for the potential energy per unit of area of the shell, we obtain

$$a^{-1}\int Wrdr=\frac{4mn\sigma^2h}{m+n}\left\{1+\frac{2m}{m+n}\frac{h^2}{a^2}+\frac{4mn}{(m+n)^2}\frac{h^2}{a^2}\left(\frac{1+\varpi}{1-\varpi}\right)^2\right\}\ldots\ldots(40)$$

The term in h agrees, as might have been expected, with (1)*. But, when the approximation is carried so far as to include h^3, (40) depends upon ϖ as well as upon σ. If the normal forces are limited to one surface, $\varpi=0$, or $\varpi=\infty$. In either case

$$(1-\varpi)^2/(1+\varpi)^2=1,$$

and
$$a^{-1}\int Wrdr=\frac{4mn\sigma^2h}{m+n}\left\{1+\frac{2m}{m+n}\frac{h^2}{a^2}+\frac{4mn}{(m+n)^2}\frac{h^2}{a^2}\right\}.\quad\ldots\ldots(41)$$

The energy involved in a given extension of the middle surface is thus the same, whether the necessary normal force be an internal pressure or an external traction; but the case is otherwise if the forces be distributed. When the work is equally divided between the two surfaces, so that there is (for example) a pressure upon the internal surface and a traction upon the external surface, $\varpi=-1$; and

$$a^{-1}\int Wrdr=\frac{4mn\sigma^2h}{m+n}\left\{1+\frac{2m}{m+n}\frac{h^2}{a^2}\right\}.\quad\ldots\ldots\ldots\ldots\ldots(42)$$

It will be seen that, in order to give rise to this discrepancy, it is not necessary to suppose the introduction of surface forces more powerful than are actually required to maintain the deformation. This instance is sufficient to show that the potential energy of deformation cannot, in general, be expressed in terms of extensions and changes of curvature of the middle surface, when it is necessary to include terms of order h^3, without further information as to the manner in which the surface forces are applied. According to Mr Love's results†, the expression for the energy in the present problem should reduce to its first term; whereas (40) indicates that there is no manner of application of the surface forces by which such a result could be brought about.

We will now abandon the restriction to $D=0$. It will then be possible to find a deformation such that, not only is there no impressed force upon the interior of the shell, but also none upon either of the surfaces. Under these circumstances the stresses between contiguous parts must reduce themselves to a simple couple.

* ϖ has there a different meaning from that belonging to it in (40). In (1) $\varpi=0$, $\sigma_2=0$, for the purposes of the present problem.

† *Loc. cit.* equations (12), (18). {December, 1889. I have been reminded by the Secretary that in the investigation of Mr Love it is expressly supposed (p. 504) that no surface tractions are applied. But the absence of normal forces would, as it appears to me, be equivalent to a limitation upon the generality of the middle surface extensions, σ_1, σ_2.}

From (6), (29), (30), we find

$$P = m(e+f) + n(e-f) = 2mA - 2nBr^{-2} + D\left\{m + \frac{2mn \log r}{m+n}\right\}. \quad \text{...(43)}$$

If $P=0$, both when $r=r_1$ and when $r=r_2$, the values of A, B, in terms of D, are

$$A = -D\left\{\frac{1}{2} + \frac{n}{m+n}\frac{r_2^{-2}\log r_1 - r_1^{-2}\log r_2}{r_2^{-2} - r_1^{-2}}\right\}, \quad \text{...........(44)}$$

$$B = \frac{mD}{m+n}\frac{\log(r_2/r_1)}{r_2^{-2} - r_1^{-2}}. \quad \text{.....................................(45)}$$

These values, substituted in (23), (24), (25), (26), determine a definite type of deformation, satisfying the conditions that there shall be no internal or surface forces, and that the strains shall be independent of θ, and this without any supposition limiting the thickness of the shell.

From the expression for Q in terms of e and f, or, more readily, by means of (11), we may verify that

$$\int_{r_1}^{r_2} Q\,dr = 0. \quad \text{...............................(46)}$$

In order to apply these results to a thin shell, we write, as before,

$$r_1 = a - h, \qquad r_2 = a + h;$$

thus

$$A = -D\left\{\frac{1}{2} + \frac{n}{m+n}\left(\log a + \frac{1}{2} + \frac{h^2}{6a^2}\right)\right\}, \quad \text{..............(47)}$$

$$B = -\frac{mD}{m+n}\frac{a^2}{2}\left(1 - \frac{5h^2}{3a^2}\right). \quad \text{...........................(48)}$$

Corresponding to these, from (29), (30),

$$e+f = \frac{nD}{m+n}\left\{2\log\frac{r}{a} - \frac{h^2}{3a^2}\right\}, \quad \text{.....................(49)}$$

$$e-f = -\frac{mD}{m+n}\left\{1 - \frac{a^2}{r^2}\left(1 - \frac{5h^2}{3a^2}\right)\right\}; \quad \text{..............(50)}$$

or, if $r = a + \rho$,

$$e+f = \frac{nD}{m+n}\left\{\frac{2\rho}{a} - \frac{\rho^2}{a^2} - \frac{h^2}{3a^2}\right\}, \quad \text{.....................(51)}$$

$$e-f = \frac{-mD}{m+n}\left\{\frac{2\rho}{a} - \frac{3\rho^2}{a^2} + \frac{5h^2}{3a^2}\right\}. \quad \text{.................(52)}$$

The strains e, f both vanish approximately when $r = a$. By (6),

$$P = \frac{2mnD}{m+n}\left\{\frac{\rho^2}{a^2} - \frac{h^2}{a^2}\right\}, \qquad Q = \frac{4mnD}{m+n}\left\{\frac{\rho}{a} - \frac{\rho^2}{a^2} + \frac{h^2}{3a^2}\right\}. \quad \text{...(53, 54)}$$

We will now calculate the potential energy of deformation. From (28), (51), (52),

$$W = \frac{mn^2 D}{(m+n)^2}\left\{\frac{2\rho^2}{a^2} - \frac{2\rho^3}{a^3} - \frac{2\rho h^2}{3a^3}\right\} + \frac{m^2 n D}{(m+n)^2}\left\{\frac{2\rho^2}{a^2} - \frac{6\rho^3}{a^3} + \frac{10\rho h^2}{3a^2}\right\};$$

so that, for the potential energy per unit of area, we get

$$a^{-1}\int W r dr = \int_{-h}^{+h} W\left(1+\frac{\rho}{a}\right) d\rho = \frac{4mnD^2}{3(m+n)}\frac{h^3}{a^2}, \quad \text{.........(55)}$$

the next term involving h^5.

In order to connect this with the change of curvature of the middle surface, we require the expression for u. From (25),

$$u = -Dr\left\{\frac{1}{2} + \frac{n}{m+n}\left(\log a + \frac{1}{2} + \frac{h^2}{6a^2}\right)\right\}$$
$$- \frac{mDr^{-1}a^2}{2(m+n)}\left(1 - \frac{5h^2}{3a^2}\right) + \frac{nD}{m+n} r \log r; \quad \text{..............(56)}$$

so that the value of u at the middle surface ($r = a$) is, approximately,

$$u = -aD. \quad \text{.................................(57)}$$

Now $a + u$ is the radius of curvature of the middle surface after deformation, or $\delta\rho_1 = u$. Thus

$$\left(\delta\frac{1}{\rho_1}\right)^2 = \frac{u^2}{a^4} = \frac{D^2}{a^2}.$$

The expression for the energy per unit area of surface is thus

$$a^{-1}\int W r dr = \frac{4mnh^3}{3(m+n)}\left(\delta\frac{1}{\rho_1}\right)^2,$$

in agreement with (2); for in the present application

$$\delta(1/\rho_2) = 0, \qquad \tau = 0.$$

It is evident that the rigorous solution from which we started is available for continuing the approximation, should it be thought desirable to retain higher powers of h.

The solution of the problem of the bending of a cylindrical shell, here put forward, favours then the idea that it is only when the middle surface of a curved shell remains unextended that it is possible to express the potential energy to the order h^3 in terms merely of the extensions and curvatures of the middle surface.

163.

ON ACHROMATIC INTERFERENCE-BANDS.

[*Philosophical Magazine*, XXVIII. pp. 77—91, 189—206, 1889.]

Introduction.

WHEN there is interference of light, the width of the resulting bands, measured for example from darkness to darkness, is usually a function of the colour of the light employed. Thus, in the case of Fresnel's well-known interference-experiment, in which light reflected from two slightly inclined mirrors illuminates a screen, the width of the bands is proportional to the wave-length of the light. In order that a considerable number of bands may be visible, it is necessary that the light be highly homogeneous; otherwise it is impossible that the various band-systems can fit one another over the necessary range. If the light could be supposed to be absolutely homogeneous, there would be no limit to the number of observable bands: and, what is especially to be remarked, there would be nothing by which one band could be distinguished from another,—in particular there could be no central band recognizable. When, on the other hand, the light is white, there may be a central band at which all the maxima of brightness coincide; and this band, being white, may be called the achromatic band. But the *system* of bands is not usually achromatic. Thus, in Fresnel's experiment the centre of symmetry fixes the position of the central achromatic band, but the system is far from achromatic. Theoretically there is not even a single place of darkness, for there is no point where there is complete discordance [opposition] of phase for all kinds of light. In consequence, however, of the fact that the range of sensitiveness of the eye is limited to less than an "octave," the centre of the first dark band on either side is sensibly black; but the existence of even one band is due to selection, and the formation of several visible bands is favoured by the capability of the retina to make chromatic distinctions within the range of vision. After two or three alternations the bands become highly coloured*; and, as the overlapping of the various elementary systems increases, the colours fade away, and the field of view assumes a uniform appearance.

There are, however, cases where it is possible to have systems of achromatic bands. For this purpose it is necessary, not merely that the maxima of illumination should coincide at some one place, but *also* that the widths of the bands should be the same for the various colours. The independence of colour, as we shall see, may be absolute; but it will probably be more convenient not to limit the use of the term so closely. The focal length of the ordinary achromatic object-glass is not entirely independent of colour. A similar use of the term would justify us in calling a system of bands achromatic, when the width of the elementary systems is a maximum or a minimum for some ray very near the middle of the spectrum, or, which comes to the same, has equal values for two rays of finitely different refrangibility. The outstanding deviation from complete achromatism, according to the same analogy, may be called the *secondary* colour.

The existence of achromatic systems was known to Newton†, and was insisted upon with special emphasis by Fox Talbot‡; but singularly little attention appears to have been bestowed upon the subject in recent times. In the article "Wave Theory" (*Encyc. Brit.* 1888 [Vol. III. p. 61]) I have discussed a few cases, but with too great brevity. It may be of interest to resume the consideration of these remarkable phenomena, and to detail some observations which I have made, in part since the publication of the "Encyclopædia" article. A recent paper by M. Mascart§ will also be referred to.

Fresnel's Bands.

In this experiment the two sources of light which are regarded as interfering with one another must not be independent; otherwise there could be no fixed phase-relation. According to Fresnel's original arrangement the sources O_1, O_2 are virtual images of a single source, obtained by reflexion in two mirrors. The mirrors may be replaced by a bi-prism. Or, as in Lloyd's form of the experiment, the second source may be obtained from the first by reflexion from a mirror placed at a high degree of obliquity. The screen upon which the bands are conceived to be thrown is parallel to O_1O_2, at distance D.

* The series of colours thus arising are calculated, and exhibited in the form of a curve upon the colour diagram, in a paper "On the Colours of Thin Plates," *Edinb. Trans.* 1887. [Vol. II. p. 498.]

† *Optics*, Book II.

‡ *Phil. Mag.* [3] IX. p. 401 (1836).

§ "On the Achromatism of Interference," *Comptes Rendus*, March 1889; *Phil. Mag.* [5] XXVII. p. 519.

If A be the point of the screen equidistant from O_1, O_2, and P a neighbouring point, then approximately

$$O_1P - O_2P = \sqrt{\{D^2 + (u + \tfrac{1}{2}b)^2\}} - \sqrt{\{D^2 + (u - \tfrac{1}{2}b)^2\}} = ub/D,$$

where

$$O_1O_2 = b, \qquad AP = u.$$

Thus, if λ be the wave-length, the places where the phases are accordant are determined by

$$u = n\lambda D/b, \quad \text{.................................(1)}$$

n being an integer representing the order of the band. The linear width of the bands (from bright to bright, or from dark to dark) is thus

$$\Lambda = \lambda D/b. \quad \text{..................................(2)}$$

The degree of homogeneity necessary for the approximate perfection of the nth band may be found at once from (1) and (2). For, if du be the change in u corresponding to the change $d\lambda$, then

$$du/\Lambda = n\, d\lambda/\lambda. \quad \text{.................................(3)}$$

Now clearly du must be a small fraction of Λ, so that $d\lambda/\lambda$ must be many times smaller than $1/n$, if the darkest places are to be sensibly black. But the phenomenon will be tolerably well marked, if the proportional range of wave-length do not exceed $1/(2n)$, provided, that is, that the distribution of illumination over this range be not concentrated towards the extreme parts.

So far we have supposed the sources at O_1, O_2 to be mathematically small. In practice the source is an elongated slit, whose direction requires to be carefully adjusted to parallelism with the reflecting surface, or surfaces. By this means an important advantage is obtained in respect of brightness without loss of definition, as the various parts of the aperture give rise to coincident systems of bands.

The question of the admissible *width* of the slit requires careful consideration. We will suppose in the first place that the lights issuing from the various parts of the aperture are without permanent phase-relation, as when the slit is backed immediately by a flame, or by the incandescent carbon of an electric lamp. Regular interference can then only take place between lights coming from *corresponding* parts of the two images; and a distinction must be drawn between the two ways in which the images may be situated relatively to one another. In Fresnel's experiment, whether carried out with mirrors or with bi-prism, the corresponding parts of the images are on the same side; that is, the right of one corresponds to the right of the other, and the left of one to the left of the other. On the other hand, in Lloyd's arrangement the reflected image is reversed relatively to the original source: the two outer edges corresponding, as also the two inner. Thus in the first arrangement the bands due to various parts of the slit differ merely by a lateral shift, and the condition of distinctness is simply that the [projection of the*] width of the slit be a small proportion of the width of the bands. From this it follows as a corollary that the limiting width is independent of the order of the bands under examination. It is otherwise in Lloyd's method. In this case the centres of the systems of bands are the same, whatever part of the slit be supposed to be operative, and it is the distance apart of the images (b) that varies. The bands corresponding to the various parts of the slit are thus upon different scales, and the resulting confusion must increase with the order of the bands. From (1) the corresponding changes in u and b are given by

$$du = -n\lambda D\, db/b^2;$$

so that

$$du/\Lambda = -n\, db/b. \qquad (4)$$

If db represents twice the width of the slit, (4) gives a measure of the resulting confusion in the bands. The important point is that the slit must be made narrower as n increases, if the bands are to retain the same degree of distinctness.

If the various parts of the width of the slit do not act as independent sources of light, a different treatment would be required. To illustrate the extreme case, we may suppose that the waves issuing from the various elements of the width are all in the same phase, as if the ultimate source were a star situated a long distance behind. It would then be a matter of indifference whether the images of the slit, acting as proximate sources of interfering light, were reversed relatively to one another, or not. It is, however, unnecessary to dwell upon this question, inasmuch as the conditions supposed are unfavourable to brightness, and therefore to be avoided in practice. The better to understand this, let us suppose that the slit is backed by the sun, and is so narrow that, in spite of the sun's angular magnitude, the luminous vibration is sensibly the same at all parts of the width. For this purpose the width must not exceed $\frac{1}{20}$ millim.† By hypothesis, the appearance presented to an eye close to the slit and looking backwards towards the sun will be the same as if the source of light were reduced to a point coincident with the sun's centre. The meaning of this is that, on account of the narrowness of the aperture, a point would appear dilated by diffraction until its apparent diameter became a large multiple

* [1901. Compare Walker, *Phil. Mag.* XLVI. p. 477, 1898.

In the case of the spectroscope, when resolving power is important, the width of the slit must evidently not exceed λ/a, where λ = wave-length and a = horizontal aperture (Vol. I. p. 420). This is the condition that the aperture of the instrument should just embrace the central diffraction fringe (from darkness to darkness) formed by light passing the slit aperture. Since full resolving power requires the cooperation of all parts of the aperture, we may conclude that an even narrower slit than that above specified is desirable.]

† Verdet's *Leçons d'Optique physique*, t. I. p. 106.

of that of the sun. Now it is evident that in such a case the brightness may be enhanced by increasing the sun's apparent diameter, as can always be done by optical appliances. Or, which would probably be more convenient in practice, we may obtain an equivalent result by so designing the experiment that the slit does not require to be narrowed to the point at which the sun's image begins to be sensibly dilated by diffraction. The available brightness is then at its limit, and would be no greater, even were the solar diameter increased. The practical rule is that, when brightness is an object, slits backed by the sun should not be narrowed to much less than half a millimetre.

Lloyd's Bands.

Lloyd's experiment deserves to be more generally known, as it may be performed with great facility and without special apparatus. Sunlight is admitted horizontally into a darkened room through a slit situated in the window-shutter, and at a distance of 15 or 20 feet is received at nearly grazing incidence upon a vertical slab of plate glass. The length of the slab in the direction of the light should not be less than 2 or 3 inches, and for some special observations may advantageously be much increased. The bands are observed on a plane through the hinder vertical edge of the slab by means of a hand magnifying-glass of from 1 to 2 inch focus. The obliquity of the reflector is of course to be adjusted according to the fineness of the bands required.

From the manner of their formation it might appear that under no circumstances could more than half the system be visible. But, according to Airy's principle*, the bands may be displaced if examined through a prism. In practice all that is necessary is to hold the magnifier somewhat excentrically. The bands may then be observed gradually to detach themselves from the mirror, until at last the complete system is seen, as in Fresnel's form of the experiment.

If we wish to observe interference under high relative retardation, we must either limit the light passing the first slit to be approximately homogeneous, or (after Fizeau and Foucault) transmit a narrow width of the band-system itself through a second slit, and subsequently analyse the light into a spectrum. In the latter arrangement, which is usually the more convenient when the original light is white, the bands seen are of a rather artificial kind. If, apart from the heterogeneity of the light, the original bands are well formed, and if the second slit be narrow enough, the spectrum will be marked out into bands; the bright places corresponding to the kinds of light for which the original bands would be bright, and the black places to the kinds of light for which the original bands would be black. The

* See below.

condition limiting the width of the second slit is obviously that it be but a moderate fraction of the width of a band (Λ).

If it be desired to pass along the entire series of bands up to those of a high order by merely traversing the second slit in a direction perpendicular to that of the light, a very long mirror is necessary. But when the second slit is in the region of the bands of highest order (that is, near the external limit of the field illuminated by both pencils), only the more distant part of the mirror is really operative; and thus, even though the mirror be small, bands of high order may be observed, if the second slit be carried backwards, keeping it of course all the time in the narrow doubly-illuminated field. In one experiment the distance from the first slit to the (3-inch) reflector was 27 feet, while the second slit was situated behind at a further distance of 4 feet. The distance (b) between the first slit and its image in the reflector (measured at the window) was about 13 inches.

As regards the spectroscope it was found convenient to use an arrangement with detached parts. The slit and collimating lens were rigidly connected, and stood upon a long and rigid box, which carried also the mirror. The narrowness of the bands in which this slit is placed renders it imperative to avoid the slightest relative unsteadiness or vibration of these parts. The prisms, equivalent to about four of 60°, and the observing telescope were upon another stand at a little distance behind the box which supported the rest of the apparatus.

Under these conditions it was easy to observe bands in the spectrum whose width (from dark to dark) could be made as small as the interval between the D lines; but for this purpose the first slit had to be rather narrow, and the direction of its length accurately adjusted, so as to give the greatest distinctness. Since the wave-lengths of the two D lines differ by about $\frac{1}{1000}$ part, spectral bands of this degree of closeness imply interference with a retardation of 1000 periods.

Much further than this it was not easy to go. When the bands were rather more than twice as close, the necessary narrowing of the slits began to entail a failing of the light, indicating that further progress would be attained with difficulty.

Indeed, the finiteness of the illumination behind the first slit imposes of necessity a somewhat sudden limit to the observable retardation. In this respect it is a matter of indifference at what angle the reflector be placed. If the angle be made small, so that the reflexion is very nearly grazing, the bands are upon a larger scale, and the width of the second slit may be increased, but in a proportional degree the width of the first slit must be reduced.

The relation of the width of the second slit to the angle of the mirror may be conveniently expressed in terms of the appearance presented to an eye placed close behind the former. The smallest angular distance which the slit, considered as an aperture, can resolve, is expressed by the ratio of the wave-length of light (λ) to the width (w_2) of the slit. Now, in order that this slit may perform its part tolerably well, w_2 must be less than $\frac{1}{2}\Lambda$; so that, by (2),

$$\lambda/w_2 > 2b/D. \quad \text{.................................(5)}$$

The width must therefore be less than the half of that which would just allow the resolution of the two images (subtending the angle b/D) as seen by an eye behind. In setting up the apparatus this property may be turned to account as a test.

The existence of a limit to n, dependent upon the intrinsic brightness of the sun, may be placed in a clearer light by a rough estimate of the illumination in the resulting spectrum; and such an estimate is the more interesting on account of the large part here played by diffraction. In most calculations of brightness it is tacitly assumed that the ordinary rules of geometrical optics are obeyed.

Limit to Illumination.

The narrowness of the second slit would not in itself be an obstacle to the attainment of full spectrum brightness, were we at liberty to make what arrangements we pleased behind it. In illustration of this, two extreme cases may be considered of a slit illuminated by ordinary sunshine. First, let the width w_2 be great enough not sensibly to dilate the solar image; that is, let w_2 be much greater than λ/s, where s denotes in circular measure the sun's apparent diameter (about 30 minutes). In this case the light streams through the slit according to the ordinary law of shadows, and the pupil (of diameter p) will be filled with light if situated at a distance exceeding d*, where

$$p/d = s. \quad \text{...................................(6)}$$

At this distance the apparent width of the slit is w_2/d, or $w_2 s/p$; and the question arises whether it lies above or below the ocular limit λ/p, that is, the smallest angular distance that can be resolved by an aperture p. The answer is in the affirmative, because we have already supposed that $w_2 s$ exceeds λ. The slit has thus a visible width, and it is seen backed by undiffracted sunshine. If a spectrum be now formed by the use of dispersion sufficient to give a prescribed degree of purity, it is as bright as is possible with the sun as ultimate source, and would be no brighter even were the solar diameter increased, as it could in effect be by the use of a burning-glass throwing a solar image upon the slit. The employment of a telescope in the formation of the spectrum gives no means of escape from this conclusion. The precise definition of the brightness of any part of the resulting spectrum

* About 30 inches [76 cm.].

would give opportunity for a good deal of discussion; but for the present purpose it may suffice to suppose that, if the spectrum is to be divided into n distinguishable parts, so that its angular width is n times the angular width of the slit, the apparent brightness is of order $1/n$ as compared with that of the sun.

Under the conditions above supposed the angular width of the slit is in excess of the ocular limit, and the distance might be increased beyond d without prejudice to the brilliancy of the spectrum. As the angular width decreases, so does the angular dispersion necessary to attain a given degree of purity. But this process must not be continued to the point where w_2/d approaches the ocular limit. Beyond that limit it is evident that no accession of purity would attend an increase in d under given dispersion. Accordingly the dispersion could not be reduced, if the purity is to be maintained; and the brightness necessarily suffers. It must always be a condition of full brightness that the angular width of the slit exceed the ocular limit.

Let us now suppose, on the other hand, that w_2 is so small that the image of the sun is dilated to many times s, or that w_2 is much less than λ/s. The divergence of the light is now not s, but λ/w_2; and, if the pupil be just immersed,

$$p/d = \lambda/w_2.$$

The angular width of the slit w_2/d is thus equal to λ/p, that is, it coincides with the ocular limit. The resulting spectrum necessarily falls short of full brightness, for it is evident that further brightness would attend an augmentation of the solar diameter, up to the point at which the dilatation due to diffraction is no longer a sensible fraction of the whole. In comparison with full brightness the actual brightness is of order $w_2 s/\lambda$; or, if the purity required is represented by n, we may consider the brightness of the spectrum relatively to that of the sun to be of order $w_2 s/(n\lambda)$.

In the application of these considerations to Lloyd's bands we have to regard the narrow slit w_2 as illuminated, not by the sun of diameter s, but by the much narrower source allowed by the first slit, whose angular width is w_1/D. On this account the reduction of brightness is at least $w_1/(sD)$. If w_1 be so narrow as itself to dilate the solar image, a further reduction would ensue; but this could always be avoided, either by increase of D, or by the use of a burning-glass focusing the sun upon the first slit. The brightness of the spectrum of purity n from the second slit is thus of order

$$\frac{w_1}{sD} \cdot \frac{w_2 s}{n\lambda} = \frac{w_1 w_2}{n\lambda D}.$$

We have now to introduce the limitations upon w_1 and w_2. By (4) w_1 must not exceed $b/(4n)$; and by (2) w_2 must not exceed $\lambda D/(2b)$. Hence the brightness is of order

$$1/(8n^2), \qquad (7)$$

independent of s, and of the linear quantities. The fact that the brightness is inversely as the *square* of the number of bands to be rendered visible explains the somewhat sudden failure observed in experiment. If $n = 2000$, the original brightness of the sun is reduced in the spectrum some 30 million times, beyond which point the illumination could hardly be expected to remain sufficient for vision of difficult objects such as narrow bands.

In Fresnel's arrangement, where the light is reflected perpendicularly from two slightly inclined mirrors, interference of high order is obtained by the movement of one of the mirrors parallel to its plane. The increase of n does not then entail a narrowing of w_1; and bands of order n may be observed in the spectrum of light transmitted through w_2, whose brightness is proportional to n^{-1}, instead of, as before, to n^{-2}.

Achromatic Interference-Bands.

We have already seen from (3) that in the ordinary arrangement, where the source is of white light entering through a narrow slit, the heterogeneity of the light forbids the visibility of more than a few bands. The scale of the various band-systems is proportional to λ. But this condition of things, as we recognize from (2), depends upon the constancy of b, that is, upon the supposition that the various kinds of light all come from the same place. Now there is no reason why such a limitation should be imposed. If we regard b as variable, we recognize that we have only to take b proportional to λ, in order to render the band-interval (Λ) independent of the colour. In such a case the *system* of bands is achromatic, and the heterogeneity of the light is no obstacle to the formation of visible bands of high order.

These requirements are very easily met by the use of Lloyd's mirror, and of a diffraction-grating with which to form a spectrum. White light enters the dark room through a slit in the window-shutter, and falls in succession upon a grating and upon an achromatic lens, so as to form a real diffraction-spectrum, or rather series of such, in the focal plane. The central image, and all the lateral coloured images, except one, are intercepted by a screen. The spectrum which is allowed to pass is the proximate source of light in the interference experiment; and since the deviation of any colour from the central white image is proportional to λ, it is only necessary so to arrange the mirror that its plane passes through the white image in order to realize the conditions for the formation of achromatic bands.

There is no difficulty in carrying out the experiment practically. I have used the spectrum of the second order, as given by a photographed grating of 6000 lines in an inch, and a photographic portrait lens of about 6 inches focus. At a distance of about 7 feet from the spectrum the light fell upon a vertical slab of thick plate-glass 3 feet in length and a few inches high. The observer upon the further side of the slab examines the bands through a Coddington lens of somewhat high power, as they are formed upon the plane passing through the end of the slab. It is interesting to watch the appearance of the bands as dependent upon the degree in which the condition of achromatism is fulfilled. A comparatively rough adjustment of the slab in azimuth is sufficient to render achromatic, and therefore distinct, the first 20 or 30 bands. As the adjustment improves, a continually larger number becomes visible, until at last the whole of the doubly illuminated field is covered with fine lines.

In these experiments the light is white, or at least becomes coloured only towards the outer edge of the field. By means of a fine slit in the plane of the spectrum we may isolate any kind of light, and verify that the band-systems corresponding to various wave-lengths are truly superposed.

When the whole spectrum was allowed to pass, the white and black bands presented so much the appearance of a grating under the microscope that I was led to attempt to photograph them, with the view of thus forming a diffraction-grating. Gelatine plates are too coarse in their texture to be very suitable for this purpose; but I obtained impressions capable of giving spectra. Comparison with spectra from standard gratings showed that the lines were at the rate of 1200 to the inch. A width of about half an inch (corresponding to 600 lines) was covered, but the definition deteriorated in the outer half. A similar deterioration was evident on direct inspection of the bands, and was due to some imperfection in the conditions—perhaps to imperfect straightness of the slab. On one occasion the bands were seen to lose their sharpness towards the middle of the field, and to recover in the outer portion.

With respect to this construction of a grating by photography of interference-bands, a question may be raised as to whether we are not virtually copying the lines of the original grating used to form the spectrum. More may be said in favour of such a suggestion than may at first appear. For it would seem that the case would not be essentially altered if we replaced the *real* spectrum by a *virtual* one, abolishing the focusing lens, and bringing Lloyd's mirror into the neighbourhood of the grating. But then the mirror would be unnecessary, since the symmetrical spectrum upon the other side would answer the purpose as well as a reflexion of the first spectrum. Indeed, there is no escape from the conclusion that a grating capable of giving on the two sides similar spectra of any one order, without spectra of other orders or central image, must produce behind it, without other appliances and at all distances, a system of achromatic interference-fringes, which could not fail to impress themselves upon a sensitive photographic plate. But a grating so obtained would naturally be regarded as merely a copy of the first.

Another apparent anomaly may be noticed. It is found in practice that, to reproduce a grating by photography, it is necessary that the sensitive plate be brought into close contact with the original; whereas, according to the argument just advanced, no such limitation would be required.

These discrepancies will be explained if, starting from the general theory, we take into account the actual constitution of the gratings with which we can experiment. If plane waves of homogeneous light (λ) impinge perpendicularly upon a plane ($z=0$) grating, whose constitution is periodic with respect to x in the interval σ, the waves behind have the general expression

$$\begin{aligned} A_0 \cos(kat-kz) &+ A_1 \cos(px+f_1)\cos(kat-\mu_1 z) \\ &+ B_1 \cos(px+g_1)\sin(kat-\mu_1 z) \\ &+ A_2 \cos(2px+f_2)\cos(kat-\mu_2 z)+\ldots; \qquad (8) \end{aligned}$$

where

$$p=2\pi/\sigma, \qquad k=2\pi/\lambda,$$

and

$$\mu_1^2=k^2-p^2, \qquad \mu_2^2=k^2-4p^2, \qquad \&c.,$$

the series being continued as long as μ is real*. Features in the wave-form for which μ is imaginary are rapidly eliminated. For the present purpose we may limit our attention to the first three terms of the series, which represent the central image and the two lateral spectra of the first order.

When the first term occurs, as usually happens, the phenomena are complicated by the interaction of this term with the following ones, and the effect varies with z in a manner dependent upon λ. This is the ordinary case of photographic reproduction, considered in the paper referred to.* If A_0 vanish, there is no central image; but various cases may still be distinguished according to the mutual relations of the other constants. If only A_1, or only B_1, occur, we have interference-fringes. The intensity of light is (in the first case)

$$A_1^2 \cos^2(px+f), \qquad (9)$$

vanishing when

$$px+f=\tfrac{1}{2}(n+1)\pi;$$

and these fringes may be regarded as arising from the interference of the two lateral spectra of the first order,

$$\tfrac{1}{2}A_1 \cos(kat-\mu_1 z+px+f_1),$$

$$\tfrac{1}{2}A_1 \cos(kat-\mu_1 z-px-f_1).$$

As an example of only one spectrum, we may suppose

$$B_1=A_1, \qquad g_1=f_1-\tfrac{1}{2}\pi,$$

* *Phil. Mag.* March 1881 [Vol. I. p. 510]; *Enc. Brit.* "Wave Theory," p. 440 [Vol. III. p. 122].

giving

$$A_1 \cos(kat - \mu_1 z - px - f_1). \quad\quad\quad (10)$$

A photographic plate exposed to this would yield no impression, since the intensity is constant.

In order, then, that a grating may be capable of giving rise to the ideal system of interference-fringes, and thus impress itself upon a sensitive plate at any distance behind, the vibration due to it must be of the form

$$A \cos(px + f) \cos(kat - \mu_1 z). \quad\quad\quad (11)$$

It does not appear how any actual grating could effect this*. Supposing $z = 0$, we see that the amplitude of the vibration immediately behind the grating must be a harmonic function of x, while the phase is independent of x, except as regards the reversals implied in the variable sign of the amplitude. Gratings may act partly by opacity and partly by retardation, but the two effects would usually be connected; whereas the requirement here is that at two points the transmission shall be the same while the phase is reversed.

We can thus hardly regard the interference-bands obtained from a grating and Lloyd's mirror as a mere reproduction of the original ruling. As will be seen in the following paragraphs, much the same result may be got from a prism, in place of a grating; and if the light be sufficiently homogeneous to begin with, both these appliances may be dispensed with altogether.

Prism instead of Grating.

If we are content with a less perfect fulfilment of the achromatic condition, the diffraction-spectrum may be replaced by a prismatic one, so arranged that $d(\lambda/b) = 0$ for the most luminous rays. The bands are then achromatic in the same sense that the ordinary telescope is so. In this case there is no objection to a merely virtual spectrum, and the experiment may be very simply executed with Lloyd's mirror and a prism of (say) 20° held just in front of it.

The number of black and white bands to be observed is not so great as might perhaps have been expected. The lack of contrast which soon supervenes can only be due to imperfect superposition of the various component systems. That the fact is so is at once proved by observation according to the method of Fizeau; for the spectrum from a slit at a very moderate distance out is seen to be traversed by bands. If the adjustment has been properly made, a certain region in the yellow-green is uninterrupted,

* [1901. It would seem that the required conditions are satisfied by a grating composed of equal transparent parts, giving alternately a relative retardation of $\frac{1}{2}\lambda$, and too fine to allow the formation of spectra of the second and higher orders.]

while the closeness of the bands increases towards either end of the spectrum. So far as regards the red and blue rays, the original bands may be considered to be already obliterated, but so far as regards the central rays, to be still fairly defined. Under these circumstances it is remarkable that so little colour should be apparent on direct inspection of the bands. It would seem that the eye is but little sensitive to colours thus presented, perhaps on account of its own want of achromatism.

It is interesting to observe the effect of coloured glasses upon the distinctness of the bands. If the achromatism be in the green, a red or orange glass, so far from acting as an aid to distinctness, obliterates all the bands after the first few. On the other hand, a green glass, absorbing rays for which the bands are already confused, confers additional sharpness. With the aid of a red glass a large number of bands are seen distinctly, if the adjustment be made for this part of the spectrum.

A still better procedure is to isolate a limited part of the spectrum by interposed screens. For this purpose a real spectrum must be formed, as in the case of the grating above considered.

We will now inquire to what degree of approximation λ/b may be made independent of λ with the aid of a prism, taking Cauchy's law of dispersion as a basis. According to it the value of b for any ray may be regarded as made up of two parts—one constant, and one varying inversely as λ^2. We therefore write

$$\frac{\lambda}{b}=\frac{\lambda^3}{A\lambda^2-B}, \quad\text{.............................(12)}$$

where A is to be so chosen that λ/b is stationary when λ has a prescribed value, λ_0. This condition gives

$$A\lambda_0^2=3B; \quad\text{.................................(13)}$$

so that

$$\frac{\lambda/b}{\lambda_0/b_0}=\frac{2}{3\lambda_0}\,\frac{\lambda^3}{\lambda^2-\frac{1}{3}\lambda_0^2}. \quad\text{........................(14)}$$

As an example, let us suppose that the disposition is achromatic for the immediate neighbourhood of the line D, so that $\lambda_0=\lambda_D$, and inquire into the proportional variation of λ/b, when we consider the ray C. Assuming

$$\lambda_D=\cdot 58890, \qquad \lambda_C=\cdot 65618,$$

we obtain from (14)

$$\frac{\lambda/b}{\lambda_0/b_0}=1\cdot 0155.$$

The meaning of this result will be best understood if we inquire for what order (n) the bands of the C-system are shifted relatively to those of the D-system through half the band-interval. From (1)

$$\delta u=nD\,\{\lambda/b-\lambda_0/b_0\}=\tfrac{1}{2}\lambda_0 D/b_0$$

by hypothesis; so that

$$n = \frac{\frac{1}{2}\lambda_0/b_0}{\lambda/b - \lambda_0/b_0}. \quad \text{.........................(15)}$$

Thus, in the case supposed, $n = 32$. After 32 periods the black places of the C-system will coincide with the bright places of the D-system, and conversely. If no prism had been employed (b constant), a similar condition of things would have arisen when

$$n = \frac{\frac{1}{2}\lambda_0}{\lambda - \lambda_0} = 4{\cdot}2.$$

If $(\lambda - \lambda_0)$, or, as we may call it, $\delta\lambda$, be small,

$$\frac{\lambda/b - \lambda_0/b_0}{\lambda_0/b_0}$$

is of the second order in $\delta\lambda$. An analytical expression is readily obtained from (14). We have

$$\frac{\lambda/b}{\lambda_0/b_0} = \frac{1 + 3\delta\lambda/\lambda_0 + 3(\delta\lambda/\lambda_0)^2 + (\delta\lambda/\lambda_0)^3}{1 + 3\delta\lambda/\lambda_0 + \frac{3}{2}(\delta\lambda/\lambda_0)^2}$$

$$= 1 + \frac{\frac{3}{2}(\delta\lambda/\lambda_0)^2 + (\delta\lambda/\lambda_0)^3}{1 + 3\delta\lambda/\lambda_0 + \frac{3}{2}(\delta\lambda/\lambda_0)^2}$$

$$= 1 + \tfrac{3}{2}(\delta\lambda/\lambda_0)^2 - \tfrac{7}{2}(\delta\lambda/\lambda_0)^3,$$

approximately; so that, by (15),

$$n = \tfrac{1}{3}\left(\frac{\lambda_0}{\delta\lambda}\right)^2 \left\{1 + \tfrac{7}{3}\frac{\delta\lambda}{\lambda_0} + \ldots\right\}. \quad \text{........................(16)}$$

This gives the order of the band at which complete discrepance first occurs for λ_0 and $\lambda_0 + \delta\lambda$, the adjustment being made for λ_0. It is, of course, inversely proportional to the *square* of $\delta\lambda$, when $\delta\lambda$ is small.

The corresponding value of n, if no prism be used, so that b is constant, is

$$n = \tfrac{1}{2}\frac{\delta\lambda}{\lambda_0}. \quad \text{.................................(17)}$$

The effect of the prism is thus to increase the number of bands in the ratio

$$2\lambda_0 : 3\delta\lambda.$$

Airy's Theory of the White Centre.

If a system of interference-bands be examined through a prism, the central white band undergoes an abnormal displacement, which has been supposed to be inconsistent with theory. The explanation has been shown by Airy* to depend upon the peculiar manner in which the white band is

* Airy, "Remarks on Mr Potter's Experiment on Interference," *Phil. Mag.* II. p. 161 (1833).

in general formed. Thus, "Any one of the kinds of homogeneous light composing the incident heterogeneous light will produce a series of bright and dark bars, unlimited in number so far as the mixture of light from the two pencils extends, and undistinguishable in quality. The consideration, therefore, of homogeneous light will never enable us to determine which is the point that the eye immediately turns to as the centre of the fringes. What, then, is the physical circumstance that determines the centre of the fringes?

"The answer is very easy. For different colours the bars have different breadths. If, then, the bars of all colours coincide at one part of the mixture of light, they will not coincide at any other part; but at equal distances on both sides from that place of coincidence they will be equally far from a state of coincidence. If, then, we can find where the bars of all colours coincide, that point is the centre of the fringes.

"It appears, then, that the centre of the fringes is *not* necessarily the point where the two pencils of light have described equal paths, but is determined by considerations of a perfectly different kind.... The distinction is important in this and other experiments."

The effect in question depends upon the dispersive power of the prism. If v be the linear shifting, due to the prism, of the originally central band, v must be regarded as a function of λ. Measured from the original centre, the position of the nth bar is now

$$v + n\lambda D/b.$$

The coincidence of the various bright bands occurs when this quantity is as independent as possible of λ; that is, when n is the nearest integer to

$$n = -\frac{b}{D}\frac{dv}{d\lambda}; \quad \text{..............................(18)}$$

or, as Airy expresses it, in terms of the width of a band (Λ),

$$n = -\,dv/d\Lambda. \quad \text{..............................(19)}$$

The apparent displacement of the white band is thus not v simply, but

$$v - \Lambda\, dv/d\Lambda. \quad \text{..............................(20)}$$

The signs of dv and $d\Lambda$ being opposite, the abnormal displacement is in addition to the normal effect of the prism. But, since $dv/d\Lambda$, or $dv/d\lambda$, is not constant, the achromatism of the white band is less perfect than when no prism is used.

If a grating were substituted for a prism, v would vary as Λ, and the displacement (20) would vanish.

More recently the matter has engaged the attention of Cornu*, who thus formulates the general principle:—"*Dans un système de franges d'interférence produites à l'aide d'une lumière hétérogène ayant un spectre continu, il existe toujours une frange achromatique qui joue le rôle de frange centrale et qui se trouve au point de champ où les radiations les plus intenses présentent une différence de phase maximum ou minimum.*"

In Fresnel's experiment, if the retardation of phase due to an interposed plate, or to any other cause, be $F(\lambda)$, the whole relative retardation of the two pencils at the point u is

$$\phi = F(\lambda) + \frac{bu}{\lambda D}, \qquad (21)$$

and the situation of the central, or achromatic, band is determined, not by $\phi = 0$, but by $d\phi/d\lambda = 0$, or

$$u = \lambda^2 DF'(\lambda)/b\dagger. \qquad (22)$$

It is scarcely necessary to say that although the nth band may be rendered achromatic, the *system* is no more achromatic than if the prism had been dispensed with. The width of the component systems being unaltered, the manner of overlapping remains as before. The present use of the prism is of course entirely different from that previously discussed, in which by a suitable adjustment the system of bands could be achromatized.

Thin Plates.

The series of tints obtained by nearly perpendicular reflexion from thin plates of varying thickness is the same as that which occurs in Lloyd's interference experiment, or at least it would be the same if the material of the plates were non-dispersive and the reflecting power small. If t be the thickness, μ the index, α' the inclination of the ray within the plate to the normal, the relative retardation of the two rays (reckoned as a distance) is $2\mu t \cos\alpha'$, and is sensibly independent of λ.

"This state of things may be greatly departed from when the thin plate is rarer than its surroundings, and the incidence is such that α' is nearly equal to 90°; for then, in consequence of the powerful dispersion, $\cos\alpha'$ may vary greatly as we pass from one colour to another. Under these circumstances the series of colours entirely alters its character, and the bands (corresponding to a graduated thickness) may even lose their coloration, becoming sensibly black and white through many alternations‡. The general explanation of this remarkable phenomenon was suggested by Newton, but

* *Journ. d. Physique*, I. p. 293 (1882).

† *Enc. Brit.*, "Wave Theory," XXIV. p. 425 [Vol. III. p. 62].

‡ Newton's *Optics*, Book II.; Fox Talbot, *Phil. Mag.* IX. p. 401 (1836).

it does not appear to have been followed out in accordance with the wave theory.

"Let us suppose that plane waves of white light travelling in glass are incident at angle α upon a plate of air, which is bounded again on the other side by glass. If μ be the index of the *glass*, α' the angle of refraction, then $\sin\alpha' = \mu \sin\alpha$; and the retardation, expressed by the equivalent distance in air, is

$$2t \sec\alpha' - \mu\, 2t \tan\alpha' \sin\alpha = 2t \cos\alpha';$$

and the retardation in *phase* is $2t\cos\alpha'/\lambda$, λ being as usual the wave-length in air.

"The first thing to be noticed is that, when α approaches the critical angle, $\cos\alpha'$ becomes as small as we please, and that, consequently, the retardation corresponding to a given thickness is very much less than at perpendicular incidence. Hence the glass surfaces need not be so close as usual.

"A second feature is the increased brilliancy of the light. But the peculiarity which most demands attention is the lessened influence of a variation in λ upon the phase retardation. A diminution of λ of itself increases the retardation of phase, but since waves of shorter wave-length are more refrangible, this effect may be more or less perfectly compensated by the greater obliquity, and consequent diminution in the value of $\cos\alpha'$. We will investigate the conditions under which the retardation of phase is stationary in spite of a variation of λ.

"In order that $\lambda^{-1}\cos\alpha'$ may be stationary, we must have

$$\lambda \sin\alpha'\, d\alpha' + \cos\alpha'\, d\lambda = 0,$$

where (α being constant)

$$\cos\alpha'\, d\alpha' = \sin\alpha\, d\mu.$$

Thus

$$\cot^2\alpha' = -\frac{\lambda}{\mu}\frac{d\mu}{d\lambda}, \quad \text{..........................(23)}$$

giving α' when the relation between μ and λ is known.

"According to Cauchy's formula, which represents the facts very well throughout most of the visible spectrum,

$$\mu = A + B\lambda^{-2}, \quad \text{..........................(24)}$$

so that

$$\cot^2\alpha' = \frac{2B}{\lambda^2\mu} = \frac{2(\mu - A)}{\mu} \quad \text{..........................(25)}$$

If we take, as for Chance's 'extra-dense flint,'

$$B = \cdot 984 \times 10^{-10},$$

and, as for the soda-lines,

$$\mu = 1{\cdot}65, \qquad \lambda = 5{\cdot}89 \times 10^{-5},$$

we get

$$\alpha' = 79^\circ\, 30'.$$

At this angle of refraction, and with this kind of glass, the retardation of phase is accordingly nearly independent of wave-length, and therefore the bands formed, as the thickness varies, are approximately achromatic."

Perfect achromatism would be possible only under a law of dispersion*

$$\mu = A' + B'\lambda^{-2}. \quad \text{..............................(26)}$$

The above investigation, as given in the *Enc. Brit.*, was intended to apply to Talbot's manner of experimenting, and it affords a satisfactory explanation of the formation of achromatic bands. In order to realize the nearly grazing incidence, the plate of air must be bounded on one side by a prism (Fig. 1).

Fig. 1.

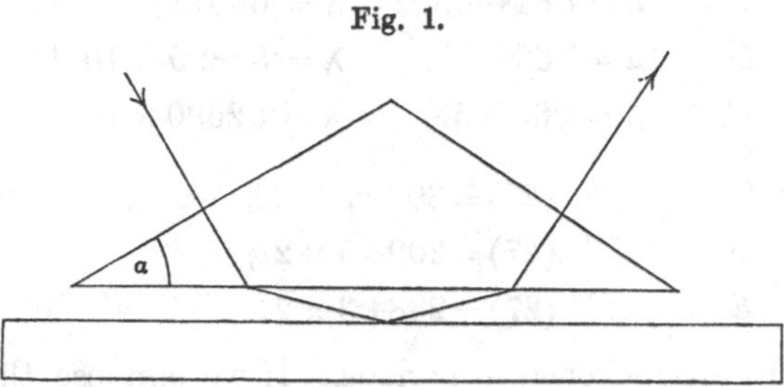

Upon this fall nearly parallel rays from a "radiant point of solar light," obtained with the aid of a lens of short focus. The bands may be observed upon a piece of ground glass held behind the prism in the reflected light, or they may be received directly upon an eyepiece.

These bands undoubtedly correspond to varying thicknesses of the plate of air, just as do the ordinary Newton's rings formed at nearly perpendicular incidence. For theoretical purposes we have the simplest conditions, if we suppose the thickness uniform, and that all the rays incident upon the plate are strictly parallel. Under these suppositions the field is uniform, the brightness for any kind of light depending upon the precise thickness in operation. If the thickness be imagined to increase gradually from zero, we are presented with a certain sequence of colours. When, however, the relation (23) is satisfied, the formation of colour is postponed, and the series commences with a number of alternations of black and white. In actual experiment it would be difficult to realize these conditions. If the surfaces

* [1901. The above formula was given in *Enc. Brit.* 1888, but at the time of publication of the present paper it was thought to be erroneous. The correctness of the original version was pointed out by Mr Preston.]

bounding the plate are inclined to one another, the various parts of the field correspond to different thicknesses; and, at any rate if the inclination be small, there is presented at one view a series of colours, constituting bands, the same as could only be seen in succession were the parallelism maintained rigorously.

The achromatism secured by (23) not being absolute, it is of interest to inquire what number of bands are to be expected. The relative retardation of phase, with which we have to deal, is $2t\cos\alpha'/\lambda$, or

$$\frac{2t\sqrt{(1-\mu^2\sin^2\alpha)}}{\lambda}. \qquad (27)$$

If this be stationary for extra-dense glass and for the line D, we have, as already mentioned, $\alpha' = 79°\,30'$, and corresponding thereto $\alpha = 36°\,34'$. Taking this as a prescribed value of α, we may compare the values of (27) for the lines C, D, E, using the data given by Hopkinson*, viz.:—

$$C, \quad \mu = 1{\cdot}644866, \quad \lambda = {\cdot}65618 \times 10^{-4},$$
$$D, \quad \mu = 1{\cdot}650388, \quad \lambda = {\cdot}58890 \times 10^{-4},$$
$$E, \quad \mu = 1{\cdot}657653, \quad \lambda = {\cdot}52690 \times 10^{-4}.$$

We find

$$\text{for } C \quad (27) = 3036{\cdot}9 \times 2t,$$
$$D \quad (27) = 3094{\cdot}5 \times 2t,$$
$$E \quad (27) = 2984{\cdot}3 \times 2t.$$

These retardations are reckoned in periods. If we suppose that the retardation for the C-system is just half a period less than for the D-system, we have $57{\cdot}6 \times 2t = \frac{1}{2}$; so that $t = \frac{1}{230}$ centim. Thus about 27 periods of the D-bands correspond to $26\frac{1}{2}$ of the C-bands.

If the range of refrangibility contemplated be small, the calculation may conveniently be conducted algebraically. According to Cauchy's law we may replace (27) by

$$\frac{2t\sqrt{(1-\mu^2\sin^2\alpha)(\mu-A)}}{\sqrt{B}}. \qquad (28)$$

Setting $\mu = \mu_0 + \delta\mu$, we have approximately

$$(1-\mu^2\sin^2\alpha)(\mu-A) = (1-\mu_0^2\sin^2\alpha)(\mu_0-A)$$
$$+\,\delta\mu\,\{(1-\mu_0^2\sin^2\alpha - 2\mu_0\sin^2\alpha)(\mu_0-A)\}$$
$$-\,(\delta\mu)^2\,\{3\mu_0 - A\}\sin^2\alpha + \dots.$$

If α be so chosen that the value of (28) is stationary for μ_0, the term of the first order in $\delta\mu$ vanishes, and we obtain finally as the approximate value of (28)

$$\frac{2t\sin\alpha\,(\mu_0-A)\sqrt{(2\mu_0)}}{\sqrt{B}}\cdot\left\{1-\frac{(3\mu_0-A)(\delta\mu)^2}{4\mu_0(\mu_0-A)^2}\right\}. \qquad (29)$$

* *Proc. Roy. Soc.* June 1877.

If now the circumstances be such that n periods of the μ_0 system correspond to $n - \frac{1}{2}$ of the μ system,

$$\frac{1}{n} = \frac{(3\mu_0 - A)(\delta\mu)^2}{2\mu_0(\mu_0 - A)^2}, \qquad \text{.........................(30)}$$

in which the ratio of $(3\mu_0 - A)$ to $2\mu_0$ does not differ much from unity. In the application to extra-dense flint the simplified formula

$$n = (\mu_0 - A)^2/(\mu - \mu_0)^2 \qquad \text{.......................(31)}$$

gives very nearly the same result as that previously found. The number of bands which approximately coincide is inversely as the square of the range of refrangibility included.

It must not be overlooked that the preceding investigation, though satisfactory so far as it goes, is somewhat special on account of the assumption that the angle of incidence (α) upon the plate of air is the same for the various colours. If the rays are parallel before they fall upon the prism, they cannot remain parallel unless the incidence upon the first surface be perpendicular. There is no reason why this should not be the case; but it is tantamount to a restriction upon the angle of the prism, since α is determined by the achromatic condition. If the angle of the prism be other than α, the required condition will be influenced by the separation of the colours upon first entering the glass. Although the general character of the phenomenon is not changed, it may be well to give the calculation applicable to all angles of prism, as was first done by M. Mascart.

Denoting, as before, by α, α' the angles of incidence and refraction upon the plate of air, let β', β be the angles of incidence and refraction at the first surface of the prism (Fig. 2), whose angle is A. Then, if Δ, equal to $n\lambda$, be the retardation,

$$\Delta = n\lambda = 2t \cos \alpha', \qquad \text{.........................(32)}$$

as before; while the relations among the angular quantities are:—

$$\sin \alpha' = \mu \sin \alpha, \qquad \alpha + \beta = A, \qquad \sin \beta' = \mu \sin \beta. \text{...(33, 34, 35)}$$

Fig. 2.

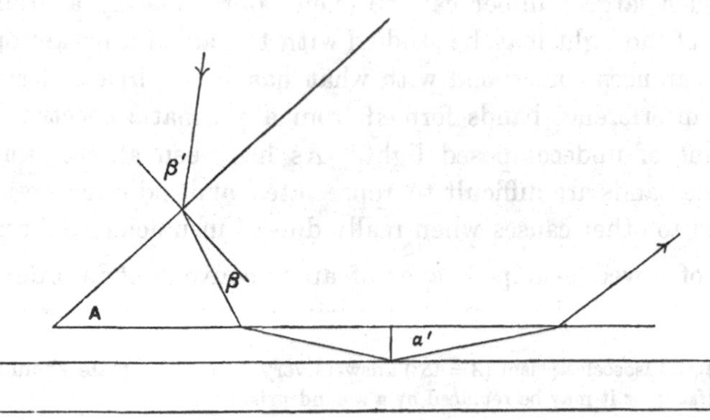

We have now to inquire under what conditions Δ/λ, or n, will be stationary, in spite of a variation of λ, if β' be constant. Thus

$$\lambda \sin \alpha' d\alpha' + \cos \alpha' d\lambda = 0,$$

while

$$\cos \alpha' d\alpha' = d\mu \sin \alpha + \mu \cos \alpha \, d\alpha,$$

$$0 = d\alpha + d\beta,$$

$$0 = d\mu \sin \beta + \mu \cos \beta \, d\beta.$$

Accordingly,

$$-\frac{\cot \alpha' d\lambda}{\lambda} \cos \alpha' = d\mu \sin \alpha + \mu \cos \alpha \, d\alpha$$

$$= d\mu \sin \alpha + \cos \alpha \tan \beta \, d\mu = \sin A \, d\mu/\cos B\,;$$

so that

$$\cot^2 \alpha' = -\frac{\lambda d\mu}{\mu d\lambda} \frac{\sin A}{\sin \alpha \cos \beta} \quad \text{.............................(36)}$$

is the condition that n should be stationary. In the more particular case considered above, $\beta' = 0$, $\beta = 0$, $\alpha = A$.

These bands, which I should have been inclined to designate after Talbot, were it not that his name is already connected with another very remarkable system of bands, are readily observed. For the "radiant point of solar light" we may substitute, if more convenient, that of the electric arc. A small hole in a piece of metal held close to the arc allows sufficient light to pass if the bands are observed without the intervention of a diffusing-screen. At a distance of say 20 feet the nearly parallel rays fall upon the prism* and plate, which should be mounted in such a fashion that the pressure may be varied, and that the whole may be readily turned in azimuth. The coloured bands are best seen when the surfaces are nearly parallel and pretty close. It is best to commence observations under these conditions. When the achromatic azimuth has been found, the interval may be increased. If it is desired to see a large number of bands, a strip of paper may be interposed between the surfaces along one edge, so as to form a plate of graduated thickness. Talbot speaks of from 100 to 200 achromatic bands; but I do not think any such large number can be even approximately achromatic. The composition of the light may be studied with the aid of a pocket spectroscope, and the appearances correspond with what has been already described under the head of interference-bands formed from a prismatic spectrum in place of the usual line of undecomposed light. As has been already remarked, the colours of fine bands are difficult to appreciate; and indistinctness is liable to be attributed to other causes when really due to insufficient achromatism.

The use of a wedge-shaped layer of air is convenient in order to obtain a simultaneous view of a large number of bands; but it must not be overlooked

* A right-angled isosceles prism ($A = 45°$) answers very well. The plate should be blackened at the hind surface; or it may be replaced by a second prism.

that it involves some departure from theoretical simplicity. The proper development of the light due to any thickness requires repeated reflexions to and fro within the layer, and at a high degree of obliquity this process occupies a considerable width. If the band-interval be too small, complications necessarily ensue, which are probably connected with the fact that the appearance of the bands changes somewhat according to the distance from the reflecting combination at which they are observed.

Herschel's Bands.

In the system of bands above discussed, substantially identical (I believe) with those observed by Talbot, all the rays of a given colour are refracted under constant angles, the variable element being the thickness of the plate of air. A system in many respects quite distinct was described by W. Herschel, and has recently been discussed by M. Mascart*. In this case the combination of prism and plate remains as before, but the thickness of the film of air is considered to be constant, the alternations constituting the bands being dependent upon the varying angles at which the light (even though of given colour) is refracted. In order to see these bands all that is necessary is to view a source of light presenting a large angle, such as the sky, by reflexion in the layer of air. They are formed a little beyond the limit of total reflexion. They are broad and richly coloured if the layer of air be thin, but as the thickness increases they become finer, and the colour is less evident.

The theoretical condition of constant thickness is better satisfied if (after Mascart) we place the layer of air in the focus of a small radiant point (*e.g.* the electric arc) as formed by an achromatic lens of wide angle. In this case the area concerned may be made so small that the thickness in operation can scarcely vary, and the ideal Herschel's bands are seen depicted upon a screen held in the path of the reflected light. It will of course be understood that bands may be observed of an intermediate character, in the formation of which both thickness and incidence vary. Herschel's observations relate to one particular case—that of constant thickness; Talbot's to the other especially simple case of constant angle of incidence.

From our present point of view there is, however, one very important distinction between the two systems of bands. The one system is achromatic, and the other is not. In order to understand this, it is necessary to follow in greater detail the theory of Herschel's bands.

We will commence by supposing that the light is homogeneous (λ constant), and inquire into the law of formation of the bands, t being given. The same equations, (32) &c., apply as before, and also Fig. 2, if we suppose

* *Loc. cit.*; also *Traité d'Optique*, tom. I. Paris, 1889.

the course of the rays reversed, so that the direction of the *emergent* ray is determined by β'. The question to be investigated is the relation of β' to n, and to the other data of the experiment.

The band of zero order ($n = 0$) occurs when $\alpha' = 90°$, that is at the limit of total reflexion. The corresponding values of α, β, and β' may be determined in succession from (33), (34), (35). The value of α' for the nth band is given immediately by (32). For the width of the band, corresponding to the change of n into $n+1$, we have

$$\lambda = -2t \sin \alpha' d\alpha',$$

and from the other equations,

$$\cos \alpha' d\alpha' = \mu \cos \alpha \, d\alpha,$$

$$d\alpha + d\beta = 0,$$

$$\cos \beta' d\beta' = \mu \cos \beta \, d\beta;$$

so that the apparent width of the nth band is given by

$$d\beta' = \frac{n\lambda^2}{4t^2} \frac{\cos \beta}{\cos \beta' \cos \alpha \sin \alpha'}. \quad \text{......................(37)}$$

In the neighbourhood of the limit of total reflexion $\sin \alpha'$ is nearly equal to unity, and the factors $\cos \beta$, $\cos \beta'$, $\cos \alpha$ vary but slowly with the order of the band and also with the wave-length. Hence the width of the nth band is approximately proportional to the order, to the square of the wave-length, and to the inverse square of the thickness.

This series of bands, commencing at the limit of total reflexion, and gradually increasing in width, are easily observed with Herschel's apparatus by the aid of a soda-flame. In order to increase the field of view, the flame may be focused upon the layer of air by a wide-angled lens. The eye should be adjusted for distant objects, and the thickness of the layer should be as uniform as possible. For the latter purpose the glass surfaces may be pressed against two strips of rather thin paper, interposed along two opposite edges.

We have now to consider what happens when the source of light is white. According to Airy's principle the centre of the system is to be found where there is coincidence of bands of order n, in spite of a variation of λ. This is precisely the question already dealt with in connexion with the other system of bands, and the answer is embodied in (36). About the achromatic centre thus determined will the visible bands be grouped.

And now the question arises, Are these bands achromatic? Certainly not. M. Mascart, to whom is due equation (37), appears to me to misinterpret it when he concludes that the bands are approximately achromatic*. At the central band n is the same for the various colours. Consequently the widths of the various systems *at this place* are approximately proportional to λ^2. It will be seen that, so far from the system being achromatic, it is much less so than the ordinary system of interference-bands, or of Newton's rings, in which the width is proportional to the *first* power of λ. And this theoretical conclusion appears to me to be in harmony with observation.

At first sight it may appear strange that an achromatic centre should be possible with bands proportional to λ^2. The explanation depends upon the fact that the limit of total reflexion, where the bands commence, is itself a function of λ.

The apparent width of the visible bands depends upon t, but is not, as might erroneously be supposed, proportional to t^{-2}. For this purpose n in (37) must be regarded as a function of t. In fact, by (32), if α' be given, n varies as t/λ; so that, in estimating the influence of t, other circumstances remaining unaltered, the width is proportional to t^{-1}. Hence, as the interval between the surfaces increases, the bands become finer, but the centre does not shift, nor is there any change in their number as limited by the advent of chromatic confusion.

Effect of a Prism upon Newton's Rings.

If Newton's rings are examined through a prism, some very remarkable phenomena are exhibited, described in his 24th observation†.

"When the two object-glasses were laid upon one another, so as to make the rings of the colours appear, though with my naked eye I could not discern above 8 or 9 of these rings, yet by viewing them through a prism I have seen a far greater multitude, insomuch that I could number more than 40, besides many others which were so very small and close together that I could not keep my eye steady on them severally so as to number them, but by their extent I have sometimes estimated them to be more than a hundred. And I believe the experiment may be improved to the discovery of far greater numbers; for they seem to be really unlimited, though visible only so far as they can be separated by the refraction, as I shall hereafter explain.

"But it was but one side of these rings—namely, that towards which the refraction was made—which by that refraction was rendered distinct; and the other side became more confused than when viewed by the naked eye,

* *Traité d'Optique*, t. I. p. 451. "On s'explique ainsi que la largeur apparente des franges voisines de la frange achromatique soit à peu près indépendante de la longueur d'onde dans une ouverture angulaire notable et qu'on en distingue un grand nombre."

† *Opticks.* See also Place, *Pogg. Ann.* CXIV. p. 504 (1861).

insomuch that there I could not discern above 1 or 2, and sometimes none of those rings, of which I could discern 8 or 9 with my naked eye. And their segments or arcs, which on the other side appeared so numerous, for the most part exceeded not the third part of a circle. If the refraction was very great, or the prism very distant from the object-glasses, the middle part of those arcs became also confused, so as to disappear and constitute an even whiteness, while on either side their ends, as also the whole arcs furthest from the centre, became distincter than before, appearing in the form as you see them designed in the fifth figure [Fig. 3]."

Fig. 3.

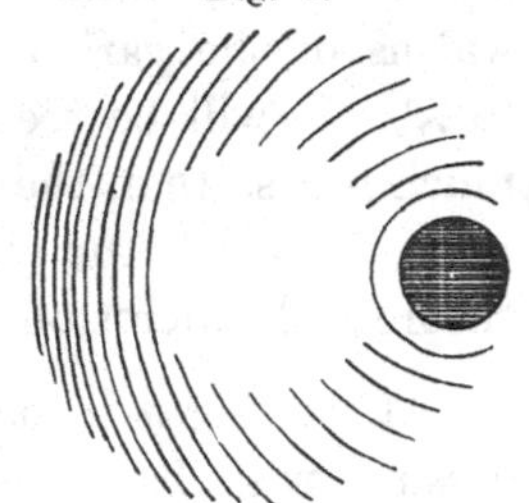

"The arcs, where they seemed distinctest, were only black and white successively, without any other colours intermixed. But in other places there appeared colours, whose order was inverted by the refraction in such manner that if I first held the prism very near the object-glasses, and then gradually removed it further off towards my eye, the colours of the 2nd, 3rd, 4th, and following rings shrunk towards the white that emerged between them, until they wholly vanished into it at the middle of the arc, and afterwards emerged again in a contrary order. But at the ends of the arcs they retained their order unchanged."

"I have sometimes so laid one object-glass upon the other, that to the naked eye they have all over seemed uniformly white, without the least appearance of any of the coloured rings; and yet, by viewing them through a prism, great multitudes of these rings have discovered themselves. And in like manner, plates of Muscovy glass, and bubbles of glass blown at a lamp-furnace, which were not so thin as to exhibit any colours to the naked eye, have through the prism exhibited a great variety of them ranged irregularly up and down in the form of waves. And so bubbles of water, before they began to exhibit their colours to the naked eye of a bystander, have appeared through a prism, girded about with many parallel and horizontal rings; to produce which effect it was necessary to hold the prism parallel, or very nearly parallel, to the horizon, and to dispose it so that the rays might be refracted upwards."

Newton was evidently much struck with these "so odd circumstances," and he explains the occurrence of the rings at unusual thicknesses as due to the dispersing power of the prism. The blue system being more refracted than the red, it is possible, under certain conditions, that the nth blue ring may be so much displaced relatively to the corresponding red ring as *at one part of the circumference* to compensate for the different diameters. White and black stripes may thus be formed in a situation where, without the prism, the mixture of colours would be complete, so far as could be judged by the eye.

The simplest case that can be considered is when the "thin plate" is bounded by plane surfaces inclined to one another at a small angle. Without the prism, the various systems coincide at the bar of zero order. The width of the bands is constant for each system, and in passing from one system to another is proportional to λ. Regarded through a prism of small angle whose refracting edge is parallel to the intersection of the bounding surfaces of the plate, the various systems no longer coincide for zero order; but by drawing back the prism, it will always be possible so to adjust the effective dispersing power as to bring the nth bars to coincidence for any two assigned colours, and therefore approximately for the entire spectrum. The formation of the achromatic band, or rather central black bar, depends indeed upon precisely the same principles as the fictitious shifting of the centre of a system of Fresnel's bands when viewed through a prism.

In this example the formation of visible rings at unusual thicknesses is easily understood; but it gives no explanation of the increased numbers observed by Newton. The width of the bands for any colour is proportional to λ, as well after the displacement by the prism as before. The manner of overlapping of two systems whose nth bars have been brought to coincidence is unaltered; so that the succession of colours in white light, and the number of perceptible bands, is much as usual.

In order that there may be an *achromatic system* of bands, it is necessary that the width of the bands near the centre be the same for the various colours. As we have seen, this condition cannot be satisfied when the plate is a true wedge; for then the width for each colour is proportional to λ. If, however, the surfaces bounding the plate be *curved*, the width for each colour varies at different parts of the plate, and it is possible that the blue bands from one part, when seen through the prism, may fit the red bands from another part of the plate. Of course, when no prism is used, the sequence of colours is the same whether the boundaries of the plate be straight or curved.

For simplicity we will first suppose that the surfaces are still cylindrical, so that the thickness is a function of but one coordinate x, measured in the direction of refraction. If we choose the point of nearest approach as the origin of x, the thickness may be taken to be

$$t = a + bx^2, \qquad (38)$$

a being thus the least distance between the surfaces. The black of the nth order for wave-length λ occurs when

$$\tfrac{1}{2}n\lambda = a + bx^2; \qquad (39)$$

so that the width (δx) of the band at this place (x) is given by

$$\tfrac{1}{2}\lambda = 2bx\,\delta x,$$

or

$$\delta x = \lambda/4bx. \qquad (40)$$

Substituting for x from (38), we obtain, as the width of the band of nth order for any colour,

$$\delta x = \frac{\lambda}{4\sqrt{b}\,.\,\sqrt{(\frac{1}{2}n\lambda - a)}} \quad \text{.....................(41)}$$

It will be seen that, while at a *given part of the plate* the width is proportional to λ, the width *for the* n*th order* is a different function dependent upon a. It is with the latter that we are concerned when, by means of the prism, the nth bars have been brought to coincidence.

If the glasses be in contact, as is usually supposed in the theory of Newton's rings, $a = 0$; and therefore, by (41), $\delta x \propto \lambda^{\frac{1}{2}}$, or the width of the band of the nth order varies as the square root of the wave-length, instead of as the first power. Even in this case the overlapping and subsequent obliteration of the bands is much retarded by the use of the prism; but the full development of the phenomenon demands that a should be finite. Let us inquire what is the condition in order that the width of the band of the nth order may be stationary, as λ varies. By (41) it is necessary that the variation of $\lambda^2/(\frac{1}{2}n\lambda - a)$ should vanish. Hence

$$2\lambda(\tfrac{1}{2}n\lambda - a) - \tfrac{1}{2}n\lambda^2 = 0,$$

or

$$a = \tfrac{1}{4}n\lambda. \quad \text{.................................(42)}$$

The thickness of the plate where the nth band for λ is formed being $\frac{1}{2}n\lambda$, equation (42) may be taken as signifying that the thickness must be half due to curvature and half to imperfect contact at the place of nearest approach. If this condition be satisfied, the achromatism of the nth band, effected by the prism, carries with it the achromatism of a large number of neighbouring bands*.

We will return presently to the consideration of the spherically curved glasses used by Newton, and to the explanation of some of the phenomena which he observed; but in the meantime it will be convenient to state the theory of straight bands in a more analytical form.

Analytical Statement.

If the coordinate ξ represent the situation of the nth band, of wave-length λ, then, in any case of straight bands, ξ may be regarded as a function of n and λ, or, conversely, n (not necessarily integral) may be regarded as a function of ξ and λ. If we write

$$n = \phi(\xi, \lambda), \quad \text{.............................(43)}$$

* *Enc. Brit.*, "Wave Theory," xxiv. p. 428 (1888). [Vol. III. p. 72.]

and expand by Taylor's theorem,

$$n - n_0 = \frac{d\phi}{d\xi_0}\delta\xi + \frac{d\phi}{d\lambda_0}\delta\lambda + \tfrac{1}{2}\frac{d^2\phi}{d\xi_0^2}(\delta\xi)^2 + \frac{d^2\phi}{d\xi_0 d\lambda_0}\delta\xi\,\delta\lambda + \tfrac{1}{2}\frac{d^2\phi}{d\lambda_0^2}(\delta\lambda)^2 + \ldots, \quad (44)$$

where

$$n_0 = \phi(\xi_0, \lambda_0). \quad \ldots\ldots(45)$$

The condition for an achromatic band at ξ_0, λ_0 is

$$\frac{d\phi}{d\lambda_0} = 0; \quad \ldots\ldots(46)$$

and, further, the condition for an achromatic system at this place is

$$\frac{d^2\phi}{d\xi_0 d\lambda_0} = 0. \quad \ldots\ldots(47)$$

If these conditions are both satisfied, n becomes very approximately a function of ξ only throughout the region in question.

In several cases considered in the present paper, the functional relation is such that

$$n = \xi \,.\, \psi(\lambda), \quad \ldots\ldots(48)$$

$\psi(\lambda)$ denoting a function of λ only. The expansion may then be written

$$n - n_0 = \xi\,\{\psi(\lambda_0) + \psi'(\lambda_0)\,\delta\lambda + \tfrac{1}{2}\psi''(\lambda_0)(\delta\lambda)^2 + \ldots\}. \quad \ldots\ldots(49)$$

The line $\xi = 0$ is here of necessity perfectly achromatic. If there be an achromatic system,

$$\psi'(\lambda_0) = 0;$$

and when this condition is satisfied, the whole field is achromatic, so long as $(\delta\lambda)^2$ can be neglected.

If the width of the bands be a function of λ only, n is of the form

$$n = \xi \,.\, \psi(\lambda) + \chi(\lambda), \quad \ldots\ldots(50)$$

more general than that just considered (48), though of course less general than (43). The condition for an achromatic line is

$$\frac{dn}{d\lambda} = \xi\psi'(\lambda_0) + \chi'(\lambda_0) = 0, \quad \ldots\ldots(51)$$

and for an achromatic system,

$$\frac{d^2n}{d\xi\,d\lambda} = \psi'(\lambda_0) = 0; \quad \ldots\ldots(52)$$

so that, for an achromatic system, ψ' and χ' must both vanish.

Curved Interference-Bands.

If the bands are not straight, n must be regarded as a function of a second coordinate η, as well as of ξ and λ. In the equation

$$n = \phi(\xi, \eta, \lambda), \quad \text{.............................(53)}$$

if we ascribe a constant value to λ, we have the relation between ξ, η corresponding to any prescribed values of n—that is, the forms of the interference-bands in homogeneous light. If the light be white, the bands are in general confused; but those points are achromatic for which

$$\frac{dn}{d\lambda_0} = 0. \quad \text{................................(54)}$$

This is a relation between ξ and η defining a curve, which we may call the achromatic curve, corresponding in some respects to the achromatic line of former investigations, where n is independent of η. There is, however, a distinction of some importance. When n is a function of ξ and λ only, the achromatic line is also an achromatic *band*; that is, n remains constant as we proceed along it. But under the present less restricted conditions n is not constant along (54). The achromatic curve is not an achromatic band; and, indeed, achromatic bands do not exist in the same development as before. They must be regarded as infinitely short, following the lines $n =$ constant, but existent only at the intersection of these with (54). Practically a small strip surrounding (54) may be regarded as an achromatic region in which are visible short achromatic bands, crossing the strip at an angle dependent upon the precise circumstances of the case.

The application of this theory to the observations of Newton presents no difficulty. The thickness of the layer of air at the point x, y, measured from the place of closest approach, is

$$t = a + b(x^2 + y^2); \quad \text{..........................(55)}$$

and since $t = \frac{1}{2}n\lambda$, the relation of n to x, y, and λ is

$$\tfrac{1}{2}n = a\lambda^{-1} + b\lambda^{-1}(x^2 + y^2). \quad \text{.......................(56)}$$

This equation defines the system of bands when the combination is viewed directly. The achromatic curve determined by (54) is

$$a + b(x^2 + y^2) = 0,$$

and is wholly imaginary if a and b are both positive and finite. Only when $a = 0$, that is when the glasses touch, is there an achromatic point $x = 0$, $y = 0$.

When a prism is brought into operation, we may suppose that each homogeneous system is shifted as a whole parallel to x by an amount variable from one homogeneous system to another. If the apparent coordinates be ξ, η, we may write

$$\xi = x - f(\lambda), \qquad \eta = y. \quad \text{......................(57)}$$

Using these in (56), we obtain as the characteristic equation of the rings viewed through a prism,

$$n = \frac{a + b\{\xi + f(\lambda)\}^2 + b\eta^2}{2\lambda}. \quad \text{......................(58)}$$

The equation of the achromatic curve is then, by (54),

$$\{\xi + f(\lambda_0) - \lambda_0 f'(\lambda_0)\}^2 + \eta^2 = \lambda_0^2 \{f'(\lambda_0)\}^2 - a/b, \quad \text{........(59)}$$

which represents a *circle*, whose centre is situated upon the axis of ξ.

If the glasses are in contact ($a = 0$), the locus is certainly real, and passes through the point

$$\xi + f(\lambda_0) = 0, \qquad \eta = 0;$$

that is, the image with rays of wave-length λ_0 of the point of contact ($x = 0$, $y = 0$). The radius of the circle is $\lambda_0 f'(\lambda_0)$, and increases with the dispersive power of the prism. The other point where the circle meets the axis,

$$x = 2\lambda_0 f'(\lambda_0), \qquad y = 0,$$

marks the place where the bands, being parallel to the achromatic curve, attain a special development. It is that which we should have found by an investigation in which the curvature of the band-systems is ignored.

If a be supposed to increase from zero, other conditions remaining unaltered, the radius of the achromatic circle decreases, while the centre maintains its position. The two places where the circle crosses the axis are thus upon the same side of the image of $x = 0$, $y = 0$. When a is such that

$$a/b = \lambda_0^2 \{f'(\lambda_0)\}^2, \quad \text{..........................(60)}$$

the circle shrinks into a point, whose situation is defined by

$$x = \xi + f(\lambda_0) = \lambda_0 f'(\lambda_0), \qquad y = \eta = 0. \quad \text{..............(61)}$$

Since there are two coincident achromatic points upon the axis, the condition is satisfied for an achromatic *system*. By (60), (61),

$$a/b = x^2,$$

so that

$$t = a + bx^2 = 2a. \quad \text{..............................(62)}$$

This is the same result as was found before (42) by the simpler treatment of the question in which points along the axis were alone considered.

If a exceed the value specified in (60), the achromatic curve becomes wholly imaginary*.

* Compare Mascart, *Traité d'Optique*, t. I. p. 435.

164.

ON BELLS*.

[*Philosophical Magazine,* XXIX. pp. 1—17, 1890.]

THE theory of the vibrations of bells is of considerable difficulty. Even when the thickness of the shell may be treated as very small, as in the case of air-pump receivers, finger-bowls, claret glasses, &c., the question has given rise to a difference of opinion. The more difficult problem presented by church bells, where the thickness of the metal in the region of the *sound-bow* (where the clapper strikes) is by no means small, has not yet been attacked. A complete theoretical investigation is indeed scarcely to be hoped for; but one of the principal objects of the present paper is to report the results of an experimental examination of several church bells, in the course of which some curious facts have disclosed themselves.

In practice bells are designed to be symmetrical about an axis, and we shall accordingly suppose that the figures are of revolution, or at least differ but little from such. Under these circumstances the possible vibrations divide themselves into classes, according to the number of times the motion repeats itself round the circumference. In the gravest mode, where the originally circular boundary becomes elliptical, the motion is once repeated, that is it occurs *twice.* The number of nodal meridians, determined by the points where the circle intersects the ellipse, is *four*, the meridians corresponding (for example) to longitudes 0° and 180° being reckoned separately. In like manner we may have 6, 8, 10... nodal meridians, corresponding to 3, 4, 5... cycles of motion. A class of vibrations is also possible which are symmetrical about the axis, the motion at any point being either in or perpendicular to the meridional plane. But these are of no acoustical importance.

* [1901. Some of the results of this investigation had been communicated to the British Association. (See *Report* for 1889, p. 491.)]

The meaning here attached to the word *nodal* must be carefully observed. The meridians are not nodal in the sense that there is no motion, but only that there is no motion *normal to the surface.* This can be best illustrated by the simplest case, that of an infinitely long thin circular cylinder vibrating in two dimensions*. The graver vibrations are here purely flexural, the circumference remaining everywhere unstretched during the motion. If we fix our attention upon one mode of vibration of n cycles, the motion at the surface is usually both radial and tangential. There are, however, $2n$ points distributed at equal intervals where the motion is purely tangential, and other $2n$ points, bisecting the intervals of the former, where the motion is purely radial. There are thus no places of complete rest; but the first set of points, or the lines through them parallel to the axis, are called nodal, in the sense that there is at these places no normal motion.

The two systems of points have important relations to the place where the vibrations are excited. "When a bell-shaped body is sounded by a blow, the point of application of the blow is a place of maximum normal motion of the resulting vibrations, and the same is true when the vibrations are excited by a violin-bow, as generally in lecture-room experiments. Bells of glass, such as finger-glasses, are, however, more easily thrown into regular vibration by friction with the wetted finger carried round the circumference. The pitch of the resulting sound is the same as that elicited by a tap with the soft part of the finger; but inasmuch as the tangential motion of a vibrating bell has been very generally ignored, the production of sound in this manner has been felt as a difficulty. It is now scarcely necessary to point out that the effect of the friction is in the first instance to excite tangential motion, and that the point of application of the friction is the place where the tangential motion is greatest, and therefore where the normal motion vanishes†."

When the symmetry is complete, the system of nodal meridians has no fixed position, and may adapt itself so as to suit the place at which a normal blow is delivered. If the point of application of the blow be conceived to travel round a circle symmetrical with respect to the axis (say, for brevity, a circle of *latitude*) the displacement will make no difference to the vibration considered as a whole, but the effect upon an observer who retains a fixed position will vary. If the bell be situated in an open space, or if the ear of the observer be so close that reflexions are relatively unimportant, the sound disappears as nodes pass by him, swelling to a maximum when the part nearest to the ear is one of the places of maximum normal motion, which for brevity we will call *loops.* In listening to a particular note it would thus be

* *Theory of Sound,* § 232.

† *Theory of Sound,* § 234. **That the rubbing finger and the violin-bow must be applied at different points in order to obtain the same vibration was known to Chladni.**

possible to determine the number of nodal meridians by watching the variations of intensity which occur as the place of the blow travels round a circle of latitude.

In practice the symmetry is seldom so complete that this account of the matter is sufficient. Theoretically the slightest departure from symmetry will in general render determinate the positions of the nodal systems. For each number n of cycles, there is one determinate mode of vibration with $2n$ nodes and $2n$ intermediate loops, and a second determinate mode in which the nodes and loops of the first mode exchange functions. Moreover the frequencies of the vibrations in the two modes are slightly different.

In accordance with the general theory, the vibrations of the two modes, as dependent upon the situation and magnitude of the initiating blow, are to be considered separately. The vibrations of the first mode will be excited, unless the blow occur at a node of this system; and in various degrees, reaching a maximum when the blow is delivered at a loop. The intensity, as appreciated by an observer, depends also upon the position of his ear, and will be greatest when a loop is immediately opposite. As regards the vibrations of the second mode, they reach a maximum when those of the first mode disappear, and conversely.

Thus in the case of n cycles, there are $2n$ places where the first vibration is not excited and $2n$ places, midway between the former, where the second vibration is not excited. At all $4n$ places the resulting sound is free from beats. In all other cases both kinds of vibration are excited, and the sound will be affected by beats. But the prominence of the beats depends upon more than one circumstance. The intensities of the two vibrations will be equal when the place of the blow is midway between those which give no beats. But it does not follow that the audible beats are then most distinct. The condition to be satisfied is that the intensities shall be equal *as they reach the ear*, and this will depend upon the situation of the observer as well as upon the vigour of the vibrations themselves. Indeed, by suitably choosing the place of observation it would be theoretically possible to obtain beats with perfect silences, wherever (in relation to the nodal systems) the blow may be delivered.

There will now be no difficulty in understanding the procedure adopted in order to fix the number of cycles corresponding to a given tone. If, in consequence of a near approach to symmetry, beats are not audible, they are introduced by suitably loading the vibrating body. By tapping cautiously round a circle of latitude the places are then investigated where the beats disappear. But here a decision must not be made too hastily. The inaudibility of the beats may be favoured by an unsuitable position of the ear, or of the mouth of the resonator in connexion with the ear. By travelling round, a situation is soon found where the observation can be made with the best advantage. In the neighbourhood of the place where the blow is being tried there is a loop of the vibration which is most excited and a (coincident) node of the vibration which is least excited. When the ear is opposite to a node of the first vibration, and therefore to a loop of the second, the original inequality is redressed, and distinct beats may be heard even although the deviation of the blow from a nodal point may be very small. The accurate determination in this way of two consecutive places where no beats are generated is all that is absolutely necessary. The ratio of the entire circumference of the circle of latitude to the arc between the points represents $4n$, that is four times the number of cycles. Thus, if the arc between consecutive points proved to be 45°, we should infer that we are dealing with a vibration of two cycles—the one in which the deformation is elliptical. As a greater security against error, it is advisable in practice to determine a larger number of points where no beats occur. Unless the deviation from symmetry be considerable, these points should be uniformly distributed along the circle of latitude*.

In the above process for determining nodes we are supposed to hear distinctly the tone corresponding to the vibration under investigation. For this purpose the beats are of assistance in directing the attention; but with the more difficult subjects, such as church bells, it is advisable to have recourse to resonators. A set of Helmholtz's pattern, manufactured by Koenig, are very convenient. The one next higher in pitch to the tone under examination is chosen and tuned by advancing the finger across the aperture. Without the security afforded by resonators, the determination of the octave is in my experience very uncertain. Thus pure tones are often estimated by musicians an octave too low.

Some years ago I made observations upon the tones of various glass bells, of which the walls were tolerably thin. A few examples may be given:—

I. *c′*, *e″*♭, *c‴*♯.

II. *a*, *c″*♯, *b″*.

III. *f′*♯, *b″*.

The value of n for the gravest tone is 2, for the second 3, and for the third 4. On account of the irregular shape and thickness only a very rough comparison with theory is possible; but it may be worth mention that for a thin uniform hemispherical bell the frequencies of the three slowest vibrations should be in the ratios

$$1 : 2{\cdot}8102 : 5{\cdot}4316;$$

* The bells, or gongs, as they are sometimes called, of striking clocks often give disagreeable beats. A remedy may be found in a suitable rotation of the bell about its axis.

so that the tones might be

$$c, \quad f'\sharp, \quad f'', \quad \text{approximately.}$$

More recently, through the kindness of Messrs Mears and Stainbank, I have had an opportunity of examining a so-called hemispherical metal bell, weighing about 3 cwt. A section is shown in Fig. 1. Four tones could be plainly heard,

$$e\flat, \quad f'\sharp, \quad e'', \quad b'',$$

the pitch being taken from a harmonium. The gravest tone has a long duration. When the bell is struck by a hard body, the higher tones are

Fig. 1.

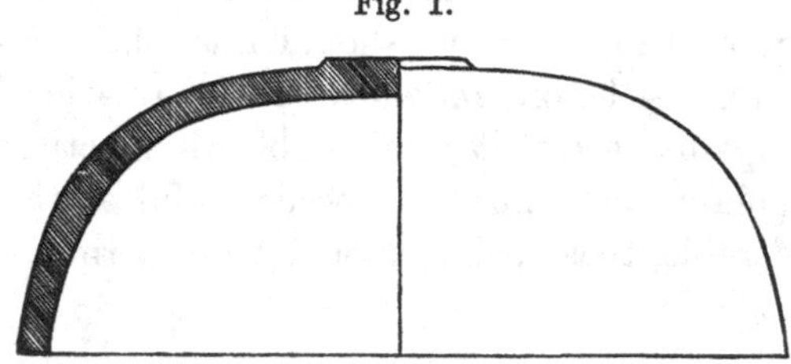

at first predominant, but after a time they die away, and leave $e\flat$ in possession of the field. If the striking body be soft, the original preponderance of the higher elements is less marked.

By the method above described there was no difficulty in showing that the four tones correspond respectively to $n = 2, 3, 4, 5$. Thus for the gravest tone the vibration is elliptical with 4 nodal meridians, for the next tone there are 6 nodal meridians, and so on. Tapping along a meridian showed that the sounds became less clear as the edge was departed from, and this in a continuous manner with no suggestion of a nodal circle of latitude.

A question, to which we shall recur in connexion with church bells, here suggests itself. Which of the various coexisting tones characterizes the pitch of the bell as a whole? It would appear to be the third in order, for the founders give the pitch as E nat.

My first attempts upon church bells were made in September 1879, upon the second bell (reckoned from the highest) of the Terling peal; and I was much puzzled to reconcile the pitch of the various tones, determined by resonators, with the effective pitch of the bell, when heard from a distance in conjunction with the other bells of the peal. There was a general agreement that the five notes of the peal were

$$f\sharp, \quad g\sharp, \quad a\sharp, \quad b, \quad c\sharp,$$

according to harmonium pitch, so that the note of the second bell was b. A tone of pitch $a\sharp$ could be heard, but at that time nothing coincident with b or its octaves. Subsequently, in January 1880, the b was found among the tones of the bell, but at much higher pitch than had been expected. The five gravest tones were determined to be

$$d', \quad a'\sharp, \quad d'', \quad g''\sharp+, \quad b'';$$

so that the nominal note of the bell agreed with the fifth component tone, and with no graver one. The octaves are here indicated by dashes in the usual way, the c' immediately below the d' being the middle c of the musical scale.

Attempts were then made to identify the modes of vibration corresponding to the various tones, but with only partial success. By tapping round the sound-bow it appeared that the minima of beats for d' occurred at intervals equal to $\frac{1}{8}$ of the circumference, indicating that the deformation in this mode was elliptical ($n = 2$), as had been expected. In like manner $g''\sharp$ gave $n = 3$; but on account of the difficulty of experimenting in the belfry, the results were not wholly satisfactory, and I was unable to determine the modes for the other tones. One observation, however, of importance could be made. All five tones were affected with beats, from which it was concluded that none of them could be due to symmetrical vibrations, as, till then, had been thought not unlikely.

Nothing further worthy of record was effected until last year, when I obtained from Messrs Mears and Stainbank the loan of a 6-cwt. bell. Hung in the laboratory at a convenient height, and with freedom of access to all parts of the circumference, this bell afforded a more convenient subject for experiment, and I was able to make the observations by which before I had been baffled. Former experience having shown me the difficulty of estimating the pitch of an isolated bell, I was anxious to have the judgment of the founders expressed in a definite form, and they were good enough to supply me with a fork tuned to the pitch of the bell. By my harmonium the fork is d''.

By tapping the bell in various places with a hammer or mallet, and listening with resonators, it was not difficult to detect 6 tones. They were identified with the following notes of the harmonium*:—

$$\underset{(4)}{e'}, \quad \underset{(4)}{c''}, \quad \underset{(6)}{f''+}, \quad \underset{(6)}{b''\flat}, \quad \underset{(8)}{d'''}, \quad f'''.$$

As in the former case, the nominal pitch is governed by the fifth component tone, whose pitch is, however, an octave higher than that of the representative fork. It is to be understood, of course, that each of the 6 tones in the above series is really double, and that in some cases the components of a pair differ sufficiently to give rise to somewhat rapid beats.

* In comparisons of this kind the observer must bear in mind the highly compound character of the notes of a reed instrument. It is usually a wise precaution to ascertain that a similar effect is not produced by the octave (or twelfth) above.

The sign + affixed to f'' indicates that the tone of the bell was decidedly sharp in comparison with the note of the instrument.

I now proceeded to determine, as far as possible, the characters of the various modes of vibration by observations upon the dependence of the sounds upon the place of tapping in the manner already described. By tapping round a circle of latitude it was easy to prove that for (each of the approximately coincident tones of) e' there were 4 nodal meridians. Again, on tapping along a meridian to find whether there were any nodal circles of latitude, it became evident that there were none such. At the same time differences of intensity were observed. This tone is more fully developed when the blow is delivered about midway between the crown and rim of the bell than at other places.

The next tone is c''. Observation showed that for this vibration also there are four, and but four, nodal meridians. But now there is a well-defined nodal circle of latitude, situated about a quarter of the way up from the rim towards the crown. As heard with the resonator, this tone disappears when the blow is accurately delivered at some point of this circle, but revives with a very small displacement on either side. The nodal circle and the four meridians divide the surface into segments, over each of which the normal motion is of one sign.

To the tone f'' correspond 6 nodal meridians. There is no well-defined nodal circle. The sound is indeed very faint, when the tap is much removed from the sound-bow; it was thought to fall to a minimum when the tap was about halfway up.

The three graver tones are heard loudly from the sound-bow. But the next in order, $b''\flat$, is there scarcely audible, unless the blow be delivered to the rim itself in a tangential direction. The maximum effect occurs at about halfway up. Tapping round the circle, we find that there are 6 nodal meridians.

The fifth tone, d''', is heard loudly from the sound-bow, but soon falls off when the locality of the blow is varied, and in the upper three-fourths of the bell it is very faint. No distinct circular node could be detected. Tapping round the circumference showed that there were here 8 nodal meridians.

The highest tone recorded, f''', was not easy of observation, and I did not succeed in satisfying myself as to the character of the vibration. The tone was perhaps best heard when the blow was delivered at a point a little below the crown.

All the above tones, except f'', were tolerably close in pitch to the corresponding notes of the harmonium.

Although the above results seemed perfectly unambiguous, I was glad to have an opportunity of confirming them by examination of another bell. This was afforded by a loan of a bell cast by Taylor, of Loughborough, and destined for the church of Ampton, Suffolk, where it now hangs. Its weight is somewhat less than 4 cwt., and the nominal pitch is *d*. The observations were entirely confirmatory of the results obtained from Messrs Mears's bell. The tones were

$$\underset{(4)}{e'\flat-2},\quad \underset{(4)}{d''-6},\quad \underset{(6)}{f''+4},\quad \underset{(6)}{b''\flat\text{—}b''},\quad \underset{(8)}{d'''},\quad g''';$$

the correspondence between the order of the tone and the number of nodal meridians being as before. In the case of d'' there was the same well-defined nodal circle. The highest tone, g''', was but imperfectly heard, and no investigation could be made of the corresponding mode of vibration.

In the specification of pitch the numerals following the note indicate by how much the frequency for the bell differed from that of the harmonium. Thus the gravest tone $e'\flat$ gave 2 beats per second, and was flat. When the number exceeds 3, it is the result of somewhat rough estimation and cannot be trusted to be quite accurate. Moreover, as has been explained, there are in strictness two frequencies under each head, and these often differ sensibly. In the case of the 4th tone, $b''\flat$—b'' means that, as nearly as could be judged, the pitch of the bell was midway between the two specified notes of the harmonium.

The sounds of bells may be elicited otherwise than by blows. Advantage may often be taken of the response to the notes of the harmonium, to the voice, or to organ-pipes, sounded in the neighbourhood. In these cases the subsequent resonance of the bell has the character of a pure tone. Perhaps the most striking experiment is with a tuning-fork. A massive $e'\flat$ (e' on the $c'=256$ scale) fork, tuned with wax, and placed upon the waist of the Ampton bell, called forth a magnificent resonance, which lasted for some time after removal and damping of the fork. The sound is so utterly unlike that usually associated with bells that an air of mystery envelops the phenomenon. The fork may be excited either by a preliminary blow upon a pad (in practice it was the bent knee of the observer), or by bowing when in contact with the bell. In either case the adjustment of pitch should be very precise, and it is usually necessary to distinguish the two nearly coincident tones of the bell. One of these is to be chosen, and the fork is to be held near a loop of the corresponding mode of vibration. In practice the simplest way to effect the tuning is to watch the course of things after the vibrating fork has been brought into contact with the bell. When the tuning is good the sound swells continuously. Any beats that are heard must be gradually slowed down by adjustment of wax, until they disappear.

Mears, 1888	Ampton, 1888	Belgian Bell	Terling (5), Osborn, 1783	Terling (4), Mears, 1810	Terling (3), Graye, 1623	Terling (2), Gardiner, 1723	Terling (1), Warner, 1863
Actual Pitch by Harmonium.							
e′	e′♭ − 2	d′ − 4	g − 3	a + 3	a♯ + 3	d′ − 6	d′ + 2
c″	d″ − 6	c″♯—d″	g′ − 4	g′♯ − 4	a′ + 6	a′♯ − 5	b′ + 2
f″ +	f″ + 4	f″ + 1	a′ + 6	b′ + 6	c″♯ + 4	d″ + 8	e″
b″♭	b″♭—b″	a″ − 6	d″ − 3	d″♯—e″	e″ + 6	g″♯ + (10)	g″♯ + 4
d‴	d‴	……	f″♯ − 2	g″♯ − 6	a″♯	b″ + 2	c‴♯ + 3
f‴	g‴						
Pitch referred to fifth tone as c.							
d	c♯ − 2		c♯ − 3	c♯ + 3	c + 3	e♭ − 6	c♯ + 2
b♭	c − 6		c♯ − 4	c − 4	b♭ + 6	b − 5	b♭ + 2
e♭ +	c♭ + 4		e♭ + 6	e♭ + 6	e♭ + 4	e♭ + 8	e♭
a♭	a♭—a		a♭ − 3	g—g♯	f♯ + 6	a + 8	g + 4
c	c		c − 2	c − 6	c	c + 2	c + 3

Observations upon the two bells in the laboratory having settled the modes of vibration corresponding to the five gravest tones, other bells of the church pattern can be sufficiently investigated by simple determinations of pitch. I give in tabular form results of this kind for a Belgian bell, kindly placed at my disposal by Mr Haweis, and for the five bells of the Terling peal. For completeness' sake the Table includes also the corresponding results for the two bells already described.

It will be seen that in every case where the test can be applied, it is the fifth tone in order which agrees with the nominal pitch of the bell. The reader will not be more surprised at this conclusion than I was, but there seems to be no escape from it. Even apart from estimates of pitch, an examination of the tones of the bells of the Terling peal proves that it is only from the third and fifth tones that a tolerable diatonic scale can be constructed. Observations in the neighbourhood of bells do not suggest any special predominance of the fifth tone, but the effect is a good deal modified by distance.

It has been suggested, I think by Helmholtz, that the aim of the original designers of bells may have been to bring into harmonic relations tones which might otherwise cause a disagreeable effect. If this be so, the result cannot be considered very successful. A glance at the Table shows that in almost every case there occur intervals which would usually be counted intolerable, such as the false octave. Terling (5) is the only bell which avoids this false interval between the two first tones; but the improvement here shown in this respect still leaves much to be desired, when we consider the relation of these two tones to the fifth tone, and the nominal pitch of the bell. Upon the assumption that the nominal pitch is governed by that of the fifth tone, I have exhibited in the second part of the above Table the relationship in each case of the various tones to this one.

One of my objects in this investigation having been to find out, if possible, wherein lay the difference between good and bad bells, I was anxious to interpret in accordance with my results the observations of Mr Haweis, who has given so much attention to the subject. The comparison is, however, not free from difficulty. Mr Haweis says*:—"The true Belgian bell when struck a little above the rim gives the dominant note of the bell; when struck two-thirds up it gives the third; and near the top the fifth; and the 'true' bell is that in which the third and fifth (to leave out a multitude of other partials) are heard in right relative subordination to the dominant note."

If I am right in respect of the dominant note, the *third* spoken of by Mr Haweis must be the minor third (or, rather, major sixth) presented by

* *Times*, October 29, 1878.

the tone third in order, which it so happens is nearly the same interval in all cases. The only *fifth* which occurs is that of the tone fourth in order. Thus, according to Mr Haweis's views, the best bell in the series would be Terling (1), for which the minor chord of the last three tones is nearly true. It must be remarked, however, that the tone fourth in order is scarcely heard in the normal use of the bell, so that its pitch can hardly be of importance directly, although it may afford a useful criterion of the character of the bell as a whole. It is evident that the first and second tones of Terling (1) are quite out of relation with the higher ones. If the first could be depressed a semitone and the second raised a whole tone, harmonic relations would prevail throughout.

Judging from the variety presented in the Table, it would seem not a hopeless task so to construct a bell that all the important tones should be brought into harmonic relation; but it would require so much tentative work that it could only be undertaken advantageously by one in connexion with a foundry. As to what advantage would be gained in the event of success, I find it difficult to form an opinion. All I can say is that the dissonant effect of the inharmonious intervals actually met with is less than one would have expected from a musical point of view; although the fact is to a great extent explained by Helmholtz's theory of dissonance.

One other point I will touch upon, though with great diffidence. If there is anything well established in theoretical acoustics it is that the frequencies of vibration of similar bodies formed of similar material are inversely as the linear dimensions—a law which extends to *all* the possible modes of vibration. Hence, if the dimensions are halved, all the tones should rise in pitch by an exact octave. I have been given to understand, however, that bells are not designed upon this principle of similarity, and that the attempt to do so would result in failure. It is just possible that differences in cooling may influence the hardness, and so interfere with the similarity of corresponding parts, in spite of uniformity in the chemical composition of the metal; but this explanation does not appear adequate. Can it be that when the scale of a bell is altered it is desirable at the same time to modify the relative intensities, or even the relative frequencies, of the various partials?

Observations conducted about ten years ago upon the manner of bending of bell-shaped bodies—waste-paper baskets and various structures of flexible material—led me to think that these shapes were especially stiff as regards the principal mode of bending (with four nodal meridians) to forces applied normally and near the rim, and that possibly one of the objects of the particular form adopted for bells might be to diminish the preponderance of the gravest tone. To illustrate this I made calculations, according to the theory of the paper already alluded to, of the deformation by pure bending of thin shells in the form of hyperboloids of revolution, and in certain composite forms built up of cylinders and cones so as represent approximately the actual shape of bells. In the case of the hyperboloid of one sheet (Fig. 2), completed by a crown in the form of a circular disk through the centre, and extending across the aperture, it appeared that there was no nodal circle for $n = 2$. The investigation is appended to this paper.

The composite forms, Figs. 4 and 5, represent the actual bell (Fig. 3*) as nearly as may be. At the top is a circular disk, and to this is attached

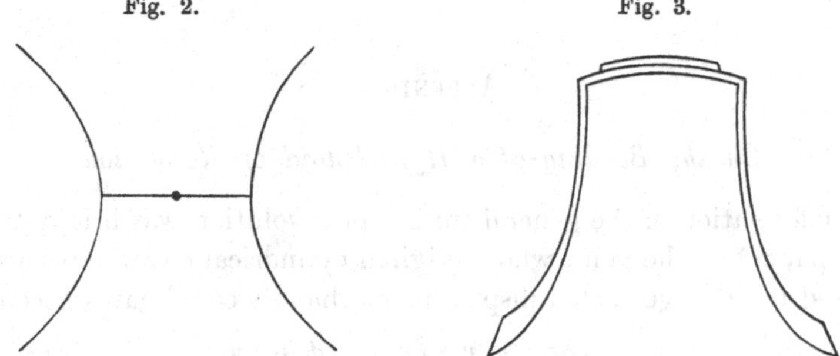
Fig. 2. Fig. 3.

a cylindrical segment. The expanding part of the bell is represented by one (Fig. 4), or with better approximation by two (Fig. 5), segments of cones. The calculations are too tedious to be reproduced here, but the results are shown upon the figures. In both cases there is a circular node N for $n = 2$, not far removed from the rim, and in Fig. 5 very nearly at the place which represents the sound-bow of an actual bell. In the latter case there is a node N' for $n = 3$ near the middle of the intermediate conical segment.

The nodal circle for $n = 2$ has been verified experimentally upon a bell constructed of thin sheet zinc in the form of Fig. 5. The gravest note, G♯, and the corresponding mode of vibration, could be investigated exactly in the

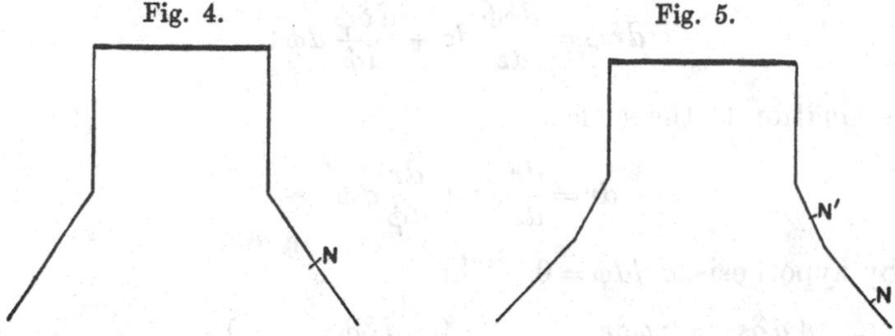

Fig. 4. Fig. 5.

manner already described. In each mode of this kind there were four nodal meridians, and a very well defined nodal circle. The situation of this circle was not quite so low as according to calculation; it was almost exactly in the middle of the lower conical segment. By merely handling the model it was

* Copied from Zamminer, *Die Musik und die musikalischen Instrumente.* Giessen, 1855.

easy to recognize that it was stiff to forces applied at N, but flexible higher up, in the neighbourhood of N'.

It is clear that the actual behaviour of a church bell differs widely from that of a bell infinitely thin; and that this should be the case need not surprise us when we consider the actual ratio of the thickness at the sound-bow to the interval between consecutive nodal meridians. I think, however, that the form of the bell does really tend to render the gravest tone less prominent.

APPENDIX.

On the Bending of a Hyperboloid of Revolution.

The deformation of the general surface of revolution was briefly treated in a former paper*. The point whose original cylindrical coordinates are z, r, ϕ, is supposed to undergo such a displacement that its coordinates become

$$z + \delta z, \qquad r + \delta r, \qquad \phi + \delta\phi.$$

The altered value $(ds + d\delta s)$ of the element of length traced upon the surface is given by

$$(ds + d\,\delta s)^2 = (dz + d\,\delta z)^2 + (r + \delta r)^2\,(d\phi + d\,\delta\phi)^2 + (dr + d\,\delta r)^2.$$

Hence, if the displacement be such that the element is unextended,

$$dz\,d\delta z + r^2 d\phi\,d\delta\phi + r\,\delta r\,(d\phi)^2 + dr\,d\delta r = 0.$$

Now

$$d\delta z = \frac{d\delta z}{dz}\,dz + \frac{d\delta z}{d\phi}\,d\phi,$$

$$d\delta r = \frac{d\delta r}{dz}\,dz + \frac{d\delta r}{d\phi}\,d\phi,$$

$$d\delta\phi = \frac{d\delta\phi}{dz}\,dz + \frac{d\delta\phi}{d\phi}\,d\phi;$$

and by the equation to the surface

$$dr = \frac{dr}{dz}\,dz + \frac{dr}{d\phi}\,d\phi,$$

in which, by hypothesis, $dr/d\phi = 0$. Thus

$$(dz)^2\left\{\frac{d\delta z}{dz} + \frac{dr}{dz}\frac{d\delta r}{dz}\right\} + (d\phi)^2\left\{r^2\frac{d\delta\phi}{d\phi} + r\,\delta r\right\}$$
$$+\, dz\,d\phi\left\{\frac{d\delta z}{d\phi} + r^2\frac{d\delta\phi}{dz} + \frac{dr}{dz}\frac{d\delta r}{d\phi}\right\} = 0.$$

* "On the Infinitesimal Bending of Surfaces of Revolution," *Proc. Math. Soc.* XIII. p. 4 (1881). [Vol. I. p. 551.]

If the displacement be of such a character that no line traced upon the surface is altered in length, the coefficients of $(dz)^2$, $(d\phi)^2$, $dz\,d\phi$, in the above equation, must vanish separately, so that

$$\frac{d\delta z}{dz}+\frac{dr}{dz}\frac{d\delta r}{dz}=0, \qquad (1)$$

$$r\frac{d\delta\phi}{d\phi}+\delta r=0, \qquad (2)$$

$$\frac{d\delta z}{d\phi}+r^2\frac{d\delta\phi}{dz}+\frac{dr}{dz}\frac{d\delta r}{d\phi}=0. \qquad (3)$$

From these, by elimination of δr,

$$\frac{d\delta z}{dz}-\frac{dr}{dz}\frac{d}{dz}\left(r\frac{d\delta\phi}{d\phi}\right)=0, \qquad (4)$$

$$\frac{d\delta z}{d\phi}+r^2\frac{d\delta\phi}{dz}-r\frac{dr}{dz}\frac{d^2d\phi}{d\phi^2}=0; \qquad (5)$$

from which again, by elimination of δz,

$$\frac{d}{dz}\left(r^2\frac{d\delta\phi}{dz}\right)-r\frac{d^2r}{dz^2}\frac{d^2\delta\phi}{d\phi^2}=0. \qquad (6)$$

For the purposes of the present problem we may assume that $\delta\phi$ varies as $\cos s\phi$, or as $\sin s\phi$; thus,

$$r^2\frac{d}{dz}\left(r^2\frac{d\delta\phi}{dz}\right)+s^2r^3\frac{d^2r}{dz^2}\delta\phi=0 \qquad (7)$$

is the equation by which the form of $\delta\phi$ as a function of z is to be determined.

When application is made to the hyperboloid of one sheet

$$\frac{r^2}{a^2}-\frac{z^2}{b^2}=1 \qquad (8)$$

we find, since

$$r\frac{dr}{dz}=\frac{a^2z}{b^2}, \qquad r^3\frac{d^2r}{dz^2}=\frac{a^4}{b^2},$$

$$r^2\frac{d}{dz}\left(r^2\frac{d\delta\phi}{dz}\right)+\frac{s^2a^4}{b^2}\delta\phi=0. \qquad (9)$$

The solution of this equation is expressed by an auxiliary variable χ, such that

$$z=b\tan\chi, \qquad r=a\sec\chi \qquad (10)$$

in the form

$$\delta\phi=A\cos s\chi+B\sin s\chi. \qquad (11)$$

In order to verify this it is only necessary to observe that by (10)

$$r^2\frac{d}{dz}=\frac{a^2}{b}\frac{d}{d\chi}.$$

We will now apply this solution to an inextensible surface formed by half the hyperboloid and a crown stretching across in the plane of symmetry $z=0$ (Fig. 2). The deformation of this crown can take place only in the direction perpendicular to its plane, so that $\delta r=0$, $\delta\phi=0$. These conditions must apply also to the hyperboloid at the place of attachment to the crown. Hence $\delta\phi$ must vanish with z, or, which is the same, with χ. Accordingly $A=0$ in (11); and dropping the constant multiplier we may take as the solution

$$\delta\phi = \sin s\chi \cos s\phi, \quad \text{.........................(12)}$$

and in correspondence therewith by (2) and (3)

$$\delta r = sr \sin s\chi \sin s\phi \quad \text{.......................................(13)}$$

$$\delta z = -\frac{a^2}{b}\{\cos s\chi + s\tan\chi \sin s\chi\}\sin s\phi. \quad \text{...............(14)}$$

It is evident from these equations that, whatever may be the value of s, there is no circle of latitude over which both $\delta\phi$ (or δr) and δz vanish*. Hence there can be no circular nodal line in the absolute sense. But just as there are meridians ($\sin s\phi=0$) on which the *normal* motion vanishes, so there may be nodal circles in this more limited sense. The condition to be satisfied is obviously

$$\delta r/\delta z = dr/dz;$$

or in the present case

$$\sin 2\chi + 2s\tan s\chi\,(\sin^2\chi + b^2/a^2) = 0. \quad \text{..............(15)}$$

In this equation the range of χ is from 0 to $\frac{1}{2}\pi$; and thus there can be solutions only when $\tan s\chi$ is negative.

In the case $s=2$ the equation reduces to

$$1 + 2\sin^2\chi + 4b^2/a^2 = 0,$$

which can never be satisfied.

When $s=3$, the roots, if any, must lie between $\chi=30°$ and $\chi=60°$. A more detailed consideration shows that there is but one root, and that it occurs when χ is a little short of 60°.

* A corresponding proposition may be proved more generally, that is without limitation to the hyperboloid.

165.

THE CLARK STANDARD CELL.

[*The Electrician*, Jan. 1890, p. 285.]

In order to expedite the settlement of any open questions respecting Clark cells, I send the following remarks upon the Paper of Prof. Carhart, reprinted in *The Electrician*, p. 271, from the *Philosophical Magazine*, entitled "An Improved Standard Clark Cell with Low Temperature Coefficient." [*Phil. Mag.* XXVIII. p. 420, 1889.]

In the first place Prof. Carhart appears to me rather to exaggerate the inconvenience arising from temperature changes in a Clark cell of ordinary construction. The coefficient is about ·00077 per degree cent., so that an uncertainty of a whole degree, affecting the E.M.F. by less than $\frac{1}{1000}$, would hardly be of practical importance. The sensitiveness to temperature is in fact only about the double of that of German-silver resistance coils. In a suitable situation, and with the most ordinary care, the temperature would not be uncertain to more than one or two tenths. I have found it possible to work even closer than this in a room (next the roof), far from specially suitable, without any particular precautions; but if desired, it is easy to reduce the uncertainty under this head by some such plan as embedding the cell, with a thermometer bulb, in a vessel of sand.

The really serious question is whether the temperature coefficient itself is liable to important variation without assignable cause. If it be uncertain whether the proper coefficient is ·00077, or, as in Prof. Carhart's cells, ·00039, the utility of the standard would indeed be seriously compromised.

Undoubtedly the lower coefficient would be an advantage in itself, if it could be obtained without loss in other respects. The principal feature insisted upon by Prof. Carhart is the separation of the zinc from the mercurous salt; but my experience is totally opposed to the view that the lower coefficient can thus be secured. The separation actually occurred in a large number of the cells upon which I experimented, especially in those of the H pattern*, where the mercury and mercurous salt occupied one leg, and

* *Phil. Trans.* 1884. [Vol. II. p. 315.]

an amalgam of zinc the other. The arrangement is shown in the figure. That these cells have practically the same temperature coefficient as others in which the paste touches the zinc, is proved by Table XIII. of my second Paper*. I am thus at a loss to explain the low temperature coefficient of Prof. Carhart's cells, unless indeed upon the supposition that his solutions were not throughout *saturated with zinc sulphate*. In this case the coefficient is just what might have been expected, for I found from two cells of this description the coefficient ·00038. It may be remarked that the **H** form is safer in this respect than those in which the zinc is at the top of the liquid, especially when removed from the paste; for the part of the liquid where saturation is of importance is that in contact with the zinc. At the top of the column the salt may easily become deficient, when the temperature rises, even though there be plenty of undissolved crystals below. The objections to unsaturated solutions are discussed in my Papers. They turn upon the difficulty of preparing a standard solution, and upon the liability to change with evaporation.

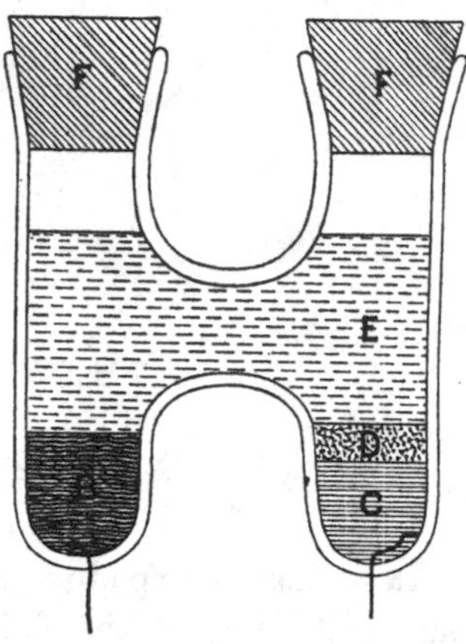

H-Pattern Cell.

B, Amalgam of Zinc; C, Pure Mercury; D, Mercurous Sulphate; E, Saturated Solution of Zinc Sulphate; F, Corks. At the bottom of each leg of the cell a platinum wire, sealed through the glass, is shown.

I quite agree with Prof. Carhart as to the importance of pure mercury. And there is undoubtedly something yet to be done in respect of the mercurous sulphate. I may remark that the sample used by me did not turn yellow when treated simply with zinc solution, but only when rubbed up also with *zinc carbonate*. In the cells made by Mr M. Evans the paste was quite white, but this did not prevent the variation with temperature being the same as in other cases. (Table XIV., T_1, T_3.)

I once came across a sample of mercurous sulphate with which it was difficult to prepare satisfactory cells. When rubbed up with zinc carbonate and zinc sulphate solution it turned dark green instead of yellow. Until the question is further elucidated I should be disposed to avoid a sample which behaved in this way.

* *Phil. Trans.* 1885, § 40. [Vol. II. p. 453.]

166.

ON THE VIBRATIONS OF AN ATMOSPHERE.

[*Philosophical Magazine*, XXIX. pp. 173—180, 1890.]

IN order to introduce greater precision into our ideas respecting the behaviour of the Earth's Atmosphere, it seems advisable to solve any problems that may present themselves, even though the search for simplicity may lead us to stray rather far from the actual question. It is proposed here to consider the case of an atmosphere composed of gas which obeys Boyle's law, viz. such that the pressure is always proportional to the density. And in the first instance we shall neglect the curvature and rotation of the earth, supposing that the strata of equal density are parallel planes perpendicular to the direction in which gravity acts.

If p, σ be the equilibrium pressure and density at the height z, then

$$\frac{dp}{dz} = -\sigma g; \qquad (1)$$

and by Boyle's law,

$$p = a^2\sigma, \qquad (2)$$

where a is the velocity of sound. Hence

$$\frac{d\sigma}{\sigma\, dz} = -\frac{g}{a^2}, \qquad (3)$$

and

$$\sigma = \sigma_0 e^{-gz/a^2}, \qquad (4)$$

where σ_0 is the density at $z = 0$. According to this law, as is well known, there is no limit to the height of the atmosphere.

Before proceeding further, let us pause for a moment to consider how the density at various heights would be affected by a small change of temperature, altering a to a', the whole quantity of air and therefore the pressure p_0

at the surface remaining unchanged. If the dashes relate to the second state of things, we have

$$\sigma = \sigma_0 e^{-gz/a^2}, \qquad \sigma' = \sigma_0' e^{-gz/a'^2},$$

$$p = p_0 e^{-gz/a^2}, \qquad p' = p_0 e^{-gz/a'^2},$$

while

$$a^2 \sigma_0 = a'^2 \sigma_0'.$$

If $a'^2 - a^2 = \delta a^2$, we may write approximately

$$\frac{p' - p}{p_0} = \frac{\delta a^2}{a^2} \frac{gz}{a^2} e^{-gz/a^2}.$$

The alteration of pressure vanishes when $z = 0$, and also when $z = \infty$. The maximum occurs when $gz/a^2 = 1$, that is when $p = p_0/e$. But relatively to σ, $(p' - p_0)$ increases continually with z.

Again, if ρ denote the proportional variation of density,

$$\rho = \frac{\sigma' - \sigma}{\sigma} = \frac{a^2}{a'^2}(e^{-gz/a'^2 + gz/a^2} - 1).$$

If $a'^2 > a^2$, ρ is negative when $z = 0$, and becomes $+\infty$ when $z = \infty$. The transition $\rho = 0$ occurs when $gz/a^2 = 1$, that is at the same place where $p' - p$ reaches a maximum.

In considering the small vibrations, the component velocities at any point are denoted by u, v, w, the original density σ becomes $(\sigma + \sigma\rho)$, and the increment of pressure is δp. On neglecting the squares of small quantities the equation of continuity is

$$\sigma \frac{d\rho}{dt} + \sigma \frac{du}{dx} + \sigma \frac{dv}{dy} + \sigma \frac{dw}{dz} + w \frac{d\sigma}{dz} = 0;$$

or by (3),

$$\frac{d\rho}{dt} + \frac{du}{dx} + \frac{dv}{dy} + \frac{dw}{dz} - \frac{gw}{a^2} = 0. \quad \text{......................(5)}$$

The dynamical equations are

$$\frac{d\delta p}{dx} = -\sigma \frac{du}{dt}, \qquad \frac{d\delta p}{dy} = -\sigma \frac{dv}{dt}, \qquad \frac{d\delta p}{dz} = -g\sigma\rho - \sigma \frac{dw}{dt};$$

or by (3), since $\delta p = a^2 \sigma \rho$,

$$a^2 \frac{d\rho}{dx} = -\frac{du}{dt}, \qquad a^2 \frac{d\rho}{dy} = -\frac{dv}{dt}, \qquad a^2 \frac{d\rho}{dz} = -\frac{dw}{dt}. \quad \text{.........(6)}$$

We will consider first the case of one dimension, where u, v vanish, while ρ, w are functions of z and t only. From (5) and (6),

$$\frac{d\rho}{dt} + \frac{dw}{dz} - \frac{gw}{a^2} = 0, \quad \text{..............................(7)}$$

$$a^2 \frac{d\rho}{dz} = -\frac{dw}{dt}; \quad \text{..............................(8)}$$

or by elimination of ρ,

$$\frac{1}{a^2}\frac{d^2w}{dt^2}=\frac{d^2w}{dz^2}-\frac{g}{a^2}\frac{dw}{dz}. \qquad (9)$$

The right-hand member of (9) may be written

$$\left(\frac{d}{dz}-\frac{g}{2a^2}\right)^2 w-\frac{g^2}{4a^4}w,$$

and in this the latter term may be neglected when the variation of w with respect to z is not too slow. If λ be of the nature of the wave-length, dw/dz is comparable with w/λ; and the simplification is justifiable when a^2 is large in comparison with $g\lambda$, that is when the velocity of sound is great in comparison with that of gravity-waves (as upon water) of wave-length λ. The equation then becomes

$$\frac{d^2w}{dt^2}=a^2\left(\frac{d}{dz}-\frac{g}{2a^2}\right)^2 w;$$

or, if

$$w=We^{\frac{1}{2}gz/a^2}, \qquad (10)$$

$$d^2W/dt^2=a^2.d^2W/dz^2; \qquad (11)$$

the ordinary equation of sound in a uniform medium. Waves of the kind contemplated are therefore propagated without change of type except for the effect of the exponential factor in (10), indicating the increase of motion as the waves pass upwards. This increase is necessary in order that the same amount of energy may be conveyed in spite of the growing attenuation of the medium. In fact $w^2\sigma$ must retain its value, as the waves pass on.

If w vary as e^{int}, the original equation (9) becomes

$$\frac{d^2w}{dz^2}-\frac{g}{a^2}\frac{dw}{dz}+\frac{n^2w}{a^2}=0. \qquad (12)$$

Let m_1, m_2 be the roots of

$$m^2-\frac{g}{a^2}m+\frac{n^2}{a^2}=0,$$

so that

$$m=\frac{g\pm\sqrt{(g^2-4n^2a^2)}}{2a^2}; \qquad (13)$$

then the solution of (12) is

$$w=Ae^{m_1z}+Be^{m_2z}, \qquad (14)$$

A and B denoting arbitrary constants in which the factor e^{int} may be supposed to be included.

The case already considered corresponds to the neglect of g^2 in the radical of (13), so that

$$m=\frac{g\pm 2nai}{2a^2}$$

and

$$w\,e^{-\frac{1}{2}gz/a^2}=A\,e^{in(t+z/a)}+B\,e^{in(t-z/a)}. \qquad (15)$$

A wave propagated upwards is thus

$$w = e^{\frac{1}{2}gz/a^2} \cos n(t - z/a), \quad \text{.......................(16)}$$

and there is nothing of the nature of reflexion from the upper atmosphere.

A stationary wave would be of type

$$w = e^{\frac{1}{2}gz/a^2} \cos nt \sin(nz/a), \quad \text{.....................(17)}$$

w being supposed to vanish with z. According to (17), the energy of the vibration is the same in every wave-length, not diminishing with elevation. The viscosity of the rarefied air in the upper regions would suffice to put a stop to such a motion, which cannot therefore be taken to represent anything that could actually happen.

When $2na < g$, the values of m from (13) are real, and are both positive. We will suppose that m_1 is greater than m_2. If w vanish with z, we have from (14) as the expression of the stationary vibration

$$w = \cos nt\,(e^{m_1z} - e^{m_2z}), \quad \text{.........................(18)}$$

which shows that w is of one sign throughout. Again by (8)

$$a^2\rho = n \sin nt \left\{ \frac{e^{m_1z}}{m_1} - \frac{e^{m_2z}}{m_2} \right\}. \quad \text{.....................(19)}$$

Hence $d\rho/dz$, proportional to w, is of one sign throughout; ρ itself is negative for small values of z, and positive for large values, vanishing once when

$$e^{(m_1-m_2)z} = m_1/m_2. \quad \text{.........................(20)}$$

When n is small, we have approximately

$$m_1 = \frac{g}{a^2} - \frac{n^2}{g}, \qquad m_2 = \frac{n^2}{g}, \quad \text{.......................(21)}$$

so that ρ vanishes when

$$e^{gz/a^2} = \frac{g^2}{n^2a^2}, \quad \text{..............................(22)}$$

or by (4) when

$$\sigma/\sigma_0 = n^2a^2/g^2. \quad \text{.............................(23)}$$

Below the point determined by (23) the variation of density is of one sign and above it of the contrary sign. The integrated variation of density, represented by $\int_0^\infty \sigma\rho\, dz$, vanishes, as of course it should do.

It may be of interest to give a numerical example of (23). Let us suppose that the period is one hour, so that in C.G.S. measure $n = 2\pi/3600$. We take $a = 33 \times 10^4$, $g = 981$. Then $\sigma/\sigma_0 = \frac{1}{290}$; showing that even for this moderate period the change of sign does not occur until a high degree of rarefaction is reached.

In discarding the restriction to one dimension, we may suppose, without real loss of generality, that $v = 0$, and that u, w, ρ are functions of x and z

only. Further, we may suppose that x occurs only in the factor e^{ikx}; that is, that the motion is periodic with respect to x in the wave-length $2\pi/k$; and that as before t occurs only in the factor e^{int}. Equations (5), (6) then become

$$in\rho + iku + dw/dz - gw/a^2 = 0, \quad\quad (24)$$

$$a^2 k\rho = -nu, \quad\quad (25)$$

$$a^2\, d\rho/dz = -inw; \quad\quad (26)$$

from which if we eliminate u, w, we get

$$\frac{d^2\rho}{dz^2} - \frac{g}{a^2}\frac{d\rho}{dz} + \left(\frac{n^2}{a^2} - k^2\right)\rho = 0, \quad\quad (27)$$

an equation which may be solved in the same form as (12).

One obvious solution of (27) is of importance. If $d\rho/dz = 0$, so that $w = 0$, the equations are satisfied by

$$n^2 = k^2 a^2. \quad\quad (28)$$

Every horizontal stratum moves alike, and the *proportional* variation of density (ρ) is the same at all levels. The possibility of such a motion is evident beforehand, since on account of the assumption of Boyle's law the velocity of sound is the same throughout.

In the application to meteorology, the shortness of the more important periods of the vertical motion suggests that an "equilibrium theory" of this motion may be adequate. For vibrations like those of (28) there is no difficulty in taking account of the earth's curvature. For the motion is that of a simple spherical sheet of air, considered in my book upon the *Theory of Sound*, § 333. If r be the radius of the earth, the equation determining the frequency of the vibration corresponding to the harmonic of order h is

$$n^2 r^2 = h(h+1)\, a^2, \quad\quad (29)$$

the actual frequency being $n/2\pi$. If τ be the period, we have

$$\tau = \frac{2\pi r}{a\sqrt{\{h(h+1)\}}}. \quad\quad (30)$$

For $h = 1$, corresponding to a swaying of the atmosphere from one side of the earth to the opposite,

$$\tau_1 = \frac{2\pi r}{a\sqrt{2}}, \quad\quad (31)$$

and in like manner for $h = 2$,

$$\tau_2 = \frac{2\pi r}{\sqrt{6}} = \frac{\tau_1}{\sqrt{3}}. \quad\quad (32)$$

To reduce these results to numbers we may take for the earth's quadrant $\frac{1}{2}\pi r = 10^9$ cm.; and if we take for a the velocity of sound at 0° as ordinarily observed, or as calculated upon Laplace's theory, viz. 33×10^3 cm./sec., we shall find

$$\tau_1 = \frac{4 \times 10^9}{\sqrt{2} \times 33 \times 10^3} \text{ seconds} = 23{\cdot}8 \text{ hours.}$$

On the same basis,

$$\tau_2 = 13{\cdot}7 \text{ hours.}$$

It must, however, be remarked that the suitability of this value of a is very doubtful, and that the suppositions of the present paper are inconsistent with the use of Laplace's correction to Newton's theory of sound propagation. In a more elaborate treatment a difficult question would present itself as to whether the heat and cold developed during atmospheric vibrations could be supposed to remain undissipated. It is evidently one thing to make this supposition for sonorous vibrations, and another for vibrations of about 24 hours period. If the dissipation were neither very rapid nor very slow in comparison with diurnal changes—and the latter alternative at least seems improbable—the vibrations would be subject to the damping action discussed by Stokes*.

In any case the near approach of τ_1 to 24 hours, and of τ_2 to 12 hours, may well be very important. Beforehand the diurnal variation of the barometer would have been expected to be much more conspicuous than the semi-diurnal. The relative magnitude of the latter, as observed at most parts of the earth's surface, is still a mystery, all the attempted explanations being illusory. It is difficult to see how the operative forces can be mainly semi-diurnal in character; and if the effect is so, the readiest explanation would be in a near coincidence between the natural period and 12 hours. According to this view the semi-diurnal barometric movement should be the same at the sea-level all round the earth, varying (at the equinoxes) merely as the square of the cosine of the latitude, except in consequence of local disturbances due to want of uniformity in the condition of the earth's surface.

* *Phil. Mag.* [4] I. p. 305, 1851. *Theory of Sound*, § 247.

167.

ON THE TENSION OF RECENTLY FORMED LIQUID SURFACES.

[*Proceedings of the Royal Society*, XLVII. pp. 281—287, 1890.]

It has long been a mystery why a few liquids, such as solutions of soap and saponine, should stand so far in advance of others in regard to their capability of extension into large and tolerably durable laminæ. The subject was specially considered by Plateau in his valuable researches, but with results which cannot be regarded as wholly satisfactory. In his view the question is one of the ratio between capillary tension and superficial viscosity. Some of the facts adduced certainly favour a connexion between the phenomena attributed to the latter property and capability of extension; but the "superficial viscosity" is not clearly defined, and itself stands in need of explanation.

It appears to me that there is much to be said in favour of the suggestion of Marangoni* to the effect that both capability of extension and so-called superficial viscosity are due to the presence upon the body of the liquid of a coating, or pellicle, composed of matter whose inherent capillary force is less than that of the mass. By means of variations in this coating, Marangoni explains the indisputable fact that in vertical soap films the effective tension is different at various levels. Were the tension rigorously constant, as it is sometimes inadvertently stated to be, gravity would inevitably assert itself, and the central parts would fall 16 feet in the first second of time. By a self-acting adjustment the coating will everywhere assume such thickness as to afford the necessary tension, and thus any part of the film, considered without distinction of its various layers, is in equilibrium. There is nothing, however, to prevent the interior layers of a moderately thick film from draining down. But this motion, taking place as it were between two fixed walls, is comparatively slow, being much impeded by ordinary fluid viscosity.

In the case of soap, the formation of the pellicle is attributed by Marangoni to the action of atmospheric carbonic acid, liberating the fatty

* *Nuovo Cimento*, Vols. v. vi. 1871—72, p. 239.

acid from its combination with alkali. On the other hand, Sondhauss* found that the properties of the liquid, and the films themselves, are better conserved when the atmosphere is excluded by hydrogen; and I have myself observed a rapid deterioration of very dilute solutions of oleate of soda when exposed to the air. In this case a remedy may be found in the addition of caustic potash. It is to be observed, moreover, that, as has long been known, the capillary forces are themselves quite capable of overcoming weak chemical affinities, and will operate in the direction required.

A strong argument in favour of Marangoni's [general] theory is afforded by his observation†, that within very wide limits the superficial tension of soap solutions, as determined by capillary tubes, is almost independent of the strength. My purpose in this note is to put forward some new facts tending strongly to the same conclusion.

It occurred to me that, if the low tension of soap solutions as compared with pure water was due to a coating, the formation of this coating would be a matter of time, and that a test might be found in the examination of the properties of the liquid surface immediately after its formation. The experimental problem here suggested may seem difficult or impossible; but it was, in fact, solved some years ago in the course of researches upon the Capillary Phenomena of Jets‡. A jet of liquid issuing under moderate pressure from an elongated, *e.g.*, elliptical, aperture perforated in a thin plate, assumes a chain-like appearance, the complete period (λ), corresponding to two links of the chain, being the distance travelled over by a given part of the liquid in the time occupied by a complete transverse vibration of the column about its cylindrical configuration of equilibrium. Since the phase of vibration depends upon the time elapsed, it is always the same at the same point in space, and thus the motion is *steady* in the hydrodynamical sense, and the boundary of the jet is a fixed surface. Measurements of λ under a given head, or velocity, determine the time of vibration, and from this, when the density of the liquid and the diameter of the column are known, follows in its turn the value of the capillary tension (T) to which the vibrations are due. *Cæteris paribus*, $T \propto \lambda^{-2}$; and this relation, which is very easily proved, is all that is needed for our present purpose. If we wish to see whether a moderate addition of soap alters the capillary tension of water, we have only to compare the wave-lengths λ in the two cases, using the same aperture and head. By this method the liquid surface may be tested before it is $\frac{1}{100}$ second old.

Since it was necessary to be able to work with moderate quantities of liquid, the elliptical aperture had to be rather fine, about 2 mm. by 1 mm.

* *Pogg. Ann.* Ergänzungsband VIII. 1878, p. 266.

† *Pogg. Ann.* Vol. CXLIII. 1871, p. 342. The original pamphlet dates from 1865.

‡ *Roy. Soc. Proc.* May 15, 1879. [Vol. I. p. 377.]

The reservoir was an ordinary flask, 8 cm. in diameter, to which was sealed below as a prolongation a (1 cm.) tube bent at right angles (Figs. 1, 2). The

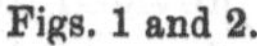
Figs. 1 and 2.

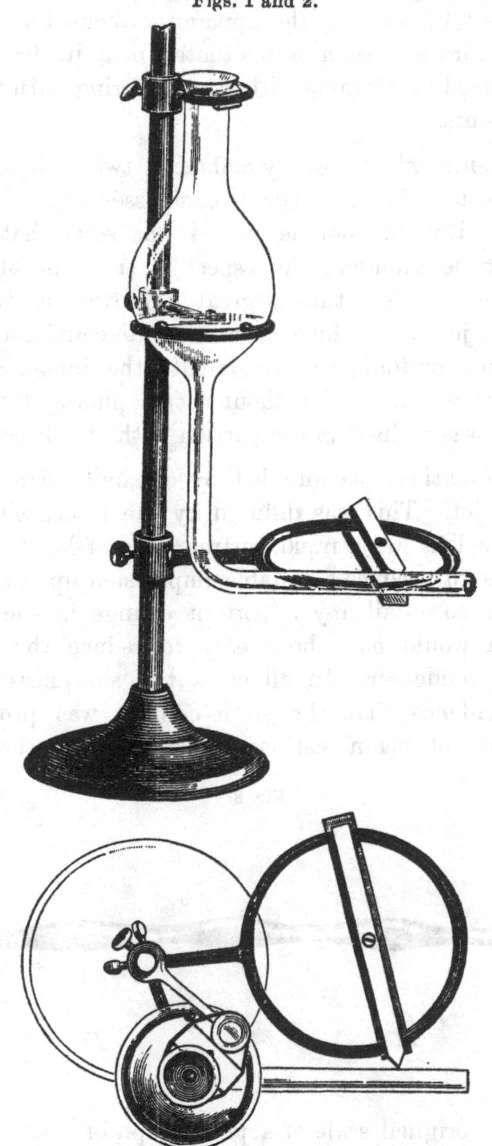

aperture was perforated in thin sheet brass, attached to the tube by cement. It was about 15 cm. below the mark, near the middle of the flask, which defined the position of the free surface at the time of observation.

The arrangement for bringing the apparatus to a fixed position, designed upon the principles laid down by Sir W. Thomson, was simple and effective.

The body of the flask rested on three protuberances from the ring of a retort stand, while the neck was held by an india-rubber band into a V-groove attached to an upper ring. This provided five contacts. The necessary sixth contact was effected by rotating the apparatus about its vertical axis until the delivery tube bore against a stop situated near its free end. The flask could thus be removed for cleaning without interfering with the comparability of various experiments.

The measurements, which usually embraced two complete periods, could be taken pretty accurately by a pair of compasses with the assistance of a magnifying glass. But the double period was somewhat small (16 mm.), and the little latitude admissible in respect to the time of observation was rather embarrassing. It was thus a great improvement to take magnified photographs of the jet, upon which measurements could afterwards be made at leisure. In some preliminary experiments the image upon the ground glass of the camera was utilised without actual photography. Even thus a decided advantage was realised in comparison with the direct measurements.

Sufficient illumination was afforded by a candle flame situated a few inches behind the jet. This was diffused by the interposition of a piece of ground glass. The lens was a rapid portrait lens of large aperture, and the ten seconds needed to produce a suitable impression upon the gelatine plate was not so long as to entail any important change in the condition of the jet. Otherwise, it would have been easy to reduce the exposure by the introduction of a condenser. In all cases the sharpness of the resulting photographs is evidence that the sixth contact was properly made, and thus that the scale of magnification was strictly preserved. Fig. 3 is a

Fig. 3.

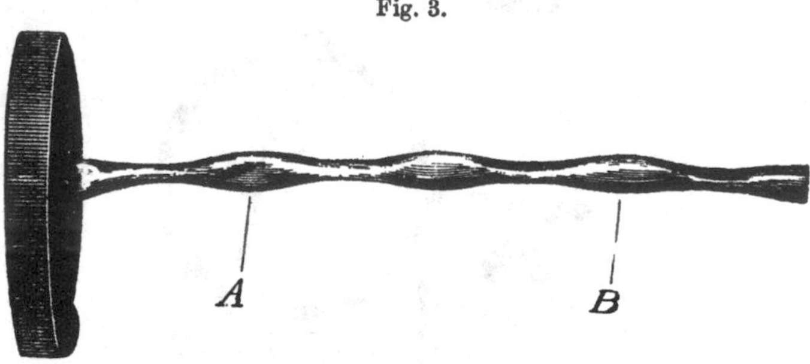

reproduction on the original scale of a photograph of a water jet taken upon 9th November. The distance recorded as 2λ is between the points marked A and B, and was of course measured upon the original negative. On each occasion when various liquids were under investigation, the photography of the water jet was repeated, and the results agreed well.

After these explanations it will suffice to summarise the actual measurements upon oleate of soda in tabular form. The standard solution contained

1 part of oleate in 40 parts of water, and was diluted as occasion required*. All lengths are given in millimetres.

	Water	Oleate 1/40	Oleate 1/80	Oleate 1/400	Oleate 1/4000
2λ	40·0	45·5	44·0	39·0	39·0
h	31·5	11·0	11·0	11·0	23·0

In the second row h is the rise of the liquid in a capillary tube, carefully cleaned before each trial with strong sulphuric acid and copious washing. In the last case, relating to oleate solution $\frac{1}{4000}$, the motion was sluggish and the capillary height but ill-defined. It will be seen that even when the capillary height is not much more than one-third of that of water, the wave-lengths differ but little, indicating that, at any rate, the greater part of the lowering of tension due to oleate requires time for its development. According to the law given above, the ratio of tensions of the newly-formed surfaces for water and oleate ($\frac{1}{80}$) would be merely as 6 : 5†.

Whether the slight differences still apparent in the case of the stronger solutions are due to the formation of a sensible coating in less than $\frac{1}{100}$ second, cannot be absolutely decided; but the probability appears to lie in the negative. No distinct differences could be detected between the first and second wave-lengths; but this observation is, perhaps, not accurate enough to settle the question. It is possible that a coating may be formed on the surface of the glass and metal, and that this is afterwards carried forward.

As a check upon the method, I thought it desirable to apply it to the comparison of pure water and dilute alcohol, choosing for the latter a mixture of 10 parts by volume of strong (not methylated) alcohol with 90 parts water. The results were as follows:—

$$2\lambda\,(\text{water}) = 38{\cdot}5, \qquad 2\lambda\,(\text{alcohol}) = 46{\cdot}5,$$
$$h\,(\text{water}) = 30{\cdot}0, \qquad h\,(\text{alcohol}) = 22{\cdot}0;$$

but it may be observed that they are not quite comparable with the preceding for various reasons, such as displacements of apparatus and changes of temperature. It is scarcely worth while to attempt an elaborate reduction of these numbers, taking into account the differences of specific gravity in the two cases; for, as was shown in the former paper, the observed values of λ are complicated by the departure of the vibrations from isochronism,

* Although I can find no note of the fact, I think I am right in saying that large bubbles could be blown with the weakest of the solutions experimented upon.

† Curiously enough, I find it already recorded in my note-book of 1879, that λ is not influenced by the addition to water of soap sufficient to render impossible the rebound of colliding jets. [Vol. I. p. 375.]

when, as in the present experiments, the deviation from the circular section is moderately great. We have—

$$(46{\cdot}5/38{\cdot}5)^2 = 1{\cdot}46, \qquad 30/22 = 1{\cdot}36;$$

and these numbers prove, at any rate, that the method of wave-lengths is fully competent to show a change in tension, provided that the change really occurs at the first moment of the formation of the free surface.

In view of the great extensibility of saponine films it seemed important to make experiments upon this material also. The liquid employed was an infusion of horse chestnuts of specific gravity 1·02, and, doubtless, contained other ingredients as well as saponine. It was capable of giving large bubbles, even when considerably diluted (6 times) with water. Photographs taken on November 23rd gave the following results:—

$$2\lambda\,(\text{water}) = 39{\cdot}2, \qquad 2\lambda\,(\text{saponine}) = 39{\cdot}5,$$
$$h\,(\text{water}) = 30{\cdot}5, \qquad h\,(\text{saponine}) = 20{\cdot}7.$$

Thus, although the capillary heights differ considerably, the tensions at the first moment are almost equal. In this case then, as in that of soap, there is strong evidence that the lowered tension is the result of the formation of a pellicle.

Though not immediately connected with the principal subject of this communication, it may be well here to record that I find saponine to have no effect inimical to the rebound after mutual collision of jets containing it. The same may be said of gelatine, whose solutions froth strongly. On the other hand, a very little soap or oleate usually renders such rebound impossible, but this effect appears to depend upon *undissolved* greasy matter. At least the drops from a nearly vertical fountain of *clear* solution of soap were found not to scatter*. The rebound of *jets* is, however, a far more delicate test than that of *drops*. A fountain of strong saponine differs in appearance from one of water; but this effect is due rather to the superficial viscosity, which retards, or altogether prevents, the resolution into drops.

The failure of rebound when jets or drops containing milk or undissolved soap come into collision has not been fully explained; but it is probably connected with the disturbance which must arise when a particle of grease from the interior reaches the surface of one of the liquid masses.

P.S.—I have lately found that the high tension of recently formed surfaces of soapy water was deduced by A. Dupré†, as long ago as 1869, from some experiments upon the vertical rise of fine jets. Although this method is less direct than that of the present paper, M. Dupré must be considered, I think, to have made out his case. It is remarkable that so interesting an observation should not have attracted more attention.

* *Roy. Soc. Proc.* June 15, 1882. [Vol. II. p. 103.]

† *Théorie Mécanique de la Chaleur*, Paris, 1869.

168.

MEASUREMENTS OF THE AMOUNT OF OIL NECESSARY IN ORDER TO CHECK THE MOTIONS OF CAMPHOR UPON WATER.

[*Proceedings of the Royal Society*, XLVII. pp. 364—367, March, 1890.]

THE motion upon the surface of water of small camphor scrapings, a phenomenon which had puzzled several generations of inquirers, was satisfactorily explained by Van der Mensbrugghe* as due to the diminished surface-tension of water impregnated with that body. In order that the rotations may be lively, it is imperative, as was well shown by Mr Tomlinson, that the utmost cleanliness be observed. It is a good plan to submit the internal surface of the vessel to a preliminary treatment with strong sulphuric acid. A touch of the finger is usually sufficient to arrest the movements by communicating to the surface of the water a film of grease. When the surface-tension is thus lowered, the differences due to varying degrees of dissolved camphor are no longer sufficient to produce the effect.

It is evident at once that the quantity of grease required is excessively small, so small that under the ordinary conditions of experiment it would seem likely to elude our methods of measurement. In view, however, of the great interest which attaches to the determination of molecular magnitudes, the matter seemed well worthy of investigation; and I have found that by sufficiently increasing the water surface the quantities of grease required may be brought easily within the scope of a sensitive balance.

In the present experiments the only grease tried is olive oil. It is desirable that the material which is to be spread out into so thin a film should be insoluble, involatile, and not readily oxidised, requirements which greatly limit the choice.

* *Mémoires Couronnés* (4to) of the Belgian Academy, Vol. XXXIV. 1869.

Passing over some preliminary trials, I will now describe the procedure by which the density of the oil film necessary for the purpose was determined. The water was contained in a sponge-bath of extra size, and was supplied to a small depth by means of an india-rubber pipe in connexion with the tap. The diameter of the circular surface thus obtained was 84 cm. (33″). A short length of fine platinum wire, conveniently shaped, held the oil. After each operation it was cleaned by heating to redness, and counterpoised in the balance. A small quantity of oil was then communicated, and determined by the difference of readings. Two releasements of the beam were tried in each condition of the wire, and the deduced weights of oil appeared usually to be accurate to $\frac{1}{20}$ milligram at least. When all is ready, camphor scrapings are deposited upon the water at two or three places widely removed from one another, and enter at once into vigorous movement. At this stage the oiled extremity of the wire is brought cautiously down so as to touch the water. The oil film advances rapidly across the surface, pushing before it any dust or camphor fragments which it may encounter. The surface of the liquid is then brought into contact with all those parts of the wire upon which oil may be present, so as to ensure the thorough removal of the latter. In two or three cases it was verified by trial that the residual oil was incompetent to stop camphor motions upon a surface including only a few square inches.

The manner in which the results are exhibited will be best explained by giving the details of the calculation for a single case, *e.g.*, the second of December 17. Here 0·81 milligram of oil was found to be very nearly enough to stop the movements. The volume of oil in cubic centimetres is deduced by dividing 0·00081 by the sp. gr., viz., 0·9. The surface over which this volume of oil is spread is

$$\tfrac{1}{4}\pi \times 84^2 \text{ square centimetres};$$

so that the thickness of the oil film, calculated as if its density were the same as in more normal states of aggregation, is

$$\frac{0{\cdot}00081}{0{\cdot}9 \times \frac{1}{4}\pi \times 84^2} = \frac{1{\cdot}63}{10^7}\text{cm.},$$

or 1·63 micro-millimetres. Other results, obtained as will be seen at considerable intervals of time, are collected in the Table. For convenience of comparison they are arranged, not in order of date, but in order of densities of film.

The sharpest test of the quantity of oil appeared to occur when the motions were nearly, but not quite, stopped. There may be some little uncertainty as to the precise standard indicated by "nearly enough," and it may have varied slightly upon different occasions. But the results are quite distinct, and under the circumstances very accordant. The thickness of oil required to take the life out of the camphor movements lies between one and two millionths of a millimetre, and may be estimated with some precision at 1·6 micro-millimetre. Preliminary results from a water surface of less area are quite in harmony.

A Sample of Oil, somewhat decolourised by exposure.

Date	Weight of oil	Calculated thickness of film	Effect upon camphor fragments
Dec. 17...	0·40 mg.	0·81	No distinct effect.
Jan. 11...	0·52	1·06	Barely perceptible.
Jan. 14...	0·65	1·32	Not quite enough.
Dec. 20...	0·78	1·58	Nearly enough.
Jan. 11...	0·78	1·58	Just enough.
Dec. 17...	0·81	1·63	Just about enough.
Dec. 18...	0·83	1·68	Nearly enough.
Jan. 22...	0·84	1·70	About enough.
Dec. 18...	0·95	1·92	Just enough.
Dec. 17...	0·99	2·00	All movements very nearly stopped.
Dec. 20...	1·31	2·65	Fully enough.
A fresh Sample.			
Jan. 28...	0·63	1·28	Barely perceptible.
Jan. 28...	1·06	2·14	Just enough.

For purposes of comparison it will be interesting to note that the thickness of the black parts of soap films was found by Messrs Reinold and Rücker to be 12 micro-millimetres.

An important question presents itself as to how far these water surfaces may be supposed to have been clean to begin with. I believe that all ordinary water surfaces are sensibly contaminated; but the agreement of the results in the Table seems to render it probable that the initial film was not comparable with that purposely contributed. Indeed, the difficulties of the experiments proved to be less than had been expected. Even a twenty-four hours' exposure to the air of the laboratory* does not usually render a water surface unfit to exhibit the camphor movements.

The thickness of the oil films here investigated is of course much below the range of the forces of cohesion; and thus the tension of the oily surface

* In the country.

may be expected to differ from that due to a complete film, and [calculated] by addition of the tensions of a water-oil surface and of an oil-air surface. The precise determination of the tension of oily surfaces is not an easy matter. A capillary tube is hardly available, as there would be no security that the degree of contamination within the tube was the same as outside. Better results may be obtained from the rise of liquid between two parallel plates. Two such plates of glass, separated at the corners by thin sheet metal, and pressed together near the centre, dipped into the bath. In one experiment of this kind the height of the water when clean was measured by 62. When a small quantity of oil, about sufficient to stop the camphor motions, was communicated to the surface of the water, it spread also over the surface included between the plates, and the height was depressed to 48. Further additions of oil, even in considerable quantity, only depressed the level to 38.

The effect of a small quantity of oleate of soda is much greater. By this agent the height was depressed to 24, which shows that the tension of a surface of soapy water is much less than the combined tensions of a water-oil and of an oil-air surface. According to Quincke, these latter tensions are respectively 2·1 and 3·8, giving by addition 5·9; that of a water-air surface being 8·3. When soapy water is substituted for clean, the last number certainly falls to less than half its value, and therefore much below 5·9.

169.

FOAM.

[*Proceedings of the Royal Institution*, XIII. pp. 85—97, March, 1890.]

WHEN I was turning over in my mind the subject for this evening, it occurred to me to take as the title of the lecture, "Froth." But I was told that a much more poetical title would be "Foam," as it would so easily lend itself to appropriate quotations. I am afraid, however, that I shall not be able to keep up the poetical aspect of the subject very long; for one of the things that I shall have most to insist upon is that foaming liquids are essentially impure, contaminated—in fact, dirty. Pure liquids will not foam. If I take a bottle of water and shake it up, I get no appreciable foam. If, again, I take pure alcohol, I get no foam. But if I take a mixture of water with 5 per cent. of alcohol there is a much greater tendency. Some of the liquids we are most familiar with as foaming, such as beer or ginger-beer, owe the conspicuousness of the property to the development of gas in the interior, enabling the foaming property to manifest itself; but of course the two things are quite distinct. Dr Gladstone proved this many years ago by showing that beer from which all the carbonic acid had been extracted *in vacuo* still foamed on shaking up. I now take another not quite pure but strong liquid, acetic acid, and from it we shall get no more foam than we did from the alcohol or the water. The bubbles, as you see, break up instantaneously. But if I take a weaker acid, the ordinary acid of commerce, there is more, though still not much, tendency to foam. But with a liquid which for many purposes may be said to contain practically no acetic acid at all, seeing that it consists of water with but one-thousandth part of acid, the tendency is far stronger; and we get a very perceptible amount of foam. These tests with the alcohol and acetic acid are sufficient to illustrate the principle that the property of foaming depends on contamination. In pure ether we have a liquid from which the bubbles break even more quickly than from alcohol or water. They are gone in a moment. In some experiments I made at home I found that water containing a small proportion of ether foamed freely; but on attempting two or three days ago to repeat the experiment, I was surprised to find a result very different. I have here some water containing a very small fraction of ether, about 1/240th part. If I shake it up, it scarcely foams at all; but another mixture made in the same proportion from another sample shows more tendency to foam. This is rather curious, because both ethers were supposed to be of the same quality; but one had been in the laboratory longer than the other, and perhaps contained more greasy matter in solution.

Another liquid which foams freely is water impregnated with camphor. Camphor dissolves sparingly; but a minute quantity of it quite alters the characteristics of water in this respect. Another substance, very minute quantities of which communicate the foaming property to water, is glue or gelatine. This liquid contains only 3 parts in 100,000 of gelatine, but it gives a froth entirely different from that of pure water. Not only are there more bubbles, but the duration of the larger bubbles is quite out of proportion. This sample contains 5 parts in 100,000, nearly double as much; but even with but 1 part in 100,000, the foaming property is so evident as to suggest that it might in certain cases prove valuable for indicating the presence of minute quantities of impurities. I have been speaking hitherto of those things which foam slightly. They are not to be compared with, say, a solution of soap in water, which, as is well known to everybody, froths very vigorously. Another thing comparable to soap, but not so well known, is saponine. It may be prepared from horse chestnuts by simply cutting them in small slices and making an infusion with water. A small quantity of this infusion added to water makes it foam strongly. The quantity required to do this is even less than in the case of soap; so the test is more delicate. It is well known that rivers often foam freely. That is no doubt due to the effect of saponine or some analogous substance*. Sea-water foams, but not, I believe, on account of the saline matter it contains; for I have found that even a strong solution of pure salt does not foam much. I believe it has been shown that the foaming of sea-water, often so conspicuous, is due to something extracted from seaweeds during the concussion which takes place under the action of breakers.

Now let us consider for a moment what is the meaning of foaming. A liquid foams when its films have a certain durability. Even in the case of pure water, alcohol, and ether, these films exist. If a bubble rises, it is covered for a moment by a thin film of the liquid. This leads us to consider the properties of liquid films in general. One of their most important and striking properties is their tendency to contract. Such surfaces may be regarded as being in the condition of a stretched membrane, as of india-rubber,—only with this difference, that the tendency to contract never ceases.

* [1901. Compare "Experiments upon Surface-Films," *Phil. Mag.* XXXIII. p. 370, 1892.]

We may show that by blowing a small soap bubble, and then removing the mouth. The air is forced back again by the pressure exerted on the bubble by the tension of the liquid. This ancient experiment suffices to prove conclusively that liquid films exercise tension.

A prettier form of the same experiment is due to Van der Mensbrugghe, who illustrated liquid tension by means of a film in which he allowed to float a loop of fine silk, tied in a knot. As long as the interior of the loop, as well as the exterior, is occupied by the liquid film, it shows no tendency to take any particular shape: but if, by insertion of, say, a bit of blotting paper, the film within the loop be ruptured, then the tension of the exterior film is free to act, and the thread flies instantaneously into the form of a circle, in consequence of the tendency of the exterior surface to become as small as possible. The exterior part is now occupied by the soap film, and the interior is empty [shown]. Many other illustrations of this property of liquids might be given, but time does not permit.

In the soap film, as in the films which constitute ordinary foam, each thin layer of liquid has two surfaces; each tends to contract; but in many cases we have only one such surface to consider, as when a drop of rain falls through the air. Again, suppose that we have three materials in contact with one another,—water, oil, and air. There are three kinds of surfaces separating the three materials, one separating water and oil, another oil and air, and a third surface separating the water from the air. These three surfaces all exert a tension, and the shape of the mass of oil depends upon the relative magnitudes of the tensions. As I have drawn it here (Fig. 1), it is implied that the tension of the water-air surface is less than the sum of the other two tensions—those of the water-oil surface and the air-oil surface; because the two latter acting obliquely balance the former. It is only under such conditions that the equilibrium of the three materials as there drawn in contact with one another is possible. If the tension of the surface separating water and air exceeded the sum of the other two, then the equilibrium as depicted would be impossible. The water-air tension, being greater, would assert its superiority by drawing out the edge of the lens, and the oil would tend to spread itself more and more over the surface.

Fig. 1.

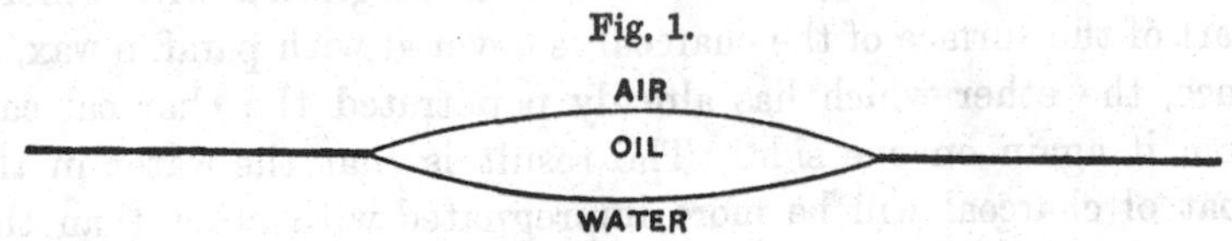

And that is what really happens. Accurate measurements, made by Quincke and others, show that the surface tension separating water and air is really greater than the sum of the two others. So oil does tend to spread upon a surface of water and air. That this is the fact, we can prove by a simple experiment. At the feet of our chairman is a large dish, containing water which at present is tolerably clean. In order to see what may happen to the surface of the water, it is dusted over with fine sulphur powder, and illuminated with the electric light. If I place on the surface a drop of water, no effect ensues; but if I take a little oil, or better still a drop of saponine, or of soap-water, and allow that to be deposited upon the middle of the surface, we shall see a great difference. The surface suddenly becomes dark, the whole of the dust being swept away to the boundary. That is the result of the spread of the film, due to the presence of the oil.

How then is it possible that we should get a lens-shaped mass of oil, as we often do, floating upon the surface of water? Seeing that the general tendency of oil is to spread over the surface of water, why does it not do so in this case? The answer is that it has already spread, and that this surface is not really a pure water surface at all, but one contaminated with oil. It is in fact only after such contamination that an equilibrium of this kind is possible. The volume of oil necessary to contaminate the surface of the water is very small, as we shall see presently; but I want to emphasise the point that, so far as we know, the equilibrium of the three surfaces in contact with one another is not possible under any other conditions. That is a fact not generally recognised. In many books you will find descriptions of three bodies in contact, and a statement of the law of the angles at which they meet; that the sides of a triangle, drawn parallel to the three intersecting surfaces, must be in proportion to the three tensions. No such equilibrium, and no such triangle, is possible if the materials are pure; when it occurs, it can only be due to the contamination of one of the surfaces. These very thin films, which spread on water, and, with less freedom, on solids also, are of extreme tenuity; and their existence, alongside of the lens, proves that the water prefers the thin film of oil to one of greater thickness. If the oil were spread out thickly, it would tend to gather itself back into drops, leaving over the surface of the water a film of less thickness than the molecular range.

One experiment by which we may illustrate some of these effects I owe to my colleague, Professor Dewar. It shows the variation in the surface tension of water, due to the presence on it of small quantities of ether. I hold in my hand masses of charcoal, which can be impregnated with ether. The greater part of the surface of the charcoal is covered with paraffin wax, and, in consequence, the ether which has already penetrated the charcoal can only escape from it again on one side. The result is that the water in the rear of this boat of charcoal will be more impregnated with ether than the part in front, so the mass of charcoal will enter into motion, and the motion will extend over a considerable interval of time. As long as the ether remains in sufficient quantity to contaminate the water in the rear, so long is there a tendency to movement of the mass. The water covered with the film of ether has less tension than the pure water in front, and the balance of tensions being upset, the mass is put in motion. If the nature of the case is such that the whole surface surrounding the solid body is contaminated, then there is no tendency to movement, the same balance in fact obtaining as if the water were pure.

Another body which we may use for this purpose is camphor. If we spread some camphor scrapings on a surface of pure water, they will, if the surface is quite clean, enter into vigorous movement, as you now see. This is because the dissolved camphor diminishes the surface tension of the water. But if I now contaminate the water with the least possible quantity of grease, the movements of the camphor will be stopped. I merely put my finger in, and you observe the effect. There is not much poetry about that! A very slight film, perfectly invisible by ordinary means, is sufficient so to contaminate the water that the effect of the dissolved camphor is no longer visible.

I was very desirous to ascertain, if possible, the actual thickness of oil necessary to produce this effect, because all data relating to molecules are, in the present state of science, of great interest. From what I have already said, you may imagine that the quantity of oil required is very small, and that its determination may be difficult. In my experiments*, I used the surface of water contained in a large sponge bath three feet in diameter. By this extension of the surface, I was able to bring the quantity of oil required within the range of a sensitive balance. In Diagram 2 [see Vol. III. p. 349], I have given a number of results obtained at various dates, showing the quantity of oil required to produce the effects recorded in the fourth column. Knowing the weight of the oil deposit, and the area of the water surface upon which it was uniformly spread, it was easy to calculate the thickness of the film. It is seen that a film of oil about $1\frac{1}{2}$ millionth of a millimetre thick is able to produce this change. I know that large numbers are not readily appreciated, and I will therefore put the matter differently. The thickness of the oil film thus determined as sufficient to stop the motions of the camphor is $\frac{1}{400}$ of the wave-length of yellow light. Another way of saying the same thing is that this thickness of oil bears to one inch the same ratio that one second of time bears to half a year.

When the movement of the camphor has been stopped by the addition of a minute quantity of oil, it is possible, by extending the water surface enclosed within the boundary, without increasing the quantity of oil, to revive the movements of the camphor; or, again, by contraction, to stop them. I can do this with the aid of a flexible boundary of thin sheet brass; and you see that the camphor recovers its activity, though a moment ago it was quite dead. It would be an interesting subject for investigation to determine what is the actual tension of an oily surface contaminated to an

* *Proc. Roy. Soc.* March, 1890.

extent just sufficient to stop the camphor movements; but it is not an easy problem. Usually we determine surface tensions by the height to which the liquids will rise in very fine tubes. Here, however, that method is not available, because if we introduce a tube into such a surface, there is no proof that the contamination of the inner surface in the tube is the same as that prevailing outside. Another method, however, may be employed which is less open to the above objection, and that is to substitute for the very fine or capillary tube, a combination of two parallel plates open at their edges. We have here two such plates of glass, kept from absolutely closing by four pieces of thin metal inserted at the corners, the plates being held close against these distance-pieces by suitable clamps. If such a combination be inserted in water, the liquid will rise above the external level, and the amount of the rise is a measure of the surface tension of the water. You see now the image on the screen. *A* is the external water surface; *B* is the height of the liquid contained between the glass plates, so that the tension may be said to be measured by the distance *AB*. If a little oil be now deposited upon the surface, it will find its way between the plates. The fall which you now see shows that the surface tension has been diminished by the oil which has found its way in. A very minute quantity will give a great effect. When the height of the pure water was measured by 62, a small quantity of oil changed the 62 into 48, and subsequent large additions of oil could only lower it to 38. But after oil has done its worst, a further effect may be produced by the addition of soap. If Mr Gordon now adds some soap, we shall find that there is a still further fall in the level, showing that the whole tension now in operation is not much more than one-third of what it was at first. This is an important point, because it is sometimes supposed that the effect of soap in diminishing the tension of water is due merely to the formation upon the surface of a layer of oil by decomposition of the soap. This experiment proves the contrary, because we find that soap can do so much more than oil. There is, indeed, something more or less corresponding to the decomposition of the soap and the formation of a superficial layer of oil. But the decomposition takes place in a very peculiar manner, and under such conditions that there is a gradual transition from the soapy liquid in the interior to the oily layer at the top, and not, as when we float a layer of oil on water, two sudden transitions, first from water to oil, and secondly from oil to air. The difference is important, because, as I showed some years ago [Vol. I. p. 234], capillary tension depends on the suddenness of change. If we suppose that the change from one liquid to another takes place by slow stages, though the final change may be as before, the capillary tension would absolutely disappear.

There is another very interesting class of phenomena due to oil films, which I hope to illustrate, though I am conscious of the difficulty of the task,—namely, the action of oil in preventing the formation of waves. From the earliest times we have records of the effect of oil in stilling waves, and all through the Middle Ages the effect was recognised, though connected with magic and fanciful explanations. Franklin, than whom, I suppose, no soberer inquirer ever existed, made the thing almost a hobby. His attention was called to it accidentally on board ship from noticing the effect on the waves caused by the greasy débris of a dinner. The captain assured him that it was due to the oil spread on the water, and for some time afterwards Franklin used to carry oil about with him, so as never to miss a chance of trying an experiment. A pond is necessary to illustrate the phenomena properly, but we shall get an idea of it by means of this trough six feet long, containing water*. Along the surface of the water we shall make an artificial wind by means of a fan†, driven by an electro-motor. In my first experiments I used wind from an organ bellows, which is not here available. Presently we shall get up a ripple, and then we will try the effect of a drop of oil put on to windward. I have now put on the drop, and you see a smooth place advancing along. As soon as the waves come up again, I will repeat the experiment. While the wind is driving the oil away, I may mention that this matter has been tested at Peterhead. Experiments were there made on a large scale to demonstrate the effect of oil in facilitating the entrance of ships into harbour in rough weather. Much advantage was gained. But here a distinction must be observed. It is not that the large swell of the ocean is damped down. That would be impossible. The action in the first instance is upon the comparatively small ripples. The large waves are not directly affected by the oil; but it seems as if the power of the wind to excite and maintain them is due to the small ripples which form on their backs, and give the wind, as it were, a better hold of them. It is only in that way that large waves can be affected. The immediate effect is on the small waves which conduce to that breaking of the large waves which from the sailor's point of view is the worst danger. It is the breaking waters which do the mischief, and these are quieted by the action of the oil.

I want to show also, though it can only be seen by those near, the return of the oil when the wind is stopped. The oil is at present driven to one end of the trough‡; when the wind stops, it will come back, because the oil film tends to spread itself uniformly over the surface. As it comes back, there will be an advancing wave of oil; and as we light the surface very obliquely by the electric lamp, there is visible on the bottom of the trough a white line, showing its progress.

* The width is 8 inches, and the depth 4 inches. The sides are of glass; the bottom and ends of wood, painted white.

† For this fan and its fittings the Institution is indebted to the liberality of the Blackman Ventilating Company.

‡ May, 1890. Any moderate quantity of oil may be driven off to leeward; but if oleate of soda be applied, the quieting effect is permanent.

Now, as to the explanation. The first attempt on the right lines was made by the Italian physicist, Marangoni. He drew attention to the importance of contamination upon the surface of the water, and to its tendency to spread itself uniformly, but for some reason which I cannot understand, he applied the explanation wrongly. More recently Reynolds and Aitken have applied the same considerations with better success. The state of the case seems to be this:—Let us consider small waves as propagated over the surface of clean water. As the waves advance, the surface of the water has to submit to periodic extensions and contractions. At the crest of a wave the surface is compressed, while at the trough it is extended. As long as the water is pure there is no force to oppose that, and the wave can be propagated without difficulty; but if the surface be contaminated, the contamination strongly resists the alternate stretching and contraction. It tends always, on the contrary, to spread itself uniformly; and the result is that the water refuses to lend itself to the motion which is required of it. The film of oil may be compared to an inextensible membrane floating on the surface of the water, and hampering its motion; and under these conditions it is not possible for the waves to be generated, unless the forces are very much greater than usual. That is the explanation of the effect of oil in preventing the formation of waves.

The all-important fact is that the surface has its properties changed, so that it refuses to submit to the necessary extensions and contractions. We may illustrate this very simply by dusting the surface of water with sulphur powder, only instead of dispersing the sulphur, as before, by the addition of a drop of oil, we will operate upon it by a gentle stream of wind projected downwards on the surface, and of course spreading out radially from the point of impact. If Mr Gordon will blow gently on the surface in the middle of the dusty region, a space is cleared*; if he stops blowing, the dust comes back again. The first result is not surprising, but why does the dusty surface come back? Such return is opposed to what we should expect from any kind of viscosity, and proves that there must be some force directly tending to produce that particular motion. It is the superior tension of the clean surface. No oil has been added here, but then no water surface is ever wholly free from contamination; there may be differences of degree, but contamination is always present to some extent. I now make the surface more dirty and greasy by contact of the finger, and the experiment no longer succeeds, because the jet of wind is not powerful enough to cleanse the place on which it impinges; the dirty surface refuses to go away, or if it goes in one direction it comes back in another.

I want now to bring to your notice certain properties of soap solutions, which, however, are not quite so novel as I thought when I first came upon

* This experiment is due to Mr Aitken. [It was exhibited by projection.]

them in my own inquiries*. If we measure by statical, or slow, methods the surface tension of soapy water, we find it very much less than that of clean water. We can prove this in a very direct manner by means of capillary tubes. Here, shown upon the screen, are two tubes of the same diameter, in which, therefore, if the liquids were the same, there would be the same elevation; one tube dips into clean water, and the other into soapy water, and the clean water rises much (nearly three times) higher than the soapy water.

Although the tension of soapy water is so much less than that of pure water when measured in this way, I had some reason to suspect that the case might be quite different if we measured the tensions immediately after the formation of the surfaces. I was led to think so by pondering on Marangoni's view that the behaviour of foaming liquids was due to the formation of a pellicle upon their surfaces; for if the change of property is due to the formation of a pellicle, it is reasonable to suppose that it will take time, so that if we can make an observation before the surface is more than say $\frac{1}{100}$ of a second old, we may expect to get a different result. That may seem an impossible feat, but there is really no difficulty about it; all that is necessary is to observe a jet of the substance in question issuing from a fine orifice. If such a jet issues from a circular orifice it will be cylindrical at first, and afterwards resolve itself into drops. If, however, the orifice is not circular, but elongated or elliptical, the jet undergoes a remarkable transformation before losing its integrity. As it issues from the elliptical orifice, it is in vibration, and trying to recover the circular form; it does so, but afterwards the inertia tends to carry it over to the other side of equilibrium. The section oscillates between the ellipse in one direction and the ellipse in the perpendicular direction. The jet thus acquires a sort of chain-like appearance, and the period of the movement, represented by the distance between corresponding points [A, B, Fig. (3), Art. 167], is a measure of the capillary tension to which these vibrations of the elliptical section about the circular form are due. A measure, then, of the wave-length of the recurrent pattern formed by the liquid gives us information as to the tension immediately after escape; and if we wish to compare the tensions of various liquids, all we have to do is to fill a vessel alternately with one liquid and another, and compare the wave-lengths in the various cases. The jet issues from a flask, to which is attached below a tubular prolongation; the aperture is made small in order that we may be able to deal with small quantities of liquid. You now see the jet upon the screen. As it issues from the orifice, it oscillates, and we can get a comparative measure of the tension by observing the distance between corresponding points (A, B).

* I here allude to the experiments of Dupré, and to the masterly theoretical discussion of liquid films by Professor Willard Gibbs.

If we were now to take out the water, and substitute for it a moderately strong solution of soap or saponine, we should find but little difference, showing that in the first moments the tension of soapy water is not very different from that of pure water. It will be more interesting to exhibit a case in which a change occurs. I therefore introduce another liquid, water containing 10 per cent. of alcohol, and you see that the wave-length is different from before. So this method gives us a means of investigating the tensions of surfaces immediately after their formation. If we calculate by known methods how long the surface has been formed before it gets to the point *B*, at which the measurement is concluded, we shall find that it does not exceed $\frac{1}{100}$ of a second.

Another important property of contaminated surfaces is what Plateau and others have described as superficial viscosity. There are cases in which the surfaces of liquids—of distilled water, for example—seem to exhibit a special viscosity, quite distinct from the ordinary interior viscosity, which is the predominant factor in determining the rate of flow through long narrow tubes. Plateau's experiment was to immerse a magnetised compass needle in water; the needle turns, as usual, upon a point, and the water is contained in a cylindrical vessel, not much larger than the free rotation of the needle requires (Fig. 4). The observation relates to the time occupied by the needle in returning to its position of equilibrium in the meridian, after having been deflected into the east and west positions, and Plateau found that in the case of water more time was required when the needle was just afloat than when it was wholly immersed, whereas in the case of alcohol the time was greater in the interior. The longer time occupied when the needle is upon the surface of water is attributed by Plateau to an excessive superficial viscosity of that body.

Fig. 4.

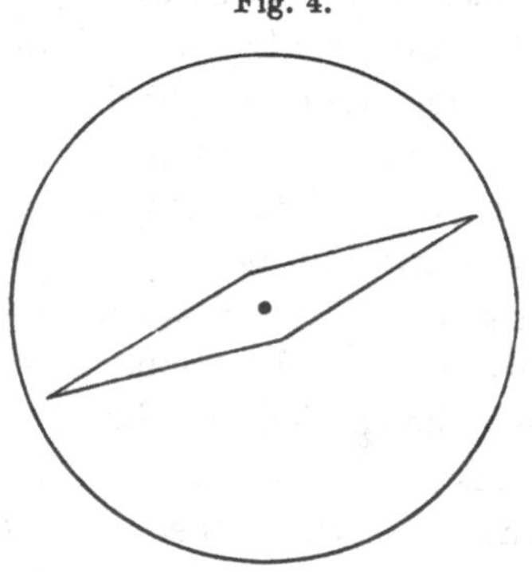

Instead of a needle, I have here a ring of brass wire (Fig. 5), floating on the surface of the water. You see upon the screen the image of the ring, as well as the surface of the water, which has been made visible by sulphur. The ring is so hung from a silk fibre that it can turn upon itself, remaining all the while upon the surface of the water. Attached to it is a magnetic needle, for the purpose of giving it a definite set, and of rotating it as required by an external magnet. On this water, which is tolerably clean, when the ring is made to turn, it leaves the dust in the interior entirely behind. That shows that the water inside the ring offers no resistance to the shearing action brought into play. The part of the surface of water immediately in contact with the ring no doubt goes round; but the movement spreads to a very little distance. The same would be observed if we added soap. But if I add some saponine, we shall find a different result, and that the behaviour of the dust in the interior of the ring is materially altered.

Fig. 5. Fig. 6.

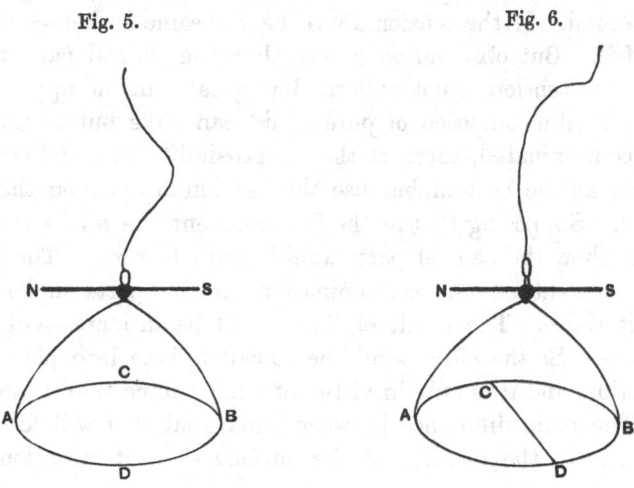

The saponine has stiffened the surface, so that the ring turns with more difficulty; and when it turns, it carries round the whole interior with it. The surface has now got a stiffness from which before it was free; but the point upon which I wish to fix your attention is that the surface of pure water does not behave in the same way. If, however, we substitute for the simple hoop another provided with a material diameter (Fig. 6), lying also in the surface of the water, then we shall find, as was found by Plateau in his experiment, that the water is carried round. In this case, it is no longer possible for the surface to be left behind, as it was with the simple hoop, unless it is willing to undergo local expansions and contractions of area. The difference of behaviour proves that what a water surface resists is not shearing, but expansions and contractions; in fact, it behaves just as a contaminated surface should do. On this supposition, it is easy to explain the effects observed by Plateau; but the question at once arises, can we believe that all water surfaces hitherto experimented upon are sensibly contaminated? and if yes, is there any means by which the contamination may be removed? I cannot in the time at my disposal discuss this question fully, but I may say that I have succeeded in purifying the surface of the water in Plateau's experiment, until it behaved like alcohol. It is therefore certain that Plateau's superficial viscosity is due to contamination, as was conjectured by Marangoni.

I must now return to the subject of foam, from which I may seem to have digressed, though I have not really done so. Why does surface contamination enable a film to exist with greater permanence than it otherwise could? Imagine a vertical soap film. Could the film continue to exist if the tension were equal at all its parts? It is evident that the film could not exist for more than a moment; for the interior part, like the others, is acted on by gravity, and, if no other forces are acting, it will fall 16 feet in a second. If the tension above be the same as below, nothing can prevent the fall. But observation proves that the central parts do not fall, and thus that the tension is not uniform, but greater in the upper parts than in the lower. A film composed of pure liquid can have but a very brief life. But if it is contaminated, there is then a possibility of a different tension at the top and at the bottom, because the tension depends on the degree of contamination. Supposing that at the first moment the film were uniformly contaminated, then the central parts would begin to drop. The first effect would be to concentrate the contamination on the parts underneath and to diminish it above. The result of that would be an increase of tension on the upper parts. So the effect would be to call a force into play tending to check the motion, and it is only in virtue of such a force that a film can have durability. The main difference between a material that will foam and one that will not, is in the liability of the surface to contamination from the interior.

I find my subject too long for my time, and must ask you to excuse the hasty explanations I have given in some parts. But I was anxious above all to show the principal experiments upon which are based the views that I have been led to entertain.

170.

ON THE SUPERFICIAL VISCOSITY OF WATER.

[*Proceedings of the Royal Society*, XLVIII. pp. 127—140, 1890.]

THE idea that liquids are endowed with a viscosity peculiar to the surface is to be found in the writings of Descartes and Rumford; but it is to Plateau that its general acceptance is due. His observations related to the behaviour of a compass needle, turning freely upon a point, and mounted in the centre of a cylindrical glass vessel of diameter not much more than sufficient to allow freedom of movement. By means of an external magnet the needle was deflected 90° from the magnetic meridian. When all had come to rest the magnet was suddenly removed, and the time occupied by the needle in recovering its position of equilibrium, or rather in traversing an arc of 85°, was noted. The circumstances were varied in two ways: first, by a change of liquid, *e.g.*, from water to alcohol; and, secondly, by an alteration in the level of the liquid relatively to the needle. With each liquid observations were made, both when the needle rested on the surface, so as to be wetted only on the under side, and also when wholly immersed to a moderate depth. A comparison of the times required in the two cases revealed a remarkable dependence upon the nature of the liquid. With water, and most aqueous solutions, the time required upon the surface was about *double* of that in the interior; whereas, with liquids of Plateau's second category, alcohol, ether, oil of turpentine, &c., the time on the surface was about *half* of the time in the interior. Of liquids in the third category (from which bubbles may be blown), a solution of soap behaved in much the same manner as the distilled water of the first category. On the other hand, solutions of albumen, and notably of saponine, exercised at their surfaces an altogether abnormal resistance.

These experiments of Plateau undoubtedly establish a special property of the surfaces of liquids of the first and third categories; but the question remains open whether the peculiar action upon the needle is to be attributed to a viscosity in any way analogous to the ordinary internal viscosity which governs the flow through capillary tubes.

In two remarkable papers*, Marangoni attempts the solution of this problem, and arrives at the conclusion that Plateau's superficial viscosity may be explained as due to the operation of causes already recognised. In the case of water and other liquids of the first category, he regards the resistance experienced by the needle as mainly the result of the deformation of the meniscuses developed at the contacts on the two sides with the liquid surface. This view does not appear to me to be sound; for a deformation of a meniscus due to inertia would not involve any dissipation of energy, nor permanent resistance to the movement. But the second suggestion of Marangoni is of great importance.

On various grounds the Italian physicist concludes that "many liquids, and especially those of Plateau's third category, are covered with a superficial pellicle; and that it is to this pellicle that they owe their great superficial viscosity." After the observations of Dupré† and myself‡, supported as they are by the theory of Professor Willard Gibbs§, the existence of the superficial pellicle cannot be doubted; and its mode of action is thus explained by Marangoni||:—"The surface of a liquid, covered by a pellicle, possesses two superficial tensions; the first, which is the weaker and in constant action, is due to the pellicle; the second is in the latent state, and comes into operation only when the pellicle is ruptured. Since the latter tension exceeds the former, it follows that any force which tends to rupture the superficial pellicle upon a liquid encounters a resistance which increases with the difference of tensions between the liquid and the pellicle." In Plateau's experiment the advancing edge of the needle tends to concentrate the superficial contamination, and the retreating edge to attenuate it; the tension in front is thus inferior to the tension behind, and a force is called into operation tending to check the vibration. On a pure surface it is evident that nothing of this sort can occur, unless it be in a very subordinate degree, as the result of difference of temperature.

There is an important distinction, discussed by Willard Gibbs, according as the contamination, to which is due the lowering of tension, is merely accidentally present upon the surface, or is derived from the body of the liquid under the normal operation of chemical and capillary forces. In the latter case, that, for example, of solutions of soap and of camphor, the changes

* *Nuovo Cimento*, Ser. 2, Vol. v. vi. Apr. 1872; *Nuovo Cimento*, Ser. 3, Vol. iii. 1878.

† *Théorie Mécanique de la Chaleur*, Paris, 1869, p. 377.

‡ "On the Tension of Recently Formed Liquid Surfaces," *Roy. Soc. Proc.* Vol. xlvii. 1890, p. 281 (*supra*). [Vol. iii. p. 341.]

§ *Connecticut Acad. Trans.* Vol. iii. Part 2, 1877—78. In my former communication I overlooked Prof. Gibbs's very valuable discussion on this subject.

|| *Nuovo Cimento*, Vol. v. vi. 1871—72, p. 260 (May, 1872).

of tension which follow an extension or contraction of the surface may be of very brief duration. After a time, dependent largely upon the amount of contaminating substance present in the body of the liquid, equilibrium is restored, and the normal tension is recovered. On the other hand, in the case of a surface of water contaminated with a film of insoluble grease, the changes of tension which accompany changes of area are of a permanent character.

It is not perfectly clear how far Marangoni regarded his principle of surface elasticity as applicable to the explanation of Plateau's observations upon distilled water; but, at any rate, he applied it to the analogous problem of the effect of oil in calming ripples. It is unfortunate that this attempt at the solution of a long-standing riddle cannot be regarded as successful. He treats the surface of the sea in its normal condition as contaminated, and therefore elastic, and he supposes that, upon an elastic surface, the wind will operate efficiently. When oil is scattered upon the sea, a non-elastic surface of oil is substituted for the elastic surface of the sea, and upon this the wind acts too locally to generate waves. It is doubtless true that an excess of oil may render a water surface again inelastic; but I conceive that the real explanation of the phenomenon is to be found by a precisely opposite application of Marangoni's principle, as in the theories of Reynolds* and Aitken†. Marangoni was, perhaps, insufficiently alive to the importance of *varying degrees* of contamination. An ordinary water surface is indeed more or less contaminated; and on that account is the less, and not the more, easily agitated by the wind. The effect of a special oiling is, in general, to increase the contamination and the elasticity dependent thereupon, and stops short of the point at which, on account of saturation, elasticity would again disappear. The more elastic surface refuses to submit itself to the local variations of area required for the transmission of waves in a normal manner. It behaves rather as a flexible but inextensible membrane would do, and, by its drag upon the water underneath, hampers the free production and propagation of waves.

The question whether the effects observed by Plateau upon the surface of distilled water are, or are not, due to contamination must, I suppose, be regarded as still undecided. Oberbeck, who has experimented on the lines of Plateau, thus sums up his discussion:—"Wir müssen daher schliessen, entweder, dass der freien Wasseroberfläche ein recht bedeutender Oberflächenwiderstand zukommt, oder dass eine reine Wasseroberfläche in Berührung mit der Luft überhaupt nicht existirt‡."

Postponing for the moment the question of the origin of "superficial viscosity," let us consider its character. A liquid surface is capable of two

* *Brit. Assoc. Rep.* 1880.

† *Edinburgh Roy. Soc. Proc.* 1882—83, Vol. XII. p. 56.

‡ *Wied. Ann.* Vol. XI. 1880, p. 650.

kinds of deformation, dilatation (positive or negative) and shearing; and the question at once presents itself, is it the former or the latter which evokes the special resistance? Towards the answer of this question Marangoni himself made an important contribution in the earlier of the memoirs cited. He found (p. 245) that the substitution for the elongated needle of Plateau of a circular disc of thin brass turning upon its centre almost obliterated the distinction between liquids of the two first categories. The ratio of the superficial to the internal viscosity was now even greater for ether than for water. From this we may infer that the special superficial viscosity of water is not called into play by the motions of the surface due to the rotation of the disc, which are obviously of the nature of shearing.

A varied form of this experiment is still more significant. I have reduced the metal in contact with the water surface to a simple (2″) ring *ABCD* of thin brass wire [Fig. (5), p. 361]. This is supported by a fine silk fibre, so that it may turn freely about its centre. To give a definite set, and to facilitate forced displacements, a magnetised sewing needle, *NS*, is attached with the aid of wax. In order to make an experiment, the ring is adjusted to the surface of water contained in a shallow vessel. When all is at rest, the surface is dusted over with a little fine sulphur* and the suspended system is suddenly set into rotation by an external magnet. The result is very distinct, and contrasts strongly with that observed by Plateau. Instead of the surface enclosed by the ring being carried round with it in its rotation, not the smallest movement can be perceived, except perhaps in the immediate neighbourhood of the wire itself. It is clear that an ordinary water surface does not appreciably resist shearing.

A very slight modification of the apparatus restores the similarity to that of Plateau. This consists merely in the addition to the ring of a material diameter of the same brass wire, *CD* [Fig. (6), p. 361]. If the experiment be repeated, the sulphur indicates that the whole water surface included within the semicircles now shares in the motion. In general terms the surface may be said to be carried round with the ring, although the motion is not that of a rigid body.

Experiments of this kind prove that what a water surface resists is not shearing, but local expansions and contractions of area, even under the condition that the total area shall remain unchanged. And this is precisely what should be expected, if the cause of the viscosity were a surface contamination. A shearing movement does not introduce any variation in the density of the contamination, and therefore does not bring Marangoni's principle into play. Under these circumstances there is no resistance.

* Sulphur seems to be on the whole the best material, although it certainly communicates some impurity to the surface. Freshly heated pumice or wood-ashes sink immediately; and probably all powders really free from grease would behave in like manner.

It remains to consider liquids of the third category in Plateau's nomenclature. The addition of a little oleate of soda does not alter the behaviour of water, at least if the surface be tolerably fresh. On the other hand, a very small quantity of saponine suffices to render the surface almost rigid. In the experiment with the simple ring the whole interior surface is carried round as if rigidly attached. A similar effect is produced by gelatine, though in a less marked degree.

In the case of saponine, therefore, it must be fully admitted that there is a superficial viscosity not to be accounted for on Marangoni's principle by the tendency of contamination to spread itself uniformly. It seems not improbable that the pellicle formed upon the surface may have the properties of a solid, rather than of a liquid. However this may be, the fact is certain that a contracting saponine surface has no definite tension alike in all directions. A sufficient proof is to be found in the well-known experiment in which a saponine bubble becomes wrinkled when the internal air is removed.

The quasi-solid pellicle on the surface of saponine would be of extreme thinness, and, even if it exist, could hardly be recognisable by ordinary methods of examination. It would moreover be capable of re-absorption into the body of liquid if unduly concentrated by contraction of surface, differing in this respect from the gross, and undoubtedly solid, pellicles which form on the surface of hard water on exposure to the atmosphere.

Two further observations relative to saponine may here find a place. The wrinkling of a bubble when the contained gas is exhausted occurs also in an atmosphere (of coal gas) from which oxygen and carbonic acid are excluded.

In Plateau's experiment a needle which is held stiffly upon the surface of a saponine solution is to a great extent released, when the surface is contaminated by grease from the finger or by a minute drop of petroleum.

To return to the case of water, it is a question of the utmost importance to decide whether the superficial viscosity of even distilled water is, or is not, due to contamination with a film of foreign matter capable of lowering the tension. The experiments of Oberbeck would appear to render the former alternative very improbable; but, on the other hand, if the existence of the film be once admitted, the observed facts can be very readily explained. The question is thus reduced to this: Can we believe that the water surface in Plateau's apparatus is almost of necessity contaminated with a greasy film? The argument which originally weighed most with me, in favour of the affirmative answer, is derived from the experiments of Quincke upon mercury. It is known that, contrary to all analogy, a drop of water does not ordinarily spread upon the surface of mercury. This is certainly due to contamination with a greasy film; for Professor Quincke* found that it was possible so to

* *Pogg. Ann.* Vol. CXXXIX. 1870, p. 66.

prepare mercury that water would spread upon it. But the precautions required are so elaborate that probably no one outside Professor Quincke's laboratory has ever witnessed what must nevertheless be regarded as the normal behaviour of these two bodies in presence of one another. The bearing of this upon the question under discussion is obvious. If it be so difficult to obtain a mercury surface which shall stand one test of purity, why may it not be equally difficult to prepare a water surface competent to pass another?

The method by which I have succeeded in proving that Plateau's superficial viscosity is really due to contamination consists in the preparation of a pure surface exhibiting quite different phenomena; and it was suggested to me by an experiment of Mr Aitken*. This observer found that, if a gentle stream of air be directed vertically downwards upon the surface of water dusted over with fine powder, a place is cleared round the point of impact. It may be added that on the cessation of the wind the dust returns, showing that the tension of the bared spot exceeds that of the surrounding surface.

The apparatus, shown in Figs. 3, 4, is constructed of sheet brass. The circular part, which may be called the *well*, has the dimensions given by

Fig. 3.

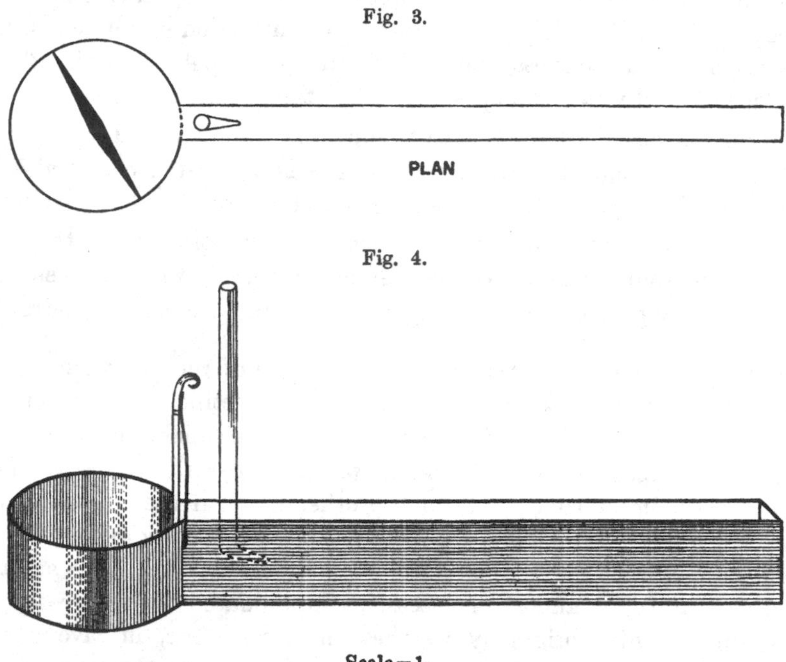

Fig. 4.

Scale = $\frac{1}{2}$.

Plateau. The diameter is 11 cm., and the depth 6 cm. The needle is 10 cm. long, 7 mm. in breadth at the centre, and about 0·3 mm. thick. It is

* *Loc. cit.* p. 69.

suspended at a height of 2½ cm. above the bottom of the vessel. So far there is nothing special; but in connexion with the well there is a rectangular trough, or tail-piece, about 2½ cm. broad and 20 cm. long. Between the two parts a sliding door may be inserted, by which the connexion is cut off, and the circular periphery of the well completed. The action of the apparatus depends upon a stream of wind, supplied from an acoustic bellows, and discharged from a glass nozzle, in a direction slightly downwards, so as to strike the water surface in the tail-piece at a point a little beyond the door. The effect of the wind is to carry any greasy film towards the far end, and thus to purify the near end of the tail-piece. When the door is up, this effect influences also the water surface in the well upon which the jet does not operate directly. For, if the tension there be sensibly less than that of the neighbouring surface in the tail-piece, an outward flow is generated, and persists as long as the difference of tensions is sensible. The movements of the surface are easily watched if a little sulphur be dusted over; when the water in the well has been so far cleansed that but little further movement is visible, the experiment may be repeated without changing the water by contaminating the surface with a little grease from the finger or otherwise. In this way the surface may be freed from an insoluble contamination any number of times, the accumulation of impurity at the far end of the tail-piece not interfering with the cleanness of the surface in the well.

Another device that I have usually employed facilitates, or at any rate hastens, the cleansing process. When the operation is nearly complete, the movement of the surface becomes sluggish on account of the approximate balance of tensions. At this stage the movement may be revived, and the purification accelerated, by the application of heat to the bottom of the well at the part furthest removed from the tail-piece. It may, perhaps, be thought that convection currents might be substituted altogether for wind; but in my experience it is not so. Until a high degree of purity is attained, the operation of convection currents does not extend to the surface, being resisted by the film according to Marangoni's principle.

When the apparatus was designed, it was hoped that the door could be made a sufficiently good fit to prevent the return of the greasy film into the well; but experience showed that this could not be relied upon. It was thus necessary to maintain the wind during the whole time of observation. The door was, however, useful in intercepting mechanical disturbance.

A very large number of consistent observations have been recorded. The return of the needle, after deflection to 90°, is timed over an arc of 60°, viz., from 90° to 30°, and is assisted by a fixed steel magnet acting in aid of the earth's magnetism. A metronome, beating three times per second, facilitates the time measurement. As an example, I may quote some observations made on April 11.

The apparatus was rinsed and carefully filled with distilled water. In this state the time was 12 (beats). After blowing for a while there was a reduction to 10, and after another operation to 8. The assistance of convection currents was then appealed to, and the time fell to $6\frac{3}{4}$, and after another operation to 6. This appeared to be the limit. The door was then opened, and the wind stopped, with the result that the time rose again to 12. More water was then poured in until the needle was drowned to the depth of about half an inch. Under these conditions the time was $6\frac{3}{4}$.

It will be seen, that while upon the unprepared surface the time was nearly twice as great as in the interior, upon the purified surface the time was somewhat less than in the interior.

For the sake of comparison, precisely similar observations were made upon the same day with substitution for water of methylated alcohol. Before the operation of wind the time was 5; after wind, 5; on repetition, still 5. Nor with the aid of convection currents could any reduction be effected. When the needle was drowned, the time rose to $7\frac{1}{4}$. The alcohol thus presents, as Plateau found, a great contrast with the unprepared water; but comparatively little with the water after treatment by wind and heat.

An even more delicate test than the time of vibration is afforded by the behaviour of the surface of the liquid towards the advancing edge of the needle. In order to observe this, it is necessary to have recourse to motes, but all superfluity should be avoided. In a good light it is often possible to see a few motes without any special dusting over. In my experience, an unprepared water surface always behaves in the manner described by Plateau; that is, it takes part in rotation of the needle, almost from the first moment. Under the action of wind a progressive change is observed. After a time the motes do not begin their movement until the needle has described a considerable arc. At the last stages of purification, a mote, situated upon a radius distant 30° or 40° from the initial direction of the needle, retains its position almost until struck; behaving, in fact, exactly as Plateau describes for the case of alcohol. I fancied, however, that I could detect a slight difference between alcohol and water even in the best condition, in favour of the former. With a little experience it was easy to predict the "time" from observations upon motes; and it appeared that the last degrees of purification told more upon the behaviour of the motes than upon the time of describing the arc of 60°. It is possible, however, that a different range from that adopted might have proved more favourable in this respect.

The special difficulties under which Plateau experimented are well known, and appealed strongly to the sympathies of his fellow workers; but it is not necessary to refer to them in order to explain the fact that the water surfaces that he employed were invariably contaminated. Guided by a knowledge of the facts, I have several times endeavoured to obtain a clean surface without the aid of wind, but have never seen the time less than 10. More often it is 12, 13, or 14. It is difficult to decide upon the source of the contamination. If we suppose that the greasy matter is dissolved, or at any rate suspended in the body of the liquid in a fine state of subdivision, it is rather difficult to understand the comparative permanence of the cleansed surfaces. In the case of distilled water, the condition will usually remain without material change for several minutes. On the other hand, with tap water (from an open cistern), which I have often used, although there is no difficulty in getting a clean surface, there is usually a more rapid deterioration on standing. The progressive diminution of the tension of well-protected water surfaces observed by Quincke* is most readily explained by the gradual formation of a greasy layer composed of matter supplied from the interior, and present only in minute quantity; although this view did not apparently commend itself to Quincke himself. If we reject the supposition that the greasy layer is evolved from the interior of the liquid, we must admit that the originally clean free surface, formed as the liquid issues from a tap, is practically certain to receive contamination from the solid bodies with which it comes into contact. The view, put forward hypothetically by Oberbeck, that contamination is almost instantly received from the atmosphere is inconsistent with the facts already mentioned.

Some further observations, made in the hope of elucidating this question, may here be recorded. First, as to the effect of soap, or rather oleate of soda. A surface of distilled water was prepared by wind and heat until the time was 5½, indicating a high degree of purity. The door being closed, so as to isolate the two parts of the surface, and the wind being maintained all the while, a few drops of solution of oleate were added to the water in the tail-piece. With the aid of gentle stirring, the oleate found its way, in a few minutes, under the door, and reached the surface of the water in the well. The time gradually rose to 13, 14, 15; and no subsequent treatment with wind and heat would reduce it again below 12. In this case there can be no doubt that the contamination comes from the interior, and is quickly renewed if necessary; not, however, so quickly that the tension is constant in spite of extension, or the surface would be free from superficial viscosity.

In like manner, the time upon the surface of camphorated distilled water could not be reduced below 10, and the behaviour of motes before the advancing needle was quite different to that observed upon a clean surface. A nearly saturated solution of chloride of sodium could not be freed from superficial viscosity; while, on the other hand, an addition of ⅓ per cent. of alcohol did not modify the behaviour of distilled water.

The films of grease that may be made evident in Plateau's apparatus are attenuated in the highest degree. In a recent paper† I have estimated

* *Pogg. Ann.* Vol. CLX. 1877, p. 580.

† *Supra*, p. 364. [Vol. III. p. 347.]

the thickness of films of olive oil competent to check the movements of camphor fragments as from one to two micro-millimetres; but these films are comparatively coarse. For example, there was never any difficulty in obtaining from tap-water surfaces upon which camphor was fully active without the aid of wind or special arrangements. I was naturally desirous of instituting a comparison between the quantities necessary to check camphor movements and the more minute ones which could be rendered manifest by Plateau's needle; but the problem is of no ordinary difficulty. A direct weighing of the contamination is out of the question, seeing that the quantity of oil required in the well of the apparatus, even to stop camphor, would be only $\frac{1}{84}$ mg.

The method that I have employed depends upon the preparation of an ethereal solution of olive oil, with which clean platinum surfaces are contaminated. It may be applied in two ways. Either we may rely upon the composition of the solution to calculate the weight of oil remaining upon the platinum after evaporation of the solvent, or we may determine the relative quantities of solution required to produce the two sorts of effects. In the latter case we are independent of the precise composition of the solution, and more especially of the question whether the ether may be regarded as originally free from dissolved oil of an involatile character. In practice, both methods have been used.

The results were not quite so regular as had been hoped, the difficulty appearing to be that the oil left by evaporation upon platinum was not completely transferred to the water surface when the platinum was immersed, even although the operation was performed slowly, and repeated two or three times. On the other hand, there was no difficulty in cleansing a large surface of platinum by ignition in the flame of a spirit lamp, so that it was absolutely without perceptible effect upon the movement of the needle over a purified water surface.

The first solution that was used contained 7 mgs. of oil in 50 c.c. of ether. The quantities of solution employed were reckoned in drops, taken under conditions favourable to uniformity, and of such dimensions that 100 drops measured 0·6 c.c. The following is an example of the results obtained:—On April 25, the apparatus was rinsed out and recharged with distilled water. Time = 13. After purification of surface by wind and heat, $5\frac{1}{2}$, rising, after a considerable interval, to 6. After insertion of a large plate of platinum, recently heated to redness, time unchanged. A narrow strip of platinum, upon which, after a previous ignition, three drops of the ethereal solution had been evaporated, was then immersed, with the result that the time was at once increased to $8\frac{1}{2}$. In subsequent trials two drops never failed to produce a distinct effect. Special experiments, in which the standard ether was tested after evaporation upon platinum, showed that nearly the whole of the effect was due to the oil purposely dissolved.

The determination of the number of drops necessary to check the movements of camphor upon the same surface seemed to be subject to a greater irregularity. In some trials 20 drops sufficed; while in others 40 or 50 drops were barely enough. There seems to be no doubt that the oil is left in a rather unfavourable condition*, very different from that of the compact drop upon the small platinum surface of former experiments; and the appearance of the platinum on withdrawal from the water often indicates that it is still greasy. Under these circumstances it is clearly the smaller number that should be adopted; but we are safe in saying that $\frac{1}{15}$ of the oil required to check camphor produces a perceptible effect upon the time in Plateau's experiment, and still more upon the behaviour of the surface before the advancing needle, as tested by observation of motes. At this rate the thickness at which superficial viscosity becomes sensible in Plateau's apparatus is about $\frac{1}{10}$ of a micro-millimetre, or about $\frac{1}{6000}$ of the wave-length of yellow light.

A tolerably concordant result is obtained from a direct estimate of the smaller quantity of oil, combined with the former results for camphor, which were arrived at under more favourable conditions. The amount of oil in two drops of the solution is about 0·0017 mg. This is the quantity which suffices to produce a visible effect upon the needle. On the large surface of water of the former experiments the oil required to check camphor was about 1 mg. In order to allow for the difference in area, this must be reduced 64 times, or to 0·016 mg. According to this estimate the ratio of thicknesses for the two classes of effects is about as 10 : 1.

Very similar results were obtained from experiments with an ethereal solution of double strength, one drop of which, evaporated as before, upon platinum, produced a distinct effect upon the time occupied by the needle in traversing the arc from 90° to 30°.

I had expected to find a higher ratio than these observations bring out between the thicknesses required for the two effects. The ratio 15 : 1 does not give any too much room for the surfaces of ordinary tap water, such as were used in the bath observations upon camphor, between the purified surfaces on the one side and those oiled surfaces upon the other which do not permit the camphor movements.

It thus became of interest to inquire in what proportion the film originally present upon the water in the bath experiments requires to be concentrated in order to check the motion of camphor fragments. This information may be obtained, somewhat roughly it is true, by dusting over a patch of the water surface in the centre of the bath. When a weighed drop of oil is deposited in the patch, it drives the dust nearly to the edge,

* It should be stated that the evaporation of the ether, and of the dew which was often visible, was facilitated by the application of a gentle warmth.

and the width of the annulus is a measure of the original impurity of the surface. When the deposited oil is about sufficient to check the camphor movements, we may infer that the original film bears to the camphor standard a ratio equal to that of the area of the annulus to the whole area of the bath. Observations of this kind indicated that a concentration of about six times would convert the original film into one upon which camphor would not freely rotate.

Another method by which this problem may be attacked depends upon the use of flexible solid boundary. This was made of thin sheet brass, and is deposited upon the bath in its expanded condition, so as to enclose a considerable area. Upon this surface camphor rotates, but the movement may be stopped by the approximation of the walls of the boundary. The results obtained by this method were of the same order of magnitude.

If these conclusions may be relied upon, it will follow that the initial film upon the water in the bath experiments is not a large multiple of that at which superficial viscosity tends to disappear. At the same time, the estimate of the total quantity of oil which must be placed upon a really pure surface in order to check the movements of camphor must be somewhat raised, say, from 1·6 to 1·9 micro-millimetre. It must be remembered, however, that on account of the want of definiteness in the effects, these estimates are necessarily somewhat vague. By a modification of Plateau's apparatus, or even in the manner of taking the observations, such as would increase the extent of surface from which the film might be accumulated before the advancing edge of the needle, it would doubtless be possible to render evident still more minute contaminations than that estimated above at one-tenth of a micro-millimetre.

Postscript, June 4.—In order to interpret with safety the results obtained by Plateau, I thought it necessary to follow closely his experimental arrangements; but the leading features of the phenomenon may be well illustrated without any special apparatus. For this purpose, the needle of the former experiments may be mounted upon the surface of water contained to a depth of 1 or 2 inches in a large flat bath. Ordinary cleanliness being observed, the motes lying in the area swept over by the needle are found to behave much as described by Plateau. Moreover, the motion of the needle, under the action of the magnet used to displace it, is decidedly sluggish. In order to purify the surface, a hoop of thin sheet brass is placed in the bath, so as to isolate a part including the needle. The width of the hoop must of course exceed the depth of the water, and that to an extent sufficient to allow of manipulation without contact of the fingers with the water. If the hoop be deposited in its contracted state, and be then opened out, the surface contamination is diminished in the ratio of the areas. By this simple device there is no difficulty in obtaining a highly purified surface, upon which motes lie quiescent, almost until struck by the oscillating needle. In agreement with what has been stated above, an expansion of three or four times usually sufficed to convert the ordinary water surface into one upon which superficial viscosity was tending to disappear.

I propose to make determinations of the actual tensions of surfaces contaminated to various degrees; but in the meantime it is evident that the higher degrees of purity do not imply much change of tension*. In the last experiment upon a tolerably pure surface, if we cause the needle to oscillate rapidly backwards and forwards through a somewhat large angle, we can clear away the contamination from a certain area. This contamination will of course tend to return, but observation of motes shows that the process is a rather slow one.

The smallness of the forces at work must be the explanation of the failure to clean the surface in Plateau's apparatus by mere expansion. For this experiment the end wall was removed from the tail-piece (Fig. 3), and a large flexible hoop substituted. By this means, it was hoped that when the whole was placed in the bath it would be possible, by mere expansion of the hoop, to obtain a clean surface in the well. The event proved, however, that the purification did not proceed readily beyond the earlier stages, unless the passage of the contamination through the long channel of the tail-piece was facilitated by wind.

* [1901. Miss Pockels' experiments (*Nature*, XLIII. p. 437, 1891) show that this is an understatement. See also Rayleigh, *Phil. Mag.* XLVIII. p. 331, 1899.]

171.

ON HUYGENS'S GEARING IN ILLUSTRATION OF THE INDUCTION OF ELECTRIC CURRENTS*.

[*Philosophical Magazine*, XXX. pp. 30—32, 1890.]

As a mechanical model of the electric machinery at work in the induction of currents, Maxwell employed differential gearing; and an apparatus on this principle, designed by him, is in use at the Cavendish Laboratory. Wishing to show something similar in a recent course of lectures, and not having differential gearing at my disposal, I designed more than one combination of pulleys, the action of which should be analogous to that of electric currents. These eventually resolved themselves into Huygens's gearing, invented, I believe, in connexion with the winding of clocks. As this apparatus is easier to understand than differential gearing, and the parts of which it is composed are more likely to be useful for general purposes in a laboratory, I have thought that it might be worth while to give a description, accompanied by an explanation of the mode of action.

Two similar pulleys, A, B, turn upon a piece of round steel fixed horizontally†. Over these is hung an endless cord, and the two bights carry similar pendent pulleys, C, D, from which again hang weights, E, F. The weight of the cord being negligible, the system is devoid of potential energy; that is, it will balance, whatever may be the vertical distance between C and D.

Since either pulley A, B may turn independently of the other, the system is capable of two independent motions. If A, B turn in the same direction and with the same velocity, one of the pendent pulleys C, D rises, and the other falls. If, on the other hand, the motions of A, B are equal and opposite, the axes of the pendent pulleys and the attached weights remain at rest.

* Read before the Physical Society on May 16, 1890.

† Light wooden laths, variously coloured and revolving with the pulleys, render the movements evident at a distance.

In the electrical analogy the rotatory velocity of A corresponds to a current in a primary circuit, that of B to a current in a secondary. If when all is at rest the rotation of A be suddenly started, by force applied at the handle or otherwise, the inertia of the masses, E, F, opposes their sudden movement, and the consequence is that the pulley B turns *backwards*, *i.e.*, in the opposite direction to the rotation imposed upon A. This is the current induced in a secondary circuit when an electromotive force begins to act in the primary. In like manner, if A having been for some time in uniform movement suddenly stops, B enters into motion in the direction of the former movement of A. This is the secondary current on the break of the current in the primary circuit.

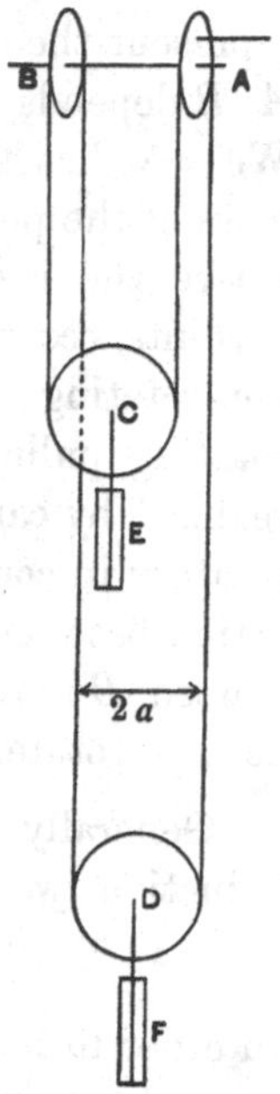

It must be borne in mind that in the absence of friction there is nothing to correspond with electrical resistance, so that the conductors must be looked upon as perfect. The frictions which actually enter do not follow the same laws as electrical resistances, and only very imperfectly represent them. However, the frictions which oppose the rotations of A and B have a general effect of the right sort; but the rotations of C and D, corresponding to dielectric machinery, should be as free as possible.

The effect of a condenser, to which the terminals of one of the circuits is joined, would be represented by a spiral spring (as in a watch) attached to the corresponding pulley, the stiffness of the spring being inversely as the capacity of the condenser. The absence of the spring, or (which comes to the same thing) the indefinite decrease of its stiffness, corresponds to infinite electrical capacity, or to a simply closed circuit.

The equations which express the mechanical properties of the system are readily found, and are precisely the same as those applicable in the electrical problem. Since the potential energy vanishes, everything turns upon the expression for the kinetic energy. If x and y denote the circumferential velocities, in the same direction, of the pulleys A, B where the cord is in contact with them, $\frac{1}{2}(x+y)$ is the vertical velocity of the pendent pulleys. Also $\frac{1}{2}(x-y)$ is the circumferential linear velocity of C, D, due to rotation, at the place where the cord engages. If the diameter be here $2a$, the angular velocity is $(x-y)/2a$. Thus, if M be the total mass of each pendent pulley and attachment, Mk^2 the moment of inertia of the revolving parts, the whole kinetic energy corresponding to each is

$$\tfrac{1}{2}M\left\{\frac{(x+y)^2}{4}+\frac{k^2}{a^2}\frac{(x-y)^2}{4}\right\}.$$

For the energy of the whole system we should have the double of this, and, if it were necessary to include them, terms proportional to x^2 and y^2 to represent the energy of the fixed pulleys. The reaction between the pulleys A, B depends upon the presence of a term xy in the expression of the energy. We see that this would disappear if $k^2 = a^2$; as would happen if the whole mass of the pendent pulleys and attachments were concentrated in the circles where the cord runs. The case discussed above, as analogous to electric currents, occurs when $k^2 < a^2$, a condition that will be satisfied, even without non-rotating attachments, if the cord run near the circumference of the rotating pulleys. The opposite state of things, in which $k^2 > a^2$, would be realized by carrying out masses beyond the groove, and thus increasing the rotatory in comparison with the translatory inertia. In this case the mutual action between A and B is reversed. If when all is at rest A be suddenly started, B moves forward in the *same* direction. Otherwise C and D would have to rotate, and this in their character of fly-wheels they oppose.

Generally, if L, N be the coefficients of self-induction, and M of mutual induction, we have (constant factors being omitted)

$$L = N = a^2 + k^2, \qquad M = a^2 - k^2.$$

In order to imitate the case of two circuits coiled together in close proximity throughout, we must have in the mechanical model $k^2 = 0$; that is, the rotatory inertia of the pendent pulleys must be negligible in comparison with the translatory inertia. Also the energy of the fixed pulleys, not included in the above expressions, must be negligible. If these conditions be satisfied, a sudden rotation imposed upon A generates an *equal* and opposite motion in B.

172.

THE BOURDON GAUGE.

[*Nature*, XLII. p. 197, 1890*.]

ALLOW me to suggest to such of your readers as are interested in this subject the following experiment. Cut out of cardboard two annular strips, each of somewhat more than a quadrant, the inner radius being, say, 7 inches, and the outer radius 9 inches. Along the middle of each strip—that is, along the circle of 8 inches radius—cut the boards half through, so as to render them flexible, and then join the two strips together with gum paper at the inner and outer edges. In this way we obtain a curved tube whose section is a rhombus, and whose curvature is connected with the magnitudes of the angles of the rhombus. The manipulation of such a tube gives definiteness to one's ideas, and enables one to recognize that internal pressure, tending to augment the included volume, and therefore to make the section square, must also cause the curvature of the axis to approach a definite associated value. In this case the deformations are practically by bending, principally indeed at the hinges; and I cannot doubt that in its main features the mechanism of an ordinary Bourdon gauge may be looked at in the same light.

* **[1901. See Vol. III. p. 230. The present note had reference to a discussion upon the theory of the Bourdon gauge which appeared in *Nature*.]**

173.

ON DEFECTIVE COLOUR VISION.

[*British Association Report (Leeds)*, pp. 728—729, 1890.]

THE existence of a defect is probably most easily detected in the first instance by Holmgren's wool test; but this method does not decide whether the vision is truly dichromic. For this purpose we may fall back upon Maxwell's colour discs. Dichromic vision allows a match between any four colours, of which black may be one. Thus we may find 64 green + 36 blue = 61 black + 39 white—a neutral matched by a green-blue. But this is apparently not the most searching test. The above match was in fact made by an observer whose vision I have reason to think is not truly dichromic, for he was unable to make a match among the four colours red, green, blue, black. The nearest approach appeared to be 100 red = 8 green + 7 blue + 85 black, but was pronounced far from satisfactory. An observer with dichromic vision, present at the same time, made without difficulty 82 red + 18 blue = 22 green + 78 black—a bright crimson against a very dark green.

It would usually be very unsafe to conclude that a colour-blind person is incapable of making a match because he thinks himself so. But, in the present instance, repeated trials led to the same result, while other matches, almost equally forced in my estimation, were effected without special difficulty. It looked as though the third colour sensation, presumably red, was defective, but not absolutely missing. When a large amount of white was present, matches could be made in spite of considerable differences in the red component, but when red light was nearly isolated its distinctive character became apparent.

This view of the matter was confirmed by experiments with my colour box, in which, by means of double refraction, a mixture of spectral red and green can be exhibited in juxtaposition with spectral yellow (*Nature*, Nov. 17, 1881) [Vol. I. p. 545]. A match to normal vision requires, of course, that (by rotation of the nicol) the red and green should be mixed in the right proportions; and secondly, that (by adjustment of gas) the brightness of the spectral yellow should be brought to the right point. An observer whose vision is dichromic does not require the first adjustment; any mixture of red and green, or even the red and green unmixed, can be matched against the yellow. In the present case, however, although the green could be matched satisfactorily against the yellow, the red could not. The construction of the instrument allowed the point to be investigated at which the match began to fail. Pure green corresponding to 0, and pure red to 25, the match with yellow began to fail when the setting reached about 17. Normal vision required a setting of about 11.

Truly dichromic vision may be thus exhibited in a diagram. If we take red, green, blue, as angular points of a triangle, there is a point upon the plane which represents darkness. Any colours which lie upon a line through this point differ only in brightness. Maxwell determined the point by comparison of colour-blind matches with his own normal ones. It seems preferable to use the colour-blind matches only, as may be done as follows:—From the match between red, green, blue, and black, the position of black on the diagram may be at once determined, and for most purposes would represent darkness sufficiently well. A match between white and the principal colours will then fix its [white's] position relatively to the fundamental points. A line joining black and white is the *neutral* line; all colours that lie on one side of it are warm, like yellow; all that lie upon the other side are cold. The point representing darkness will lie upon the neutral line and a little beyond black. The diagram sketched depends upon the following matches obtained from an observer, whom Holmgren would call green-blind:—

Red	Green	Blue	Black	White	Yellow
−82·0	+21·8	−18·0	+78·2	0	0
+57·2	−100	+ 4·8	0	+38·0	0
0	+96·0	+ 4·0	−53·0	−47·0	0
−100	0	+ 5·0	+78·7	0	+16·3

174.

INSTANTANEOUS PHOTOGRAPHS OF WATER JETS.

[*British Association Report* (*Leeds*), p. 752, 1890.]

THESE photographs were taken by the light of the electric spark. A battery of Leyden jars was charged by a Wimshurst machine, and discharged itself between brass balls, held half an inch apart, in the optical lantern. By means of a large condenser a good proportion of light was concentrated upon the lens of the camera. The jet of water, regularized by a tuning-fork, fell in front of the condenser, and was focused upon the photographic plate.

In the absence of anything to diffuse the light, the pictures are simple shadows, such as have been obtained without any optical appliances by Mr Bell and Mr Boys. The only detail is due to the lens-like action of the jets and the drops into which it is resolved. This arrangement is quite sufficient to illustrate the behaviour of electrified jets. But the interposition of a plate of ground glass close to the condenser effects a great improvement in the pictures by bringing out half-tones, and the results printed on aristotype paper are now very good. The only difficulty is that due to loss of light. In some of the experiments it was found advantageous to diminish the diffusion by slightly oiling the ground glass.

The degree of instantaneity required depends upon circumstances. In some cases the outlines would have lost their sharpness had the exposure exceeded $\frac{1}{10000}$ second. It is probable that the actual duration of the principal illumination was decidedly less than this.

175.

ON THE TENSION OF WATER SURFACES, CLEAN AND CONTAMINATED, INVESTIGATED BY THE METHOD OF RIPPLES*.

[*Philosophical Magazine*, XXX. pp. 386—400, Nov. 1890.]

NUMEROUS and varied phenomena prove that the tension of a water surface is lowered by the presence of even a trace of grease. In the case of olive-oil, a film whose calculated thickness is as low as 2 micro-millimetres is sufficient to entirely alter the properties of the surface in relation to fragments of camphor floating thereupon. It seemed to me of importance for the theory of capillarity to ascertain with some approach to precision the tensions of greasy surfaces; and in a recent paper† I gave some results applicable to the comparison of a clean surface with one just greasy enough to stop the camphor movements and also with one saturated with olive-oil. The method employed was that depending upon the rise of liquid between parallel plates of glass; and I was not satisfied with it, not merely on account of the roughness of the measurement, but also because the interpretation of the result depends upon the assumption that the angle of contact with the glass is zero. In the opinion of Prof. Quincke, whose widely extended researches in this field give great weight to his authority, this assumption is incorrect even in the case of pure liquids, and, as it seemed to me, is still less to be trusted in its application to contaminated surfaces, the behaviour of which is still in many respects obscure. I was thus desirous of checking my results by a method independent of the presence of a solid body.

The solution of the problem was evidently to be found in the observation of ripples, as proposed by Prof. Tait, upon the basis of Sir W. Thomson's theory. Thomson has shown that when the wave-length is small, the

* Read September 6 before Section A of the British Association at Leeds. [*Brit. Ass. Rep.* p. 746, 1890.]

† *Proc. Roy. Soc.* March 1890, Vol. XLVII. p. 367. [Vol. III. p. 350.]

vibration depends principally upon capillary tension; so that a knowledge of corresponding wave-lengths and periods will lead to a tolerably accurate estimate of tension.

Besides some early observations of my own*, made for the most part for another purpose, I had before me the work of Matthiessen†, who has compared Thomson's formulæ with observation over a wide range of wave-length. The results are calculated on the basis of an assumed surface-tension, and are exhibited as a comparison of calculated and observed wave-lengths. On the whole the agreement is fair; but the accuracy attained seemed to be insufficient for the purpose which I had in view. As will presently appear, an error in the wave-length is multiplied about three times in the tension deduced from it, so that, in a reversal of Matthiessen's calculations, the errors would appear much magnified.

Quite recently Mr Michie Smith has published an account of experiments made by Thomson's method for the determination of the tension of mercury. Some anomalies were met with; and it seems not improbable that the vibrations observed were in some cases an octave below those of the vibrating source‡.

When it is remembered that Thomson's theory is one of infinitely small vibrations, it will be seen that for my purpose it was necessary above all things that the amplitude of vibration should be very moderate. The sub-octave vibrations of Faraday are especially to be avoided as almost necessarily of large amplitude. At the same time the limitation is not without its inconvenience. One of the great difficulties of the experiment is to see the waves properly, and this is much increased when the vibrations are extremely small.

In considering the problem thus presented, it occurred to me that it was essentially the same as that so successfully solved by Foucault in relation to the figuring of optical surfaces. The undisturbed surface of liquid is an accurate plane. The waves upon it may be regarded as deviations from optical truth, and may be made evident in the same way as any other deviations from truth in a reflecting surface. Guided by this idea, I was able to work with waves of which nothing whatever was to be seen by ordinary observation of the surface over which they were travelling.

In the application of Foucault's method it is necessary that light from a radiant point, after reflexion from the surface under test, should be brought to a focus, in the immediate neighbourhood of which is placed the eye of the

* "On the Crispations of Fluid resting upon a Vibrating Support," *Phil. Mag.* July 1883. [Vol. II. p. 212.]

† *Wied. Ann.* XXXVIII. p. 118 (1889).

‡ Faraday, *Phil. Trans.* 1831. See also Rayleigh, *Phil. Mag.* April and July 1883. [Vol. II. pp. 188, 212.]

observer. Any small irregularities in the surface then render themselves conspicuous to the eye focused upon it. In the present case the reflector is plane, and the formation of a real image of the radiant requires the aid of a lens. In my experiments this was usually a large single lens of 6 inches diameter and 34 inches focus. On one occasion an achromatic telescope-lens was substituted, but the aperture was too small to include the number of waves necessary for accuracy. Although the want of achromatism was prejudicial to the appearance of the image, it is not certain that the accuracy of the determinations was impaired, at least after experience in observation had been acquired. The lens was fixed horizontally near the floor, a few inches above the surface of the water under examination. The radiant point, a very small gas-flame, was situated in the principal focal plane, but a little on one side of the axis of symmetry, so that the image formed after reflexion from the water and a double passage through the lens might be a little separated from the source. For greater convenience reflecting strips of looking-glass were introduced at angles of 45°, or thereabouts, so that the initial and final directions of the rays were horizontal.

The smallness of the disturbance is not the only obstacle to its visibility. Even with Foucault's arrangement for viewing minute departures from planeness, nothing could usually be seen of the waves here employed without a further device necessary on account of the rapidity with which all phases are presented in succession. A clear view of the waves must be an intermittent one, isoperiodic with the vibrations themselves, and may be obtained in the manner first described by Plateau. In the present case it was found simplest to render the light itself intermittent. Close in front of the small gas-flame was placed a vibrating blade of tin-plate rigidly attached to the extremity of the prong of a large tuning-fork, and so situated that once during each vibration the light was intercepted by the interposition of the blade. The vibrations of the fork were maintained electromagnetically in the usual manner, and the intermittent current furnished by the interrupter fork was utilized, as in Helmholtz's vowel-sound experiments, to excite a second, in unison with itself. The second fork generated waves in the dish of water by means of a dipper attached to its lower prong.

When the action is regular, the vibrations of the two forks are strictly isochronous, even though the natural periods may differ somewhat*. The view presented to the observer is then perfectly steady, and corresponds to one particular phase of the vibration, or rather, since the illumination is not

* A dirty condition of the mercury sometimes leads to the failure of several successive contacts. During the interval the vibrations of the second fork, being unconstrained, take place in their natural period. In this way a phase-discrepancy may set in, to be subsequently corrected when the regular contacts are re-established. Such a state of things is to be avoided as distracting to the eye, and unfavourable to accurate observation.

instantaneous, to an average of phases in the neighbourhood of a particular one.

Even in the case of a perfectly regular train of waves, the appearance will depend upon the precise position occupied by the eye. It is evident that the light most diverted from its course is that reflected from the shoulders of the waves—the points midway between the troughs and crests, for it is here that the slope of the surface is greatest. Thus if the eye be moved laterally outwards from the focal point, until all light has nearly disappeared, the residual illumination will mark out the instantaneous positions of one set of shoulders, all other parts of the complete wave remaining dark. This is one of the most favourable positions for observation. If the deviation from the focal point be in the opposite direction, the other set of shoulders will be seen bright.

The aspect of the waves was not always equally pleasing. Sometimes the formation of stationary waves, due to reflexions, interfered with regularity. A readjustment of the walls of the vessel relatively to the dipper would then often effect an improvement. The essential thing is that there should be no ambiguity in the wave pattern over the measured part of the field. It would occasionally happen that in certain positions of the eye a change of phase would occur in the middle of the field, so that the bright bands in one part were the continuation of the dark bands of another part. Near the transition

Fig. 1.

the bands would appear confused, a sufficient indication that no measurement must be attempted. On the other hand, it is not necessary that the contrast between the dark and bright parts should be very great. Indeed the measuring marks were better seen when no part of the field was very dark.

Fig. 1 gives a general idea of the appearance of the field. On the right is seen a paper with a notched edge, the use of which was to facilitate the counting. The measuring arrangement was something like a beam compass. Stout brass wires, attached to a bar of iron, were shaped at their ends like bradawls, and the edges were placed parallel to the crests of the waves. In order to avoid residual parallax, the rod was so supported that the edges were in close proximity to the water surface.

In many of the experiments the distance between the edges was set beforehand, *e.g.* to 10 cm., and was not altered when the wave-lengths varied with the deposition of grease. The number of wave-lengths included was determined by counting, and estimation of tenths. Usually the discrepancy between Mr Gordon's estimation and my own did not exceed a single tenth, and in a large proportion of cases there was no difference. Probably the mean of our readings would rarely be wrong by more than $\frac{1}{20}$ of a wave-length, when the pattern was well seen. In the experiments specially directed to the determination of the tension of a clean surface, it was found advisable to work with an unknown distance; otherwise the recollection of previous results interfered with the independence of the estimates.

It is probable that somewhat greater accuracy in single measurements might have been attained had the distance been adjustable by a smooth motion within reach of the observer. Each measuring edge might then have been set to the most favourable position, that is, to the centre of a bright band. The frequent removal of the apparatus for comparison with a scale would, however, be rather objectionable; and it was thought doubtful whether any final gain would accrue in the mean of several observations.

Some trouble was experienced from the communication of vibration through unintended channels. In order to prevent the direct influence of the interrupter fork upon the liquid surface, it was found advantageous to isolate it from the floor by supporting it upon a shelf carried upon the walls across a corner of the laboratory. On one occasion it was noticed that the waves were visible without the aid of the arrangement for making the light intermittent. This was traced to a tremor of one of the mirrors, supported upon the same shelf as the interrupter fork. Such a method of rendering the waves visible is objectionable, since it destroys the definition of the measuring points. The tremor was eliminated by the introduction of rubber tubing under the stand of the interrupter.

During the experiments on greasy surfaces one pair of forks only was employed. The frequency of the interrupter was about 42 per second, so that the intermittent current could be used to excite a fork of about 126. The beats between this and a standard Koenig fork of 128 were counted at intervals, and found to be sufficiently constant. The pitch of the standard has been verified by myself*, and at the temperature of the laboratory may be taken with sufficient accuracy to be 128. If we take the number of beats per minute at 98, we have for the frequency of the interrupter

$$f = \tfrac{1}{3}\left(128 - \frac{98}{60}\right) = 42 \cdot 12.$$

In the case of clean water another pair of forks of about 128 was employed as a check. The number of beats was 184 per minute, and

$$f = 128 - \frac{184}{60} = 124 \cdot 9.$$

The water was contained in a shallow 12″ × 10″ porcelain dish; and before commencing observations its surface was purified with the aid of an expansible hoop of thin sheet brass. The width of the hoop is greater than the depth of water, and it is deposited in the dish so as to include the dipper, but otherwise in as contracted a condition as possible. It is then opened out to its maximum area with the effect of attenuating many times the thickness of the greasy film, which no amount of preliminary cleaning seems able to obviate. It not unfrequently happened that the first attempt to get a clean surface was a partial failure, but a repetition of the operation was usually successful. It seems as if impurity attaches itself to the brass so obstinately that only contact with a clean water surface will remove it.

In the earlier experiments the waves were generated by a dipper of circular section, a closed tube of glass, somewhat like a test-tube. The measurements were quite satisfactory, but I felt doubts as to a possible influence of curvature upon wave-length. In order to avoid any risk of this kind, and to render the waves straight from the commencement, a straight horizontal edge of glass plate, about 2½ inches long, was afterwards substituted, and worked very satisfactorily. It is not necessary or desirable that the dipper should pass in and out of the water. In most cases the vibrations employed were very small, and the edge of the dipper was immersed throughout.

The purity of the water surface could be judged by the result of the observation of the number of wave-lengths; the smallest number corresponding to the purest surface. But it soon became apparent that a more delicate test was to be found in the general appearance of the wave pattern. Upon a clean surface there is a strong tendency to irregularity, dependent no doubt upon reflexions, which become more important when the propagation is very free. In order to meet this, it was often found necessary to weaken the vibrations of the secondary fork, either by putting it more out of tune with the primary, or by shifting its magnet to a less favourable position, or,

* *Phil. Trans.* p. 316, 1883. [Vol. II. p. 177.]

finally, by shunting the current across. A slight trace of grease would then render itself evident by a damping down of the waves before any change could be observed in the wave-length. After a little experience with the forks in a given state of adjustment, a momentary glance at the pattern was sufficient to enable one to recognize the condition of the surface.

The interpretation of the observations depends upon the following formula, due to Thomson:—

Let U = velocity of propagation, λ = wave-length, τ = periodic time, ρ = density, T = superficial tension, h = depth of water; then (Basset's *Hydrodynamics*, Vol. II. p. 177)

$$U^2 = \frac{\lambda^2}{\tau^2} = \left(\frac{g\lambda}{2\pi} + \frac{2\pi T}{\rho\lambda}\right)\tanh\frac{2\pi h}{\lambda};$$

so that to find T we have

$$T = \frac{\rho\lambda^3}{2\pi\tau^2}\coth\frac{2\pi h}{\lambda} - \frac{g\lambda^2\rho}{4\pi^2}.$$

In the present experiments the effect of the limitation of depth is negligible. We have $h = 1{\cdot}8$ cm., and for the greatest value of λ about ·7 cm. Now

$$\coth\frac{2\pi h}{\lambda} = \frac{1 + e^{-4\pi h/\lambda}}{1 - e^{-4\pi h/\lambda}} = 1 + 2e^{-4\pi h/\lambda},$$

approximately, when h is relatively large; so that

$$\coth(2\pi h/\lambda) = 1 + 2e^{-30} = 1,$$

with abundant accuracy. Again, in the case of water we have $\rho = 1$; and thus

$$T = \frac{\lambda^3}{2\pi\tau^2} - \frac{g\lambda^2}{4\pi^2}.$$

which is the formula by which the calculation of T is to be made. The second term will be found to be small in comparison with the first, so that approximately T varies as λ^3. A one-per-cent. error in the estimation of λ will therefore involve one of three per cent. in the deduced value of T. In many of the experiments about 15 waves were included between the marks. An error of $\frac{1}{10}$ of a wave is thus 1 in 150, leading to a two-per-cent. error in T. We may expect the final mean value to be correct to less than one per cent., but we must not be surprised if individual results show discrepancies of two per cent.

An example (August 2) will now be given in which the surface of clean water was greased with oleic acid. The dish after rinsing was filled with water drawn from a tap in connexion with a cistern supplied mainly by rain water, and placed in position. On expansion of the brass hoop, the number of waves included between the measuring points was estimated to be 13·7, 13·8 by the two observers. A piece of paper was then greased with oleic acid, and with this a platinum wire, previously cleaned by ignition, was wiped. On introduction of part of the wire into the water contained within the hoop, the number of waves rose to 15·4, 15·3. Upon this surface camphor scrapings were found to be quite dead, so that the mark had been overshot.

The dish was then refilled. Upon expansion the number of waves upon the clean surface was 13·7, 13·7. On contamination with a little oleic acid, 14·8, 14·8. Camphor was now moderately active. More oleic was added. Readings were now 15·4, 15·4, and camphor was quite dead.

The point to be fixed evidently lay between 14·8 and 15·4. A fresh surface was taken, and on addition of a little oleic the readings were 14·8, 14·8. Camphor was then tried and found moderately active. Reading still 14·8. A little more oleic added; readings 15·1, 15·1; camphor scrapings were now "nearly dead." More oleic; 15·2, 15·2; camphor "very nearly dead." More oleic; 15·4, 15·4; "not absolutely dead." More oleic; 15·5, 15·5; camphor "absolutely dead." The temperature of the water was 63° F.

On a previous occasion (July 29) accordant results had been obtained. Clean water 13·7, 13·7. Oleic added; 15·0, 15·0; camphor nearly dead. More oleic; 15·2, 15·25; camphor very nearly dead. Oleic; 15·55, 15·6; camphor dead. On both days the distance over which the waves were measured was 9·20 cm.

It may be well to exhibit in full the calculation for the clean water:—

$$\begin{array}{ll} \log 9{\cdot}2 &= \cdot 9638 \\ \log 13{\cdot}7 &= 1{\cdot}1367 \\ \hline \log \lambda &= \bar{1}{\cdot}8271 \\ & \quad 3 \\ \hline \log \lambda^3 &= \bar{1}{\cdot}4813 \\ \log 42{\cdot}12 &= 1{\cdot}6245 \\ & \quad 1{\cdot}6245 \\ \hline & \quad 2{\cdot}7303 \\ \log 2\pi &= \cdot 7981 \\ \hline \log 85{\cdot}5 &= 1{\cdot}9322 \end{array} \qquad \begin{array}{ll} \log g &= 2{\cdot}9917 \\ \log \lambda &= \bar{1}{\cdot}8271 \\ & \quad \bar{1}{\cdot}8271 \\ \hline & \quad 2{\cdot}6459 \\ \log 4\pi^2 &= 1{\cdot}5962 \\ \hline \log 11{\cdot}2 &= 1{\cdot}0497 \end{array}$$

$$\text{Finally,} \quad T = 85{\cdot}5 - 11{\cdot}2 = 74{\cdot}3.$$

If we take as the reading when the camphor is nearly dead 15·2, we find in like manner

$$T = 62{\cdot}7 - 9{\cdot}1 = 53{\cdot}6.$$

After this example a summary of results may suffice. The interest attaching to the determination of the tension of a clean surface led me to strive after a higher degree of accuracy than perhaps would otherwise have been necessary. The following table contains the results obtained with both forks:—

Date	Distance	Frequency	Tension	Water	Temp. F.	Remarks
1890					°	
June 23......	9·05	40·9	72·3	Tap	......	
25......	4·12	40·9	74·5	„	73	Telescope lens
26......	11·70	40·9	73·7	„	66	
30......	11·27	42·12	74·0	„	61	
July 1......	9·96	42·12	73·2	„	61	
2......	9·96	42·12	74·7	„	62	Strip dipper introduced
4......	9·96	42·12	74·7	Distilled	64	
8......	9·96	42·12	74·7	„	60	
25......	10·00	42·12	74·2	Tap	65	
25......	9·20	42·12	75·2	„	65	
28......	9·20	42·12	74·3	„	68	
28......	9·20	42·12	74·3	Distilled	68	
29......	9·20	42·12	74·3	„	......	
Aug. 2......	9·20	42·12	74·3	Tap	63	
July 23......	10·00	124·9	74·1	Tap	65	
23......	9·49	124·9	73·2	„	66	
23......	8·13	124·9	73·5	Distilled	66	

The mean result with the graver fork is $T=74{\cdot}2$; and with the quicker one $T=73{\cdot}6$. The discrepancy of nearly one per cent. marks the limit of accuracy. It should be remarked that some of the consecutive results where no variation occurred in the distance between the points cannot be regarded as quite independent.

On several occasions distilled water proved a less satisfactory subject than tap water. The surface seemed more unwilling to become and remain clean. Sometimes after expansion a notable increase of readings would occur in the course of a few minutes without assignable cause.

I was very anxious to satisfy myself that in the surfaces experimented upon by the wave method a high degree of purity was really attained. In the experiments of July 28 a Plateau needle vibrating upon a portable stand was introduced. *After* the examination by the method of waves, the dish was brought out into a good light, and the quality of the surface tested by observation of the motion of motes when the needle lying upon it was caused to vibrate by an external magnet*. In making the necessary arrangements

* "On the Superficial Viscosity of Water," *Proc. Roy. Soc.* June 1890, Vol. XLVIII. p. 139. [Vol. III. p. 374.]

there was some risk of introducing contamination, so that the discovery of an unclean surface would prove nothing definite. If, however, the behaviour of the surface under the needle test was good, it could be inferred with confidence that the measured waves were not affected by impurity. On two occasions the test succeeded fairly well.

The observations with the 128 fork were rather difficult, the waves being about twice as close as in the other case. In the calculation of results it appears, as was to be expected, that the importance of the second term, due to gravity, is diminished. Thus for July 22,

$$T = 76{\cdot}5 - 2{\cdot}4 = 74{\cdot}1.$$

The general result that at temperatures such as 65° (18° C.) the tension of clean water surfaces is about 74·0 C.G.S. absolute units of force per centimetre seems entitled to considerable confidence. It agrees with some former observations* of my own upon the transverse vibrations of jets, as has been remarked by Mr Worthington†. Some interesting experiments upon the vibrations of falling drops by Lenard‡ point also in the same direction. On the other hand it deviates largely from the higher value, about 81, which Prof. Quincke thinks the most probable. The deviation from 81 is certainly not due to contamination. It has been explained that great care was taken in this respect during the present experiments; and in the jet method the surfaces are probably the purest attainable. The method favoured by Quincke depends upon the measurement of large flat bubbles confined under the horizontal surface of a solid body. In default of experience I must leave it to others to judge whether a systematic error due to optical or other causes could enter here. Mr Worthington contends that some of Quincke's deductions from his measurements require correction for curvature perpendicular to the meridional plane. To this and other criticisms Prof. Quincke has replied§.

Experimenters upon capillary tubes have generally been led to adopt the lower value, but here the interpretation involves an assumption that the angle of contact θ is zero. What these measurements give in the first instance is $T\cos\theta$; so that if $\theta = 30°$, or thereabouts, the higher value of T is the one really indicated. This is the view adopted by Quincke, who in an important series of observations|| has shown that the edge angle between water and glass has frequently a considerable value dependent upon the impurity of glass surfaces, even when carefully cleaned by ordinary methods. But I confess that the argument does not appear to me conclusive. The angles recorded are maximum angles. If after a drop has been deposited

* *Proc. Roy. Soc.* Vol. XXIX. p. 71, 1879. [Vol. I. p. 387.]

† *Phil. Mag.* Vol. XX. p. 51, 1885.

‡ *Wied. Ann.* Bd. XXX. (1887).

§ *Ibid.* XXVII. p. 219 (1886).

|| *Ibid.* Vol. II. p. 145, 1877.

some of the liquid is drawn off, the angle may be diminished almost to zero. Observations upon capillary heights correspond surely to the latter condition of things, for no experimenter measures the gradual rise of liquid in a dry tube. I am disposed to think that the assumption $\theta = 0$ is legitimate, and thus that the lower value of T is really supported by experiments of this class.

Leaving now the results for pure surfaces, let us pass on to those found for water contaminated with grease up to the point where the camphor scrapings were judged to be "very nearly dead." It must be remembered that the additions of oil were discontinuous, and that the point could not always be hit with precision. On any one day it is possible to set up a fairly precise standard of what one means by "very nearly dead"; but the standard is liable to vary in one's own mind, and is of course impossible to communicate to another. Too much importance therefore must not be ascribed to exact agreement or the failure of it. On one day experiments were made by varying the areas enclosed within the hoop. Thus, if the motions were a little too lively, they could be deadened to the required point by contraction of the area and consequent concentration of grease. This procedure was not so convenient as had been hoped, in consequence of the mechanical disturbance attending a motion of the hoop. In all cases an observation, for the most part recorded in the previous table, was made first upon a clean surface, so as to ensure that the contamination was all of the kind intended. The results are collected in the annexed table:—

Date	Water	Oil	Tension	Remarks
June 30......	Tap	Olive-oil	53·7	
July 1......	„	„	51·1	
2......	„	„	52·1	
4......	Distilled	„	53·0	
7......	„	„	53·0	Not quite independent
11......	„	„	53·0	„ „ „
29......	Tap	Oleic Acid	53·6	
Aug. 2......	„	„	53·6	
2......	„	Olive-oil	52·4	
2......	„	„	52·4	Another sample

The tension of the surface when the camphor movements are just stopping may thus be reckoned at 53·0 C.G.S., or about 72 per cent. of that of a clean surface. There is some reason to infer that the tension is the same whatever kind of grease be used. In the last experiment the sample of oil was one of which it was necessary to take decidedly more than usual (in the ratio of about 3 : 2) in order to stop the motions. This was proved by the balance in the manner described in a former paper. I have other grounds for thinking it probable that the tension does not depend upon the kind of oil, and hope to investigate the matter further by a more appropriate method*.

On several occasions the effect of large additions of oil was tried. The limit did not appear to be very definite; for a second and even a third drop gave a sensible indication. The results were June 30, 38·8; July 1, 40·3; July 7, 41·0; July 8, 41·7; July 26, 38·9; mean 40·1. They relate to olive-oil; and it is possible that the largeness of the quantity required to approach the limit depends upon the heterogeneous character of the substance.

Two observations were made of the effect of additions of oleate of soda to distilled water. When the limit seemed nearly attained final readings were taken with the results: July 8, 25·3; July 11, 24·6; mean 25·0. It will be seen that the tension is lowered very much further by soap than by oil.

The principal results of the present experiments may be thus summarized. The tension of a water surface, reckoned in C.G.S. measure, is in the various cases:—

Clean	74·0	100
Greasy to the point where the camphor motions nearly cease	53·0	72
Saturated with olive-oil	41·0	54
Saturated with oleate of soda	25·0	34

In the last column the tensions are exhibited as fractions of that of a clean surface.

POSTSCRIPT, *Sept.* 19.

It appeared probable that the tension of otherwise pure water saturated with camphor would be the same as that of greasy water upon which camphor fragments were just dead; and before the above paper was written I had already attempted to examine this point. The experiment, however, did not succeed. The camphorated water had decidedly too much tension (wave-number 14·7 instead of 15·5), but on the other hand the liquid was clearly not saturated, inasmuch as fresh camphor scrapings were lively upon an expanded surface. I have recently returned to the subject with water which has stood in contact with excess of camphor for more than a month.

Sept. 15. Fresh clean water. Expanded 13·7, 13·7. Motes still. Olive-oil added, 15·2, 15·2; camphor fragments moderately active. More oil, 15·3, 15·3; camphor nearly dead. More oil, 15·5, 15·5; camphor dead. Fragments of camphor and motes quite still.

* [*Phil. Mag.* XXXIII. p. 366, 1892.]

The saturated solution of camphor was now substituted. Surface expanded; 15·5, 15·5. Expanded, 15·5. This number could not be reduced by any number of expansions of the surface.

It was observed that the surface was usually in motion, as evidenced by an irregular drift of motes and camphor fragments. The latter had no individual motion, all neighbouring particles moving together. The effect is probably due to local evaporation of camphor and accompanying increase of tension. Associated with this was a fluctuation backwards and forwards of the number of waves, such as was never observed with pure, or simply greasy, water.

We are thus justified in the conclusion that saturated solution of camphor has the same tension as is found for greasy water when camphor fragments are just dead. When the saturated solution was diluted with about an equal volume of water, the wave-number was reduced to 14·7. In these experiments the distance between the points was 9·20 cm., and the frequency was 42·12, so that the observations are directly comparable with those in the example calculated at length.

The comparison of tensions for clean and camphorated water may also be effected by the method of capillary heights. Some observations by Mr Gordon gave the following:—

Clean water	7·94,	7·91,	7·92
Water changed.................	7·92,	7·90,	7·90
Saturated camphor	5·63,	5·68,	5·65
Clean water	7·97,	7·90,	7·92
Water changed.................	7·94,	7·96,	7·93
Saturated camphor	5·62,	5·63,	5·66

Thus, as a mean, capillary height for clean water is 7·93 cm., and for water saturated with camphor 5·64 cm. The ratio of these is ·71.

Observations by myself upon the same tube, but read in a somewhat different manner, gave

Clean water...........	8·04,	8·03,	8·04,	8·05.		
Water changed	8·02,	8·02.				
Camphorated water...	5·77,	5·80,	5·79,	5·80,	5·80,	5·83.

As means we may take 8·03 cm. and 5·80 cm., giving for the ratio ·71, as before.

The ratio of tensions thus found agrees remarkably well with that deduced from the observations upon ripples, viz. ·72. It will be remembered that the latter might be expected to be somewhat higher, as corresponding with a condition of things where camphor fragments were *nearly*, but not quite, dead.

October 8.—I take this opportunity of recording that a film of grease, insufficient to check the motion of camphor fragments, exercises a marked influence upon the reflexion of light from the surface of water in the neighbourhood of the polarizing angle. In the case of a clean surface and at the Brewsterian angle, the reflexion of light polarized perpendicularly to the plane of incidence appears to vanish, in accordance with the formula of Fresnel.

[1901. This subject is further treated in *Phil. Mag.* XXXIII. p. 1, 1892; Vol. III. of present collection, Art. 185 below.]

176.

ON THE THEORY OF SURFACE FORCES.

[*Philosophical Magazine*, XXX. pp. 285—298, 456—475, 1890.]

SINCE the time of Young the tendency of a liquid surface to contract has always been attributed to the mutual attraction of the parts of the liquid, acting through a very small range,—to the same forces in fact as those by which the cohesion of liquids and solids is to be explained. It is sometimes asserted that Laplace was the first to look at the matter from this point of view, and that Young contented himself with calculations of the *consequences* of superficial tension. Such an opinion is entirely mistaken, although the authority of Laplace himself may be quoted in its favour*. In the introduction to his first paper†, which preceded the work of Laplace, Young writes:—"It will perhaps be more agreeable to the experimental philosopher, although less consistent with the strict course of logical argument, to proceed in the first place to the comparison of this theory [of superficial tension] with the phenomena, and to inquire afterwards for its foundation in the ultimate properties of matter." This he attempts to do in Section VI., which is headed *Physical Foundation of the Law of Superficial Cohesion.* The argument is certainly somewhat obscure; but as to the character of the "physical foundation" there can be no doubt. "We may suppose the particles of liquids, and probably those of solids also, to possess that power of repulsion, which has been demonstrably shown by Newton to exist in aëriform fluids, and which varies in the inverse ratio of the distance of the particles from each other. In air and vapours this force appears to act uncontrolled; but in liquids it is overcome by a cohesive force, while the particles still retain a power of moving freely in all directions...It is simplest

* *Méc. Cél. Supplément au Xe livre*, 1805:—"Mais il n'a pas tenté, comme Segner, de dériver ces hypothèses, de la loi de l'attraction des molécules, décroissante avec une extrême rapidité; ce qui était indispensable pour les réaliser."

† "On the Cohesion of Fluids," *Phil. Trans.* 1805.

to suppose the force of cohesion nearly or perfectly constant in its magnitude, throughout the minute distance to which it extends, and owing its apparent diversity to the contrary action of the repulsive force which varies with the distance."

Although nearly a century has elapsed, we are still far from a satisfactory theory of these reactions. We know now that the pressure of gases cannot be explained by a repulsive force varying inversely as the distance, but that we must appeal to the impacts of colliding molecules*. There is every reason to suppose that the molecular movements play an important part in liquids also; and if we leave them out of account, we can only excuse ourselves on the ground of the difficulty of the subject, and with full recognition that a theory so founded is probably only a first approximation to the truth. On the other hand, the progress of science has tended to confirm the views of Young and Laplace as to the existence of a powerful attraction operative at short distances. Even in the theory of gases it is necessary, as Van der Waals has shown, to appeal to such a force in order to explain their condensation under increasing pressure in excess of that indicated by Boyle's law, and explicable by impacts. Again, it would appear that it is in order to overcome this attraction that so much heat is required in the evaporation of liquids.

If we take a statical view of the matter, and ignore the molecular movements†, we must introduce a repulsive force to compensate the attraction. Upon this point there has been a good deal of confusion, of which even Poisson cannot be acquitted. And yet the case seems simple enough. For consider the equilibrium of a spherical mass of mutually attracting matter, free from external force, and conceive it divided by an ideal plane into hemispheres. Since the hemispheres are at rest, their total action upon one another must be zero, that is, no force is transmitted across the interface. If there be attraction operative across the interface, it must be precisely compensated by repulsion. This view of the matter was from the first familiar to Young, and he afterwards gave calculations, which we shall presently notice, dependent upon the hypothesis that there is a constant attractive force operative over a limited range and balanced by a repulsive force of suitable intensity operative over a different range. In Laplace's theory, upon the other hand, no mention is made of repulsive forces, and it would appear at first as if the attractive forces were left to perform the impossible feat of balancing themselves. But in this theory there is introduced a pressure which is really the representative of the repulsive forces.

* The argument is clearly set forth in Maxwell's lecture "On the Dynamical Evidence of the Molecular Constitution of Bodies" (*Nature*, Vol. XI. p. 357, 1875. [Maxwell's *Scientific Papers*, Vol. II. p. 418]).

† Compare Worthington, "On Surface Forces in Fluids," *Phil. Mag.* XVIII. p. 334 (1884).

It may be objected that if the attraction and repulsion must be supposed to balance one another across any ideal plane of separation, there can be no sense, or advantage, in admitting the existence of either. This would certainly be true if the origin and law of action of the forces were similar, but such is not supposed to be the case. The inconclusiveness of the objection is readily illustrated. Consider the case of the earth, conceived to be at rest. The two halves into which it may be divided by an ideal plane do not upon the whole act upon one another; otherwise there could not be equilibrium. Nevertheless no one hesitates to say that the two halves attract one another under the law of gravitation. The force of the objection is sometimes directed against the pressure, denoted by K, which Laplace conceives to prevail in the interior of liquids and solids. How, it is asked, can there be a pressure, if the whole force vanishes? The best answer to this question may be found in asking another—Is there a pressure in the interior of the earth?

It must no doubt be admitted that in availing ourselves of the conception of pressure we are stopping short of a complete explanation. The mechanism of the pressure is one of the things that we should like to understand. But Laplace's theory, while ignoring the movements and even the existence of molecules, cannot profess to be complete; and there seems to be no inconsistency in the conception of a continuous, incompressible liquid, whose parts attract one another, but are prevented from undergoing condensation by forces of infinitely small range, into the nature of which we do not further inquire. All that we need to take into account is then covered by the ordinary idea of pressure. However imperfect a theory developed on these lines may be, and indeed must be, it presents to the mind a good picture of capillary phenomena, and, as it probably contains nothing not needed for the further development of the subject, labour spent upon it can hardly be thrown away.

Upon this view the pressure due to the attraction measures the cohesive force of the substance, that is the tension which must be applied in order to cause rupture. It is the quantity which Laplace denoted by K, and which is often called the molecular pressure. Inasmuch as Laplace's theory is not a molecular theory at all, this name does not seem very appropriate. Intrinsic pressure is perhaps a better term, and will be employed here. The simplest method of estimating the intrinsic pressure is by the force required to break solids. As to liquids, it is often supposed that the smallest force is adequate to tear them asunder. If this were true, the theory of capillarity now under consideration would be upset from its foundations, but the fact is quite otherwise. Berthelot* found that water could sustain a tension of about

* *Ann. de Chimie*, xxx. p. 232 (1850). See also Worthington, *Brit. Assoc. Report*, 1888, p. 583.

50 atmospheres applied directly, and the well-known phenomenon of retarded ebullition points in the same direction. For if the cohesive forces which tend to close up a small cavity in the interior of a superheated liquid were less powerful than the steam-pressure, the cavity must expand, that is the liquid must boil. By supposing the cavity infinitely small, we see that ebullition must necessarily set in as soon as the steam* pressure exceeds that intrinsic to the liquid. The same method may be applied to form a conception of the intrinsic pressure of a liquid which is not superheated. The walls of a moderately small cavity certainly tend to collapse with a force measured by the constant surface-tension of the liquid. The pressure in the cavity is at first proportional to the surface-tension and to the curvature of the walls. If this law held without limit, the consideration of an infinitely small cavity shows that the intrinsic pressure would be infinite in all liquids. Of course the law really changes when the dimensions of the cavity are of the same order as the range of the attractive forces, and the pressure in the cavity approaches a limit, which is the intrinsic pressure of the liquid. In this way we are forced to admit the reality of the pressure by the consideration of experimental facts which cannot be disputed.

The first estimate of the intrinsic pressure of water is doubtless that of Young. It is 23,000 atmospheres, and agrees extraordinarily well with modern numbers. I propose to return to this estimate, and to the remarkable argument which Young founded upon it.

The first great advance upon the theory of Young and Laplace was the establishment by Gauss of the principle of surface-energy. He observed that the existence of attractive forces of the kind supposed by his predecessors leads of necessity to a term in the expression of the potential energy proportional to the surface of the liquid, so that a liquid surface tends always to contract, or, what means precisely the same thing, exercises a tension. The argument has been put into a more general form by Boltzmann†. It is clear that all molecules in the interior of the liquid are in the same condition. Within the superficial layer, considered to be of finite but very small thickness, the condition of all molecules is the same which lie at the same very small distance from the surface. If the liquid be deformed without change in the total area of the surface, the potential energy necessarily remains unaltered; but if there be a change of area the variation of potential energy must be proportional to such change.

A mass of liquid, left to the sole action of cohesive forces, assumes a spherical figure. We may usefully interpret this as a tendency of the surface

* If there be any more volatile impurity (*e.g.*, dissolved gas) ebullition must occur much earlier.

† *Pogg. Ann.* CXLI. p. 582 (1870). See also Maxwell's *Theory of Heat*, 1870; and article "Capillarity," *Enc. Brit.* [Maxwell's *Scientific Papers*, Vol. II. p. 541.]

to contract; but it is important not to lose sight of the idea that the spherical form is the result of the endeavour of the parts to get *as near to one another* as is possible*. A drop is spherical under capillary forces for the same reason that a large gravitating mass of (non-rotating) liquid is spherical.

In the following sketch of Laplace's theory we will commence in the manner adopted by Maxwell†. If f be the distance between two particles m, m', the cohesive attraction between them is denoted in Laplace's notation by $mm'\phi(f)$, where $\phi(f)$ is a function of f which is insensible for all sensible values of f, but which becomes sensible and even enormously great, when f is exceedingly small.

"If we next introduce a new function of f and write

$$\int_f^\infty \phi(f)\,df = \Pi(f), \quad\text{.............................(1)}$$

then $mm'\Pi(f)$ will represent (1) the work done by the attractive force on the particle m, while it is brought from an infinite distance from m' to the distance f from m'; or (2) the attraction of a particle m on a narrow straight rod resolved in the direction of the length of the rod, one extremity of the rod being at a distance f from m, and the other at an infinite distance, the mass of unit of length of the rod being m'. The function $\Pi(f)$ is also insensible for sensible values of f, but for insensible values of f it may become sensible and even very great."

"If we next write

$$\int_z^\infty \Pi(f)\,f\,df = \psi(z), \quad\text{.........................(2)}$$

then $2\pi m\sigma\psi(z)$ will represent (1) the work done by the attractive force while a particle m is brought from an infinite distance to a distance z from an infinitely thin stratum of the substance whose mass per unit of area is σ; (2) the attraction of a particle m placed at a distance z from the plane surface of an infinite solid whose density is σ."

The intrinsic pressure can now be found immediately by calculating the mutual attraction of the parts of a large mass which lie on opposite sides of an imaginary plane interface. If the density be σ, the attraction between the whole of one side and a layer upon the other, distant z from the plane and of thickness dz, is $2\pi\sigma^2\psi(z)dz$, reckoned per unit of area. The expression for the intrinsic pressure is thus simply

$$K = 2\pi\sigma^2 \int_0^\infty \psi(z)\,dz. \quad\text{.........................(3)}$$

* See Sir W. Thomson's lecture on "Capillary Attraction" (*Proc. Roy. Inst.* 1886), reprinted in *Popular Lectures and Addresses*.

† *Enc. Brit.*, "Capillarity." [Maxwell's *Scientific Papers*, Vol. II. p. 541.]

In Laplace's investigation σ is supposed to be unity. We may call the value which (3) then assumes K_0, so that

$$K_0 = 2\pi \int_0^\infty \psi(z)\, dz. \qquad (4)$$

The expression for the superficial tension is most readily found with the aid of the idea of superficial energy, introduced into the subject by Gauss. Since the tension is constant, the work that must be done to extend the surface by one unit of area measures the tension, and the work required for the generation of any surface is the product of the tension and the area. From this consideration we may derive Laplace's expression, as has been done by Dupré* and Thomson†. For imagine a small cavity to be formed in the interior of the mass and to be gradually expanded in such a shape that the walls consist almost entirely of two parallel planes. The distance between the planes is supposed to be very small compared with their ultimate diameters, but at the same time large enough to exceed the range of the attractive forces. The work required to produce this crevasse is twice the product of the tension and the area of one of the faces. If we now suppose the crevasse produced by direct separation of its walls, the work necessary must be the same as before, the initial and final configurations being identical; and we recognize that the tension may be measured by half the work that must be done per unit of area against the mutual attraction in order to separate the two portions which lie upon opposite sides of an ideal plane to a distance from one another which is outside the range of the forces. It only remains to calculate this work.

If σ_1, σ_2 represent the densities of the two infinite solids, their mutual attraction at distance z is per unit of area

$$2\pi\sigma_1\sigma_2 \int_z^\infty \psi(z)\, dz, \qquad (5)$$

or $2\pi\sigma_1\sigma_2\,\theta(z)$, if we write

$$\int_z^\infty \psi(z)\, dz = \theta(z). \qquad (6)$$

The work required to produce the separation in question is thus

$$2\pi\sigma_1\sigma_2 \int_0^\infty \theta(z)\, dz; \qquad (7)$$

and for the tension of a liquid of density σ we have

$$T = \pi\sigma^2 \int_0^\infty \theta(z)\, dz. \qquad (8)$$

The form of this expression may be modified by integration by parts. For

$$\int \theta(z)\, dz = \theta(z) \,.\, z - \int z \frac{d\theta(z)}{dz}\, dz = \theta(z) \,.\, z + \int z \psi(z)\, dz.$$

* *Théorie Mécanique de la Chaleur* (Paris, 1869).

† "Capillary Attraction," *Proc. Roy. Inst.*, Jan. 1886. Reprinted, *Popular Lectures and Addresses*, 1889.

Since $\theta(0)$ is finite, proportional to K, the integrated term vanishes at both limits, and we have simply

$$\int_0^\infty \theta(z)\, dz = \int_0^\infty z\psi(z)\, dz, \qquad\text{.........................(9)}$$

and

$$T = \pi\sigma^2 \int_0^\infty z\psi(z)\, dz. \qquad\text{.........................(10)}$$

In Laplace's notation the second member of (9), multiplied by 2π, is represented by H.

As Laplace has shown, the values for K and T may also be expressed in terms of the function ϕ, with which we started. Integrating by parts, we get by means of (1) and (2),

$$\int \psi(z)\, dz = z\psi(z) + \tfrac{1}{3} z^3 \Pi(z) + \tfrac{1}{3}\int z^3 \phi(z)\, dz,$$

$$\int z\psi(z)\, dz = \tfrac{1}{2} z^2\psi(z) + \tfrac{1}{8} z^4 \Pi(z) + \tfrac{1}{8}\int z^4 \phi(z)\, dz.$$

In all cases to which it is necessary to have regard the integrated terms vanish at both limits, and we may write

$$\int_0^\infty \psi(z)\, dz = \tfrac{1}{3}\int_0^\infty z^3\phi(z)\, dz, \qquad \int_0^\infty z\psi(z)\, dz = \tfrac{1}{8}\int_0^\infty z^4\phi(z)\, dz; \;\text{...(11)}$$

so that

$$K_0 = \frac{2\pi}{3}\int_0^\infty z^3\phi(z)\, dz, \qquad T_0 = \frac{\pi}{8}\int_0^\infty z^4\phi(z)\, dz. \;\text{.........(12)}$$

A few examples of these formulæ will promote an intelligent comprehension of the subject. One of the simplest suppositions open to us is that

$$\phi(f) = e^{-\beta f}. \qquad\text{............................(13)}$$

From this we obtain

$$\Pi(z) = \beta^{-1} e^{-\beta z}, \qquad \psi(z) = \beta^{-3}(\beta z + 1)\, e^{-\beta z}. \;\text{.........(14)}$$

$$K_0 = 4\pi\beta^{-4}, \qquad T_0 = 3\pi\beta^{-5}. \;\text{.........................(15)}$$

The range of the attractive force is mathematically infinite, but practically of the order β^{-1}, and we see that T is of higher order in this small quantity than K. That K is in all cases of the fourth order and T of the fifth order in the range of the forces is obvious from (12) without integration.

An apparently simple example would be to suppose $\phi(z) = z^n$. From (1), (2), (4) we get

$$\Pi(z) = -\frac{z^{n+1}}{n+1}, \qquad \psi(z) = \frac{z^{n+3}}{n+3\,.\,n+1},$$

$$K_0 = \left.\frac{2\pi z^{n+4}}{n+4\,.\,n+3\,.\,n+1}\right|_0^\infty. \qquad\text{.......................(16)}$$

The intrinsic pressure will thus be infinite whatever n may be. If $n+4$ be positive, the attraction of infinitely distant parts contributes to the result; while if $n+4$ be negative, the parts in immediate contiguity act with infinite power. For the transition case, discussed by Sutherland*, of $n+4=0$, K_0 is also infinite. It seems therefore that nothing satisfactory can be arrived at under this head.

As a third example we will take the law proposed by Young, viz.

$$\left.\begin{array}{ll}\phi(z)=1 & \text{from } z=0 \text{ to } z=a,\\ \phi(z)=0 & \text{from } z=a \text{ to } z=\infty;\end{array}\right\} \quad\text{............(17)}$$

and corresponding therewith,

$$\left.\begin{array}{ll}\Pi(z)=a-z & \text{from } z=0 \text{ to } z=a,\\ \Pi(z)=0 & \text{from } z=a \text{ to } z=\infty,\end{array}\right\} \quad\text{............(18)}$$

$$\left.\begin{array}{ll}\psi(z)=\tfrac{1}{2}a(a^2-z^2)-\tfrac{1}{3}(a^3-z^3) & \\ & \text{from } z=0 \text{ to } z=a,\\ \psi(z)=0 & \text{from } z=a \text{ to } z=\infty.\end{array}\right\} \quad\text{............(19)}$$

Equations (12) now give

$$K_0=\frac{2\pi}{3}\int_0^\infty z^3dz=\frac{\pi a^4}{6}, \qquad T_0=\frac{\pi}{8}\int_0^a z^4dz=\frac{\pi a^5}{40}. \quad\text{......(20, 21)}$$

The numerical results differ from those of Young†, who finds that "*the contractile force is one-third of the whole cohesive force of a stratum of particles, equal in thickness to the interval to which the primitive equable cohesion extends,*" viz. $T=\frac{1}{3}aK$; whereas according to the above calculation $T=\frac{3}{20}aK$. The discrepancy seems to depend upon Young having treated the attractive force as operative in one direction only.

In his Elementary Illustrations of the Celestial Mechanics of Laplace‡, Young expresses views not in all respects consistent with those of his earlier papers. In order to balance the attractive force he introduces a repulsive force, following the same law as the attractive except as to the magnitude of the range. The attraction is supposed to be of constant intensity C over a range c, while the repulsion is of intensity R, and is operative over a range r. The calculation above given is still applicable, and we find that

$$K=\frac{\pi}{6}(c^4C-r^4R), \qquad T=\frac{\pi}{40}(c^5C-r^5R). \quad\text{............(22)}$$

In these equations, however, we are to treat K as vanishing, the specification of the forces operative across a plane being supposed to be complete. Hence, as Young finds, we must take

$$c^4C=r^4R; \quad\text{..................................(23)}$$

and accordingly

$$T=\frac{\pi c^4C(c-r)}{40}. \quad\text{..........................(24)}$$

At this point I am not able to follow Young's argument, for he asserts (p. 490) that "the existence of such a cohesive tension proves that the mean sphere of

* *Phil. Mag.* XXIV. p. 113 (1887).

† *Enc. Brit.; Collected Works*, Vol. I. p. 461.

‡ 1821. *Collected Works*, Vol. I. p. 485.

action of the repulsive force is more extended than that of the cohesive: a conclusion which, though contrary to the tendency of some other modes of viewing the subject, shows the absolute insufficiency of all theories built upon the examination of one kind of corpuscular force alone." According to (24) we should infer, on the contrary, that if superficial tension is to be explained in this way, we must suppose that $c > r$.

My own impression is that we do not gain anything by this attempt to advance beyond the position of Laplace. So long as we are content to treat fluids as incompressible, there is no objection to the conception of intrinsic pressure. The repulsive forces which constitute the machinery of this pressure are probably intimately associated with actual compression, and cannot advantageously be treated without enlarging the foundations of the theory. Indeed it seems that the view of the subject represented by (23), (24), with c greater than r, cannot consistently be maintained. For consider the equilibrium of a layer of liquid at a free surface A of thickness AB equal to r. If the void space beyond A were filled up with liquid, the attractions and repulsions across B would balance one another; and since the action of the additional liquid upon the parts below B is wholly attractive, it is clear that in the actual state of things there is a finite repulsive action across B, and a consequent failure of equilibrium.

I now propose to exhibit another method of calculation, which not only leads more directly to the results of Laplace, but allows us to make a not unimportant extension of the formulæ to meet the case where the radius of a spherical cavity is neither very large nor very small in comparison with the range of the forces.

The density of the fluid being taken as unity, let V be the potential of the attraction, so that

$$V = \iiint \Pi(f)\, dx\, dy\, dz, \qquad \text{.......................(25)}$$

f denoting the distance of the element of the fluid $dx\,dy\,dz$ from the point at which the potential is to be reckoned. The hydrostatic equation of pressure is then simply $dp = dV$; or, if A and B be any two points,

$$p_B - p_A = V_B - V_A. \qquad \text{..........................(26)}$$

Suppose, for example, that A is in the interior, and B upon a plane surface of the liquid. The potential at B is then exactly one half of that at A, or $V_B = \frac{1}{2} V_A$; so that

$$p_A - p_B = \tfrac{1}{2} V_A = 2\pi \int_0^{\frac{1}{2}\pi} \int_0^\infty \Pi(f) f^2\, df \sin\theta\, d\theta = 2\pi \int_0^\infty \Pi(f) f^2 df.$$

Now $p_A - p_B$ is the intrinsic pressure K_0; and thus

$$K_0 = 2\pi \int_0^\infty \Pi(f) f^2\, df = \frac{2\pi}{3} \int_0^\infty \phi(f) f^3\, df,$$

as before.

Again, let us suppose that the fluid is bounded by concentric spherical surfaces, the interior one of radius r being either large or small, but the exterior one so large that its curvature may be neglected. We may suppose that there is no external pressure, and that the tendency of the cavity to collapse is balanced by contained gas. Our object is to estimate the necessary internal pressure.

Fig. 1.

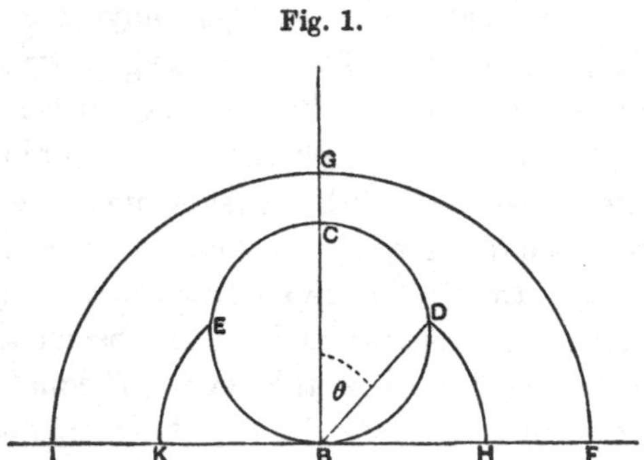

In the figure $BDCE$ represents the cavity, and the pressure required is the same as that of the fluid at such a point as B. [A is supposed to lie upon the external surface.] Since $p_A = 0$, $p_B = V_B - V_A$. Now V_A is equal to that part of V_B which is due to the infinite mass lying below the plane BF. Accordingly the pressure required (p_B) is the potential at B due to the fluid which lies above the plane BF. Thus

$$p_B = \iiint \Pi(f)\, dx\, dy\, dz,$$

where the integrations are to be extended through the region above the plane BF which is external to the sphere $BDCE$. On the introduction of polar coordinates the integral divides itself into two parts. In the first from $f=0$ to $f=2r$ the spherical shells (*e.g.* DH) are incomplete hemispheres, while in the second part from $f=2r$ to $f=\infty$ the whole hemisphere (*e.g.* IGF) is operative. The spherical area DH, divided by f^2,

$$= 2\pi \int_\theta^{\frac{1}{2}\pi} \sin\theta\, d\theta = 2\pi \cos\theta = \pi f / r.$$

The area $IGF = 2\pi f^2$.

Thus, dropping the suffix B, we get the unexpectedly simple expression

$$p = \frac{\pi}{r} \int_0^{2r} \Pi(f) f^3\, df + 2\pi \int_{2r}^{\infty} \Pi(f) f^2\, df. \qquad (27)$$

If $2r$ exceed the range of the force, the second integral vanishes and the first may be supposed to extend to infinity. Accordingly

$$p = \frac{\pi}{r} \int_0^{\infty} \Pi(f) f^3\, df = \frac{2}{r} \times \frac{\pi}{8} \int_0^{\infty} f^4 \phi(f)\, df, \qquad (28)$$

in accordance with the value (12) already given for T_0. We see then that, if the curvature be not too great, the pressure in the cavity can be calculated as if it were due to a constant tension tending to contract the surface. In the other extreme case where r tends to vanish, we have ultimately

$$p = 2\pi \int_0^\infty \Pi(f) f^2\, df = K_0.$$

In these extreme cases the results are of course well known; but we may apply (27) to calculate the pressure in the cavity when its diameter is of the order of the range. To illustrate this we may take a case already suggested, in which $\phi(f) = e^{-\beta f}$, $\Pi(f) = \beta^{-1} e^{-\beta f}$. Using these, we obtain on reduction,

$$p = 2\pi\beta^{-4}\left\{\frac{3}{\beta r} - e^{-2\beta r}\left(2\beta r + 4 + \frac{3}{\beta r}\right)\right\}. \quad \text{..........(29)}$$

From (29) we may fall back upon particular cases already considered. Thus, if r be very great,

$$p = \frac{2}{r} \times 3\pi\beta^{-5};$$

and if r be very small, $p = 4\pi\beta^{-4}$, in agreement with (15).

In a recent memoir* Fuchs investigates a second approximation to the tension of curved surfaces, according to which the pressure in a cavity would consist of two terms; the first (as usual) directly as the curvature, the second subtractive, and proportional to the cube of the curvature. This conclusion does not appear to harmonize with (27), (29), which moreover claim to be exact expressions. It may be remarked that when the tension depends upon the curvature, it can no longer be identified with the work required to generate a unit surface. Indeed the conception of surface-tension appears to be appropriate only when the range is negligible in comparison with the radius of curvature.

The work required to generate a spherical cavity of radius r is of course readily found in any particular case. It is expressed by the integral

$$\int_0^r p \cdot 4\pi r^2 \cdot dr. \quad \text{..........................(30)}$$

As a second example we may consider Young's supposition, viz. that the force is unity from 0 to a, and then altogether ceases. In this case by (18), $\Pi(f)$ absolutely vanishes, if $f > a$; so that if the diameter of the cavity at all exceed a, the internal pressure is given rigorously by

$$p = \frac{2}{r} \times \frac{\pi}{8} \int_0^a f^4 \phi(f)\, df = \frac{2}{r} \times \frac{\pi a^5}{40} \quad \text{..................(31)}$$

* *Wien. Ber.* Bd. xcviii. Abth. ii. *a*, Mai 1889.

When, on the other hand, $2r < a$, we have

$$p = \frac{\pi}{r}\int_0^{2r}(a-f)f^3\,df + 2\pi\int_{2r}^{a}(a-f)f^2\,df$$

$$= \pi\left\{\frac{a^4}{6} - \frac{4}{3}ar^3 + \frac{8}{5}r^4\right\}, \qquad\qquad (32)$$

coinciding with (31) when $2r = a$. If $r = 0$, we fall back upon $K_0 = \pi a^4/6$.

We will now calculate by (30) the work required to form a cavity of radius equal to $\frac{1}{2}a$. We have

$$4\pi\int_0^{\frac{1}{2}a} p\,.\,r^2\,dr = \frac{\pi^2 a^7}{4}\left(\frac{1}{18} + \frac{1}{35}\right).$$

The work that would be necessary to form the same cavity, supposing the pressure to follow the law (31) applicable when $2r > a$, is

$$\int_0^{\frac{1}{2}a} \frac{2}{r}\cdot\frac{\pi a^5}{40}\,.\,4\pi r^2\,dr = \frac{\pi^2 a^7}{40}.$$

The work required to generate a cavity for which $2r > a$ is therefore less than if the ultimate law prevailed throughout by the amount

$$\frac{\pi^2 a^7}{4}\left(\frac{1}{10} - \frac{1}{18} - \frac{1}{35}\right) = \frac{\pi^2 a^7}{4\,.\,9\,.\,7}. \qquad\qquad (33)$$

We may apply the same formulæ to compare the pressures at the centre and upon the surface of a spherical mass of fluid, surrounded by vacuum. If the radius be r, we have at the centre

$$V = 4\pi\int_0^r f^2\,\Pi(f)\,df,$$

and at the surface

$$V = 2\pi\int_0^{2r}\left(1 - \frac{f}{2r}\right)f^2\,\Pi(f)\,df;$$

so that the excess of pressure at the centre is

$$4\pi\int_0^r f^2\Pi(f)\,df - 2\pi\int_0^{2r} f^2\Pi(f)\,df + \frac{\pi}{r}\int_0^{2r} f^3\Pi(f)\,df. \qquad (34)$$

If r exceed the range of the forces, (34) becomes

$$2\pi\int_0^\infty f^2\Pi(f)\,df + \frac{\pi}{r}\int_0^\infty f^3\Pi(f)\,df = K + \frac{2T}{r}, \qquad (35)$$

as was to be expected. As the curvature increases from zero, there is at first a rise of pressure. A maximum occurs when r has a particular value, of the order of the range. Afterwards a diminution sets in, and the pressure approaches zero, as r decreases without limit.

If the surface of fluid, not acted on by external force, be of variable curvature, it cannot remain in equilibrium. For example, at the pole of an oblate ellipsoid of revolution the potential will be greater than at the equator, so that in order to maintain equilibrium an external polar pressure would be needed. An extreme case is presented by a rectangular mass, in which the potential at an edge is only one half, and at a corner only one [quarter], of that general over a face.

When the surface is other than spherical, we cannot obtain so simple a general expression as (34) to represent the excess of internal over superficial pressure; but an approximate expression analogous to (35) is readily found.

The potential at a point upon the surface of a convex mass differs from that proper to a plane surface by the potential of the meniscus included between the surface and its tangent plane. The equation of the surface referred to the normal and principal tangents is approximately

$$2z = x^2/R_1 + y^2/R_2,$$

R_1, R_2 being the radii of curvature. The potential, at the origin, of the meniscus is thus

$$V = \iint \Pi(f)\, zf\, df\, d\theta,$$

where $f^2 = x^2 + y^2$; and

$$\int_0^{2\pi} z\, d\theta = \int \left(\frac{f^2\cos^2\theta}{2R_1} + \frac{f^2\sin^2\theta}{2R_2}\right) d\theta = \frac{\pi f^2}{2}\left(\frac{1}{R_1} + \frac{1}{R_2}\right).$$

Accordingly

$$V = \frac{\pi}{2}\left(\frac{1}{R_1} + \frac{1}{R_2}\right)\int_0^\infty f^3\Pi(f)\, df = \frac{T}{R_1} + \frac{T}{R_2}.$$

The excess of internal pressure above that at the superficial point in question is thus

$$K + \frac{T}{R_1} + \frac{T}{R_2}, \quad \text{.............................(36)}$$

in agreement with (35).

For a cylindrical surface of radius r, we have simply

$$K + T/r. \quad \text{.................................(37)}$$

Returning to the case of a plane surface, we know that upon it $V = K$, and that in the interior $V = 2K$. At a point P (Fig. 2) just within the surface, the value of V cannot be expressed in terms of the principal quantities K and T, but will depend further upon the precise form of the function Π. We can, however, express the value of $\int V dz$, where z is measured inwards along the normal, and the integration extends over the whole of the superficial layer where V differs from $2K$.

Fig. 2.

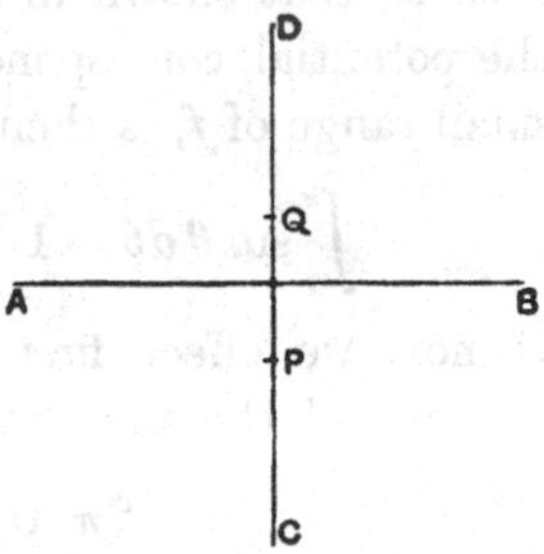

It is not difficult to recognize that this integral must be related to T. For if Q be a point upon the normal equidistant with P from the surface AB, the potential at Q due to fluid below AB is the same as the potential at P due to imaginary fluid above AB. To each of these add the potential of the lower fluid at P. Then the sum of the potentials at P and Q due to the lower fluid is equal to the potential at P due to both fluids, that is to the constant $2K$. The deficiency of potential at a point P near the plane surface of a fluid, as compared with the potential in the interior, is thus the same as the potential at an external point Q, equidistant from the surface. Now it is evident that $\int V_Q\,dz$ integrated upwards along the normal represents the work per unit of area that would be required to separate a continuous fluid of unit density along the plane AB and to remove the parts beyond the sphere of influence, that is, according to the principle of Dupré, $2T$. We conclude that the deficiency in $\int V_P\,dz$, integrated along the normal inwards, is also $2T$; or that

$$\int_0^z V_P\,dz = 2K\,.\,z - 2T, \qquad\qquad (38)$$

z being large enough to include the whole of the superficial stratum. The pressure p at any point P is given by $p = V_P - K$, so that

$$\int_0^z p\,dz = K\,.\,z - 2T. \qquad\qquad (39)$$

We may thus regard $2T$ as measuring the total deficiency of pressure in the superficial stratum.

The argument here employed is of course perfectly satisfactory; but it is also instructive to investigate the question directly, without the aid of the idea of superficial tension, or energy, and this is easily done.

In polar coordinates the potential at any point P is expressed by

$$V_P = 2\pi \iint \Pi\,(f)\,f^2 \sin\theta\, d\theta\, df,$$

the integrations extending over the whole space ACB (Fig. 3). If the distance EP, that is z, exceed the range of the forces, every sphere of radius f, under consideration, is complete, and $V_P = 2K$. But in the integration with respect to z incomplete spheres have to be considered, such as that shown in the figure. The value of the potential, corresponding to a given infinitely small range of f, is then proportional to

Fig. 3.

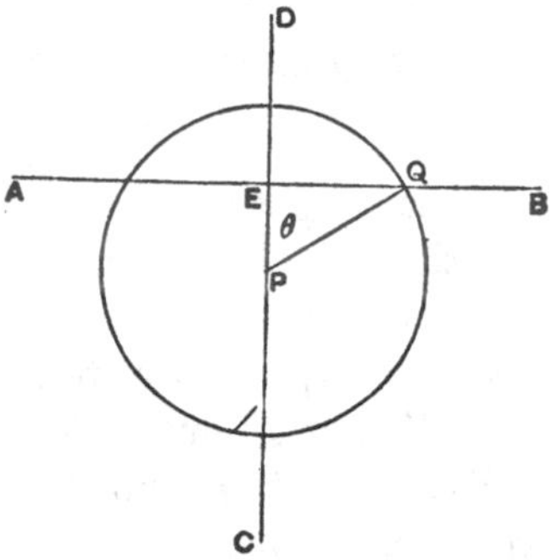

$$\int_\theta^\pi \sin\theta\, d\theta = 1 + \cos\theta = 1 + z/f.$$

If now we effect first the integration with respect to z, we have as the element of the final integral,

$$2\pi\Pi(f)f^2\,df\left\{\int_0^f (1 + z/f)\,dz + \int_f^z 2\,dz\right\},$$

or

$$2\pi\Pi\,(f)\,f^2\,df(2z - \tfrac{1}{2}f);$$

and thus, on the whole,

$$\int_0^z V_P dz = z \,.\, 4\pi \int_0^\infty \Pi(f) f^2 df - \pi \int_0^\infty \Pi(f) f^3 df$$
$$= z \,.\, 2K - 2T, \quad \text{as before.}$$

An application of this result to a calculation of the pressure operative between the two halves of an isolated sphere will lead us to another interpretation of T. The pressure in the interior is $K + 2T/r$, r being the radius; and this may be regarded as prevailing over the whole of the diametral dividing plane, subject to a correction for the circumferential parts which are near the surface of the fluid. If the radius r increase without limit, the correction will be the same per unit of length as that investigated for a plane surface. The whole pressure between the two infinite hemispheres is thus

$$\pi r^2 (K + 2T/r) - 2T \,.\, 2\pi r, \quad \text{or} \quad \pi r^2 K - T \,.\, 2\pi r. \quad \text{........(40)}$$

This expression measures equally the attraction between the two hemispheres, which the pressure is evoked to balance. If the fluid on one side of the diametral plane extended to infinity, the attraction upon the other hemisphere, supposed to retain its radius r, would be $\pi r^2 K$ simply; so that the second term $T \,.\, 2\pi r$ may be considered to represent the deficiency of attraction due to the absence of the fluid external to one hemisphere. Regarding the matter in two dimensions, we recognize T as the attraction per unit of length perpendicular to the plane of the paper of the fluid occupying (say) the first quadrant XOY (Fig. 4) upon the fluid in the third quadrant $X'OY'$, the attraction being resolved in one or other of the directions OX, OY. In its actual direction, bisecting the angle XOY the attraction will be of course $\sqrt{2} \,.\, T$.

Fig. 4.

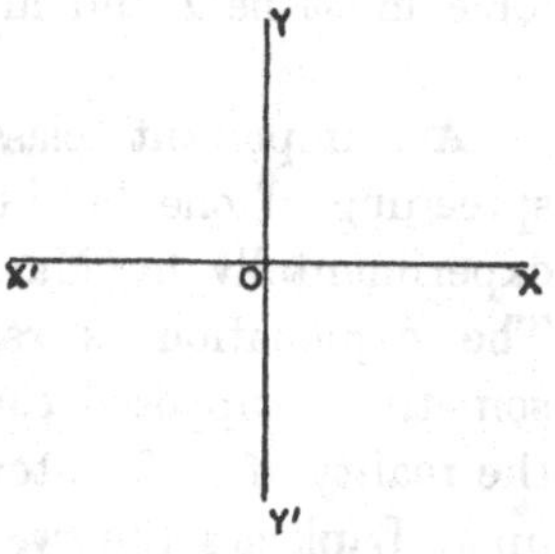

We will now suppose that the sphere is divided by a plane AB (Fig. 5), which is not diametral, but such that the angle $BAO = \theta$; $AO = r$, $AB = 2\rho$. In the interior of the mass, and generally along the section AB, $V = 2K$. On the surface of the sphere, and therefore along the circumference of AB, $V = K - 2T/r$. When V was integrated along the normal, from a plane surface inwards, the deficiency was found to be $2T$. In the present application the integration is along the oblique line AB, and the deficiency will be $2T \sec\theta$. Hence when r and ρ increase without limit, we may take as the whole pressure over the area AB

Fig. 5.

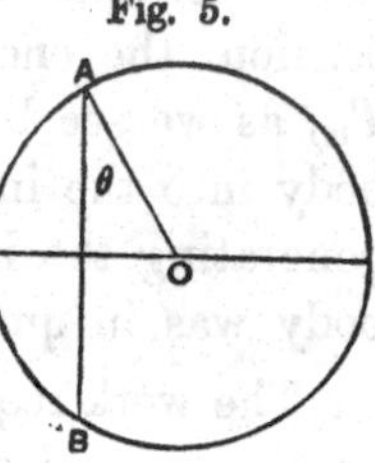

$$\pi\rho^2 (K + 2T/r) - 2\pi\rho \,.\, 2T \sec\theta = \pi\rho^2 K - 2\pi\rho\,(2T \sec\theta - T\cos\theta).$$

The deficiency of attraction perpendicular to AB is thus for each unit of perimeter

$$2T \sec\theta - T\cos\theta, \quad\quad(41)$$

and this we may think of as applicable in two dimensions (Fig. 6) to each unit of length. When $\theta = 0$, (41) reduces to T.

The term $T\cos\theta$ in the expression for the total pressure appears to have its origin in the curvature of the surface, only not disappearing when the curvature vanishes, in consequence of the simultaneous increase without limit of the area over which the pressure is reckoned. If we consider only a distance AB, which, though infinite in comparison with the range of the attraction, is infinitely small in comparison with the radius of curvature, $T\cos\theta$ will disappear from the expression for the pressure, though it must necessarily remain in the expression for the attraction. The pressure acting across a section AB proceeding inwards from a plane surface AE of a fluid is thus inadequate to balance the attraction of the two parts. It must be aided by an external force perpendicular to AB of magnitude $T\cos\theta$; and since the imaginary section AB may be made at any angle, we see that the force must be T and must act along AE.

Fig. 6.

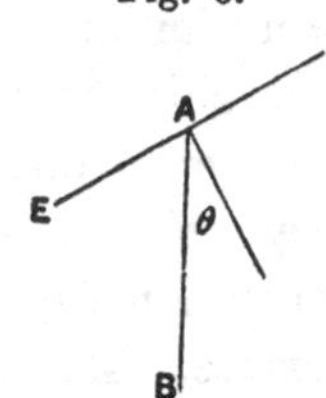

An important class of capillary phenomena are concerned with the spreading of one liquid upon the surface of another, a subject investigated experimentally by Marangoni, Van der Mensbrugghe, Quincke, and others. The explanation is readily given in terms of surface-tension; and it is sometimes supposed that these phenomena demonstrate in a special manner the reality of surface-tension, and even that they are incapable of explanation upon Laplace's theory, which dealt in the first instance with the capillary pressures due to curvature of surfaces*.

In considering this subject, we have first to express the dependence of the tension at the interface of two bodies in terms of the forces exercised by the bodies upon themselves and upon one another, and to effect this we cannot do better than follow the method of Dupré. If T_{12} denote the interfacial tension, the energy corresponding to unit of area of the interface is also T_{12}, as we see by considering the introduction (through a fine tube) of one body into the interior of the other. A comparison with another method of generating the interface, similar to that previously employed when but one body was in question, will now allow us to evaluate T_{12}.

The work required to cleave asunder the parts of the first fluid which lie on the two sides of an ideal plane passing through the interior, is per unit

* Van der Mensbrugghe, "Essai sur la Théorie Mécanique de la Tension Superficielle, &c." *Bulletins de l'Acad. roy. de Belgique*, 3me série, t. IX. No. 5, 1885, p. 12. Worthington, *Phil. Mag.* Oct. 1884, p. 364.

of area $2T_1$, and the free surface produced is two units in area. So for the second fluid the corresponding work is $2T_2$. This having been effected, let us now suppose that each of the units of area of free surface of fluid (1) is allowed to approach normally a unit of area of (2) until contact is established. In this process work is gained which we may denote by $4T'_{12}$, $2T'_{12}$ for each pair. On the whole, then, the work expended in producing two units of interface is $2T_1 + 2T_2 - 4T'_{12}$, and this, as we have seen, may be equated to $2T_{12}$. Hence

$$T_{12} = T_1 + T_2 - 2T'_{12} \qquad (42)$$

If the two bodies are similar, $T_1 = T_2 = T'_{12}$; and $T_{12} = 0$, as it should do.

Laplace does not treat systematically the question of interfacial tension, but he gives incidentally in terms of his quantity H a relation analogous to (42).

If $2T'_{12} > T_1 + T_2$, T_{12} would be negative, so that the interface would of itself tend to increase. In this case the fluids must mix. Conversely, if two fluids mix, it would seem that T'_{12} must exceed the mean of T_1 and T_2; otherwise work would have to be *expended* to effect a close alternate stratification of the two bodies, such as we may suppose to constitute a first step in the process of mixture*.

The value of T'_{12} has already been calculated (7). We may write

$$T'_{12} = \pi\sigma_1\sigma_2 \int_0^\infty \theta(z)\,dz = \tfrac{1}{8}\pi\sigma_1\sigma_2 \int_0^\infty z^4\phi(z)\,dz\,; \qquad (43)$$

and in general the functions θ, or ϕ, must be regarded as capable of assuming different forms. Under these circumstances there is no limitation upon the values of the interfacial tensions for three fluids, which we may denote by T_{12}, T_{23}, T_{31}. If the three fluids can remain in contact with one another, the sum of any two of the quantities must exceed the third, and by Neumann's rule the directions of the interfaces at the common edge must be parallel to the sides of a triangle, taken proportional to T_{12}, T_{23}, T_{31}. If the above-mentioned condition be not satisfied, the triangle is imaginary, and the three fluids cannot rest in contact, the two weaker tensions, even if acting in full concert, being incapable of balancing the strongest. For instance, if $T_{31} > T_{12} + T_{23}$, the second fluid spreads itself indefinitely upon the interface of the first and third fluids.

Fig. 7.

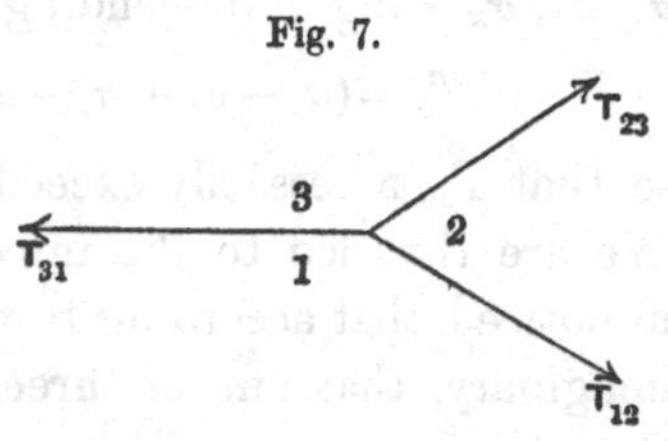

The experimenters who have dealt with this question, Marangoni, Van der Mensbrugghe, Quincke, have all arrived at results inconsistent with the reality of Neumann's triangle. Thus Marangoni says†;—"Die gemeinschaft-

* Dupré, *loc. cit.* p. 372. Thomson, *Popular Lectures*, p. 53.

† *Pogg. Ann.* CXLIII. p. 348, 1871 (1865). It was subsequently shown by Quincke that mercury is not really an exception.

liche Oberfläche zweier Flüssigkeiten hat eine geringere Oberflächenspannung als die Differenz der Oberflächenspannung der Flüssigkeiten selbst (mit Ausnahme des Quecksilbers)." Three pure bodies (of which one may be air) cannot accordingly remain in contact. If a drop of oil stands in lenticular form upon a surface of water, it is because the water-surface is already contaminated with a greasy film.

On the theoretical side the question is open until we introduce some limitation upon the generality of the functions. By far the simplest supposition open to us is that the functions are the same in all cases, the attractions differing merely by coefficients analogous to densities in the theory of gravitation. This hypothesis was suggested by Laplace, and may conveniently be named after him. It was also tacitly adopted by Young, in connexion with the still more special hypothesis which Young probably had in view, namely that the force in each case was constant within a limited range, the same in all cases, and vanished outside that range.

As an immediate consequence of this hypothesis we have from (3)

$$K = K_0\sigma^2, \qquad T = T_0\sigma^2, \quad \text{......................(44, 45)}$$

where K_0, T_0 are the same for all bodies.

But the most interesting results are those which Young* deduced relative to the interfacial tensions of three bodies. By (12), (43),

$$T'_{12} = \sigma_1\sigma_2 T_0; \quad \text{..............................(46)}$$

so that by (42), (45),

$$T_{12} = (\sigma_1 - \sigma_2)^2 T_0. \quad \text{..........................(47)}$$

According to (47), the interfacial tension between any two bodies is proportional to the square of the difference of their densities. The densities σ_1, σ_2, σ_3 being in descending order of magnitude, we may write

$$T_{31} = (\sigma_1 - \sigma_2 + \sigma_2 - \sigma_3)^2 T_0 = T_{12} + T_{23} + 2(\sigma_1 - \sigma_2)(\sigma_2 - \sigma_3) T_0;$$

so that T_{31} necessarily exceeds the sum of the other two interfacial tensions. We are thus led to the important conclusion, so far as I am aware hitherto unnoticed, that according to this hypothesis Neumann's triangle is necessarily imaginary, that one of three fluids will always spread upon the interface of the other two.

Another point of importance may be easily illustrated by this theory, viz. the dependency of capillarity upon abruptness of transition. "The reason why the capillary force should disappear when the transition between two liquids is sufficiently gradual will now be evident. Suppose that the transition from 0 to σ is made in two equal steps, the thickness of the intermediate layer of density $\frac{1}{2}\sigma$ being large compared to the range of the molecular forces, but small in comparison with the radius of curvature. At

* *Works*, Vol. I. p. 463.

each step the difference of capillary pressure is only one quarter of that due to the sudden transition from 0 to σ, and thus altogether half the effect is lost by the interposition of the layer. If there were three equal steps, the effect would be reduced to one third, and so on. When the number of steps is infinite, the capillary pressure disappears altogether*."

According to Laplace's hypothesis the whole energy of any number of contiguous strata of liquids is least when they are arranged in order of density, so that this is the disposition favoured by the attractive forces. The problem is to make the sum of the interfacial tensions a minimum, each tension being proportional to the square of the difference of densities of the two contiguous liquids in question. If the order of stratification differ from that of densities, we can show that each step of approximation to this order lowers the sum of tensions. To this end consider the effect of the abolition of a stratum σ_{n+1}, contiguous to σ_n and σ_{n+2}. Before the change we have

$$(\sigma_n - \sigma_{n+1})^2 + (\sigma_{n+1} - \sigma_{n+2})^2,$$

and afterwards $(\sigma_n - \sigma_{n+2})^2$. The second *minus* the first, or the increase in the sum of tensions, is thus

$$2(\sigma_n - \sigma_{n+1})(\sigma_{n+1} - \sigma_{n+2}).$$

Hence, if σ_{n+1} be intermediate in magnitude between σ_n and σ_{n+2}, the sum of tensions is increased by the abolition of the stratum; but, if σ_{n+1} be not intermediate, the sum is decreased. We see, then, that the removal of a stratum from between neighbours where it is out of order and its introduction between neighbours where it will be in order is doubly favourable to the reduction of the sum of tensions; and since by a succession of such steps we may arrive at the order of magnitude throughout, we conclude that this is the disposition of minimum tensions and energy.

So far the results of Laplace's hypothesis are in marked accordance with experiment; but if we follow it out further, discordances begin to manifest themselves. According to (47)

$$\sqrt{T_{31}} = \sqrt{T_{12}} + \sqrt{T_{23}}, \quad \text{.........................(48)}$$

a relation not verified by experiment. What is more, (47) shows that according to the hypothesis T_{12} is necessarily positive; so that, if the preceding argument be correct, no such thing as mixture of two liquids could ever take place.

But although this hypothesis is clearly too narrow for the facts, it may be conveniently employed in illustration of the general theory. In extension of (25) the potential at any point may be written

$$V = \iiint \sigma \Pi(f)\, dx\, dy\, dz, \quad \text{.....................(49)}$$

and the hydrostatical equation of equilibrium is

$$dp = \sigma\, dV. \quad \text{..................................(50)}$$

* "Laplace's Theory of Capillarity," *Phil. Mag.* October 1883, p. 315. [Vol. II. p. 234.]

By means of the potential we may prove, independently of the idea of surface tension, that three fluids cannot rest in contact. Along the surface of contact of any two fluids the potential must be constant. Otherwise, there would be a tendency to circulation round a circuit of which the principal parts are close and parallel to the surface, but on opposite sides. For in the limit the variation of potential will be equal and opposite in the two parts of the circuit, and the resulting forces at corresponding points, being proportional also to the densities, will not balance. It is thus necessary to equilibrium that there be no force at any point; that is, that the potential be constant along the whole interface.

Fig. 8.

It follows from this that if three fluids can rest in contact, the potential must have the same constant value on all the three intersecting interfaces. But this is clearly impossible, the potential on each being proportional to the sum of the densities of the two contiguous fluids, as we see by considering places sufficiently removed from the point of intersection.

According to Laplace's hypothesis, then, three fluids cannot rest in contact; but the case is altered if one of the bodies be solid. It is necessary, however, that the quality of solidity attach to the body of intermediate density. For suppose, for example (Fig. 9), that the body of greatest density, σ_1, is solid, and that fluids of densities σ_2, σ_3 touch it and one another. It is now no longer necessary that the potential be constant along the interfaces (1, 2), (1, 3); but only along the interface (3, 2). The potential at a distant point of this interface may be represented by $\sigma_2 + \sigma_3$. But at the point of intersection the potential cannot be so low as this, being at least equal to $\sigma_1 + \sigma_3$, even if the angle formed by the two faces of (2) be evanescent. By this and similar reasoning it follows that the conditions of equilibrium cannot be satisfied, unless the solid be the body of intermediate density σ_2.

Fig. 9.

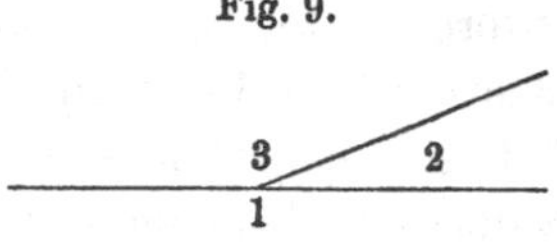

One case where equilibrium is possible admits of very simple treatment. It occurs when $\sigma_2 = \frac{1}{2}(\sigma_1 + \sigma_3)$, and the conditions are satisfied by supposing (Fig. 10) that the fluid interface is plane and perpendicular to the solid wall. At a distance from O the potential is represented by $\sigma_1 + \sigma_3$; and the same value obtains at a point P, near O, where the sphere of influence cuts into (2). For the areas of spherical surface lost by (1) and (3) are equal, and are replaced by equal areas of (2); so that if the above condition between the densities holds good, the potential is constant all the way up to O. The sub-case, where $\sigma_3 = 0$, $\sigma_2 = \frac{1}{2}\sigma_1$, was given by Clairaut.

Fig. 10.

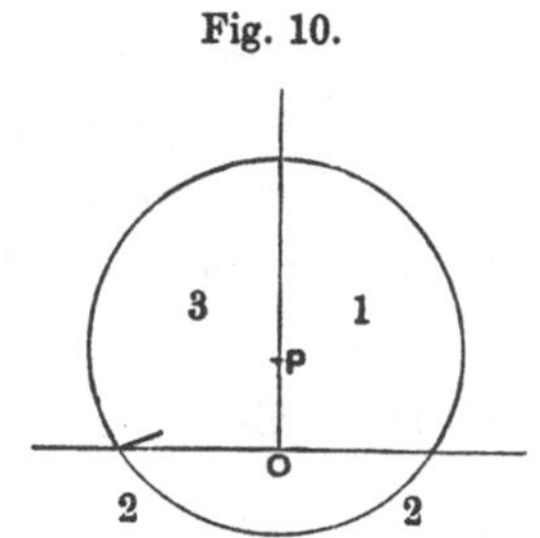

If the intermediate density differ from the mean of the other two, the problem is less simple; but the general tendency is easily recognized. If, for example, $\sigma_2 > \frac{1}{2}(\sigma_1 + \sigma_3)$, it is evident that along a perpendicular interface the potential would increase as O is approached. To compensate this the interface must be inclined, so that, as O is approached, σ_1 loses its importance relatively to σ_3. In this case therefore the angle between the two faces of (1) must be acute.

The general problem was treated by Young by means of superficial tensions, which must balance when resolved parallel to the surface of the solid, though not in the perpendicular direction. In this way Young found at once

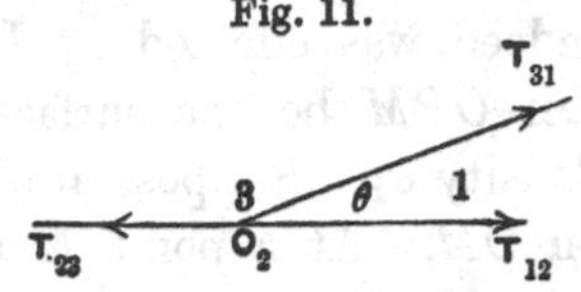

$$T_{31}\cos\theta + T_{12} = T_{23}; \quad \text{.........................(51)}$$

or rather, in terms of the more special hypothesis,

$$(\sigma_1 - \sigma_3)^2\cos\theta + (\sigma_1 - \sigma_2)^2 = (\sigma_2 - \sigma_3)^2. \quad \text{..............(52)}$$

From this we deduce

$$\cos\theta = \frac{2\sigma_2 - \sigma_1 - \sigma_3}{\sigma_1 - \sigma_3}, \quad \text{.........................(53)}$$

in agreement with what we found above for a special case. The equation may also be written

$$\sigma_1\cos^2\tfrac{1}{2}\theta + \sigma_3\sin^2\tfrac{1}{2}\theta = \sigma_2; \text{.........................(54)}$$

or if, as we may suppose without real loss of generality, $\sigma_3 = 0$,

$$\sigma_1\cos^2\tfrac{1}{2}\theta = \sigma_2, \quad \text{.............................(55)}$$

a form given by Laplace. In discussing the equation (53) with $\sigma_3 = 0$, Young* remarks:—"Supposing the attractive density of the solid to be very small, the cosine will approach to -1, and the angle of the liquid to two right angles; and on the other hand, when σ_2 becomes equal to σ_1, the cosine will be 1, and the angle will be evanescent, the surface of the liquid coinciding in direction with that of the solid. If the density σ_2 be still further increased, the angle cannot undergo any further alteration, and the excess of force will only tend to spread the liquid more rapidly on the solid, so that a thin film would always be found upon its surface, unless it were removed by evaporation, or unless its formation were prevented by some unknown circumstance which seems to lessen the intimate nature of the contact of liquids with solids."

The calculation of the angle of contact upon these lines is thus exceedingly simple, but I must admit that I find some difficulty in forming a definite conception of superficial tension as applied to the interface of a solid and a fluid. It would seem that interfacial *tension* can only be employed in

* *Works*, Vol. I. p. 464. I have introduced an insignificant change in the notation.

such cases as the immediate representative of interfacial *energy*, as conceived by Gauss. This principle, applied to a hypothetical displacement in which the point of meeting travels along the wall, leads with rigour to the required result.

In view of the difficulties which have been felt upon this subject, it seems desirable to show that the calculation of the angle of contact can be made without recourse to the principle of interfacial tension or energy. This indeed was effected by Laplace himself, but his process is very circuitous. Let OPM be the surface of fluid (σ_1) resting against a solid wall ON of density σ_2. Suppose also that $\sigma_3 = 0$, and that there is no external pressure on OM. At a point M at a sufficient distance from O the curvature must be uniform (or the potential could not be constant), and we will suppose it to be zero. It would be a mistake, however, to think that the surface can be straight throughout up to O. This we may recognize by consideration of the potential at a point P just near enough to O for the sphere of influence to cut the solid. As soon as this occurs, the potential would begin to vary by substitution of σ_2 for σ_1, and equilibrium would fail. The argument does not apply if $\theta = \frac{1}{2}\pi$.

Fig. 12.

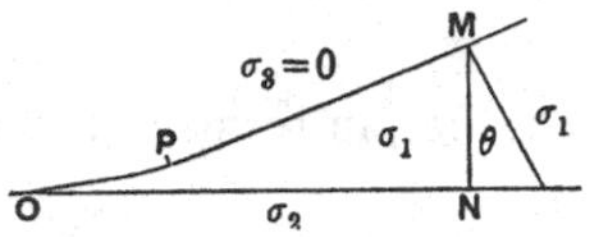

We may attain the object in view by considering the equilibrium of the fluid MNO, or rather of the forces which tend to move it parallel to ON. Of pressures we have only to consider that which acts across MN, for on OM there is no pressure, and that on ON has no component in the direction considered. Moreover, the solid σ_2 below ON exercises no attraction parallel to ON. Equilibrium therefore demands that the pressure operative across MN shall balance the horizontal attraction exercised upon OMN by the fluid σ_1 which lies to the right of MN. The evaluation of the attraction in such cases has been already treated. It is represented by $MN.\sigma_1^2 K_0$, subject to corrections for the ends at M and N. The correction for M is by (41) $\sigma_1^2 T_0 (2 \sec\theta - \cos\theta)$, and for N it is $\sigma_1^2 T_0$. On the whole the attraction in question is therefore

$$\sigma_1^2 \{MN.K_0 - 2T_0 \sec\theta + T_0 \cos\theta - T_0\}.$$

We have next to consider the pressure. In the interior of MN, we have $\sigma_1^2 K_0$; but the whole pressure $MN.\sigma_1^2 K_0$ is subject to corrections for the ends. The correction for M we have seen to be $2\sigma_1^2 T_0 \sec\theta$. In the neighbourhood of N the potential, and therefore the pressure, is influenced by the solid. If σ_2 were zero, the deficiency would be $2\sigma_1^2 T_0$. If σ_2 were equal to σ_1, there would be no deficiency. Under the actual circumstances the deficiency is accordingly

$$2\sigma_1 (\sigma_1 - \sigma_2) T_0;$$

so that the expression for the total pressure operative across MN is

$$\sigma_1 \{MN . \sigma_1 K_0 - 2\sigma_1 T_0 \sec\theta - 2(\sigma_1 - \sigma_2) T_0\}.$$

If we now equate the expressions for the pressure and the resolved attraction, we find as before

$$\sigma_1 (1 - \cos\theta) = 2(\sigma_1 - \sigma_2).$$

In connexion with edge-angles it may be well here to refer to a problem, which has been the occasion of much difference of opinion—that of the superposition of several liquids in a capillary tube. Laplace's investigation led him to the conclusion that the whole weight of liquid raised depends only upon the properties of the lowest liquid. Thereupon Young* remarks:—"This effect may be experimentally illustrated by introducing a minute quantity of oil on the surface of the water contained in a capillary tube, the joint elevation, instead of being increased as it ought to be according to Mr Laplace, is very conspicuously diminished; and it is obvious that since the capillary powers are represented by the squares of the density of oil and of its difference from that of water, their sum must be less than the capillary power of water, which is proportional to the square of the sum of the separate quantities."

But the question is not to be dismissed so summarily. That Laplace's conclusion is sound, upon the supposition that *none of the liquids wets the walls of the tube,* may be shown without difficulty by the method of energy. In a hypothetical displacement the work done against gravity will balance the work of the capillary forces. Now it is evident that the liquids, other than the lowest, contribute nothing to the latter, since the relation of each liquid to its neighbours and to the walls of the tube is unaltered by the displacement. The only effect of the rise is that a length of the tube before in contact with air is replaced by an equal length in contact with the lowest liquid. The work of the capillary forces is the same as if the upper liquids did not exist, and therefore the total weight of the column supported is independent of these liquids.

The case of Young's experiment, in which oil stands upon water in a glass tube, is not covered by the foregoing reasoning. The oil must be supposed to wet the glass, that is to insinuate itself between the glass and air, so that the upper part of the tube is covered to a great height with a very thin layer of oil. The displacement here takes place under conditions very different from before. As the column rises, no new surface of glass is touched by oil, while below water replaces oil. The properties of the oil are thus brought into play, and Laplace's theorem does not apply.

* *Works*, Vol. I. p. 463.

That theory indicates the almost indefinite rise of a liquid like oil in contact with a vertical wall of glass is often overlooked, in spite of Young's explicit statement quoted above. It may be of interest to look into the question more narrowly on the basis of Laplace's hypothesis.

If we include gravity in our calculations, the hydrostatic equation of equilibrium is

$$p = \text{const.} + \sigma V - g\rho z, \quad \text{......................(56)}$$

where z is measured upwards, and V denotes as before the potential of the cohesive forces. Along the free surface of the liquid the pressure is constant, so that

$$\sigma V = \sigma^2 K_0 + g\rho z, \quad \text{........................(57)}$$

z being reckoned from a place where the liquid is deep and the surface plane.

At a point upon the surface, whose distance from the wall exceeds the range of the forces,

$$\sigma V = K + T\left(\frac{1}{R_1} + \frac{1}{R_2}\right); \quad \text{......................(58)}$$

or, if we take the problem in two dimensions,

$$\sigma V = K + T/R, \text{............................(59)}$$

where R is the radius of curvature, and K, T denote the intrinsic pressure and tension proper to the liquid and proportional to σ^2. Upon this equation is founded the usual calculation of the form of the surface.

When the point under consideration is nearer to the wall than the range of the forces, the above expression no longer applies. The variation of V on the surface of the thin layer which rises above the meniscus is due not to variations of curvature, for the curvature is here practically evanescent, but to the inclusion within the sphere of influence of the more dense matter constituting the wall. If the attraction be a simple function of the distance, such as those considered above in illustrative examples, the thickness of the layer diminishes constantly with increasing height. The limit is reached when the thickness vanishes, and the potential attains the value due simply to the solid wall. This potential is $\sigma' K_0$, the intrinsic pressure within the wall being $\sigma'^2 K_0$; so that if we compare the point above where the layer of fluid disappears with a point below upon the horizontal surface, we find

$$g\rho z = \sigma(\sigma' - \sigma) K_0. \quad \text{........................(60)}$$

By this equation is given the total head of liquid in contact with the wall; and, as was to be expected, it is enormous.

The height of the meniscus itself in a very narrow tube wetted by the liquid is obtained from (57), (58). If R be the radius of curvature at the centre of the meniscus,

$$g\rho z = 2T/R; \quad \text{............................(61)}$$

and R may be identified with the radius of the tube, for under the circumstances supposed the meniscus is very approximately hemispherical.

The calculation of the height by the method of energy requires a little attention. The simplest displacement is an equal movement upwards of the whole body of liquid, including the layer above the meniscus. In this case the work of the cohesive forces depends upon the substitution of liquid for air in contact with the tube, and therefore not merely upon the interfacial tension between liquid and air, as (61) might lead us to suppose. The fact is that in this way of regarding the subject the work which compensates that of the cohesive forces is not simply the elevation against gravity of the column (z), but also an equal elevation of the very high, though very thin, layer situated above it. The complication thus arising may be avoided by taking the hypothetical displacement so that the thin layer does not accompany the column (z). In this case the work of the cohesive forces depends upon a reduction of surface between liquid and air simply, without reference to the properties of the walls, and (61) follows immediately.

Laplace's integral K was, as we have seen, introduced originally to express the intrinsic pressure, but according to the discovery of Dupré* it is susceptible of another and very important interpretation. "Le travail de désagrégation totale d'un kilogramme d'un corps quelconque égale le produit de l'attraction au contact par le volume, ou, ce qui équivaut, le travail de désagrégation totale de l'unité de volume égale l'attraction au contact." *Attraction au contact* here means what we have called intrinsic pressure. The following reasoning is substantially that of Dupré.

We have seen (2) that $2\pi m\sigma\psi(z)$ represents the attraction of a particle m placed at distance z from the plane surface of an infinite solid whose density is σ. The work required to carry m from $z = 0$ to $z = \infty$ is therefore

$$2\pi m\sigma \int_0^\infty \psi(z)\, dz = m\sigma K_0,$$

by (4); so that the work necessary to separate a superficial layer of thickness dz from the rest of the mass and to carry it beyond the range of the attraction is $\sigma^2 dz K_0$. The complete disaggregation of unit of volume into infinitesimal slices demands accordingly an amount of work represented by $\sigma^2 K_0$, or K. The work required further to separate the infinitesimal slices into component filaments or particles and to remove them beyond the range of the mutual attraction is negligible in the limit, so that K is the total work of complete disaggregation.

A second law formulated by Dupré is more difficult to accept. "Pour un même corps prenant des volumes variés, le travail de désagrégation restant

* *Théorie Mécanique de la Chaleur*, 1869, p. 152.

Van der Waals gives the same result in his celebrated essay of 1873.—German Translation, 1881, p. 31.

à accomplir est proportionel à la densité ou en raison inverse du volume." The argument is that the work remaining to be done upon a given mass at any stage of the expansion is proportional first to the square of the density, and secondly to the actual volume, on the whole therefore inversely as the volume. The criticism that I am inclined to make here is that Dupré's theory attempts either too little or too much. If we keep strictly within the lines of Laplace's theory the question here discussed cannot arise, because the body is supposed to be incompressible. That bodies are in fact compressible may be so much the worse for Laplace's theory, but I apprehend that the defect cannot be remedied without a more extensive modification than Dupré attempts. In particular, it would be necessary to take into account the work of compression. We cannot leave the attractive forces unbalanced; and the work of the repulsive forces can only be neglected upon the hypothesis that the compressibility itself is negligible. Indeed it seems to me, that a large part of Dupré's work, important and suggestive as it is, is open to a fundamental objection. He makes free use of the two laws of thermodynamics, and at the same time rests upon a molecular theory which is too narrow to hold them. One is driven to ask what is the real nature of this heat, of which we hear so much. It seems hopeless to combine thermodynamics with a merely statical view of the constitution of matter.

On these grounds I find it difficult to attach a meaning to such a theorem as that enunciated in the following terms*:—"La dérivée partielle du travail mécanique interne prise par rapport au volume égale l'attraction par mètre carré qu'exercent l'une sur l'autre les deux parties du corps situées des deux côtés d'une section plane," viz. the intrinsic pressure. In the partial differentiation the volume is supposed to vary and the temperature is supposed to remain constant. The difficulty of the first part of the supposition has been already touched upon; and how in a fundamental theory can we suppose temperature to be constant without knowing what it is? It is possible, however, that some of these theorems may be capable of an interpretation which shall roughly fit the facts, and it is worthy of consideration how far they may be regarded as applicable to matter in a state of extreme cold.

With respect to the value of K, Young's estimate of 23,000 atmospheres for water has already been referred to. It is not clear upon what basis he proceeded, but a chance remark suggests that it may have been upon the assumption that cohesion was of the same order of magnitude in liquids and solids. Against this, however, it may be objected that the estimate is unduly high. Even steel is scarcely capable of withstanding a tension of 23,000 atmospheres.

* *Loc. cit.* p. 47.

So far as I am aware, the next estimates of K are those of Dupré. One of them proceeds upon the assumption that for rough purposes K may be identified with the mechanical equivalent of the heat rendered latent in the evaporation of the liquid, that in fact evaporation may be regarded as a process of disaggregation in which the cohesive forces have to be overcome. This view appears to be substantially sound. If we take the latent heat of water as 600°, we find for the work required to disintegrate one gram of water

$$600 \times 4{\cdot}2 \times 10^7 \text{ C.G.S.}$$

One atmosphere is about 10^6 C.G.S.; so that

$$K = 25{,}000 \text{ atmospheres.}$$

The estimates of his predecessors were apparently unknown to Van der Waals, who (in 1873) undertook his work mainly with the object of determining the quantity in question. He finds for water 11,000 atmospheres. The application of Clausius's equation of virial to gases and liquids is obviously of great importance; but, as it lies outside the scope of the present paper, I must content myself with referring the reader to the original memoir and to the account of it by Maxwell*.

One of the most remarkable features of Young's treatise is his estimate of the range a of the attractive force on the basis of the relation $T = \frac{1}{3}aK$. Never once have I seen it alluded to; and it is, I believe, generally supposed that the first attempt of the kind is not more than twenty years old. Estimating K at 23,000 atmospheres, and T at 3 grains per inch, Young finds† that "the extent of the cohesive force must be limited to about the 250 millionth of an inch"; and he continues, "nor is it very probable that any error in the suppositions adopted can possibly have so far invalidated this result as to have made it very many times greater or less than the truth." It detracts nothing from the merit of this wonderful speculation that a more precise calculation does not verify the numerical coefficient in Young's equation. The point is that the range of the cohesive force is necessarily of the order T/K.

But this is not all. Young continues:—"Within similar limits of uncertainty, we may obtain something like a conjectural estimate of the mutual distance of the particles of vapours, and even of the actual magnitude of the elementary atoms of liquids, as supposed to be nearly in contact with each other; for if the distance at which the force of cohesion begins is constant at the same temperature, and if the particles of steam are condensed when they approach within this distance, it follows that at 60° of Fahrenheit the distance of the particles of pure aqueous vapour is about the 250 millionth of an inch;

* *Nature*, Vol. x. p. 477 (1874). See also Vol. xi. pp. 357, 374. [Maxwell's *Scientific Papers*, Vol. ii. pp. 407, 418.]

† *Works*, Vol. i. p. 461.

and since the density of this vapour is about one sixty thousandth of that of water, the distance of the particles must be about forty times as great; consequently the mutual distance of the particles of water must be about the ten thousand millionth of an inch. It is true that the result of this calculation will differ considerably according to the temperature of the substances compared.... This discordance does not, however, wholly invalidate the general tenour of the conclusion...and on the whole it appears tolerably safe to conclude that, whatever errors may have affected the determination, the diameter or distance of the particles of water is between the two thousand and the ten thousand millionth of an inch." This passage, in spite of its great interest, has been so completely overlooked that I have ventured briefly to quote it, although the question of the size of atoms lies outside the scope of the present paper.

Another matter of great importance to capillary theory I will only venture to touch upon. When oil spreads upon water, the layer formed is excessively thin, about two millionths of a millimetre. If the layer be at first thicker, it exhibits instability, becoming perforated with holes. These gradually enlarge, until at last, after a series of curious transformations, the superfluous oil is collected in small lenses. It would seem therefore that the energy is less when the water is covered by a very thin layer of oil, than when the layer is thicker. Phenomena of this kind present many complications, for which various causes may be suggested, such as solubility, volatility, and—perhaps more important still—chemical heterogeneity. It is at present, I think, premature to draw definite physical conclusions; but we may at least consider what is implied in the preference for a thin as compared with a thicker film.

Fig. 13. Fig. 14. Fig. 15.

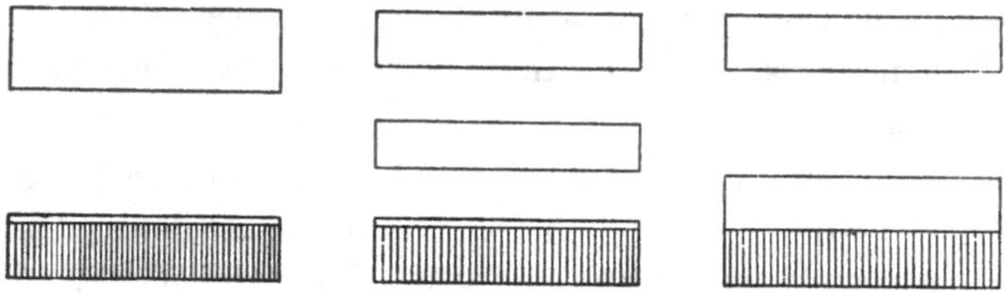

The passage from the first stage to the second may be accomplished in the manner indicated in Figs. 13, 14, 15. We begin (Fig. 13) with a thin layer of oil on water and an independent thick layer of oil. In the second stage (Fig. 14) the thick layer is split in two, also thick in comparison with the range of the cohesive forces, and the two parts are separated. In the third stage one of the component layers is brought down until it coalesces with the thin layer on water. The last state differs from the first by the substitution of a thick film of oil for a thin one in contact with the water, and we have to consider the work spent or gained in producing the change. If, as observation suggests, the last state has more energy than the first, it follows that more work is spent in splitting the thick layer of oil than is gained in the approach of a thick layer to the already oiled water. At some distances therefore, and those not the smallest, oil must be more attracted (or less repelled) by oil than by water. The reader will not fail to notice the connexion between this subject and the black of soap-films investigated by Profs. Reinold and Rücker [*Phil. Trans.* 172, p. 645, 1884].

[1901. Continuations of the present memoir under the same title will be found below, reprinted from *Phil. Mag.* XXXIII. pp. 209, 468, 1892. Reference may be made also to *Phil. Mag.* XLVIII. p. 331, 1899.]

177.

CLERK-MAXWELL'S PAPERS*.

[*Nature*, XLIII. pp. 26, 27, 1890.]

THE gratitude with which we receive these fine volumes is not unmingled with complaint. During the eleven years which have elapsed since the master left us, the disciples have not been idle, but their work has been deprived, to all appearance unnecessarily, of the assistance which would have been afforded by this collection of his works. However, it behoves us to look forward rather than backward; and no one can doubt that for many years to come earnest students at home and abroad will derive inspiration from Maxwell's writings, and will feel thankful to Mr Niven and the committee of friends and admirers for the convenient and handsome form in which they are here presented.

Under the modest title of preface, the editor contributes a sketch of Maxwell's life, which will be valued even by those who are acquainted with the larger work of Profs. Lewis Campbell and W. Garnett; and while abstaining from entering at length into a discussion of the relation which Maxwell's work bears historically to that of his predecessors, or attempting to estimate the effect which it had upon the scientific thought of the present day, he points out under the various heads what were the leading advances made.

In the body of the work the editor's additions reduce themselves to a few useful footnotes, placed in square brackets. Doubtless there is some difficulty in knowing where to stop, but the number of these footnotes might, I think, have been increased. For example, the last term in the differential equation of a stream-function symmetrical about an axis is allowed to stand with a wrong sign (Vol. I. p. 591) and on the following page the fifth term in the

* *The Scientific Papers of James Clerk-Maxwell.* Two Vols. Edited by W. D. Niven. (London: Cambridge University Press, 1890.)

expression for the self-induction of a coil should be $-\frac{1}{3}\pi \operatorname{cosec} 2\theta$, and not $-\frac{1}{3}\pi \cos 2\theta$.

To a large and enterprising group of physicists, Maxwell's name at once suggests electricity, and some, familiar with the great treatise, may be tempted to suppose that this book can contain little that is new to them. It was De Morgan, I think, who remarked that a great work often overshadows too much lesser writings of an author upon the same subject. In the present case it is true that much of the "Dynamical Theory of the Electro-magnetic Field" was subsequently embodied in the separate treatise. Nevertheless, there were important exceptions. Among these may be noticed the experimental method of determining the self-induction of a coil of wire in the Wheatstone's balance. By adjustment of resistances, the steady current through the galvanometer in the bridge is reduced to zero; but at the moment of making or breaking battery contact, an instantaneous current passes. From the magnitude of the throw thus observed, in comparison with the effect of upsetting the resistance-balance to a known extent, the self-induction can be calculated. The letter to Sir W. Grove, entitled "Experiment in Magneto-electric Induction" (Vol. II. p. 121), will also be read with interest by electricians. It gives the complete theory of what is sometimes called "electric resonance."

There can be little doubt but that posterity will regard as Maxwell's highest achievement in this field his electro-magnetic theory of light, whereby optics becomes a department of electrics. The clearest statement of his views will be found in the note appended to the "Direct comparison of Electro-static with Electro-magnetic Force" (Vol. II. p. 125). Several of the points which were then obscure have been cleared up by recent researches.

Scarcely, if at all, less important than his electrical work was the part taken by Maxwell in the development of the Dynamical Theory of Gases. Even now the difficulties which meet us here are not entirely overcome; but in the whole range of science there is no more beautiful or telling discovery than that gaseous viscosity is the same at all densities. Maxwell anticipated from theory, and afterwards verified experimentally, that the retarding effect of the air upon a body vibrating in a confined space is the same at atmospheric pressure and in the best vacuum of an ordinary air-pump.

Besides the more formal writings, these volumes include several reviews, contributed to *Nature*, as well as various lectures and addresses, all abounding in valuable suggestions, and enlivened by humorous touches. Among the most noticeable of these are the address to Section A of the British Association, the lectures on Colour-vision, on Molecules, and on Action at a Distance, and, one of his last efforts, the Rede Lecture on the Telephone. Many of the articles from the *Encyclopædia Britannica* are also of great importance, and become here for the first time readily accessible to foreigners. Under "Constitution of Bodies," ideas are put forward respecting the breaking up of but feebly stable groups of molecules, which, in the hands of Prof. Ewing, seem likely to find important application in the theory of magnetism.

A characteristic of much of Maxwell's writing is his dissatisfaction with purely analytical processes, and the endeavour to find physical interpretations for his formulæ. Sometimes the use of physical ideas is pushed further than strict logic can approve*; but those of us who are unable to follow a Sylvester in his analytical flights will be disposed to regard the error with leniency. The truth is that the limitation of human faculties often imposes upon us, as a condition of advance, temporary departure from the standard of strict method. The work of the discoverer may thus precede that of the systematizer; and the division of labour will have its advantage here as well as in other fields.

The reader of these volumes, not already familiarly acquainted with Maxwell's work, will be astonished at its variety and importance. Would that another ten years' teaching had been allowed us! The premature death of our great physicist was a loss to science that can never be repaired.

* "With all possible respect for Prof. Maxwell's great ability, I must own that to deduce purely analytical properties of spherical harmonics, as he has done, from 'Green's theorem' and the 'principle of potential energy,' seems to me a proceeding at variance with sound method, and of the same kind and as reasonable as if one should set about to deduce the binomial theorem from the laws of virtual velocities or make the rule for the extraction of the square root flow as a consequence from Archimedes' law of floating bodies." Sylvester, *Phil. Mag.* Vol. II. p. 306. 1876.

178.

ON PIN-HOLE PHOTOGRAPHY.

[*Philosophical Magazine*, XXXI. pp. 87—99, 1891.]

IT has long been known that the resolving power of lenses, however perfect, is limited, and more particularly that the capability of separating close distant objects, *e.g.* double stars, is proportional to aperture. The ground of the limitation lies in the finite magnitude of the wave-length of light (λ), and the consequent diffusion of illumination round the geometrical image of even an infinitely small radiant point. It is easy to understand the *rationale* of this process without entering upon any calculations. At the focal point itself all the vibrations proceeding from various parts of the aperture arrive in the same phase. The illumination is therefore here a maximum. But why is it less at neighbouring points in the focal plane which are all equally exposed to the vibrations from the aperture? The answer can only be that at such points the vibrations are discrepant. This discrepance can only enter by degrees; so that there must be a small region round the focus, at any point of which the phases are practically in agreement and the illumination sensibly equal to the maximum.

These considerations serve also to fix at least the order of magnitude of the patch of light. The discrepancy of phase is the result of the different distances of the various parts of the aperture from the eccentric point in question; and the greatest discrepancy is that between the waves which come from the nearest and furthest parts of the aperture. A simple calculation shows that the greatest difference of distance is expressed by $2rx/f$, where $2r$ is the diameter of the aperture, f the focal length, and x the linear eccentricity of the point under consideration. The question under discussion is at what stage does this difference of path introduce an important discrepancy of phase? It is easy to recognize that the illumination will not be greatly reduced until the extreme discrepancy of phase reaches half a wave-length. In this case

$$2x = f\lambda/2r,$$

which may be considered to give roughly the diameter of the patch of light. If there are two radiant points, the two representative patches will seriously overlap, unless the distance of their centres exceed $2x$. Supposing it to be equal to $2x$, which corresponds to an angular interval $2x/f$, we see that the double radiant cannot be resolved in the image, unless the angular interval exceed $\lambda/2r$.

Experiment* shows that the value thus roughly estimated is very near the truth for a rectangular aperture of width $2r$. If the aperture be of circular form, the resolving power is somewhat less, in the ratio of about 1·1 : 1.

It is therefore not going too far to say that there is nothing better established in optics than the limit to resolving power as proportional to aperture. On the other hand, the focal length is a matter of indifference, if the object-glass be perfect.

This is one side of the question before us. We now pass on to another, in which the focal length becomes of paramount importance.

"The function of a lens in forming an image is to compensate by its variable thickness the differences in phase which would otherwise exist between secondary waves arriving at the focal point from various parts of the aperture. If we suppose the diameter of the lens ($2r$) to be given, and its focal length (f) gradually to increase, these differences of phase at the image of an infinitely distant luminous point diminish without limit. When f attains a certain value, say f_1, the extreme error of phase to be compensated falls to $\frac{1}{4}\lambda$. Now, as I have shown on a previous occasion†, an extreme error of phase amounting to $\frac{1}{4}\lambda$, or less, produces no appreciable deterioration in the definition; so that from this point onwards the lens is useless, as only improving an image already sensibly as perfect as the aperture admits of. Throughout the operation of increasing the focal length, the resolving power of the instrument, which depends only upon the aperture, remains unchanged; and we thus arrive at the rather startling conclusion that a telescope of any degree of resolving power might be constructed without an object-glass, if only there were no limit to the admissible focal length. This last proviso, however, as we shall see, takes away almost all practical importance from the proposition.

"To get an idea of the magnitudes of the quantities involved, let us take the case of an aperture of $\frac{1}{5}$ inch, about that of the pupil of the eye. The distance f_1, which the actual focal length must exceed, is given by

$$\sqrt{\{f_1^2 + r^2\}} - f_1 = \tfrac{1}{4}\lambda;$$

* "On the Resolving Power of Telescopes," *Phil. Mag.* August 1880. [Vol. I. p. 488.]

† *Phil. Mag.* November 1879. [Vol. I. p. 415.]

so that {approximately}

$$f_1 = 2r^2/\lambda.$$

Thus, if $\lambda = \frac{1}{40000}$, $r = \frac{1}{10}$, $f_1 = 800$.

"The image of the sun thrown on a screen at a distance exceeding 66 feet, through a hole $\frac{1}{5}$ inch in diameter, is therefore at least as well defined as that seen direct. In practice it would be better defined, as the direct image is far from perfect. If the image on the screen be regarded from a distance f_1, it will appear of its natural angular magnitude. Seen from a distance less than f_1, it will appear magnified. Inasmuch as the arrangement affords a view of the sun with full definition {corresponding to aperture} and with an increased apparent magnitude, the name of a telescope can hardly be denied to it.

"As the minimum focal length increases with the square of the aperture, a quite impracticable distance would be required to rival the resolving power of a modern telescope. Even for an aperture of four inches f_1 would be five miles*."

A more practical application of these principles is to be found in landscape photography, where a high degree of definition is often unnecessary, and where a feeble illumination can be compensated by length of exposure. In a recent communication to the British Association† it was pointed out that a suitable aperture is given by the relation

$$2r^2 = f\lambda; \quad \text{.................................}(1)$$

and a photograph was exhibited in illustration of the advantage to be derived from an increase of f. The subject was a weather-cock, seen against the sky, and it was taken with an aperture of $\frac{1}{16}$ inch [inch = 2·54 cm.] and at a distance of 9 feet. The amount of detail in the photograph is not markedly short of that observable by direct vision from the actual point of view. The question of brightness was also considered. As the focal length increases, the brightness (B) in the image of a properly proportioned pin-hole camera diminishes. For

$$B \propto r^2/f^2 \propto r^2\lambda^2/r^4 \propto \lambda^2/r^2 \propto \lambda/f. \quad \text{..................}(2)$$

There will now be no difficulty in understanding why a certain aperture is more favourable than either a larger or a smaller one, when f and λ are given. If the aperture be very small, the definition is poor even if the aid of a lens be invoked. If, on the other hand, the aperture be large, the lens becomes indispensable. The size of the aperture should accordingly be increased up to the point at which the lens is sensibly missed; and this, as we have seen, will occur in the neighbourhood of the value determined by (1).

* "On Images formed without Reflection or Refraction," *Phil. Mag.* March 1881. [Vol. I. p. 513.]

† *Brit. Assoc. Report*, 1889, p. 493.

A more precise calculation can be made only upon the basis of a detailed knowledge of the distribution of light in the image.

The question of the best size of aperture for a pin-hole camera was first considered by Petzval*. His theory, though it can hardly be regarded as sound, brings out the failure of definition when the aperture is either too large or too small, and, as is very remarkable, gives (1) as the best relation between r, f, and λ. The argument is as follows:—If the hole be very small, the diameter of the patch of light representative of a luminous point is given by

$$D = f\lambda/r,$$

the measurement being made up to the first blackness in the diffraction-pattern. "This formula is only an approximate one, applicable when r is very small; in the case of a larger aperture, its diameter must be added to the value above given, that is to say,

$$D = 2r + f\lambda/r.$$

From the last formula we can at once deduce the best value for r; in other words, the size of the aperture which corresponds to the least possible value of D, and therefore to the sharpest possible image. In fact, differentiating the last expression, and setting in the ordinary manner, $dD/dr = 0$, we find at once

$$r = \sqrt{(\tfrac{1}{2}f\lambda)},$$

which corresponds to

$$D = 2\sqrt{(2f\lambda)}."$$

The assumption that intermediate cases can be represented by mere addition of the terms appropriate in the extreme cases of very large and very small apertures appears to be inadmissible.

The complete determination of the image of a radiant point as given by a small aperture is a problem in diffraction, solved only within the last years by Lommel†. In view of the practical application to pin-hole photography, I have thought that it would be interesting to adapt Lommel's results to the problem in hand, and to exhibit upon the same diagram curves showing the distribution of illumination in various cases. For the details of the investigation reference must be made to Lommel's memoir, or to the account of it in the *Encyclopædia Britannica*, Art. "Wave Theory," p. 444. But it may be well to state the results somewhat fully. [These results, having been already given—equations (1) to (19), Vol. III. pp. 135, 136—are not now repeated.]

* *Wien. Sitz. Ber.* XXVI. p. 33 (1857); *Phil. Mag.* XVII. (1859), p. 1.

† "Die Beugungserscheinungen einer kreisrunden Oeffnung und eines kreisrunden Schirmchens," *Aus den Abhandlungen der k. bayer. Akademie der Wiss.* II. Cl. XV. Bd. II. Abth. (München, 1884.)

At the central point of the image where $z = 0$, $V_0 = 1$, $V_1 = 0$,

$$C = \pi r^2 \frac{\sin \frac{1}{2} y}{\frac{1}{2} y}, \qquad S = \pi r^2 \left\{ \frac{2}{y} - \frac{\cos \frac{1}{2} y}{\frac{1}{2} y} \right\},$$

and

$$I^2 = \frac{4}{(a+b)^2} \sin^2 \left(\frac{\pi r^2}{\lambda} \frac{a+b}{2ab} \right). \qquad \text{.....................(20)}$$

In general by (10), (11),

$$C^2 + S^2 = \frac{4\pi^2 r^4}{y^2} \{U_1^2 + U_2^2\} = \pi^2 r^4 . M^2, \qquad \text{..............(21)}$$

if with Lommel we set

$$M^2 = \left(\frac{2}{y} U_1 \right)^2 + \left(\frac{2}{y} U_2 \right)^2. \qquad \text{..........................(22)}$$

Also

$$I^2 = \frac{\pi^2 r^4}{a^2 b^2 \lambda^2} . M^2. \qquad \text{..............................(23)}$$

In these formulæ U_1^2, U_2^2, and therefore by (22), (23) M^2 and I^2 are known functions of y and z. The connexion with r and ζ is given by the relations

$$y = \frac{2\pi r^2}{\lambda} \frac{a+b}{ab}, \qquad z = \frac{2\pi r \zeta}{\lambda b}. \qquad \text{.....................(24)}$$

In Lommel's memoir are given the values of M^2 for integral values of z from 0 to 12 when y has the values π, 2π, 3π, &c. If we regard a, b, λ as given, each of these Tables affords a knowledge of the distribution of illumination as a function of ζ for a certain radius of aperture by means of the two equations (24). In each case ζ is proportional to z; but in comparing one case with another we have to bear in mind that the ratio of ζ to z varies. As our object is to compare the distributions of illumination when the aperture varies, we must treat ζ, and not z, as the abscissa in our diagrams. Another question arises as to how the scale of the ordinate I^2 should be dealt with in the various cases. If we take (23) as it stands, we shall have curves corresponding to the same actual intensity of the radiant point. For some purposes this might be desirable; but in the application to photography the deficiency of illumination when the aperture is much reduced would always be compensated by increased exposure. It will be more practical to vary the scale of ordinates from that prescribed in (23), so as to render the illumination corresponding to an extended source of light, such as the sky, the same in all cases. We shall effect this by removing from the right-hand member of (23) a factor proportional to the area of aperture, proportional that is to r^2, or y. Thus for any value of y equal to $s\pi$, we shall require to plot as ordinate, not M^2 simply, but sM^2, and as abscissa, not z simply, but $z/\sqrt{s}$. The following are at once deduced from Lommel's tables III.—VI.

$y = \pi.$

z	$z/\sqrt{1} = z$	M^2
0	0	·8106 Max.
1	1	·6286
2	2	·2772
3	3	·0623
4	4	·0269
5	5	·0306
6	6	·0121
7	7	·0018
8	8	·0051
9	9	·0037
10	10	·0004
11	11	·0013
12	12	·0016
3·8317	......	·0263 Min.
4·7153	......	·0320 Max.
7·0156	......	·0018 Min.
8·3060	......	·0055 Max.
10·1735	......	·0003 Min.
11·5785	......	·0019 Max.

$y = 2\pi.$

z	$z/\sqrt{2}$	$2M^2$
0	·000	·8106 Max.
1	·707	·6316
2	1·414	·3117
3	2·121	·1560
4	2·829	·1438
5	3·536	·1077
6	4·243	·0426
7	4·950	·0200
8	5·657	·0227
9	6·364	·0125
10	7·071	·0034
11	7·778	·0053
12	8·485	·0046
3·5977	2·544	·1440 Min.
3·8317	2·710	·1442 Max.
7·0156	4·961	·0198 Min.
7·8879	5·578	·0229 Max.
10·1735	7·193	·0032 Min.
11·4135	8·070	·0059 Max.

$y = 3\pi.$

z	$z/\sqrt{3}$	$3M^2$
0	·000	·2702 Max.
1	·577	·2159
2	1·154	·1631
3	1·732	·2110
4	2·310	·2449
5	2·887	·1734
6	3·464	·0916
7	4·041	·0739
8	4·619	·0651
9	5·195	·0335
10	5·773	·0156
11	6·350	·0178
12	6·927	·0122
1·9969	1·153	·1631 Min.
3·8317	2·212	·2467 Max.
7·0156	4·050	·0737 Min.
7·0878	4·092	·0739 Max.
10·1735	5·871	·0154 Min.
11·0361	6·374	·0178 Max.

$y = 4\pi.$

z	$z/2$	$4M^2$
0	·000	·0000 Min.
1	·500	·0056
2	1·000	·0609
3	1·500	·1594
4	2·000	·1947
5	2·500	·1515
6	3·000	·1293
7	3·500	·1399
8	4·000	·1148
9	4·500	·0658
10	5·000	·0484
11	5·500	·0458
12	6·000	·0280
3·8317	1·9158	·1961 Max.
5·8978	2·9489	·1291 Min.
7·0156	3·5078	·1399 Max.
10·1735	5·0867	·0483 Min.
10·3861	5·1930	·0483 Max.

As it appeared desirable to trace the curve corresponding to a smaller value of y than any given by Lommel, I have calculated by means of (12), (13) the value of $\frac{1}{2}M^2$, that is of

$$\frac{8}{\pi^2}(U_1^2 + U_2^2),$$

corresponding to $z = 0, 1, 2, 3, 4$.

The results are as follows:

$$y = \tfrac{1}{2}\pi.$$

z	$\sqrt{2}.z$	$\frac{1}{2}M^2$
0	·000	·4748
1	1·414	·3679
2	2·828	·1590
3	4·243	·0272
4	5·657	·0041

The various curves, or rather the halves of them, are plotted in the Figure, and exhibit to the eye the distribution of light in the images corresponding to the different apertures. It is at once evident that $y = \frac{1}{2}\pi$ is too small, and that $y = 3\pi$ is too great. The only question that can arise is between $y = \pi$ and $y = 2\pi$. The latter has decidedly the higher resolving power, but the advantage is to some extent paid for in the greater diffusion of light outside the image proper. In estimating this we must remember that the amount of light is represented, not by the *areas* of the various parts of the diagrams, but by the *volumes* of the solids formed by the revolution of the curves round the axis of I^2. In virtue of the method of construction the total volume is the same in all cases. The best aperture will thus depend in some degree upon the subject to be represented; but there is every reason to think that in general $y = 2\pi$ will prove more advantageous than $y = \pi$. It will be convenient to recall that

$$y/\pi = \frac{2r^2}{\lambda}\frac{a+b}{ab};$$

or, if we write $a = \infty$, $b = f$,

$$y/\pi = 2r^2/\lambda f. \qquad\qquad (25)$$

The curve $y = \pi$ thus corresponds to (1); and we conclude that the aperture may properly be somewhat enlarged so as to make

$$r^2 = \lambda f. \qquad\qquad (26)$$

In the general case when a is finite, y/π represents four times the number of wave-lengths by which the extreme ray is retarded relatively to the central ray; for

$$\frac{\sqrt{(a^2+r^2)} + \sqrt{(b^2+r^2)} - a - b}{\lambda} = \frac{r^2}{2\lambda}\frac{a+b}{ab}, \quad \text{approximately.}$$

According to (26) the aperture should be enlarged until the retardation amounts to $\frac{1}{2}\lambda$.

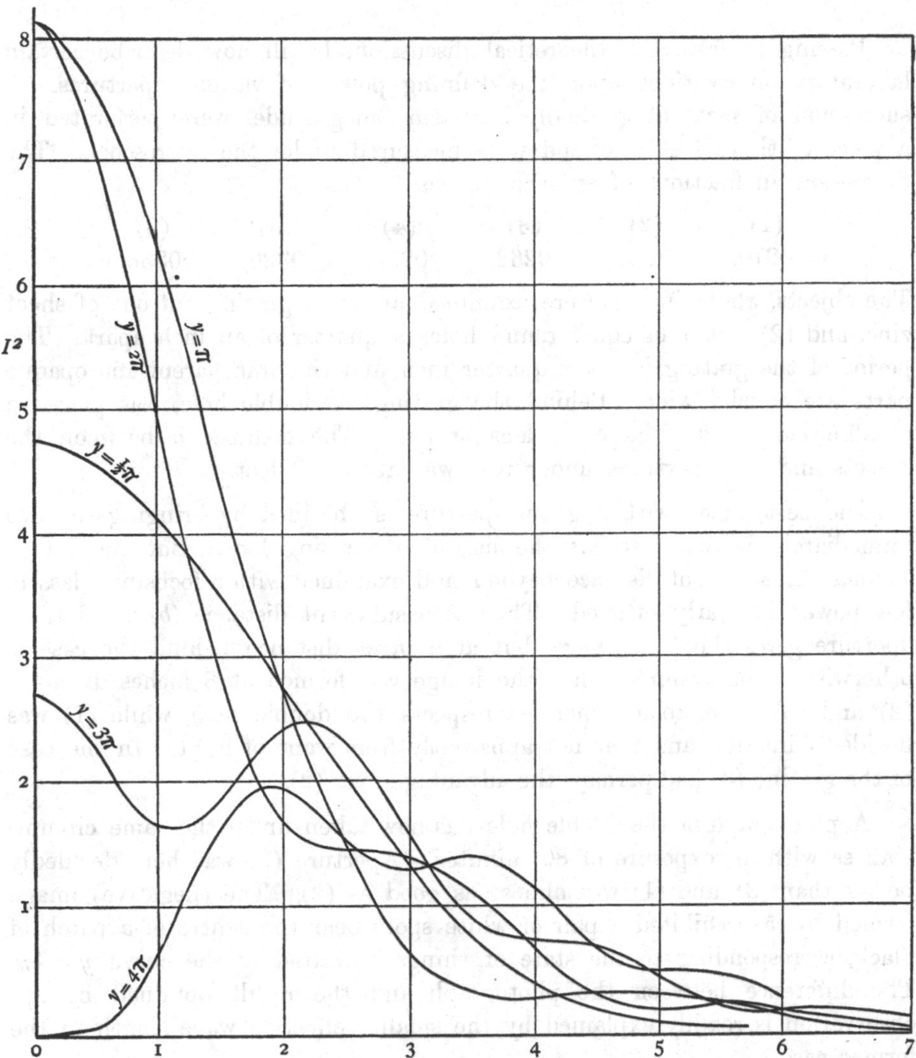

In the image of a double star the curves of brightness proper to the two components are superposed. If the components are equal, resolution will be just beginning when the distance of the geometrical images asunder is the double of the value of ζ for which I^2 has about one-half its maximum value. By inspection of the curve for $y = 2\pi$ we see that there will not be much appearance of resolution until $z/\sqrt{2} = 1{\cdot}5$. The corresponding angular interval between the two components is

$$\frac{2\zeta}{f} = \frac{1{\cdot}5 \times \sqrt{2}}{\pi} \sqrt{\left(\frac{\lambda}{f}\right)}. \qquad \text{.......................(27)}$$

This may be regarded as defining the maximum separating power as a function of λ and f.

Passing on from the theoretical discussion, I will now describe certain laboratory observations upon the defining power of various apertures. A succession of such, of gradually increasing magnitude, were perforated in a piece of thin sheet zinc, and were measured under the microscope. The diameters, in fractions of an inch, are as follows:—

(1)	(2)	(3)	(4)	(5)	(6)
·0210,	·0240,	·0262,	·0290,	·0326,	·0366.

The objects, whose images were examined, are (1) a grating cut out of sheet zinc, and (2) a pair of equal round holes a quarter of an inch apart. The period of the grating is also a quarter inch, and the transparent and opaque parts are equally wide. Behind the grating, or double hole, was placed a paraffin lamp and a large condensing lens. The distance a between the objects and the apertures under test was about 18 feet.

The best image with a given aperture is obtained by bringing the eye immediately behind, without the use of a focusing lens. But the image formed at a sufficient distance beyond, and examined with a focusing glass of low power, is nearly as good. Thus at a sufficient distance (b) the largest aperture gives the best image, but at a *given* distance behind the case is otherwise. For example, when the image was formed at 8 inches distance, (2) and (3) were about equal as respects the double hole, while (1) was decidedly inferior, and that not apparently from want of light. In the case of the grating (3) had perhaps the advantage over (2).

A photograph of the double hole was now taken under the same circumstances with an exposure of 80 minutes. Aperture (2) was here decidedly better than (3), and (1) was almost as good as (2). The (negative) image formed by (5) exhibited a pair of white spots near the centre of a patch of black, corresponding to the state of things indicated in the curve $y = 4\pi$. The difference between the photograph and the result obtained by eye observation is readily explained by the smaller effective wave-length in the former case.

The difference just spoken of is intensified when the light is white. In one experiment upon cloud-light $a = 21$ feet, $b = 10$ inches. In the resulting photograph, obtained upon an Ilford plate with an exposure of 30 minutes, the image from (2) was decidedly the best.

We may utilize the last result to calculate the relation between aperture and focus most suitable for out of door photography. We have

$$(2r)^2\left(\frac{1}{a}+\frac{1}{b}\right) = (\cdot 0240)^2\left(\frac{1}{252}+\frac{1}{10}\right) = 10^{-5} \times 5\cdot 99 \text{ inches} = 10^{-4} \times 1\cdot 52 \text{ cm}.$$

Thus, if $a = \infty$, as may usually be supposed in landscape-photography, the most favourable diameter of aperture is given by

$$(2r)^2/f = 10^{-5} \times 6{\cdot}0 \text{ inches} = 10^{-4} \times 1{\cdot}5 \text{ cm.},$$

the first number being employed if r and f are measured in inches, and the latter when the measures are in centimetres*. If $f = 12$ inches, $2r = {\cdot}027$ inch. If $f = 7 \times 12 = 84$ inches, $2r = {\cdot}071$ inch.

The experimental determination of the best value of y is more easily effected by eye observations. In order to render the wave-length more definite, an orange-red glass was employed. With $a = 18$ feet, $b = 8$ inches, the image formed by aperture (3) was judged to be decidedly the best, (2) was next, while (1) and (4) were decidedly behind. Thus we may take as the most favourable aperture $2r = {\cdot}026$ inch.

The mean wave-length of the light employed was found with the aid of a grating by a comparison with a soda flame:—

$$\text{Mean } \lambda : \lambda_D = 239 : 226;$$

so that

$$\lambda = \frac{239}{226} \times 5{\cdot}89 \times 10^{-5} = 6{\cdot}23 \times 10^{-5} \text{ centim.}$$

Hence

$$y/\pi = \frac{2r^2}{\lambda}\left(\frac{1}{a} + \frac{1}{b}\right) = \frac{({\cdot}026)^2}{2} \frac{7}{54} \frac{2{\cdot}54 \times 10^5}{6{\cdot}23} = 1{\cdot}79,$$

agreeing very well with what was expected from the curves.

If we now assume that the best value of y is 1·8, we can calculate backwards from the photographic results to find the mean λ there effective. We have

$$10^{-4} \times 1{\cdot}52 = (2r)^2 \left(\frac{1}{a} + \frac{1}{b}\right) = \frac{2\lambda y}{\pi} = 2\lambda \times 1{\cdot}8;$$

whence

$$\lambda = 4{\cdot}2 \times 10^{-5} \text{ centim.},$$

a little less than that belonging to Fraunhofer's line G.

To test the improvement of definition which according to (27) accompanies an increase of f, I have used an aperture of ·07 inch and a focal length of 7 feet. The aperture was perforated in sheet zinc, and was placed in the shutter of a room otherwise completely darkened. The subject was a group of cedars, and, being somewhat dark in the shadows, required an exposure of about an hour and a half, even in sunshine. The resulting 12 × 10-inch photographs fully bear out expectations. To appear in natural magnitude

* The effect of varying the diameter of aperture in photographic landscape work has been tested by Capt. Abney; but I am not in possession of the data as to focal length necessary for a comparison with the above.

the pictures would of course need to be held 7 feet distance from the eye; but even at 3 or 4 feet the apparent definition is sufficient. I have also taken panoramic pictures with an aperture of ·027 inch and a focal distance of 12 inches; but in this case there is nothing that could not equally well be done with an ordinary portable camera.

TERLING PLACE, WITHAM, ESSEX,
Dec. 2, 1890.

179.

SOME APPLICATIONS OF PHOTOGRAPHY.

[*Proc. Roy. Inst.* XIII. pp. 261—272, Feb. 1891; *Nature,* XLIV. pp. 249—254, 1891.]

ONE of the subjects to which I propose to invite your attention this evening is the application of instantaneous photography to the illustration of certain mechanical phenomena which pass so quickly as to elude ordinary means of observation. The expression "instantaneous photography" is not quite a defensible one, because no photography can be really instantaneous—some time must always be occupied. One of the simplest and most commonly used methods of obtaining very short exposures is by the use of movable shutters, for which purpose many ingenious mechanical devices have been invented. About two years ago we had a lecture from Prof. Muybridge, in which he showed us the application of this method—and a remarkably interesting application it was—to the examination of the various positions assumed by a horse in his several gaits. Other means, however, may be employed to the same end, and one of them depends upon the production of an instantaneous light. It will obviously come to the same thing whether the light to which we expose the plates be instantaneous, or whether by a mechanical device we allow the plate to be submitted to a continuous light for only a very short time. A good deal of use has been made in this way of what is known as the magnesium flash light. A cloud of magnesium powder is ignited, and blazes up quickly with a bright light of very short duration. Now I want to compare that mode of illumination with another, in order to be able to judge of the relative degree of instantaneity, if I may use such an expression. We will illumine for a short time a revolving disc, composed of black and white sectors; and the result will depend upon how quick the motion is as compared with the duration of the light. If the light could be truly instantaneous, it would of necessity show the disc apparently stationary.

I believe that the duration of this light is variously estimated at from one-tenth to one-fiftieth of a second; and as the arrangement that I have here is one of the slowest, we may assume that the time occupied will be about a tenth of a second. I will say the words one, two, three, and at the word three Mr Gordon will project the powder into the flame of a spirit lamp, and the flash will be produced. Please give your attention to the disc, for the question is whether the present uniform grey will be displaced by a perception of the individual black and white sectors. [Experiment.] You see the flash was *not* instantaneous enough to resolve the grey into its components.

I wish now to contrast with that mode of illumination one obtained by means of an electric spark. We have here an arrangement by which we can charge Leyden jars from a Wimshurst machine. When the charge is sufficient, a spark will pass inside a lantern, and the light proceeding from it will be condensed and thrown upon the same revolving disc as before. The test will be very much more severe; but severe as it is, I think we shall find that the electric flash will bear it. The teeth on the outside of the disc are very numerous, and we will make them revolve as fast as we can, but we shall find that under the electric light they will appear to be absolutely stationary. [Experiment.] You will agree that the outlines of the black and white sectors are seen perfectly sharp.

Now, by means of this arrangement we might investigate a limit to the duration of the spark, because with a little care we could determine how fast the teeth are travelling—what space they pass through in a second of time. For this purpose it would not be safe to calculate from the multiplying gear on the assumption of no slip. A better way would be to direct a current of air upon the teeth themselves, and make them give rise to a musical note, as in the so-called siren. From the appearance of the disc under the spark we might safely say, I think, that the duration of the light is less than a tenth of the time occupied by a single tooth in passing. But the spark is in reality much more instantaneous than can be proved by the means at present at our command. In order to determine its duration, it would be necessary to have recourse to that powerful weapon—the revolving mirror; and I do not, therefore, propose to go further into the matter to-night.

Experiments of this kind were made some twenty years ago by Prof. Rood, of New York, both on the duration of the discharge of a Leyden jar, and also on that of lightning. Prof. Rood found that the result depended somewhat upon the circumstances of the case; the discharge of a small jar being generally more instantaneous than that of a larger one. He proved that in certain cases the duration of the principal part of the light was as low as one twenty-five-millionth part of a second of time. That is a statement which probably conveys very little of its real meaning. A million seconds is about twelve days and nights. Twenty-five million seconds is nearly a year. So that the time occupied by the spark in Prof. Rood's experiment is about the same fraction of one second that one second is of a year. In many other cases the duration was somewhat greater; but in all his experiments it was well under the one-millionth part of a second. In certain cases you may have multiple sparks. I do not refer to the oscillating discharges of which Prof. Lodge gave us so interesting an account last year; Prof. Rood's multiple discharge was not of that character. It consisted of several detached overflows of his Leyden jar when charged by the Rhumkorff coil. One number mentioned for the total duration was one six-thousandth part of a second; but the individual discharges had the degree of instantaneity of which I have spoken.

It is not a difficult matter to adapt the electrical spark to instantaneous photography. We will put the lantern into its proper position, excite the electric sparks within it, causing them to be condensed by the condenser of the lantern on to the photographic lens. We will then put the object in front of the lantern-condenser, remove the cap from the lens, expose the plate to the spark when it comes, and thus obtain an instantaneous view of whatever may be going on. I propose to go through the operation of taking such a photograph presently. I will not attempt any of the more difficult things of which I shall speak, but will take a comparatively easy subject,—a stream of bubbles of gas passing up through a liquid. In order that you may see what this looks like when observed in the ordinary way, we have arranged it here for projection upon the screen. [Experiment.] The gas issues from the nozzle, and comes up in a stream, but so fast that you cannot fairly see the bubbles. If, however, we take an instantaneous picture, we shall find that the stream is decomposed into its constituent parts. We arrange the trough of liquid in front of the lantern which contains the spark-making apparatus—[Experiment]—and we will expose a plate, though I hardly expect a good result in a lecture. A photographer's lamp provides some yellow light to enable us to see when other light is excluded. There goes the spark; the plate is exposed, and the thing is done. We will develop the plate, and see what it is good for; and if it turns out fit to show, we will have it on the screen within the hour.

In the meantime, we will project on the screen some slides taken in the same way and with the same subject. [Photograph shown.] That is an instantaneous photograph of a stream of bubbles. You see that the bubbles form at the nozzle from the very first moment, contrasting in that respect with the behaviour of jets of water, projected into air. [Fig. 1, Plate I.]

The latter is our next subject. This is the reservoir from which the water is supplied. It issues from a nozzle of drawn-out glass, and at the moment of issue it consists of a cylindrical body of water. The cylindrical form is unstable, however, and the water rapidly breaks up into drops which succeed one another so rapidly that they can hardly be detected by ordinary vision. But by means of instantaneous photography the individual drops can be made evident. I will first project the jet itself on the screen, in order that you may appreciate the subject which we shall see presently represented by photography. [Experiment.] Along the first part of its length the jet of water is continuous. At a certain point it breaks into drops, but you cannot see them because of their rapidity. If we act on the jet with a vibrating body, such as a tuning-fork, the breaking into drops occurs still earlier, the drops are more regular, and assume a curious periodic appearance, investigated by Savart. I have some photographs of jets of that nature. Taken as described, they do not differ much in appearance from those obtained by Chichester Bell, and by Mr Boys. We get what we may regard as simply shadows of the jet obtained by instantaneous illumination; so that these photographs show little more than the outlines of the subject. They show a little more, on account of the lens-like action of the cylinder and of the drops. Here we have an instantaneous view of a jet similar to the one we were looking at just now. [Fig. 2, Plate I.] This is the continuous part; it gradually ripples itself as it comes along; the ripples increase; then the contraction becomes a kind of ligament connecting consecutive drops; the ligament next gives way, and we have the individual drops completely formed. The small points of light are the result of the lens-like action of the drops. [Other instantaneous views also shown.]

The pictures can usually be improved by diffusing somewhat the light of the spark with which they are taken. In front of the ordinary condensing lens of the magic lantern we slide in a piece of ground glass, slightly oiled, and we then get better pictures showing more shading. [Photograph shown.] Here is one done in that way; you would hardly believe it to be water resolved into drops under the action of a tremor. It looks more like mercury. You will notice the long ligament trying to break up into drops on its own account, but not succeeding. [Fig. 3, Plate I.]

There is another, with the ligament extremely prolonged. In this case it sometimes gathers itself into two drops. [Fig. 4, Plate I.]

[A number of photographs showing slight variations were exhibited.]

The mechanical cause of this breaking into drops is, I need hardly remind you, the surface tension or capillary force of the liquid surface. The elongated cylinder is an unstable form, and tends to become alternately swollen and contracted. In speaking on this subject I have often been embarrassed for want of an appropriate word to describe the condition in question. But a few days ago, during a biological discussion, I found that there is a recognised, if not a very pleasant, word. The cylindrical jet may

PLATE I

Fig. 1.

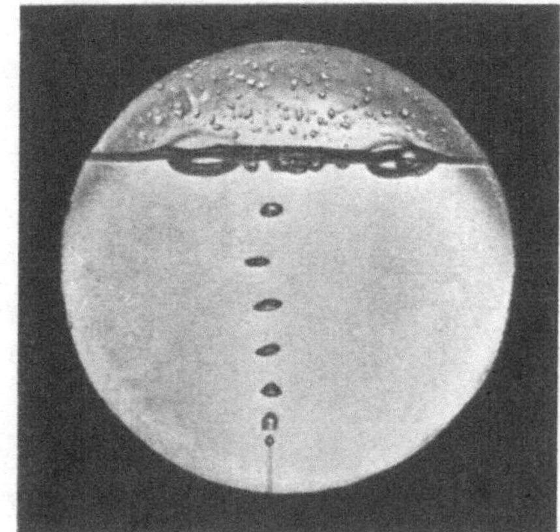

Fig. 2.

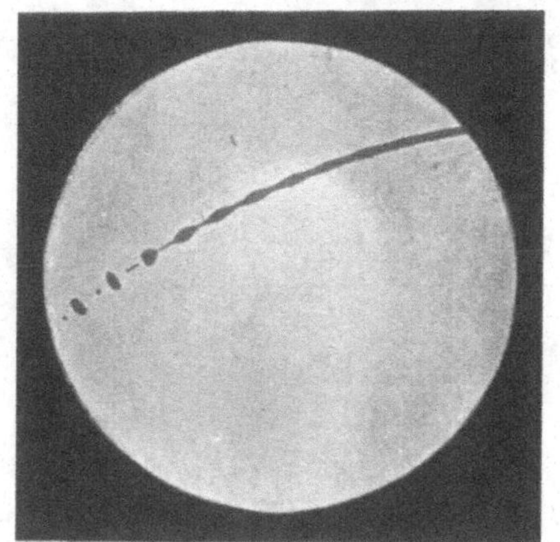

Fig. 3.

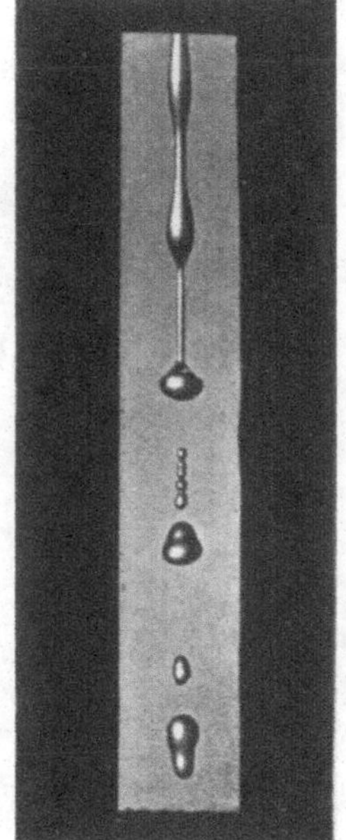

Fig. 4.

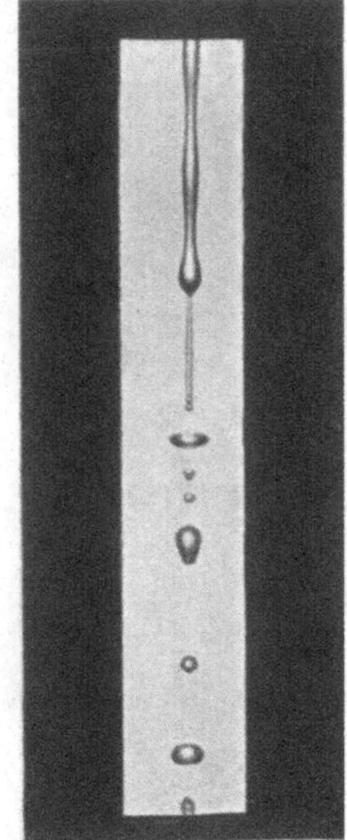

To face p. 444.

be said to become *varicose*, and the varicosity goes on increasing with time, until eventually it leads to absolute disruption.

There is another class of unstable jets presenting many points of analogy with the capillary ones, and yet in many respects quite distinct from them. I refer to the phenomena of sensitive flames. The flame, however, is not the essential part of the matter, but rather an indicator of what has happened. Any jet of fluid playing into a stationary environment is sensitive, and the most convenient form for our present purpose is a jet of coloured in uncoloured water. In this case we shall use a solution of permanganate of potash playing into an atmosphere of other water containing acid and sulphate of iron, which exercises a decolourising effect on the permanganate, and so retards the general clouding up of the whole mass by accumulation of colour. [Experiment.] Mr Gordon will release the clip, and we shall get a jet of permanganate playing into the liquid. If everything were perfectly steady, we might see a line of purple liquid extending to the bottom of the trough; but in this theatre it is almost impossible to keep anything steady. The instability to which the jet is subject now manifests itself, and we get a breaking away into clouds something like smoke from chimneys. A heavy tuning-fork, vibrating at ten to the second, acts upon it with great advantage, and regularizes the disruption. A little more pressure will increase the instability, and the jet goes suddenly into confusion, although at first, near the nozzle, it is pretty regular.

It may now be asked "What is the jet doing?" That is just the question which the instantaneous method enables us to answer. For this purpose the permanganate which we have used to make the jet visible is not of much service. It is too transparent to the photographic rays, and so it was replaced by bichromate of potash. Here the opposite difficulty arises; for the bichromate is invisible by the yellow light in which the adjustments have to be made. I was eventually reduced to mixing the two materials together, the one serving to render the jet visible to the eye and the other to the photographic plate. Here is an instantaneous picture of such a jet as was before you a moment ago, only [now] under the action of a regular vibrator. It is *sinuous*, turning first in one direction and then in the other. The original cylinder, which is the natural form of the jet as it issues from the nozzle, curves itself gently as it passes along through the water. It thus becomes sinuous, and the amount of the sinuosity increases, until in some cases the consecutive folds come into collision with one another. [Several photographs of sinuous jets were shown, two of which are reproduced in Figs. 5, 6, Plate II.]

The comparison of the two classes of jets is of great interest. There is an analogy as regards the instability, the vibrations caused by disturbance gradually increasing as the distance from the nozzle increases; but there is a great difference as to the nature of the deviation from the equilibrium condition, and as to the kind of force best adapted to bring it about. The one gives way by becoming *varicose*; the other by becoming *sinuous*. The only forces capable of producing varicosity are symmetrical forces, which act alike all round. To produce sinuosity, we want exactly the reverse—a force which acts upon the jet transversely and unsymmetrically.

I will now pass on to another subject for instantaneous photography, namely, the soap film. Everybody knows that if you blow a soap bubble it breaks—generally before you wish. The process of breaking is exceedingly rapid, and difficult to trace by the unaided eye. If we can get a soap film on this ring, we will project it upon the screen and then break it before our eyes, so as to enable you to form your own impressions as to the rapidity of the operation. For some time it has been my ambition to photograph a soap bubble in the act of breaking. I was prepared for difficulty, believing that the time occupied was less than the twentieth of a second. But it turns out to be a great deal less even than that. Accordingly this subject is far more difficult to deal with than are the jets of water or coloured liquids, which one can photograph at any moment that the spark happens to come.

There is the film, seen by reflected light. One of the first difficulties we have to contend with is that it is not easy to break the film exactly when we wish. We will drop a shot through it. The shot has gone through, as you see, but it has not broken the film! and when the film is a thick one, you may drop a shot through almost any number of times from a moderate height without producing any effect. You would suppose that the shot in going through would necessarily make a hole, and end the life of the film. The shot goes through, however, without making a hole. The operation can be traced, not very well with a shot, but with a ball of cork stuck on the end of a pin, and pushed through. A dry shot does not readily break the film; and as it was necessary for our purpose to effect the rupture in a well defined manner, here was a difficulty which we had to overcome. We found, after a few trials, that we could get over it by wetting the shot with alcohol.

We will try again with dry shot. Three shots have gone through and nothing has happened. Now we will try one wetted with alcohol, and I expect it will break the film at once. There! It has gone!

The apparatus for executing the photography of a breaking soap film will of necessity be more complicated than before, because we have to time the spark exactly with the breaking of the film. The device I have used is to drop two balls simultaneously, so that one should determine the spark and the other rupture the film. The most obvious plan was to hang iron balls to two electro-magnets [connected in series], and cause them to drop by breaking the circuit, so that both were let go at the same moment. The

PLATE II

Fig. 5.

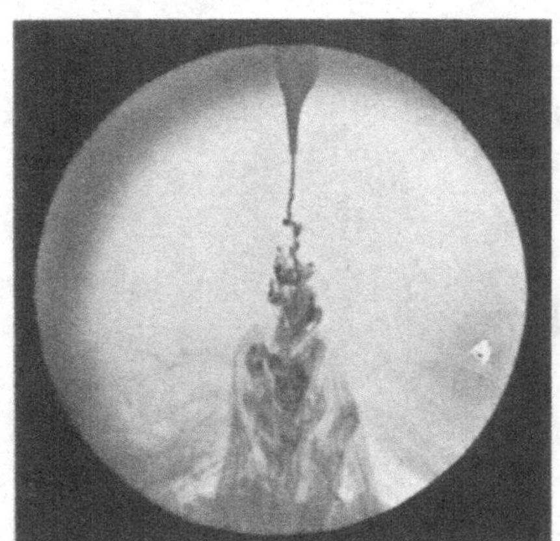

Fig. 6.

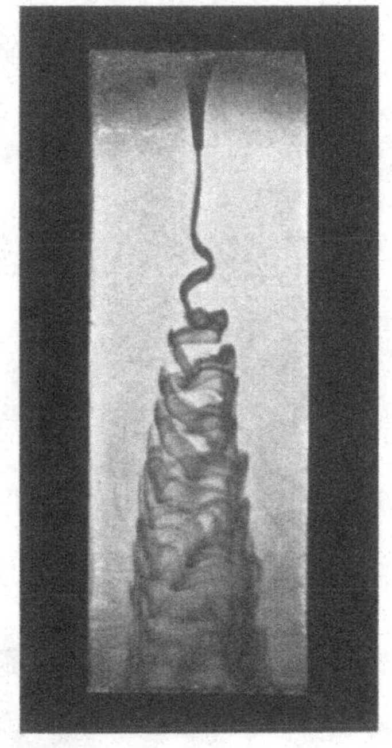

Fig. 7.

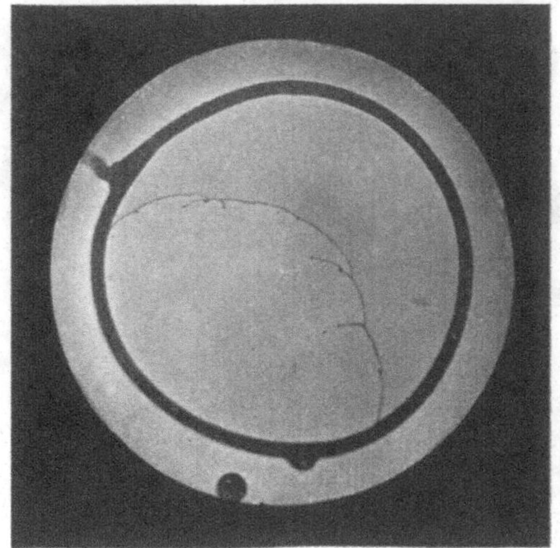

To face p. 446.

method was not quite a success, however, because there was apt to be a little hesitation in letting go the balls. So we adopted another plan. The balls were not held by electro-magnetism but by springs (Fig. 8) pressing laterally, and these were pulled off by electro-magnets. The proper moment for putting down the key and so liberating the balls, is indicated by the tap of the beam of an attracted disc electrometer as it strikes against the upper stop. One falling ball determines the spark, by filling up most of the interval between two fixed ones submitted to the necessary electric pressure. Another ball, or rather shot, wetted with alcohol, is let go at the same moment, and breaks the film on its passage through. By varying the distances dropped through, the occurrence of one event may be adjusted relatively to the other. The spark which passes to the falling ball is, however, not the one which illuminates the photographic plate. The latter occurs within the lantern, and forms part of a circuit in connexion with the *outer* coatings of the Leyden jars*, the whole arrangement being similar to

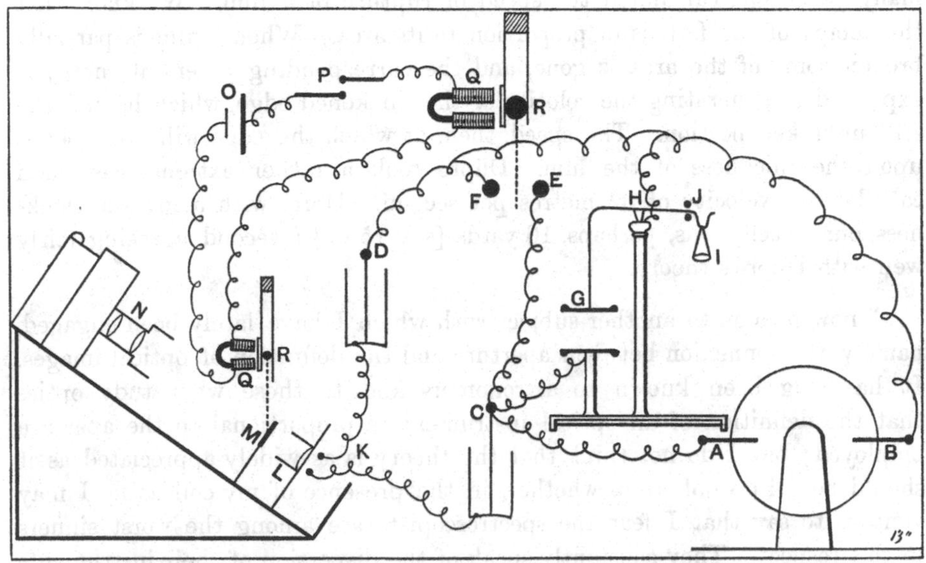

DESCRIPTION OF FIG. 8.

A, B, Electrodes of Wimshurst machine.
C, D, Terminals of interior coatings of Leyden jars.
E, F, Balls on insulating supports between which the discharge is taken.
G, Attracted disc of electrometer.
H, Knife edge.
I, Scale pan.
J, Stops limiting movement of beam.
K, Sparking balls in connexion with exterior coatings of jars. [These exterior coatings are to be joined by an imperfect conductor, such as a table.]
L, Lantern condenser.
M, Soap film.
N, Photographic camera.
O, Daniell cell.
P, Key.
Q, Electromagnets.
R, Balls.

* In practice there were two sets of three jars each.

that adopted by Prof. Lodge in his experiments upon alternative paths of discharge. Fig. 8 will give a general idea of the disposition of the apparatus. [Several photographs of breaking films were shown upon the screen; one of these is reproduced in Fig. 7, Plate II.*]

This work proved more difficult than I had expected; and the evidence of our photographs supplies the explanation, namely, that the rupture of the film is an extraordinarily rapid operation. It was found that the whole difference between being too early and too late was represented by a displacement of the falling ball through less than a diameter, viz. $\frac{1}{4}$ inch nearly. The drop was about a foot. The speed of the ball would thus be about 100 inches per second; and therefore the whole difference between being too soon and too late is represented by $\frac{1}{300}$ second. Success is impossible, unless the spark can be got to occur within the limits of this short interval.

Prof. Dewar has directed my attention to the fact that Dupré, a good many years ago, calculated the speed of rupture of a film. We know that the energy of the film is in proportion to its area. When a film is partially broken, some of the area is gone, and the corresponding potential energy is expended in generating the velocity of the thickened edge, which bounds the still unbroken portion. The speed, then, at which the edge will go depends upon the thickness of the film. Dupré took a rather extreme case, and calculated a velocity of 32 metres per second. Here, with a greater thickness, our velocity was, perhaps, 16 yards [say 15 m.] a second, agreeing fairly well with Dupré's theory.

I now pass on to another subject with which I have lately been engaged, namely, the connexion between aperture and the definition of optical images. It has long been known to astronomers and to those who study optics that the definition of an optical instrument is proportional to the aperture employed; but I do not think that the theory is as widely appreciated as it should be. I do not know whether, in the presence of my colleague, I may venture to say that I fear the spectroscopists are among the worst sinners in this respect. They constantly speak of the dispersion of their instruments as if that by itself could give any idea of the power employed. You may have a spectroscope of any degree of dispersion, and yet of resolving power insufficient to separate even the D lines. What is the reason of this? Why is it that we cannot get as high a definition as we please with a limited aperture? Some people say that the reason why large telescopes are necessary, is, because it is only by their means that we can get enough light. That may be in some cases a sufficient reason, but that it is inadequate in others will be apparent, if we consider the case of the sun. Here we do

* The appearance of the breaking bubble, as *seen* under instantaneous illumination, was first described by Marangoni and Stephanelli, *Nuovo Cimento*, 1873.

not want more light, but rather are anxious to get rid of a light already excessive. The principal *raison d'être* of large telescopes, is, that without a large aperture definition is bad, however perfect the lenses may be. In accordance with the historical development of the science of optics, the student is told that the lens collects the rays from one point to a focus at another; but when he has made further advance in the science he finds that this is not so. The truth is that we are in the habit of regarding this subject in a distorted manner. The difficulty is not to explain why optical images are imperfect, no matter how good the lens employed, but rather how it is that they manage to be as good as they are. In reality the optical image of even a mathematical point has a considerable extension; light coming from one point cannot be concentrated into another point by any arrangement. There must be diffusion, and the reason is not hard to see in a general way. Consider what happens at the mathematical focus, where, if anywhere, the light should all be concentrated. At that point all the rays coming from the original radiant point arrive in the same phase. The different paths of the rays are all rendered optically equal, the greater actual distance that some of them have to travel being compensated for, in the case of those which come through the centre, by an optical retardation due to the substitution of glass for air; so that all the rays arrive at the same time*. If we take a point not quite at the mathematical focus but near it, it is obvious that there must be a good deal of light there also. The only reason for any diminution at the second point lies in the discrepancies of phase which now occur; and these can only enter by degrees. Once grant that the image of a mathematical point is a diffused patch of light, and it follows that there must be a limit to definition. The images of the components of a close double point will overlap; and if the distance between the centres do not exceed the diameter of the representative patches of light, there can be no distinct resolution. Now their diameter varies inversely as the aperture; and thus the resolving power is directly as the aperture.

My object to-night is to show you by actual examples that this is so. I have prepared a series of photographs of a grating consisting of parallel copper wires, separated by intervals equal to their own diameter, and such that the distance from centre to centre is $\frac{1}{10}$ inch [inch = 2·54 cm.]. The grating was backed by a paraffin lamp and a large condensing lens; and the photographs were taken in the usual way, except that the lens employed was a telescopic object glass, and was stopped by a screen perforated with a narrow adjustable slit, parallel to the wires†. In each case the exposure was inversely as the aperture employed. The first [thrown upon the screen], is a picture done by an aperture of eight hundredths of an inch, and the

* On this principle we may readily calculate the focal lengths of lenses without use of the law of sines. See *Phil. Mag.* Dec. 1879. [Vol. I. p. 439.]

† The distance between the grating and the telescope lens was 12 ft. 3 in.

definition is tolerably good (Fig. 8). The next (Fig. 9), with six hundredths, is rather worse. In the third case (Fig. 10), I think that everyone can see that the definition is deteriorating; that was done by an aperture of four hundredths of an inch. The next (Fig. 11) is one done by an aperture of three hundredths of an inch, and you-can see that the lines are getting washed out. In focusing the plate for this photograph, I saw that the lines had entirely disappeared, and I was surprised, on developing the plate, to find them still visible. That was in virtue of the shorter wave-length of the light operative in photography as compared with vision. In the last example (Fig. 12), the aperture was only two-and-a-half hundredths of an inch, and the effect of the contraction has been to wash away the image altogether, although, so far as ordinary optical imperfections are concerned, the lens was acting more favourably with the smaller aperture than with the larger ones*.

This experiment may be easily made with very simple apparatus; and I have arranged that each one of my audience may be able to repeat it by means of the piece of gauze and perforated card which have been distributed. The piece of gauze should be placed against the window so as to be backed by the sky, or in front of a lamp provided with a ground-glass or opal globe. You then look at the gauze through the pin-holes. Using the smaller hole, and gradually drawing back from the gauze, you will find that you lose definition and ultimately all sight of the wires. That will happen at a distance of about 4½ feet from the gauze. If, when looking through the smaller hole, you have just lost the wires, you shift the card so as to bring the larger hole into operation, you will see the wires again perfectly.

That is one side of the question. However perfect your lens may be, you cannot get good definition if the aperture is too much restricted. On the other hand if the aperture is much restricted, then the lens is of no use, and you will get as good an image without it as with it.

I have not time to deal with this matter as I could wish, but I will illustrate it by projecting on the screen the image of a piece of gauze as formed by a narrow aperture parallel to one set of wires. There is no lens whatever between the gauze and the screen. [Experiment.] There is the image—if we can dignify it by such a name—of the gauze as formed by an aperture which is somewhat large. Now, as the aperture is gradually narrowed, we will trace the effect upon the definition of the wires parallel to it. The definition is improving; and now it looks tolerably good. But I will go on, and you will see that the definition will become bad again. Now, the aperture has been further narrowed, and the lines are getting

* [1901. The original photographs were exhibited by projection, and are now reproduced for the first time. In these reproductions the distinction between Figs. (8), (9), (10) is barely visible.]

PLATE III

Fig. 8.

Fig. 9.

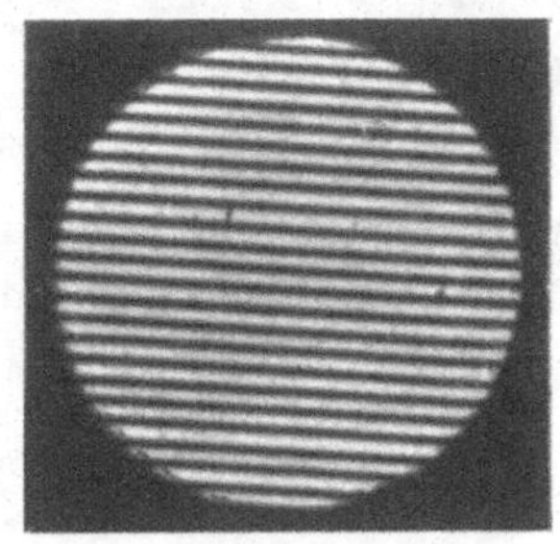

Fig. 10.

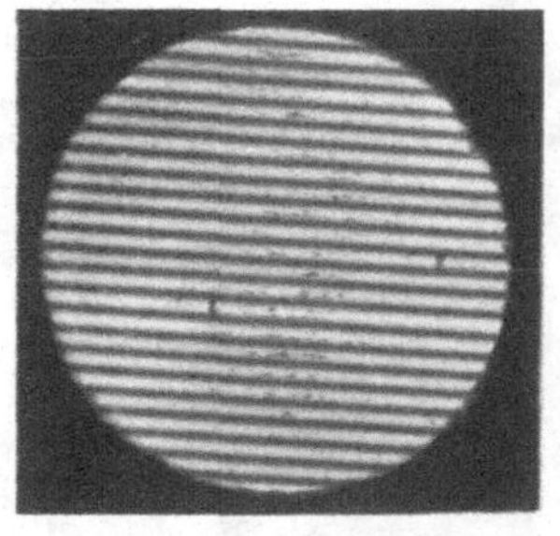

Fig. 11.

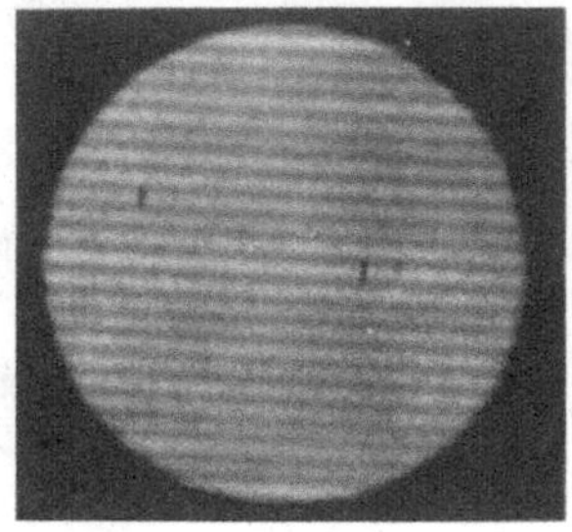

Fig. 12.

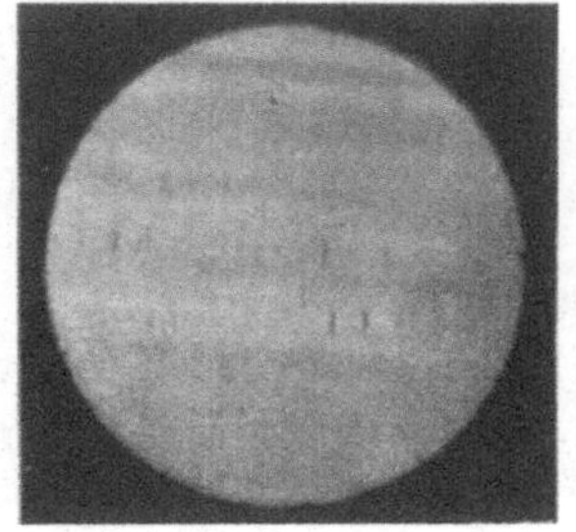

To face p. 450.

washed out. Again, a little more, and they are gone. Perhaps you may think that the explanation lies in the faintness of the light. We cannot avoid the loss of light which accompanies the contraction of aperture, but to prove that the result is not so to be explained, I will now put in a lens. This will bring the other set of wires into view, and prove that there was plenty of light to enable us to see the first set if the definition had been good enough. Too small an aperture, then, is as bad as one which is too large; and if the aperture is sufficiently small, the image is no worse without a lens than with one.

What, then, is the best size of the aperture? That is the important question in dealing with pin-hole photography. It was first considered by Petzval, of Vienna, and he arrived at the result indicated by the formula, $2r^2 = f\lambda$, where $2r$ is the diameter of the aperture, λ the wave-length of light, and f the focal length, or rather simply the distance between the aperture and the screen upon which the image is formed.

His reasoning, however, though ingenious, is not sound, regarded as an attempt at an accurate solution of the question. In fact it is only lately that the mathematical problem of the diffraction of light by a circular hole has been sufficiently worked out to enable the question to be solved. The mathematician to whom we owe this achievement is Prof. Lommel. I have adapted his results to the problem of pin-hole photography. [A series of curves* were shown, exhibiting to the eye the distribution of illumination in the images obtainable with various apertures.] The general conclusion is that the hole may advantageously be enlarged beyond that given by Petzval's rule. A suitable radius is $r = \sqrt{(f\lambda)}$.

I will not detain you further than just to show one application of pin-hole photography on a different scale from usual. The definition improves as the aperture increases; but in the absence of a lens the augmented aperture entails a greatly extended focal length. The limits of an ordinary portable camera are thus soon passed. The original of the transparency now to be thrown upon the screen was taken in an ordinary room, carefully darkened. The aperture (in the shutter) was ·07 inch, and the distance of the 12×10 plate from the aperture was 7 feet. The resulting picture of a group of cedars shows nearly as much detail as could be seen direct from the place in question.

* *Phil. Mag.* Feb. 1891. [Vol. III. p. 437.]

180.

ON THE SENSITIVENESS OF THE BRIDGE METHOD IN ITS APPLICATION TO PERIODIC ELECTRIC CURRENTS.

[*Proceedings of the Royal Society*, XLIX. pp. 203—217, 1891.]

THE most favourable conditions in the ordinary measurement of resistance have been investigated by Schwendler* and by O. Heaviside†. It is here proposed to treat the problem more generally, so as to cover the application to conductors endowed with self-induction, or combined with condensers. The receiving instrument may be supposed to be a telephone, which takes the place of the galvanometer employed in ordinary testing. In the conjugate "battery" branch a periodic electromotive force of given frequency is the origin of the currents.

Special attention will be given to the case where the branches are equal in pairs, *e.g.*, $a=c$, $b=d$ (Fig. 1). The advantages of this arrangement are important even in ordinary resistance testing, and in the generalised application are still more to be insisted upon. By mere interchange of a and c and combination of results, the equality of b and d can be verified independently of the exactitude of the ratio $a : c$.

Fig. 1.

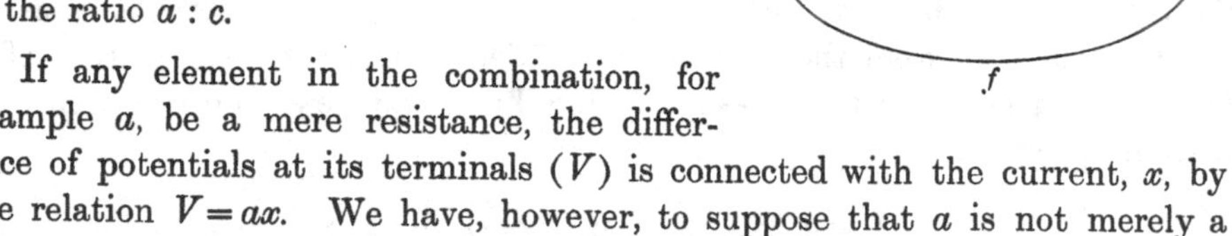

If any element in the combination, for example a, be a mere resistance, the difference of potentials at its terminals (V) is connected with the current, x, by the relation $V=ax$. We have, however, to suppose that a is not merely a

* "On the Galvanometric Resistance to be employed in Testing with Wheatstone's Diagram," *Phil. Mag.* Vol. XXXI. p. 364, 1866.

† "On the Best Arrangement of Wheatstone's Bridge for measuring a given Resistance with a given Galvanometer and Battery," *Phil. Mag.* Vol. XLV. p. 114, 1873.

resistance or even combination of such. It may include an electromagnet*, and it may be interrupted by a condenser. So long as the current is strictly harmonic, proportional to e^{ipt}, the most general possible relation between V and x is expressed by $V=(a_1+ia_2)x$, where a_1 and ia_2 are the real and imaginary parts of a complex coefficient a, and are functions of the frequency $p/2\pi$. In the particular case of a simple conductor, endowed with inductance L, a_1 represents the resistance, and a_2 is equal to pL. In general, a_1 is positive; but a_2 may be either positive, as in the above example, or negative. The latter case arises when a resistance, R, is interrupted by a condenser of capacity C. Here $a_1=R$, $a_2=-1/pC$. If there be also inductance L,

$$a_1=R, \qquad a_2=pL-1/pC.$$

Since the parts of a_2 may be either positive or negative, there is nothing to hinder its evanescence by compensation. In the above combination of an electromagnet and condenser, compensation occurs when $p^2LC=1$, that is, when the natural period with terminals connected coincides with the forced period. The combination is then equivalent to a simple resistance†; but a variation of frequency will give rise to a positive or negative a_2.

The case of two electromagnets in parallel is treated in my paper on "Forced Harmonic Oscillations‡"; and other combinations have been discussed by Mr Heaviside and myself. But the above examples will suffice to illustrate the principle that the relation of V to x is one of proportionality, and may be expressed by the single complex symbol a. We fall back at any time upon the case of mere resistance by supposing a to be real. In like manner b, c, d, e, and f are symbols expressing the electrical properties of the remaining branches.

In all electrical problems the generalised quantities a, b, &c., combine, just as they do when they represent simple resistances. Thus, if a, a' be two complex quantities representing two conductors in series, the corresponding quantity for the combination is $a+a'$. Again, if a, a' represent two conductors in parallel, the reciprocal of the resultant is given by addition of the reciprocals of a, a'. For, if the currents be x and x', corresponding to a difference of potentials V at the common terminals,

$$V=ax=a'x',$$

so that

$$x+x'=V(1/a+1/a').$$

The investigation of the currents in networks of conductors is usually treated by "Kirchhoff's rules," and this procedure may of course be adopted in the present case to determine the current through the bridge of a Wheatstone combination. But it will be more instructive to put the argument in

* An electromagnet here denotes a conductor with sensible inductance. Iron may be present if the range of magnetisation be small. *Phil. Mag.* March 1887. [Vol. II. p. 579.]

† *Theory of Sound*, § 46, Macmillan, 1877.

‡ *Phil. Mag.* May 1886. [Vol. II. p. 475.]

the form applicable to the forced vibrations of all mechanical systems which oscillate about a configuration of equilibrium.

If $p/2\pi$ represent the frequency of the vibration, the coordinates ψ_1, ψ_2, ψ_3... determining the condition of the system, and the corresponding forces Ψ_1, Ψ_2, Ψ_3... are all proportional to e^{ipt}, and the coordinates are linear functions of the forces*. For the present purpose we suppose that all the forces vanish, except the first and second. Thus ψ_1, ψ_2 are linear functions of Ψ_1 and Ψ_2, and, conversely, Ψ_1, Ψ_2 may be regarded as linear functions of ψ_1 and ψ_2. We may therefore set

$$\Psi_1 = A\psi_1 + B\psi_2, \qquad \Psi_2 = B\psi_1 + C\psi_2, \quad \text{..............(1)}$$

the coefficient of ψ_2 in the first equation being identical with that of ψ_1 in the second by the reciprocal property. The three constants A, B, C are in general complex quantities, functions of p.

In the application that we have to make of these equations, ψ_1, ψ_2, Ψ_1, Ψ_2 will represent respectively currents and electromotive forces in the battery and telephone branches of the combination. The reciprocal property may then be interpreted as follows:—If $\Psi_2 = 0$,

$$B\psi_1 + C\psi_2 = 0,$$

and

$$\psi_2 = \frac{B}{B^2 - AC}\Psi_1. \quad \text{......................(2)}$$

In like manner, if we had supposed $\Psi_1 = 0$, we should have found

$$\psi_1 = \frac{B}{B^2 - AC}\Psi_2, \quad \text{......................(3)}$$

showing that the ratio of the current in one branch to an electromotive force operative in the other is independent of the way in which the parts are assigned to the two branches.

We have now to determine the constants A, B, C in terms of the electrical properties of the system. If ψ_2 be maintained zero by a suitable force Ψ_2, the relation between ψ_1 and Ψ_1 is $\Psi_1 = A\psi_1$. In our application, A therefore denotes the (generalised) resistance to an electromotive force in the battery branch, *when the telephone branch is open*. This resistance is made up of f, the resistance in the battery branch, and of that of the conductors $a + c$, $b + d$ combined in parallel. Thus,

$$A = f + \frac{(a+c)(b+d)}{a+b+c+d}. \quad \text{......................(4)}$$

In like manner,

$$C = e + \frac{(a+b)(c+d)}{a+b+c+d}. \quad \text{......................(4')}$$

* *Theory of Sound*, Vol. I. § 107.

To determine B let us consider the force Ψ_2 which must act in e in order that the current through it (ψ_2) may be zero, in spite of the operation of Ψ_1. We have $\Psi_2 = B\psi_1$. The total current ψ_1 flows partly along the branch $a+c$ and partly along $b+d$. The current through $a+c$ is

$$\frac{\dfrac{1}{a+c}}{\dfrac{1}{a+c}+\dfrac{1}{b+d}}\psi_1 = \frac{(b+d)\psi_1}{a+b+c+d}, \qquad\text{..................(5)}$$

and that through $b+d$ is

$$\frac{(a+c)\psi_1}{a+b+c+d}. \qquad\text{..........................(6)}$$

The difference of potentials at the terminals of e, supposed to be interrupted, is thus

$$\frac{c(b+d)\psi_1 - d(a+c)\psi_1}{a+b+c+d};$$

or

$$B = \frac{bc-ad}{a+b+c+d}. \qquad\text{..........................(7)}$$

By (4), (4′), (7) the relationship of Ψ_1, Ψ_2 to ψ_1, ψ_2 is completely determined.

The problem of the bridge requires the determination of the current ψ_2, as proportional to Ψ_1, when $\Psi_2 = 0$, that is, when no electromotive force acts in the bridge itself, and the solution is given at once by simple introduction into (2) of the values A, C, B from (4), (4′), (7).

If there be an approximate balance, the expression simplifies. For $bc-ad$ is then small, and B^2 may be neglected relatively to AC in the denominator of (2). Thus, as a sufficient approximation in this case, we have

$$\frac{\psi_2}{\Psi_1} = \frac{\dfrac{ad-bc}{a+b+c+d}}{\left\{e+\dfrac{(a+b)(c+d)}{a+b+c+d}\right\}\left\{f+\dfrac{(a+c)(b+d)}{a+b+c+d}\right\}}, \qquad\text{..........(8)}$$

in agreement with the equation used by Mr Heaviside for simple resistances.

The following interpretation of the process leads very simply to the approximate form (8), and may be acceptable to readers less familiar with the general method. Let us first inquire what E.M.F. is necessary in the telephone branch to stop the current through it. If such a force acts, the conditions are, externally, the same as if the branch were open, and the current ψ_1 in the battery branch due to an E.M.F. equal to Ψ_1 in that branch is Ψ_1/A, where A is written for brevity as representing the right-hand member of (4). The difference of potential at the terminals of e, still supposed to be open, is found at once when ψ_1 is known. It is equal to

$$c\times(5) - d\times(6) = B\psi_1,$$

where B is defined by (7). In terms of Ψ_1 the difference of potentials is thus $B\Psi_1/A$. If e be now closed, the same fraction expresses the E.M.F. necessary in e in order to prevent the generation of a current in that branch.

The case that we have to deal with is when Ψ_1 acts in f, and there is no E.M.F. in e. We are at liberty, however, to suppose that two opposite forces, each of magnitude $B\Psi_1/A$, act in e. One of these, as we have seen, acting in conjunction with Ψ_1 in f, gives no current in e; so that, since electromotive forces act independently of one another, the actual current in e, closed without internal E.M.F., is simply that due to the other component. The question is thus reduced to the determination of the current in e due to a given E.M.F. in that branch.

So far the argument is rigorous; but we will now suppose that we have to deal with an approximate balance. In this case an E.M.F. in e gives rise to very little current in f, and in calculating the current in e we may suppose f to be broken. The total resistance to the force in e is then given simply by C of equation (4′), and the approximate value for ψ_2 is derived by dividing $-B\Psi_1/A$ by C, as we found in (8).

A continued application of the foregoing process gives ψ_2/Ψ_1 in the form of an infinite geometric series:

$$\psi_2/\Psi_1 = -\frac{B}{AC}\left\{1 + \frac{B^2}{AC} + \frac{B^4}{A^2C^2} + \dots\dots\right\} = \frac{B}{B^2 - AC}. \quad \dots\dots(2)$$

This is the rigorous solution already found; but the first term of the series suffices for practical purposes.

The form of (8) enables us at once to compare the effects of increments of resistance and inductance in disturbing a balance. For let $ad = bc$, and then change d to $d + d'$ where $d' = d'_1 + id'_2$. The value of ψ_2/Ψ_1 is proportional to d', and the amplitude of the vibratory current in the bridge is proportional to Mod d', that is, to $\sqrt{(d'^2_1 + d'^2_2)}$. Thus d'_1, d'_2 are equally efficacious when numerically equal.

The next application that we shall make of (8) is to the generalised form of Schwendler's problem. When all else is given, how should the telephone, or other receiving instrument, be wound in order to get the greatest effect?

If by separation of real and imaginary parts we set

$$e = e_1 + ie_2, \qquad \frac{(a+b)(c+d)}{a+b+c+d} = r_1 + ir_2, \quad \dots\dots\dots\dots\dots(9)$$

the factor in the denominator of (6) with which we are concerned becomes

$$e_1 + r_1 + i(e_2 + r_2);$$

and the square of the modulus is given by

$$\text{Mod}^2 = (e_1 + r_1)^2 + (e_2 + r_2)^2. \quad \dots\dots\dots\dots\dots\dots\dots(10)$$

In this equation e_1, r_1 are essentially positive, while e_2, r_2 may be either positive or negative. If e_1 and e_2 are both at disposal, the minimum of (10), corresponding to the maximum current, is found by making

$$e_1 = 0, \qquad e_2 = -r_2. \quad \text{.......................(11)}$$

But this is not the practical question. As in the case of simple resistances, what we have to aim at is not to render the current in the bridge a maximum, but rather the *effect* of the current. Whether the receiving instrument be a galvanometer or a telephone, we cannot in practice reduce its resistance to zero without at the same time nullifying the effect desired. We must rather regard the space available for the windings as given, and merely inquire how it may best be utilised. Now the effect required to be exalted is, *cæteris paribus*, proportional to the number of windings (m); and, if the space occupied by insulation be proportional to that occupied by copper, the resistance varies as m^2. So also does the inductance; and accordingly, if the instrument be connected to the bridge by leads sensibly devoid of resistance and inductance,

$$e_1 + ie_2 = m^2(\epsilon_1 + i\epsilon_2), \quad \text{.......................(12)}$$

where ϵ_1, ϵ_2 are independent of m. The quantity whose modulus is to be made a minimum by variation of m is thus

$$\frac{e_1 + ie_2 + r_1 + ir_2}{m} = \frac{r_1 + m^2\epsilon_1 + i(r_2 + m^2\epsilon_2)}{m};$$

and we have

$$\text{Mod}^2 = \frac{(r_1 + m^2\epsilon_1)^2 + (r_2 + m^2\epsilon_2)^2}{m^2}$$

$$= (r_1^2 + r_2^2)m^{-2} + 2(r_1\epsilon_1 + r_2\epsilon_2) + (\epsilon_1^2 + \epsilon_2^2)m^2.$$

This is a minimum by variation of m when

$$m^4 = \frac{r_1^2 + r_2^2}{\epsilon_1^2 + \epsilon_2^2},$$

or

$$\text{Mod}(r_1 + ir_2) = \text{Mod}(e_1 + ie_2). \quad \text{.....................(13)}$$

We may express this result by saying that to get the best effect the instrument must be so wound that its *impedance* is equal to that of the compound conductor $r_1 + ir_2$. If for any reason the inductances can be omitted from consideration, then the resistance of the instrument is to be made equal to r_1, in accordance with Schwendler's rule.

The case of the "battery" branch may often be treated in like manner. As Mr Heaviside has shown, if a number of cells are available for ordinary resistance testing, they should be combined, so that their resistance is equal to that (s_1) of the corresponding combination of wires in parallel. Periodic currents may be conceived to arise from the rotation of a coil in a magnetic field of given strength. If the space occupied by the windings of the coil be supposed to be given, their number m will be determined by the condition of equal impedances. Thus, if

$$\frac{(a+c)(b+d)}{a+b+c+d} = s_1 + is_2, \qquad\qquad (14)$$

$$\text{Mod}\,(f_1 + if_2) = \text{Mod}\,(s_1 + is_2), \qquad\qquad (15)$$

in analogy with (13).

The above is the solution of the problem, if the coils of the sending and receiving instruments represent the whole of their respective branches, and are limited to occupy given spaces. The inductances and resistances cannot then be varied independently. But there would often be no difficulty in escaping from this limitation. The inclusion of additional resistance, external to the instrument, can only do harm; but the case is otherwise with inductance, positive or negative. If the inductance of the instrument added to r_2, or to s_2, be positive, the total inductance may be reduced to zero by the insertion of a suitable condenser, and this without material increase of resistance. If the inductance be already negative, the remedy is not so easily carried out; but, theoretically, it is possible to add the necessary inductance without sensible increase of resistance. The greater the frequency of vibration, the more feasible does this course become. We may, therefore, without much violence, suppose that the inductances of two branches can be reduced to zero without additional resistance. Thus,

$$e_2 + r_2 = 0, \qquad f_2 + s_2 = 0; \qquad\qquad (16)$$

and the condition of maximum efficiency of the transmitting and receiving coils is then given by Schwendler's rule,

$$e_1 = r_1, \qquad f_1 = s_1. \qquad\qquad (17)$$

These suppositions form a reasonable basis for further investigation; but conclusions founded upon them will be subject to re-examination, especially in extreme cases. We may also now introduce the promised simplification,

$$a = c, \qquad b = d, \qquad\qquad (18)$$

in accordance with which (8) becomes

$$\frac{\psi_2}{\Psi_1} = \frac{d-b}{4b}\,\frac{2ab/(a+b)}{\{e+\tfrac{1}{2}(a+b)\}\,\{f+2ab/(a+b)\}}. \qquad\qquad (19)$$

Also
$$r_1 + ir_2 = \tfrac{1}{2}(a+b) = \tfrac{1}{2}(a_1+b_1) + \tfrac{1}{2}i(a_2+b_2). \qquad\qquad (20)$$

$$\begin{aligned} s_1 + is_2 &= \frac{2(a_1+ia_2)(b_1+ib_2)}{a_1+ia_2+b_1+ib_2} \\ &= 2\,\frac{(a_1+b_1)(a_1b_1-a_2b_2)+(a_2+b_2)(a_2b_1+a_1b_2)}{(a_1+b_1)^2+(a_2+b_2)^2} \\ &\quad + 2i\,\frac{(a_1+b_1)(a_2b_1+a_1b_2)-(a_2+b_2)(a_1b_1-a_2b_2)}{(a_1+b_1)+(a_2+b_2)^2}. \end{aligned} \qquad\qquad (21)$$

It may be well to examine, first, the consequences of (19), in the case of simple resistances. Here

$$r_1 = \tfrac{1}{2}(a_1 + b_1), \qquad r_2 = 0; \quad \text{.................}(22)$$

$$s_1 = 2a_1 b_1/(a_1 + b_1), \qquad s_2 = 0. \quad \text{.................}(23)$$

In accordance with the plan proposed, we are to make $e_2 = 0$, $f_2 = 0$*; $e_1 = r$, $f_1 = s_1$. Our equation then becomes

$$\frac{\psi_2}{\Psi_1} = \frac{d_1 - b_1}{8b_1(a_1 + b_1)}. \quad \text{.........................}(24)$$

Here a_1 is still at disposal, and we see that according to (24) it ought to be diminished without limit. This conclusion does not harmonize with one obtained by Mr Heaviside†. It must be observed, however, that $a_1 = 0$ is unpractical, involving, as it does, $s_1 = 0$, $f_1 = 0$. Even according to (24) there is little to be gained by diminishing a_1 below, say, $\frac{1}{2}b_1$. In this case

$$a_1 = \tfrac{1}{2}b_1, \qquad e_1 = r_1 = \tfrac{3}{4}b_1, \qquad f_1 = s_1 = \tfrac{2}{3}b_1. \quad \text{........}(25)$$

Such an arrangement as (25) may be recommended for practical use.

When b_1 is large, there may be advantage in taking a_1 relatively smaller than in the above example. In such cases we approach the limiting condition of things, and have approximately

$$e_1 = r_1 = \tfrac{1}{2}b_1, \qquad f_1 = s_1 = 2a_1, \quad \text{....................}(26)$$

$$\frac{\psi_2}{\Psi_1} = \frac{d_1 - b_1}{8b_1^2}. \quad \text{..............................}(27)$$

And the smallness of f_1 in comparison with b_1 may sometimes be a convenience.

The next remark that has to be made is that, even when the conductors, b and d, to be compared are endowed with sensible inductances (positive or negative), the problem may still, theoretically, be brought under the above head. Suppose, for example, that b, d represent nearly equal electromagnets. Their inductances may be compensated by the introduction (in series) of suitable equal condensers into these branches, so that b and d are reduced to b_1 and d_1. If then we assume a to be a simple resistance ($a_2 = 0$), the solution is as before. Two objections may here be raised. First, on the theoretical side it has not been proved to be advantageous to assume $a_2 = 0$; and, secondly, the introduction of extraneous condensers‡, even with interchange,

* These conditions require no attention in galvanometric testing with steady currents, being satisfied by $p = 0$, independently of the nature of the instrument.

† *Loc. cit.* p. 120, "In conclusion, if, to measure a certain resistance, the best resistances for the galvanometer, battery, and the three sides, a, b, c, were required, then we should have to make $a = b = c = d = e = f$."

‡ The use of condensers or electromagnets in the branches e and f stands, of course, upon a different footing.

into the branches to be accurately compared may be a complication unfavourable to success.

We will now resume the consideration of (19), supposing that

$$e = e_1 + ie_2 = r_1 - ir_2, \qquad f = f_1 + if_2 = s_1 - is_2, \quad \text{.........(28)}$$

r_1, r_2, s_1, s_2 being given by (20), (21). Thus,

$$\psi_2/\Psi_1 = \frac{d-b}{16}\frac{s_1 + is_2}{br_1 s_1}, \quad \text{.........................(29)}$$

and the question before us is how to make the modulus of the second fraction on the right a maximum by variation of a. In the denominator of this fraction r_1 and s_1 are real, and the modulus of b is $\sqrt{(b_1^2 + b_2^2)}$. For the numerator we have

$$\frac{1}{a} + \frac{1}{b} = \frac{1}{a_1 + ia_2} + \frac{1}{b_1 + ib_2} = \frac{2}{s_1 + is_2} = \frac{2(s_1 - is_2)}{s_1^2 + s_2^2},$$

so that

$$\frac{2s_1}{s_1^2 + s_2^2} = \frac{a_1}{a_1^2 + a_2^2} + \frac{b_1}{b_1^2 + b_2^2}.$$

Also from the definition of s

$$s_1^2 + s_2^2 = \frac{4(a_1^2 + a_2^2)(b_1^2 + b_2^2)}{(a_1 + b_1)^2 + (a_2 + b_2)^2};$$

so that

$$\frac{s_1^2}{s_1^2 + s_2^2} = \frac{(a_1^2 + a_2^2)(b_1^2 + b_2^2)}{(a_1 + b_1)^2 + (a_2 + b_2)^2}\left\{\frac{a_1}{a_1^2 + a_2^2} + \frac{b_1}{b_1^2 + b_2^2}\right\}^2.$$

Thus

$$\text{Mod}\,\frac{br_1 s_1}{s_1 + is_2} = \frac{(a_1 + b_1)\{a_1(b_1^2 + b_2^2) + b_1(a_1^2 + a_2^2)\}}{2\sqrt{(a_1^2 + a_2^2)}\,.\,\sqrt{\{(a_1 + b_1)^2 + (a_2 + b_2)^2\}}}, \quad \text{......(30)}$$

and this is to be made a minimum by variation of a_1, a_2.

We shall show presently that (30) can be reduced to zero; but for the moment we will so far limit the generality of a_1, a_2 as to suppose that $a_1 = xb_1$, $a_2 = xb_2$, x being real and positive.

(30) then reduces to $\frac{1}{2}b_1^2(1 + x)$; and by (29)

$$\text{Mod}\,\psi_2/\Psi_1 = \frac{\text{Mod}\,(d-b)}{8b_1^2(1+x)}. \quad \text{.....................(31)}$$

Accordingly, the maximum sensitiveness cannot be attained until x is reduced to zero, so that a_1, a_2 vanish. (31) may be regarded as a generalised form of (24), free from the limitation that $b_2 = 0$, provided a_2 be so taken that $a_2/b_2 = a_1/b_1$.

We will now suppose in (30) that a_1 and a_2 are both small, and in the first instance that b_1 is finite. We have

$$\tfrac{1}{2}b_1\sqrt{(b_1^2 + b_2^2)}\frac{a_1}{\sqrt{(a_1^2 + a_2^2)}} + \frac{\tfrac{1}{2}b_1^2}{\sqrt{(b_1^2 + b_2^2)}}\sqrt{(a_1^2 + a_2^2)}; \quad \text{.........(32)}$$

and this reduces ultimately to its first term, depending upon the *ratio* only of a_1 and a_2. The expression vanishes if $a_1 : a_2$ be small enough, so that (30) can certainly be thus reduced to zero. It is remarkable that the expression for the sensitiveness should be capable of becoming infinite by suitable choice of a_2. If we first suppose that a_2 is absolutely zero, and afterwards that a_1 diminishes without limit, the ultimate value of (32) is $\frac{1}{2}b_1\sqrt{(b_1^2+b_2^2)}$, in place of zero.

From the practical point of view, these conclusions from our equations are not particularly satisfactory. We began with certain proposals which, in ordinary cases, could be carried out; but in the end we are directed to apply them to an extreme and impossible state of things. We have found, however, in what direction we must tend in the search for sensitiveness; and useful information may be gathered from (32). In practice a_1 could not be reduced below a certain point. The question may then be asked, what is the best value of a_2, when a_1 is given? From (32) we find at once that

$$a_1^2 + a_2^2 = \frac{a_1(b_1^2 + b_2^2)}{b_1}, \qquad \text{......................(33)}$$

(32) then becoming

$$b_1\sqrt{(a_1 b_1)}. \qquad \text{..............................(34)}$$

In this case from (29)

$$\text{Mod}\,\psi_2/\Psi_1 = \frac{\text{Mod}\,(d-b)}{16 b_1 \sqrt{(a_1 b_1)}}, \qquad \text{....................(35)}$$

independent of b_2.

If we suppose in (32) that $a_2 = 0$, we have

$$\tfrac{1}{2}b_1\sqrt{(b_1^2+b_2^2)} + \frac{\frac{1}{2}b_1^2 a_1}{\sqrt{(b_1^2+b_2^2)}}. \qquad \text{....................(36)}$$

To take a numerical example, let $b_2 = 0$; and suppose $a_1 = \frac{1}{10}b_1$. Then, according to (33), $a_2 = \pm\frac{3}{10}b_1$. Also by (20), (21),

$$e_1 = \tfrac{11}{20}b_1, \qquad e_2 = \mp\tfrac{3}{20}b_1;$$

$$f_1 = \tfrac{4}{13}b_1, \qquad f_2 = \mp\tfrac{6}{13}b_1.$$

The corresponding minimum value of (32), equal to (34), is $b_1^2/\sqrt{(10)}$.

But with this value of a_1 the gain by allowing a_2 to be finite is not great. If $a_2 = 0$,

$$e_1 = \tfrac{11}{20}b_1, \qquad e_2 = 0;$$

$$f_1 = \tfrac{11}{65}b_1, \qquad f_2 = 0;$$

and the value of (32), equal to (36), is $\frac{11}{20}b_1^2$.

We see from (36) that when $a_2 = 0$ there is little to be gained by further reduction of a_1. But when a_2 is suitably chosen the gain may be worth having. Thus, in (34), if $a_1 = \frac{1}{100}b_1$, we have $\frac{1}{10}b_1^2$. Corresponding to this $a_2 = \pm\frac{1}{10}b_1$ nearly, and

$$e_1 = \tfrac{1}{2}b_1, \qquad e_2 = \mp\tfrac{1}{20}b_1;$$

$$f_1 = \tfrac{1}{25}b_1, \qquad f_2 = \mp\tfrac{1}{5}b_1.$$

These are not unreasonable proportions, and we see that the use of a_2 may be advantageous, even when the subject of measurement is a mere resistance. It will be remarked too that, except as regards e_2, f_2, the sign of a_2 is immaterial.

When the branches b, d consist of electromagnets, and still more when they consist of condensers, b_1 may be very small. If we suppose it to be zero, (30) becomes

$$\frac{a_1^2 b_2^2}{2\sqrt{(a_1^2+a_2^2)}\,.\,\sqrt{\{a_1^2+(a_2+b_2)^2\}}}. \qquad \text{.................(37)}$$

Corresponding to this from (20), (21),

$$e_1 = \tfrac{1}{2}a_1, \qquad e_2 = -\tfrac{1}{2}(a_2+b_2), \qquad \text{....................(38)}$$

$$f_1 = \frac{2a_1 b_2^2}{a_1^2+(a_2+b_2)^2}, \qquad f_2 = -\frac{2a_1^2 b_2 + 2a_2 b_2 (a_2+b_2)}{a_1^2+(a_2+b_2)^2}. \qquad \text{......(39)}$$

From (37) we see that the increase of a_2 is favourable, especially if the sign be the same as of b_2. Even if $a_2 = 0$, (37) now assuming the form

$$\frac{a_1 b_2^2}{2\sqrt{(a_1^2+b_2^2)}} \qquad \text{............................(40)}$$

can be reduced to zero by taking a_1 small enough. But of course (37) ceases to be applicable unless b_1 be small relatively to a_1. In correspondence with (40),

$$e_1 = \tfrac{1}{2}a_1, \qquad e_2 = -\tfrac{1}{2}b_2; \qquad \text{....................(41)}$$

$$f_1 = \frac{2a_1 b_2^2}{a_1^2+b_2^2}, \qquad f_2 = -\frac{2a_1^2 b_2}{a_1^2+b_2^2}. \qquad \text{..............(42)}$$

As an example of (37), suppose

$$a_1 = \tfrac{1}{4}b_2, \qquad a_2 = 4b_2.$$

Then
$$(37) = \frac{b_2^2}{640} \text{ nearly.}$$

Also approximately

$$e_1 = \tfrac{1}{8}b_2, \qquad e_2 = -\tfrac{5}{2}b_2, \qquad f_1 = \tfrac{1}{50}b_2, \qquad f_2 = -\tfrac{8}{5}b_2.$$

If b_2 represent the stiffness of a condenser, f_2 must be a positive inductance, and its magnitude, relatively to f_1, would probably constitute a difficulty.

As an example, with a_2 equal to zero, take

$$a_1 = \tfrac{1}{10} b_2, \qquad a_2 = 0.$$

Then $$(37) = (40) = \tfrac{1}{20} b_2^2 \text{ nearly},$$

and $$e_1 = \tfrac{1}{20} b_2, \qquad e_2 = -\tfrac{1}{2} b_2, \qquad f_1 = \tfrac{1}{5} b_2, \qquad f_2 = -\tfrac{1}{50} b_2.$$

So far as the general theory is concerned, it is a matter of indifference whether the indicating instrument be in the branch e, or in f. The latter corresponds to the connexions in De Sauty's method of testing condensers by means of the galvanometer. In practice, more space would probably be available for the coils of a transmitting instrument than of the receiving instrument, at least, if the latter be a telephone; and this would tell in favour of choosing that branch for the transmitter which should have the larger time constant (L/R).

To get an idea of the relative capacities, resistances, and inductances involved, we must assume a particular pitch. A frequency suitable for telephonic experiments is 1000 per second, for which $p = 2000\pi$. Thus, if the value of a_2 for a condenser of capacity C, and for an inductance L, and that of a_1 for a resistance R, are all numerically equal,

$$R = 2000\pi L = \frac{1}{2000\pi C}.$$

If R be 1 ohm, equal to 10^9 C.G.S., the corresponding capacity is $1{\cdot}6 \times 10^{-13}$ C.G.S., equal to 160 microfarads, and the corresponding inductance is $1{\cdot}6 \times 10^5$ C.G.S. Again, if C be one microfarad, equal to 10^{-15} C.G.S., R is 160 ohms, and L is $2{\cdot}5 \times 10^7$ cm.

In the preceding calculations e and f are supposed to be adjusted to the values most favourable to the effect in the receiving instrument. A question, which arises quite as often in practice, is how to make the best of given instruments. The full answer is necessarily somewhat complicated; for there could be no objection to the insertion of a condenser, for example, if the sensitiveness could be improved thereby. In what follows, however, the transmitting and receiving branches will be supposed to be fully given, so that e and f are known complex quantities; and the only question to be considered is as to the most suitable value of a, assumed to be equal to c.

For this purpose the modulus of the second fraction on the right in (19) is to be a maximum, or that of

$$(a + b + 2e)\left(\frac{1}{a} + \frac{1}{b} + \frac{2}{f}\right) \quad \text{.......................(43)}$$

is to be a minimum, by variation of a. The problem thus arising of determining the minimum modulus of a function of a complex quantity may be treated generally.

Let $$F(z) = F(x+iy) = \phi(x, y) + i\psi(x, y),$$

and let it be required to find when the modulus² of $F(z)$, viz., $\phi^2 + \psi^2$, is a minimum by variation of x, y. We have

$$\phi \frac{d\phi}{dx} + \psi \frac{d\psi}{dx} = 0, \qquad \phi \frac{d\phi}{dy} + \psi \frac{d\psi}{dy} = 0. \quad \text{............(44)}$$

And in general

$$d\phi/dx = d\psi/dy, \qquad d\phi/dy = -d\psi/dx. \quad \text{............(45)}$$

In order that (44), (45) may both obtain, we must have either $\phi^2 + \psi^2 = 0$, or else

$$d\phi/dx = 0, \qquad d\phi/dy = 0, \qquad d\psi/dx = 0. \qquad d\psi/dy = 0.$$

The latter conditions are equivalent to

$$F'(z) = 0. \quad \text{................................(46)}$$

For example, let

$$F(z) = (z + \alpha)(\beta + 1/z), \quad \text{........................(47)}$$

where α, β are complex constants.

The application of (46) gives

$$z^2 = \alpha/\beta, \quad \text{................................(48)}$$

and $$F(z) = \{1 + \sqrt{(\alpha\beta)}\}^2. \quad \text{..........................(49)}$$

We see then that the modulus of (43) will be a minimum, when

$$a^2 = \frac{b + 2e}{2/f + 1/b}, \quad \text{..........................(50)}$$

and in taking the square root the ambiguity must be so determined as to make the real part of a positive.

Equation (50) coincides with that obtained by Mr Heaviside for the case where all the quantities are real.

181.

ON VAN DER WAALS'S TREATMENT OF LAPLACE'S PRESSURE IN THE VIRIAL EQUATION: LETTERS TO PROF. TAIT.

[*Nature*, XLIV. pp. 499, 597, 1891.]

IN Part IV. of your "Foundations of the Kinetic Theory of Gases*," you take exception to the manner in which Van der Waals has introduced Laplace's intrinsic pressure K into the equation of virial. "I do not profess to be able fully to comprehend the arguments by which Van der Waals attempts to justify the mode in which he obtains the above equation. Their nature is somewhat as follows:—He repeats a good deal of Laplace's capillary work, in which the existence of a large, but unknown, internal molecular pressure is established, entirely from a statical point of view. He then gives reasons (which seem, on the whole, satisfactory from this point of view) for assuming that the magnitude of this force is as the square of the density of the aggregate of particles considered. But his justification of the introduction of the term a/v^2 into an account already closed, as it were, escapes me. He seems to treat the surface-skin of the group of particles as if it were an additional bounding-surface, exerting an additional and enormous pressure on the contents. Even were this justifiable, nothing could justify the multiplying of this term by $(v-\beta)$ instead of by v alone. But the whole procedure is erroneous. If one begins with the virial equation, one must keep strictly to the assumptions made in obtaining it, and consequently *everything* connected with molecular force, whether of attraction or of elastic resistance, must be extracted from the term $\Sigma(Rr)$."

With the last sentence all will agree; but it seemed to me when I first read Van der Waals's essay that his treatment of Laplace's pressure was satisfactory, and on reperusal it still appears to me to conform to the requirements above laid down. As the point is of importance, it may be well to

* *Ed. Trans.* Vol. XXXVI. Part 2, p. 261.

examine it somewhat closely. The question is as to the effect in the virial equation of a mutual attraction between the parts of the fluid, whose range is small compared with the dimensions of bodies, but large in comparison with molecular distances.

The problem thus presented may be attacked in two ways. The first, to which I will recur, is that followed by Van der Waals; but the second is more immediately connected with that form of the equation which you had in view in the passage above quoted.

In the notation of Van der Waals (equation 8)

$$\tfrac{1}{2}\Sigma m V^2 = \tfrac{1}{2}\Sigma f\rho - \tfrac{1}{2}\Sigma Rr\cos(R, r),$$

where V denotes the velocity of a particle m, which is situated at a distance r from the origin, and is acted upon by a force R, while (R, r) denotes the angle between the directions of R and r. The intermediate term is to be omitted if R be the total force acting upon m. It represents the effect of such forces, f, as act mutually between two particles at distances from one another equal to ρ. In the summation the force between two particles is to be reckoned once only, and the forces accounted for in the second term are, of course, to be excluded in the third term.

In the present application we will suppose all the mutual forces accounted for in the second term, and that the only external forces operative are due to the pressure of the containing vessel. No one disputes that the effect of the external pressure is given by

$$-\tfrac{1}{2}\Sigma Rr\cos(R, r) = \tfrac{3}{2}pv;$$

so that

$$\tfrac{1}{2}\Sigma m V^2 = \tfrac{3}{2}pv + \tfrac{1}{2}\Sigma\rho\phi(\rho),$$

if with Laplace we represent by $\phi(\rho)$ the force between two particles at distance ρ. The last term is now easily reckoned upon Laplace's principles. For one particle in the interior we have

$$\tfrac{1}{2}\,.\,4\pi\int_0^\infty \phi(\rho)\,\rho^3 d\rho,$$

and this, as Laplace showed*, is equal to $3K$. The second summation over the volume gives $3Kv$, but this must be halved. Otherwise each force would be reckoned twice. Hence

$$\tfrac{1}{2}\Sigma m V^2 = \tfrac{3}{2}pv + \tfrac{3}{2}Kv = \tfrac{3}{2}v(p+K),$$

showing that the effect of such forces as Laplace supposed to operate is represented by the addition to p, the pressure exerted by the walls of the vessel, of the intrinsic pressure K. In the above process the particles situated near the surface are legitimately neglected in comparison with those in the interior.

* See also *Phil. Mag.* October 1890, p. 292. [Vol. III. p. 403.]

Van der Waals's own process starts from the original form of virial equation—

$$\tfrac{1}{2}\Sigma m V^2 = -\tfrac{1}{2}\Sigma Rr \cos(R, r),$$

where R now refers to the *whole* force operative upon any particle; and it appears to me equally legitimate. For all particles in the interior of the fluid R vanishes in virtue of the symmetry, so that the reckoning is limited to a surface stratum whose thickness is equal to the range of the forces. Upon this stratum act normally both the pressure of the vessel and the attraction of the interior fluid. The integrated effect of the latter throughout the stratum is equal to the intrinsic pressure; and, on account of thinness of the stratum, it enters into the equations in precisely the same way as the external pressure exerted by the vessel. The effect of Laplace's forces is thus represented by adding K to p, in accordance with the assertion of Van der Waals.

I am in hopes that, upon reconsideration, you will be able to admit that this conclusion is correct. Otherwise, I shall wish to hear more fully the nature of your objection, as the matter is of such importance that it ought not longer to remain in doubt.

Sept. 7.

I gather from your letter of September 28 (*Nature*, October 8, p. 546) that you admit the correctness of Van der Waals's deduction from the virial equation (i) when the particles are infinitely small, in which case

$$\left(p + \frac{a}{v^2}\right) v = \tfrac{1}{3}\Sigma m V^2, \quad \text{......................(1)}$$

a representing a cohesive force, whose range is great in comparison with molecular distances; and (ii) when, in the absence of a cohesive force, the volume of the particles is small in comparison with the total volume v, in which case the virial of the repulsive forces at impact gives

$$p(v-b) = \tfrac{1}{3}\Sigma m V^2. \quad \text{..........................(2)}$$

For hard spherical masses, the value of b is four times the total volume of the sphere. But you ask, "How can the factor $(v-b)/v$, which Van der Waals introduces on the left (in the first case) in consequence of the finite diameters of the particles, be justifiably applied to the term in K (or a/v^2) as well as to that in p?"

In my first letter I desired to avoid the complication entailed by the consideration of the finite size of the particles; but it appears to me that the argument there given (after Van der Waals) suffices to answer your question. For, if the cohesive force be of the character supposed, it exercises no influence upon any particle in the interior, and is *completely* accounted for by the addition to p of a/v^2. In so far, therefore, as (2) is correct when there is no cohesive force, the effect of such is properly represented by

$$\left(p+\frac{a}{v^2}\right)(v-b)=\tfrac{1}{2}\Sigma m V^2, \quad \text{..........................(3)}$$

in which b is to be multiplied by a/v^2, as well as by p.

October 13.

[1901. Prof. Tait maintained his opinion. In a reply to the second of the above letters, dated Oct. 17, 1891 (*Nature*, XLIV. p. 628), he writes:—

"On the question of the introduction by Van der Waals of the factor $(v-b)/v$, whether or not it is applied alike to p and to K, I regret to find that our views must continue to differ. For it appears to me that when once the various terms of the virial equation have been correctly extracted from the expression $\Sigma(Rr)$, we have no right to modify any of them. There seems therefore to be no doubt whatever that the procedure in Van der Waals's sixth chapter is entirely wrong in principle:—except in so far as (in the German version) he borrows some correct expressions from Lorentz. The meanings of v and p, in the term pv of the virial equation, are (from the very beginning of the inquiry) definitely assigned as total volume and external pressure:—so that this term cannot in any way be altered. No more can the term $\Sigma mu^2/3$, or the ratio of these two terms...."]

182.

ON THE VIRIAL OF A SYSTEM OF HARD COLLIDING BODIES.

[*Nature*, XLV. pp. 80—82, 1891.]

A RECENT correspondence [see Vol. III. p. 465] has led me to examine the manner in which various authors have treated the influence of the finite size of molecules in the virial equation, and I should like to lay a few remarks upon the subject before the readers of *Nature*.

To fix the ideas, we may begin by supposing that the molecules are equal hard elastic spheres, which exert no force upon one another except at the instant of collision. By calling the molecules hard, it is implied that the collisions are instantaneous, and it follows that at any moment the potential energy of the system is negligible in comparison with the kinetic energy.

If the volume of the molecules be very small in comparison with the space they occupy, the virial of the impulsive forces may be neglected, and the equation may be written

$$pv = \tfrac{1}{3}\Sigma m V^2, \quad\text{.................................(1)}$$

where p is the pressure exerted upon the walls of the enclosure, v the volume, m the mass, and V the velocity of a molecule.

In his essay of 1873 Van der Waals took approximate account of the finite size of the molecules, using a peculiar process to which exception has been taken by Maxwell and other subsequent writers. It must be said, however, that this method has not been proved to be illegitimate, and that at any rate it led Van der Waals to the correct conclusion—

$$p(v-b) = \tfrac{1}{3}\Sigma m V^2, \quad\text{................................(2)}$$

in which b denotes four times the total volume of the spheres. In calling (2) correct, I have regard to its character as an approximation, which was sufficiently indicated by Van der Waals in the original investigation, though perhaps a little overlooked in some of the applications.

In his (upon the whole highly appreciative) review of Van der Waals's essay, Maxwell (*Nature*, Vol. x. p. 477, 1874; *Scientific Papers*, Vol. II. p. 407) comments unfavourably upon the above equation, remarking that in the virial equation v is the volume of the vessel and is not subject to correction*. "The effect of the repulsion of the molecules causing them to act like elastic spheres is therefore to be found by calculating the virial of this repulsion." As the result of the calculation he gives

$$pv = \tfrac{1}{3}\Sigma m V^2 \left\{1 - 2 \log\left(1 - 8\frac{\rho}{\sigma} + \frac{17\rho^2}{\sigma^2} - \dots\right)\right\}, \qquad \text{.............(3)}$$

where σ is the density of the molecules, and ρ the mean density of the medium, so that $\rho/\sigma = b/4v$. If we expand the logarithm in (3), we obtain as the approximate expression, when ρ/σ is small

$$pv = \tfrac{1}{3}\Sigma m V^2 (1 + 4b/v), \qquad \text{..........................(4)}$$

or, as equally approximate,

$$p(v - 4b) = \tfrac{1}{3}\Sigma m V^2, \qquad \text{..........................(5)}$$

which does *not* agree with (2).

The details of the calculation of (3) have not been published, but there can be no doubt that the equation itself is erroneous. In his paper of 1881 (*Wied. Ann.* XII. p. 127), Lorentz, adopting Maxwell's suggestion, investigated afresh the virial of the impulsive forces, and arrived at a conclusion which, to the order of approximation in question, is identical with (2). A like result has been obtained by Prof. Tait (*Edin. Trans.* XXXIII. p. 90, 1886).

It appears that, while the method has been improved, no one has succeeded in carrying the approximation beyond the point already attained by Van der Waals in 1873. But a suggestion of great importance is contained in Maxwell's equation (3), numerically erroneous though it certainly is. For, apart from all details, it is there implied that the virial of the impacts is represented by $\tfrac{1}{3}\Sigma m V^2$, multiplied by some function of ρ/σ, so that, if the volume be maintained constant, the pressure as a function of V is proportional to $\Sigma m V^2$. The truth of this proposition is evident, because we may suppose the velocities of all the spheres altered in any constant ratio, without altering the motion in any respect except the scale of time, and then the pressure will necessarily be altered in the square of that ratio.

It will be interesting to inquire how far this conclusion is limited to the suppositions laid down at the commencement. It is necessary that the collisions be instantaneous in relation, of course, to the free time. Otherwise, the similarity of the motion could not be preserved, the duration of a collision, for example, bearing a variable ratio to the free time. On the same ground,

* In connexion with this it may be worth notice that for motion *in one dimension* the form (2) is *exact*.

vibrations within a molecule are not admissible. On the other hand, the limitation to the spherical form is unnecessary, and the theorem remains true whatever be the shape of the colliding bodies. Again, it is not necessary that the shapes and sizes of the bodies be the same, so that application may be made to mixtures.

In the theory of gases $\Sigma m V^2$ is proportional to the absolute temperature; and whatever doubts may be felt in the general theory can scarcely apply here, where the potential energy does not come into question. So far, then, as a gas may be compared to our colliding bodies, the relation between pressure, volume, and temperature is

$$p = T\phi(v), \quad \dots\dots\dots(6)$$

where $\phi(v)$ is some function of the volume. When v is large, the first approximation to the form of ϕ is

$$\phi(v) = \frac{A}{v}.$$

In the case of spheres, the second approximation is

$$\phi(v) = \frac{A}{v} + \frac{Ab}{v^2},$$

where b is four times the volume of the spheres.

Thus far we have supposed that there are no forces between the bodies but the impulses on collision. Many and various phenomena require us to attribute to actual molecules an attractive force operative to much greater distances than the forces of collision, and the simplest supposition is a cohesive force such as was imagined by Young and Laplace to explain capillarity. We are thus led to examine the effect of forces whose range, though small in comparison with the dimensions of sensible bodies, is large in comparison with molecular distances. In the extreme case, the influence of the discontinuous distribution of the attractive centres disappears, and the problem may be treated by the methods of Laplace. The modification then required in the virial equation is simply to add to p* a term inversely proportional to v^2, as was proved by Van der Waals; so that (6) becomes

$$p = T\phi(v) - av^{-2}. \quad \dots\dots\dots(7)$$

According to (7) the relation between pressure and temperature is *linear*—a law verified by comparison with observations by Van der Waals, and more recently and extensively by Ramsay and Young. It is not probable, however, that it is more than an approximation. To such cases as the behaviour of water in the neighbourhood of the freezing-point it is obviously inapplicable.

* It thus appears that, contrary to the assertion of Maxwell, p *is* subject to correction. It is pretty clear that he had in view an attraction of much smaller range than that considered by Van der Waals.

In their discussions, Ramsay and Young employ the more general form

$$p = T\phi(v) + \chi(v); \qquad (8)$$

and the question arises, whether we can specify any generalization of the theoretical conditions which shall correspond to the substitution of $\chi(v)$ for av^{-2}. It would seem that, as long as the only forces in operation are of the kinds, impulsive and cohesive, above defined, the result is expressed by (7); and that if we attempt to include forces of an intermediate character, such as may very probably exist in real liquids, and must certainly exist in solids, we travel beyond the field of (8) as well as of (7). It may be remarked that the equation suggested by Clausius, as an improvement on that of Van der Waals, is not included in (8).

Returning to the suppositions upon which (7) was grounded, we see that, if the bodies be all of one *shape, e.g.* spherical, the formula contains only two constants—one determining the size of the bodies, and the second the intensity of the cohesive force; for the mean kinetic energy is supposed to represent the temperature in all cases. From this follows the theorem of Van der Waals respecting the identity of the equation for various substances, provided pressure, temperature, and volume be expressed as fractions of the critical pressure, temperature, and volume respectively. If, however, the *shape* of the bodies vary in different cases, no such conclusion can be drawn, except as a rough approximation applicable to large volumes.

183.

DYNAMICAL PROBLEMS IN ILLUSTRATION OF THE THEORY OF GASES.

[*Phil. Mag.* XXXII. pp. 424—445, 1891.]

Introduction.

THE investigations, of which a part is here presented, had their origin in a conviction that the present rather unsatisfactory position of the Theory of Gases is due in some degree to a want of preparation in the mind of readers, who are confronted suddenly with ideas and processes of no ordinary difficulty. For myself, at any rate, I may confess that I have found great advantage from a more gradual method of attack, in which effort is concentrated upon one obstacle at a time. In order to bring out fundamental statistical questions, unencumbered with other difficulties, the motion is here limited to one dimension, and in addition one set of impinging bodies is supposed to be very small relatively to the other. The simplification thus obtained in some directions allows interesting extensions to be made in others. Thus we shall be able to follow the whole process by which the steady state is attained, when heavy masses originally at rest are subjected to bombardment by projectiles fired upon them indifferently from both sides. The case of pendulums, or masses moored to fixed points by elastic attachments, is also considered, and the stationary state attained under a one-sided or a two-sided bombardment is directly calculated.

Collision Formulæ.

If u', v' be the velocities before collision, u, v after collision, of two masses P, Q, we have by the equation of energy

$$P(u'^2-u^2)+Q(v'^2-v^2)=0, \qquad (1)$$

and by the equation of momentum,

$$P(u'-u)+Q(v'-v)=0. \qquad (2)$$

From (1) and (2)

$$u' + u = v' + v, \qquad (3)$$

or, as it may be written,

$$u' - v' = v - u,$$

signifying that the relative velocity of the two masses is reversed by the collision. From (2) and (3),

$$\left.\begin{aligned}(P+Q)\,u' &= (P-Q)\,u + 2Qv \\ (P+Q)\,v' &= 2Pu + (Q-P)\,v\end{aligned}\right\}. \qquad (4)$$

As is evident from (1) and (2), we may in (4), if we please, interchange the dashed and undashed letters. Thus from the first of (4),

$$(P+Q)\,u = (P-Q)\,u' + 2Qv',$$

or

$$u' = \frac{P+Q}{P-Q}u - \frac{2Q}{P-Q}v' = u + \frac{2Q}{P-Q}(u-v'). \qquad (5)$$

In the application which we are about to make, P will denote a relatively large mass, and Q will denote the relatively small mass of what for the sake of distinction we will call a projectile. All the projectiles are equal, and in the first instance will be supposed to move in the two directions with a given great velocity. After collision with a P the projectile rebounds and disappears from the field of view. Since in the present problem we have nothing to do with the velocity of rebound, it will be convenient to devote the undashed letter v to mean the given initial velocity of a projectile. Writing also q to denote the small ratio $Q : P$, we have

$$u' = u + \frac{2q}{1-q}(u-v). \qquad (6)$$

If u and v be supposed positive, this represents the case of what we may call a favourable collision, in which the velocity of the heavy mass is increased. If the impact of the projectile be in the opposite direction, the velocity u'', which becomes u after the collision, is given by

$$u'' = u + \frac{2q}{1-q}(u+v). \qquad (7)$$

The symbol v thus denotes the velocity of a projectile without regard to sign, and (7) represents the result of an unfavourable collision.

Permanent State of Free Masses under Bombardment.

The first problem that we shall attack relates to the *ultimate* effect upon a mass P of the bombardment of projectiles striking with velocity v, and moving indifferently in the two directions. It is evident of course that the ultimate state of a particular mass is indefinite, and that a definite result can relate only to probability or statistics. The statistical method of expression being the more convenient, we will suppose that a very large number of masses are undergoing bombardment independently, and inquire what we are to expect as the ultimate distribution of velocity among them. If the number of masses for which the velocity lies between u and $u+du$ be denoted by $f(u)\,du$, the problem before us is the determination of the form of $f(u)$.

The number of masses, whose velocities lie between u and $u+du$, which undergo collision in a given small interval of time, is proportional in the first place to the number of the masses in question, that is to $f(u)\,du$, and in the second place to the relative velocity of the masses and of the projectiles. In all the cases which we shall have to consider v is greater than u, so that the chance of a favourable collision is always proportional to $v-u$, and that of an unfavourable collision to $v+u$. It is assumed that the chances of collision depend upon u in no other than the above specified ways. The number of masses whose velocities in a given small interval of time are passing, as the result of favourable collisions, from below u to above u, is thus proportional to

$$\int_{u'}^{u} f(w)\,.\,(v_1-w)\,dw, \quad\text{.........................(8)*}$$

where u' is defined by (6); and in like manner the number which pass in the same time from above u to below u, in consequence of unfavourable collisions, is

$$\int_{u}^{u''} f(w)\,.\,(v_1+w)\,dw, \quad\text{.........................(9)}$$

u'' being defined by (7). In the steady state as many must pass one way as the other, and hence the expressions (8) and (9) are to be equated. The result may be written in the form

$$v_1\left\{\int_{u'}^{u}-\int_{u}^{u''}\right\}f(w)\,dw=\int_{u'}^{u''}wf(w)\,dw. \quad\text{..............(10)}$$

Now, if q be small enough, one collision makes very little impression upon u; and the range of integration in (10) is narrow. We may therefore expand the function f by Taylor's theorem:

$$f(w)=f(u)+(w-u)\,f'(u)+\tfrac{1}{2}(w-u)^2 f''(u)+\ldots\ldots;$$

so that

$$\int f(w)\,dw=wf(u)+\tfrac{1}{2}(w-u)^2 f'(u)+\tfrac{1}{6}(w-u)^3 f''(u)+\ldots\ldots,$$

$$\left\{\int_{u'}^{u}-\int_{u}^{u''}\right\}f(w)\,dw=(2u-u'-u'')\,f(u)$$

$$-\tfrac{1}{2}\{(u'-u)^2+(u''-u)^2\}\,f'(u)+\ldots\ldots$$

$$=-\frac{4q}{1-q}\,uf(u)-\frac{4q^2}{(1-q)^2}(v^2+u^2)\,f'(u)+\text{cubes of } q. \quad\ldots(11)$$

* In the present problem $v_1=v$; but it will be convenient at this stage to maintain the distinction.

Also $$\int w\,f(w)\,dw = \int \{(w-u)+u\}\,f(w)\,dw$$
$$= \tfrac{1}{2}(w-u)^2 f(u) + \tfrac{1}{3}(w-u)^3 f'(u) + \ldots\ldots$$
$$+\,u f(u)\,\{w f(u) + \tfrac{1}{2}(w-u)^2 f'(u) + \ldots\ldots\};$$
so that
$$\int_{u'}^{u''} w f(w)\,dw = u f(u)\,.\,(u''-u')$$
$$+\,\{\tfrac{1}{2} f(u) + \tfrac{1}{3} u f(u)\}\,\{(u''-u)^2 - (u'-u)^2\} + \ldots\ldots$$
$$= \frac{4qv}{1-q}\,u f(u) + \frac{8q^2uv}{(1-q)^2}\{f(u) + u f'(u)\} + \text{cubes of } q. \quad \ldots(12)$$
As far as q^2 inclusive (10) thus becomes
$$\frac{4qv_1}{1-q}\,u\,f(u) + \frac{4q^2v_1}{(1-q)^2}(v^2+u^2)\,f'(u)$$
$$+\,\frac{4qv}{1-q}\,u f(u) + \frac{8q^2uv}{(1-q)^2}\{f(u) + u f'(u)\} = 0,$$
or $$u f(u)\,\{(1-q)\,v_1 + (1+q)\,v\} + q f'(u)\,\{v_1 v^2 + u^2(v_1 + 2v)\} = 0.$$
If $v_1 = v$, q disappears from the first term as it stands, and will do so in any case in the limit when it is made infinitely small. Moreover, in the second term u^2 is to be neglected in comparison with v^2. We thus obtain
$$u f(u)\,\{1 + v/v_1\} + qv^2 f'(u) = 0 \quad \ldots\ldots\ldots\ldots\ldots\ldots(13)$$
as the differential equation applicable to the determination of $f(u)$ when q is infinitely small. The integral is
$$qv^2 \log f(u) + \tfrac{1}{2}(1 + v_1/v)\,u^2 = \text{constant},$$
or
$$f(u) = Ae^{-hu^2}, \quad \ldots\ldots\ldots\ldots\ldots\ldots\ldots\ldots\ldots\ldots(14)$$
where
$$h = \frac{1 + v/v_1}{2qv^2}; \quad \ldots\ldots\ldots\ldots\ldots\ldots\ldots\ldots\ldots(15)$$
or, if $v_1 = v$,
$$h = 1/qv^2. \quad \ldots\ldots\ldots\ldots\ldots\ldots\ldots\ldots\ldots\ldots(16)$$
The ultimate distribution of velocities among the masses is thus a function of the energy of the projectiles and not otherwise of their common mass and velocity. The ultimate state is of course also independent of the number of the projectiles.

The form of f is that found by Maxwell. To estimate the mean value of u^2 we must divide
$$\int_{-\infty}^{+\infty} u^2 f(u)\,du \;\text{ by }\; \int_{-\infty}^{+\infty} f(u)\,du.$$
Now
$$\int u^2 e^{-u^2/qv^2}\,du = -\tfrac{1}{2}qv^2\,\{u e^{-u^2/qv^2} - \int e^{-u^2/qv^2}\,du\},$$
so that
$$\int_{-\infty}^{+\infty} u^2 e^{-u^2/qv^2}\,du = \tfrac{1}{2}qv^2 \int_{-\infty}^{+\infty} e^{-u^2/qv^2}\,du.$$

The ratio in question is thus $\frac{1}{2}qv^2$, showing that the mean kinetic energy of a mass is *one half* that of a projectile, deviating from the law of equal energies first (1845) laid down by Waterston. We must remember, however, that we have thus far supposed the velocities of the projectiles to be all equal.

The value of A in (14) may be determined as usual. If N be the whole (very great) number of masses to which the statistics relate,

$$N = \int_{-\infty}^{+\infty} f(u)\,du = A\int_{-\infty}^{+\infty} e^{-u^2/qv^2}\,du = Av\sqrt{(\pi q)};$$

so that

$$f(u)\,du = \frac{N}{v\sqrt{(\pi q)}}\,e^{-u^2/qv^2}\,du. \quad \ldots\ldots\ldots\ldots\ldots\ldots\ldots(15')$$

If we were to suppose that the chances of a favourable or unfavourable collision were independent of the actual velocity of a mass, there would still be a stationary state defined by writing $v_1 = \infty$ in (15). Under these circumstances the mean energy would be twice as great as that calculated above.

It is easy to extend our result so as to apply to the case of projectiles whose velocities are distributed according to any given law $F(v)$, of course upon the supposition that the projectiles of different velocities do not interfere with one another. We have merely to multiply by $F(v)\,dv$ and to integrate between 0 and ∞. Thus from (13) we obtain

$$2uf(u)\int_0^{+\infty} vF(v)\,dv + qf'(u)\int_0^{+\infty} v^3F(v)\,dv = 0. \quad \ldots\ldots\ldots(17)$$

If $F(v) = e^{-kv^2}$, we find

$$\int v^3e^{-kv^2}\,dv = -\frac{1}{2k}\left\{v^2e^{-kv^2} - \int e^{-kv^2}\,2v\,dv\right\},$$

so that

$$\int_0^{+\infty} v^3e^{-kv^2}\,dv = \frac{1}{k}\int_0^{+\infty} ve^{-kv^2}\,dv. \quad \ldots\ldots\ldots\ldots\ldots\ldots\ldots(18)$$

Our equation then becomes

$$2kuf(u) + qf'(u) = 0,$$

giving

$$f(u) = Ae^{-ku^2/q}. \quad \ldots\ldots\ldots\ldots\ldots\ldots\ldots\ldots\ldots\ldots(19)$$

The mean energy of the masses is $\frac{1}{2}q/k$, and this is now *equal* to the mean energy of the projectiles. We see that if the mean energy of the projectiles is given, their efficiency is greater when the velocity is distributed according to the Maxwell law than when it is uniform, and that in the former case the Waterston relation is satisfied, as was to be expected from investigations in the theory of gases.

It may perhaps be objected that the law e^{-kv^2} is inconsistent with our assumption that v is always great in comparison with u. Certainly there will be a few projectiles for which the assumption is violated; but it is pretty evident that in the limit when q is small enough, the effect of these will become negligible. Even when the velocity of the projectiles is constant, the law e^{-u^2/qv^2} must not be applied to values of u comparable with v.

The independence of the stationary state of conditions, which at first sight would seem likely to have an influence, may be illustrated by supposing that the motion of the masses is constrained to take place along a straight line, but that the direction of motion of the projectiles, striking always centrically, is inclined to this line at a constant angle θ.

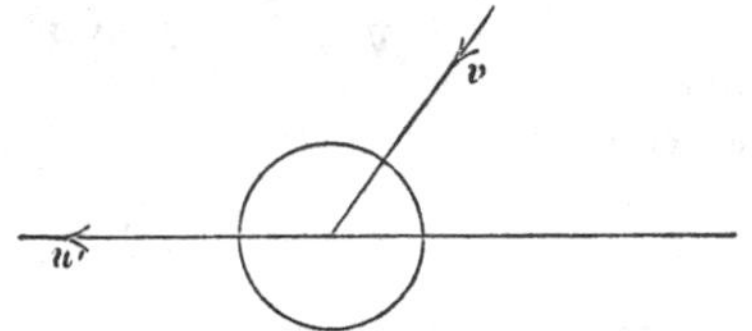

If u' be the velocity of the mass (unity) before impact, and u after impact, B the impulsive action between the mass and the projectile,

$$u - u' = B \cos \theta.$$

Also, if v, V be the velocities of the projectile (q) before and after impact,

$$q(v - V) = B;$$

so that

$$q(v - V) \cos \theta = u - u'.$$

By the equation of energy

$$u^2 - u'^2 = q(v^2 - V^2).$$

From these we find, as before,

$$u' = u - \frac{2q \cos^2 \theta}{1 - q \cos^2 \theta} \left(u - \frac{v}{\cos \theta}\right).$$

This may be regarded as a generalization of (6); and we see that it may be derived from (6) by writing $v/\cos \theta$ for v, and $q \cos^2 \theta$ for q. In applying equation (10) to determine the stationary state, we must remember that the velocity of retreat is now no longer w, but $w \cos \theta$, so that (10) becomes

$$v \left\{ \int_{u'}^{u} - \int_{u}^{u''} \right\} f(w)\, dw = \int_{u'}^{u''} w \cos \theta f(w)\, dw.$$

The entire effect of the obliquity θ is thus represented by the substitution of $v/\cos \theta$ for v, and of $q \cos^2 \theta$ for q, and since these leave qv^2 unaltered, the stationary state, determined by (15), is the same as if $\theta = 0$.

The results that we have obtained depend entirely upon the assumption that the individual projectiles are fired at random, and without distinction between one direction and the other. The significance of this may be illustrated by tracing the effect of a restriction. If we suppose that the projectiles are despatched in pairs of closely following components, we should expect that the effect would be the same as of a doubling of the mass. If, again, the components of a pair were so projected as to strike almost at the same time upon opposite sides, while yet the direction of the first was at random, we should expect the whole effect to become evanescent. These anticipations are confirmed by calculation.

By (5) the velocity u_1', which on collision becomes u, is

$$u_1' = \frac{1+q}{1-q}u \mp \frac{2q}{1-q}v;$$

so that the velocity, which after *two* consecutive collisions upon the same side becomes u, is given by

$$u_2' = \frac{1+q}{1-q}u_1 \mp \frac{2q}{1-q}v = \frac{1+2q+(q^2)}{1-2q+(q^2)}u \mp \frac{4qv}{1-2q+(q^2)}.$$

The masses which by single collisions at velocity v would ultimately produce the same effect as these pairs are therefore very approximately $2q$.

If the projectiles be distributed in pairs in such a way that the components of each strike nearly simultaneously and upon *opposite* sides,

$$\begin{aligned} u_2' &= \frac{1+q}{1-q}\left\{\frac{1+q}{1-q}u \pm \frac{2qv}{1-q}\right\} \mp \frac{2qv}{1-q} \\ &= \frac{(1+q)^2}{(1-q)^2}u \pm \frac{4q^2v}{(1-q)^2} \\ &= \frac{1+2q+(q^2)}{1-2q+(q^2)}u \pm \frac{4q\,.\,qv}{1-2q+(q^2)}; \end{aligned}$$

showing that the effect is the same as if the mass were doubled, and the velocity reduced from v to qv. Thus, when q is infinitely small, the effect is negligible in comparison with that obtained when the connexion of the components of a pair is dissolved, and each individual is projected at random.

Another Method of Investigation.

The method followed in the formation of equation (10) seems to lead most simply to the required determination of $f(u)$; but it is an instructive variation to consider directly the balance between the numbers of masses which change their velocities *from* and *to* u.

The number of masses whose velocities lie between u and $u+du$ being $f(u)\,du$, we have as the number whose velocities in a given small interval of time are expelled from the range du,

$$f(u)\,du\,(v-u) + f(u)\,du\,(v+u),$$

or

$$2vf(u)\,du.$$

This, in the steady state, is equal to the number which enter the range du from the two sides in consequence of favourable and unfavourable collisions; so that

$$f(u')(v-u')\,du' + f(u'')(v+u'')\,du'' - 2vf(u)\,du = 0. \quad \text{......(20)}$$

By (6), (7), since v is constant,

$$du' = \frac{1+q}{1-q}\,du, \qquad du'' = \frac{1+q}{1-q}\,du;$$

so that

$$\frac{1+q}{1-q}f(u').(v-u') + \frac{1+q}{1-q}f(u'').(v+u'') - 2vf(u) = 0.$$

Now

$$v - u' = \frac{1+q}{1-q}(v-u), \qquad v + u'' = \frac{1+q}{1-q}(v+u),$$

and thus

$$\frac{(1+q)^2}{(1-q)^2}\{(v-u)\,f(u') + (v+u)\,f(u'')\} - 2vf(u) = 0.$$

In this

$$f(u') = f(u) + \frac{2q(u-v)}{1-q}f'(u) + \frac{2q^2(u-v)^2}{(1-q)^2}f''(u) + \ldots\ldots$$

$$f(u'') = f(u) + \frac{2q(u+v)}{1-q}f'(u) + \frac{2q^2(u+v)^2}{(1-q)^2}f''(u) + \ldots\ldots;$$

so that

$$\frac{(1+q)^2}{(1-q)^2}\left\{2vf(u) + \frac{8qv}{1-q}uf'(u) + \frac{4q^2v^3}{(1-q)^2}f''(u)\right\} - 2vf(u) = 0,$$

or, when q is small enough,

$$8qv\{f(u) + uf'(u)\} + 4q^2v^3f''(u) = 0. \quad \text{..............(21)}$$

Accordingly

$$f(u) + uf'(u) + \tfrac{1}{2}qv^2f''(u) = 0, \quad \text{....................(22)}$$

or on integration

$$uf(u) + \tfrac{1}{2}qv^2f'(u) = C.$$

It is easy to recognize that the constant C of integration must vanish. On putting $u=0$, its value is seen to be

$$C = \tfrac{1}{2}qv^2f'(0),$$

for $f(0)$ is not infinite. Now $f(u)$ is by its nature an even function of u, so that $f'(0)$ must vanish. We thus obtain the same equation (14) of the first order as by the former process.

Progress towards the Stationary State.

Passing from the consideration of the steady state, we will now suppose that the masses are initially at rest, and examine the manner in which they acquire velocity under the impact of the projectiles. In the very early stages of the process the momentum acquired during one collision is practically independent of the existing velocity (u) of a mass, and may be taken to be $\pm 2qv$. Moreover, the chance of a collision is at first sensibly independent of u. In the present investigation we are concerned not merely, as in considering the ultimate state, with the mass and velocity of a projectile, but also with the frequency of impact. We will denote by ν the whole number of projectiles launched in both directions in the unit of time in the path of each mass. The chance of a collision for a given mass in time dt is thus represented by $\nu\,dt$. The number of collisions by which masses are expelled from the range du in time dt is $f(u)\,du\,.\,\nu\,dt$. The number which enter the range from the two sides is

$$\{f(u-2qv)+f(u+2qv)\}\,du\,.\,\tfrac{1}{2}\nu\,dt,$$

so that the excess of the number which enter the range over the number which leave is

$$\{\tfrac{1}{2}f(u-2qv)+\tfrac{1}{2}f(u+2qv)-f(u)\}\,du\,.\,\nu\,dt,$$

and this is to be equated to $\dfrac{df(u,t)\,du}{dt}\,dt$. Thus

$$\frac{df}{\nu\,dt}=\tfrac{1}{2}f(u-2qv)+\tfrac{1}{2}f(u+2qv)-f(u)=2q^2v^2\frac{d^2f}{du^2}, \quad \text{......(23)}$$

the well-known equation of the conduction of heat. When $t=0$, $f(u)$ is to be zero for all finite values of u. The Fourier solution, applicable under these conditions, is

$$f(u,t')=\frac{A}{\sqrt{t'}}\,e^{-u^2/4t'},$$

where t' is written for $2q^2v^2\nu t$. The total number of masses being N, we get to determine A

$$N=\int_{-\infty}^{+\infty}f(u,t')\,du=2\sqrt{\pi}\,.\,A\,;$$

so that

$$f(u,t')=\frac{N}{2\sqrt{(\pi t')}}\,e^{-u^2/4t'}. \quad \text{......................(24)}$$

If n be the whole number of collisions (for each mass), $n=\nu t$, and we have

$$4t'=4q^2v^2\,.\,2n. \quad \text{............................(25)}$$

If the unit of velocity be so chosen that the momentum ($2qv$) communicated at each impact is unity, (24) takes the form

$$f(u,n)=\frac{N}{\sqrt{(2\pi n)}}\,e^{-u^2/2n}, \quad \text{......................(26)}$$

which exhibits the distribution of momentum among the masses after n impacts. In this form the problem coincides with one formerly treated* relating to the composition of vibrations of arbitrary phases. It will be seen

* *Phil. Mag.* August, 1880, p. 73. [Vol. I. p. 491.]

that there is a sharp contrast between the steady state and the early stages of the variable state. The latter depends upon the *momentum* of the projectiles, and upon the number of impacts; the former involves the *energy* of the projectiles, and is independent of the rapidity of the impacts.

The mean square of velocity after any number (n) of impacts is

$$N^{-1}\int_{-\infty}^{+\infty} u^2 f(u, n)\, du = n,$$

or, if we restore $4q^2v^2$,

$$\text{mean } u^2 = n \,.\, 4q^2v^2. \qquad (27)$$

It must be distinctly understood that the solution expressed by (24), (25), (26) applies only to the first stages of the bombardment, beginning with the masses at rest. If the same state of things continued, the motion of the masses would increase without limit. But, as time goes on, two causes intervene to prevent the accumulation of motion. When the velocity of the masses becomes sensible, the chance of an unfavourable collision increases at the expense of the favourable collisions, and this consideration alone would prevent the unlimited accumulation of motion, and lead to the ultimate establishment of a steady state. But another cause is also at work in the same direction, and, as may be seen from the argument which leads to (13), with equal efficiency. The favourable collisions, even when they occur, produce less effect than the unfavourable ones, as is shown by (6) and (7).

We will now investigate the general equation, applicable not merely to the initial and final, but to all stages of the acquirement of motion. As in (20), (23) we have

$$\frac{df(u, t)\, du}{dt}\, dt = \frac{\nu\, dt}{v}\left\{\tfrac{1}{2}f(u')\,.\,(v-u')\, du' + \tfrac{1}{2}f(u'')\,.\,(v+u'')\, du'' - f(u)\,.\, v\, du\right\};$$

and thus by the same process as for (22)

$$\frac{df}{\nu\, dt} = 4q\frac{d}{du}\{uf(u)\} + 2q^2v^2\frac{d^2f}{du^2}. \qquad (28)$$

If we write, as before,

$$t' = 2q^2v^2\nu t, \quad\text{and}\quad h = 1/qv^2, \qquad (29)$$

we have

$$\frac{df}{dt'} = \frac{d^2f}{du^2} + 2h\frac{d}{du}(uf). \qquad (30)$$

Both in the case where the left side was omitted, and also when h vanished, we found that the solution was of the form

$$f = \sqrt{\phi}\,.\, e^{-\phi u^2}, \qquad (31)$$

where ϕ was constant, or a function of t' only. We shall find that the same form applies also to the more general solution. The factor $\sqrt{\phi}$ is evidently necessary in order to make $\int_{-\infty}^{+\infty} f(u)\,du$ independent of the time. By differentiation of (31),

$$df/dt' = \tfrac{1}{2}\phi^{-\frac{1}{2}}e^{-\phi u^2}(1 - 2\phi u^2)\,d\phi/dt',$$

$$d^2f/du^2 = -2\phi^{\frac{3}{2}}e^{-\phi u^2}(1 - 2\phi u^2),$$

$$f + df/du = \phi^{\frac{1}{2}}e^{-\phi u^2}(1 - 2\phi u^2);$$

so that (30) is satisfied provided ϕ is so chosen as a function of t' that

$$\tfrac{1}{2}\phi^{-\frac{1}{2}}\frac{d\phi}{dt'} = -2\phi^{\frac{3}{2}} + 2h\phi^{\frac{1}{2}},$$

or

$$\frac{1}{4\phi^2}\frac{d\phi}{dt'} = -1 + \frac{h}{\phi}.$$

Thus

$$4t' = \int \frac{d\phi^{-1}}{1 - h\phi^{-1}} = -\frac{1}{h}\log(1 - h\phi^{-1}) + \text{const.},$$

where, however, the constant must vanish, since $\phi = \infty$ corresponds to $t' = 0$. Accordingly

$$\phi = \frac{h}{1 - e^{-4ht'}}, \quad \text{..............................(32)}$$

which with (31) completes the solution.

If t' is small, (32) gives $\phi = 1/4t'$, in agreement with (24); while if t' be great, we have $\phi = h = 1/qv^2$, as in (15′).

The above solution is adapted to the case where $f(u) = 0$ for all finite values of u, when $t' = 0$. The next step in the process of generalization will be to obtain a solution applicable to the initial concentration of $f(u)$, no longer merely at zero, but at any arbitrary value of u; that is, to the case where initially all the masses are moving with one constant velocity α.

Assume

$$f = \sqrt{\phi}\,.\,e^{-\phi(u-\psi)^2}, \quad \text{..............................(33)}$$

where ϕ, ψ are functions of t' only. Substituting, as before, in (30), we find

$$\{1 - 2\phi(u-\psi)^2\}\left\{\tfrac{1}{2}\frac{d\phi}{dt'} + 2\phi^2 - 2h\phi\right\} + 2\phi^2(u-\psi)\left\{\frac{d\psi}{dt} + 2h\psi\right\} = 0;$$

so that the equation is satisfied provided

$$\tfrac{1}{2}\frac{d\phi}{dt'} + 2\phi^2 - 2h\phi = 0, \quad \text{........................(34)}$$

and

$$\frac{d\psi}{dt'} + 2h\psi = 0. \quad \text{..............................(35)}$$

The first is the same equation as we found before, and its solution is given by (32); while (35) gives

$$\psi = \alpha e^{-2ht'}. \quad \text{..................................(36)}$$

Thus (32), (33), (36) constitute the complete solution of the problem proposed, and show how the initial concentration at $u=\alpha$ passes gradually into the steady state when $t'=\infty$. In the early stages of the process

$$f(u, t')=\frac{1}{\sqrt{(4t')}}e^{-(u-\alpha)^2/4t'}; \quad \text{.......................(37)}$$

to which the factor $N/\sqrt{\pi}$ may be applied, when it is desired to represent that the whole number of masses is N. It appears that during the whole process the law of distribution is in a sense maintained, the only changes being in the value of u round which the grouping takes place, and in the degree of concentration about that value.

There will now be no difficulty in framing the expression applicable to an arbitrary initial distribution of velocity among the masses. For this purpose we need only multiply (33) by $\chi(\alpha)\,d\alpha$, and integrate over the necessary range. Thus

$$f(u, t')=\sqrt{\phi}\,.\int_{-\infty}^{+\infty}d\alpha\,\chi(\alpha)\,Exp\,\{-\phi\,(u-\alpha e^{-2ht'})^2\}, \quad \text{.........(38)}$$

ϕ being given, as usual, by (32). The limits for α are taken $\pm\infty$; but we must not forget that the restriction upon the magnitude of u requires that $\chi(u)$ shall be sensible only for values of u small in comparison with v.

When t' is small, we have from (38),

$$f(u, t')=\frac{1}{\sqrt{(4t')}}\int_{-\infty}^{+\infty}d\alpha\,\chi(\alpha)\,e^{-(u-\alpha)^2/4t'}=\sqrt{\pi}\,.\,\chi(u)$$

ultimately; so that

$$\chi(\alpha)=\frac{1}{\sqrt{\pi}}f(\alpha, 0).$$

Accordingly the required solution expressing the distribution of velocity at t' in terms of that which obtains when $t'=0$, is

$$f(u, t')=\sqrt{\frac{\phi}{\pi}}\,.\int_{-\infty}^{+\infty}d\alpha\,f(\alpha, 0)\,Exp\,\{-\phi\,(u-\alpha e^{-2ht'})^2\}. \quad \text{......(39)}$$

We may verify this by supposing that $f(u, 0)=e^{-hu^2}$, representing the steady state. The integration of (39) then shows that

$$f(u, t')=e^{-hu^2},$$

as of course should be.

An example of more interest is obtained by supposing that initially

$$f(u, 0)=e^{-h'u^2}; \quad \text{..............................(40)}$$

that is, that the velocities are in the state which would be a steady state under the action of projectiles moving with an energy different from the actual energy. In this case we find from (32), (39),

$$f(u, t')=\sqrt{\left(\frac{\phi}{\phi-h+h'}\right)}\,e^{-\frac{\phi h'u^2}{\phi-h+h'}}. \quad \text{..................(41)}$$

We will now introduce the consideration of variable velocity of projectiles into the problem of the progressive state. In (28) we must regard ν as a function of v. If we use $\nu\, dv$ to denote the number of projectiles launched in unit of time with velocities included between v and $v + dv$, (28) may be written

$$\frac{df}{dt} = 4q \int \nu\, dv \,.\, \frac{d}{du}\{uf(u)\} + 2q^2 \int \nu v^2 dv \,.\, \frac{d^2f}{du^2}, \quad \text{...........(42)}$$

which is of the same form as before. The only difference is that we now have in place of (29),

$$t' = 2q^2 t \int \nu v^2 dv, \quad \text{.................................(43)}$$

$$h = \int \nu\, dv \div q \int \nu v^2 dv. \quad \text{..........................(44)}$$

In applying these results to particular problems, there is an important distinction to be observed. By definition $\nu\, dv$ represents the number of projectiles which in the unit *time* pass a given place with velocities included within the prescribed range. It will therefore not represent the distribution of velocities in a given *space*; for the projectiles, passing in unit time, which move with the higher velocities cover correspondingly greater spaces. If therefore we wish to investigate the effect of a Maxwellian distribution of velocities among the projectiles, we are to take, not $\nu = Be^{-kv^2}$, but

$$\nu = B\, v\, e^{-kv^2}. \quad \text{..............................(45)}$$

In this case, by (18),

$$h = k/q; \quad \text{...................................(46)}$$

and, as we saw, the mean energy of a mass in the steady state is equal to the mean energy of the projectiles which at any moment of time occupy a given space. From (43),

$$t' = Bq^2 k^{-2} t. \quad \text{...............................(47)}$$

Pendulums in place of Free Masses.

We will now introduce a new element into the question by supposing that the masses are no longer free to wander indefinitely, but are moored to fixed points by similar elastic attachments. And for the moment we will assume that the stationary state is such that no change would occur in it were the bombardment at any time suspended. To satisfy this condition it is requisite that the phases of vibrations of a given amplitude should have a certain distribution, dependent upon the law of force. For example, in the simplest case of a force proportional to displacement, where the velocity u is connected with the amplitude (of velocity) r and with the phase θ by the relation $u = r \cos\theta$, the distribution must be uniform with respect to θ, so that the number of vibrations in phases between θ and $\theta + d\theta$ must be $d\theta/2\pi$

of the whole number whose amplitude is r. Thus, if r be given, the proportional number with velocities between u and $u + du$ is

$$\frac{du}{2\pi\sqrt{(r^2 - u^2)}}. \qquad\qquad (48)$$

And, in general, if r be some quantity by which the amplitude is measured, the proportional number will be of the form

$$\phi(r, u)\, du, \qquad\qquad (49)$$

where ϕ is a determinate function of r and u, dependent upon the law of vibration. If now $\chi(r)\,dr$ denote the number of vibrations for which r lies between r and $r + dr$, we have altogether for the distribution of velocities u,

$$f(u) = \int \chi(r)\, \phi(r, u)\, dr. \qquad\qquad (50)$$

If the vibrators were left to themselves, $\chi(r)$ might be chosen arbitrarily, and yet the distribution of velocity, denoted by $f(u)$, would be permanent. But if the vibrators are subject to bombardment, $f(u)$ cannot be permanent, unless it be of the form already determined. The problem of the permanent state may thus be considered to be the determination of $\chi(r)$ in (50), so as to make $f(u)$ equal to e^{-hu^2}.

We will now limit ourselves to a law of force proportional to displacement, so that the vibrations are isochronous; and examine what must be the form of $\chi(r)$ in (8) in order that the requirements of the case may be satisfied.

By (15′), if N be the whole number of vibrators,

$$\frac{N\sqrt{h}}{\sqrt{\pi}} e^{-hu^2} = \int_u^\infty \frac{\chi(r)\, dr}{2\pi\sqrt{(r^2 - u^2)}}. \qquad\qquad (51)$$

The determination of the form of χ is analogous to a well-known investigation in the theory of gases. We assume

$$\chi(r) = Ar\, e^{-hr^2}, \qquad\qquad (52)$$

where A is a constant to be determined. To integrate the right-hand member of (51), we write

$$r^2 = u^2 + \eta^2; \qquad\qquad (53)$$

so that

$$\int_u^\infty \frac{\chi(r)\, dr}{\sqrt{(r^2 - u^2)}} = A \int_0^\infty e^{-h(u^2+\eta^2)}\, d\eta = \frac{A\sqrt{\pi}}{2\sqrt{h}} e^{-hu^2}.$$

Thus

$$A = 4hN. \qquad\qquad (54)$$

The distribution of the amplitudes (of velocity) is therefore such that the number of amplitudes between r and $r + dr$ is

$$N \,.\, 4hr\, e^{-hr^2}\, dr, \qquad\qquad (55)$$

while for each amplitude the phases are uniformly distributed round the complete cycle.

The argument in the preceding paragraphs depends upon the assumption that a steady state exists, which would not be disturbed by a suspension, or relaxation, of the bombardment. Now this is a point which demands closer examination; because it is conceivable that there may be a steady state, permanent so long as the bombardment itself is steady, but liable to alteration when the rate of bombardment is increased or diminished. And in this case we could not argue, as before, that the distribution must be uniform with respect to θ.

If x denote the displacement of a vibrator at time t,

$$x = n^{-1} r \sin(nt - \theta), \qquad dx/dt = r\cos(nt-\theta).$$

When $t=0$,

$$x = -n^{-1} r \sin\theta, \qquad dx/dt = u = r\cos\theta;$$

and we may regard the amplitude and phase of the vibrator as determined by u, η where

$$u = r\cos\theta, \quad \eta = r\sin\theta.$$

Any distribution of amplitudes and phases may thus be expressed by $f(u, \eta)\, du\, d\eta$.

If we consider the effect of the collisions which may occur at $t=0$, we see that u is altered according to the laws already laid down, while η *remains unchanged.* The condition that the distribution remains undisturbed by the collisions is, as before, that, for every constant η, $f(u, \eta)$ should be of the form e^{-hu^2}, or, as we may write it,

$$f(u, \eta) = \chi(\eta)\, e^{-hu^2}.$$

But this condition is not sufficient to secure a stationary state, because, even in the absence of collisions, a variation would occur, unless $f(u, \eta)$ were a function of r, independent of θ. Both conditions are satisfied, if $\chi(\eta) = A\, e^{-h\eta^2}$, where A is a constant; so that

$$f(u, \eta)\, du\, d\eta = A\, e^{-h(u^2+\eta^2)}\, du\, d\eta = 2\pi A\, e^{-hr^2} r\, dr.$$

Under this law of distribution there is no change either from the progress of the vibrations themselves, or as the result of collisions.

The principle that the distribution of velocities in the stationary state is the same as if the masses were free is of great importance, and leads to results that may at first appear strange. Thus the mean kinetic energies of the masses is the same in the two cases, although in the one case there is an accompaniment of potential energy, while in the other there is none. But it is to be observed that nothing is here said as to the rate of progress towards the stationary condition when, for instance, the masses start from rest; and the fact that the ultimate distribution of velocities should be independent of the potential energy is perhaps no more difficult to admit than its independence of the number of projectiles which strike in a given time. One difference may, however, be alluded to in passing. In the case of the vibrators it is necessary to suppose that the collisions are instantaneous; while the result for the free masses is independent of such a limitation.

The simplicity of f in the stationary state has its origin in the independence of θ. It is not difficult to prove that this law of independence fails during the development of the vibrations from a state of rest under a vigorous bombardment. The investigation of this matter is accordingly more complicated than in the case of the free masses, and I do not propose here to enter upon it.

In a modification of the original problem of some interest even the stationary distribution is not entirely independent of phase. I refer to the case where the bombardment is from one side only, or (more generally) is less vigorous on one side than on the other. It is easy to see that a one-sided bombardment would of necessity disturb a uniform distribution of phase, even if it were already established. The permanent state is accordingly one of unequal phase-distribution, and is not, as for the symmetrical bombardment, independent of the vigour with which the bombardment is conducted.

But in one important particular case the simplicity of the symmetrical bombardment is recovered. For if the number of projectiles striking in a given time be sufficiently reduced, the stationary condition must ultimately become one of uniform phase-distribution.

Under this limitation it is easy to see what the stationary state must be. Since the ultimate distribution is uniform with respect to phase, it must be the same from whichever side the bombardment comes. Under these circumstances it could not be altered if the bombardment proceeded indifferently from both sides, which is the case already investigated. We conclude that, *provided the bombardment be very feeble*, there is a definite stationary condition, independent both of the amount of the bombardment and of its distribution between the two directions. It is of course understood that from whichever side a projectile be fired, the moment of firing is absolutely without relation to the phase of the vibrator which it is to strike.

The problem of the one-sided bombardment may also be attacked by a direct calculation of the distribution of amplitude in the stationary condition. The first step is to estimate the effect upon the amplitude of a given collision. From (6), if u' be the velocity before collision, and u after,

$$u = u' + \frac{2q}{1+q}(v - u').$$

The fraction $2q/(1+q)$ occurs as a whole, and we might retain it throughout. But inasmuch as in the final result only one power of q need be retained, it will conduce to brevity to omit the denominator at once, and take simply

$$u = u' + 2q\,(v - u'). \qquad \text{.........................(56)}$$

Thus if ρ, ϕ and r, θ be the amplitude and phase before and after collision respectively,

$$\left.\begin{aligned} r\cos\theta &= \rho\cos\phi + 2q\,(v - \rho\cos\phi), \\ r\sin\theta &= \rho\sin\phi\,; \end{aligned}\right\}\ldots\ldots\ldots\ldots\ldots\ldots(57)$$

so that

$$r^2 = \rho^2 + 4q\rho\cos\phi\,(v - \rho\cos\phi) + 4q^2\,(v - \rho\cos\phi)^2.$$

From this we require the approximate value of ρ in terms of r and ϕ. The term in q^2 cannot be altogether neglected, but it need only be retained when multiplied by v^2. The result is $\rho = r - \delta r$, where

$$\delta r = 2q\,(v\cos\phi - r\cos^2\phi) + \frac{2q^2v^2}{r}\sin^2\phi.\ \ldots\ldots\ldots\ldots(58)$$

This equation determines for a given ϕ the value of ρ which the blow converts into r. Values of ρ nearer to r will be projected across that value. The chance of a collision at ρ, ϕ is proportional to $(v - \rho\cos\phi)$. Thus if a number of vibrators in state ρ, ϕ be $F(\rho)\,d\rho\,d\phi$*, the condition for the stationary state is

$$\int_0^{2\pi} d\phi \int_\rho^r (v - \rho\cos\theta)\,F(\rho)\,d\rho = 0,\ \ldots\ldots\ldots\ldots\ldots(59)$$

the integral on the left expressing the whole number (estimated algebraically) of amplitudes which in a small interval of time pass outwards through the value r.

By expansion of $F(\rho)$ in the series

$$F(\rho) = F(r) + F'(r)\,(\rho - r) + \ldots,$$

we find

$$\int_\rho^r F(\rho)\,d\rho = F(r)\,\delta r - \tfrac{1}{2}F'(r)\,(\delta r)^2 + \text{cubes of } q,$$

$$\int_\rho^r \rho F(\rho)\,d\rho = r\,F(r)\,\delta r - \tfrac{1}{2}(\delta r)^2\,\{F(r) + rF'(r)\} + \text{cubes of } q.$$

Again from (58),

$$\int_0^{2\pi} \delta r\,d\phi = -\,qr + q^2v^2/r, \qquad \int_0^{2\pi} \cos\phi\,\delta r\,d\phi = qv,$$

$$\int_0^{2\pi} (\delta r)^2\,d\phi = 2q^2v^2, \qquad \int_0^{2\pi} \cos\phi\,(\delta r)^2\,d\phi = 0.$$

The condition for the stationary state is therefore

$$v\,\{F(r)(-\,qr + q^2v^2/r) - F'(r)\,q^2v^2\} - r\,F(r)\,qv = 0,$$

or

$$F(r)\,\{-\,2r + qv^2/r\} - F'(r)\,qv^2 = 0.$$

Thus, on integration,

$$r^2 - qv^2\log r + qv^2\log F(r) = \text{const.},\ \ldots\ldots\ldots\ldots\ldots\ldots(60)$$

or

$$F(r) = Ar\,e^{-r^2/qv^2}\ldots\ldots\ldots\ldots\ldots\ldots\ldots\ldots\ldots(61)$$

* We here assume that the bombardment is feeble.

The mean value of r^2, expressed by

$$\int_0^\infty r^3 F(r)\,dr \div \int_0^\infty r\,F(r)\,dr,$$

is qv^2; that is, the mean value of the maximum kinetic energy attained during the vibration is equal to the kinetic energy of a projectile. The mean of all the actual kinetic energies of the vibrators is the half of this; but would rise to equality with the mean energy of the projectiles, if the velocities of the latter, instead of being uniform, as above supposed, were distributed according to the Maxwellian law.

If we are content to assume the *law* of distribution, $\rho\, e^{-h\rho^2}$, leaving only the constant h to be determined, the investigation may be much simplified. Thus from (57) the gain of energy from the collision is

$$\tfrac{1}{2}r^2 - \tfrac{1}{2}\rho^2 = 2q\rho\cos\phi\,(v - \rho\cos\phi) + 2q^2v^2.$$

The chance of the collision in question is proportional to the relative velocity $(v - \rho\cos\phi)$; and in the stationary state the whole gain of energy is zero. Hence

$$\iint \rho\, e^{-h\rho^2}\, d\rho\, d\phi\, \{2q\rho\cos\phi\,(v - \rho\cos\phi)^2 + 2q^2v^3\} = 0.$$

In the integration with respect to ϕ the odd powers of $\cos\phi$ vanish. Hence

$$2qv \int_0^\infty \rho\, d\rho\, e^{-h\rho^2}\,(qv^2 - \rho^2) = 0\,;$$

so that

$$h = 1/qv^2,$$

as in (61).

184.

EXPERIMENTS IN AËRODYNAMICS*.

[*Nature*, XLV. pp. 108, 109, 1891.]

THE subject of this memoir is of especial interest at the present time, when the skill of a distinguished inventor is understood to be engaged in attacking the many practical difficulties which lie in the way of artificial flight upon a large scale. For a long time the resistance of fluids formed an unsatisfactory chapter in our treatises on hydrodynamics. According to the early suggestions of Newton, the resistances are (1) proportional to the surfaces of the solid bodies acted upon, to the densities of the fluids, and to the squares of the velocities: while (2) "the direct impulse of a fluid on a plane surface is to its absolute oblique impulse on the same surface as the square of the radius to the square of the sine of the angle of incidence." The author of the work† from which these words are quoted, in comparing the above statements with the experimental results available in his time (1822), remarks:—"(1) It is very consonant to experiment that the resistances are proportional to the squares of the velocities.... (2) It appears from a comparison of all the experiments, that the impulses and resistances are very nearly in the proportion of the surfaces.... (3) The resistances do by no means vary in the duplicate ratio of the sines of the angle of incidence." And he subsequently states that for small angles the resistances are more nearly proportional to the sines of incidence than to their squares.

It is probable that the law of velocity tended to support in men's minds the law of the square of the sine. For, if both be admitted, it follows that the resistance, normal to the surface, experienced by a plane when immersed in a stream of fluid, depends only upon the *component* of the velocity perpendicular to the surface. That the effect should be independent

* *Experiments in Aërodynamics.* By S. P. Langley. *Smithsonian Contributions to Knowledge* (Washington, 1891).

† *System of Mechanical Philosophy.* By John Robison, Vol. II., 1822.

of the component parallel to the plane seems plausible, inasmuch as this component, *if it existed alone*, would exercise no pressure: but that such a view is entirely erroneous has been long recognized by practical men, especially by those concerned in navigation.

From the law of the simple sine, enunciated by Robison, it follows at once that the pressure upon a lamina exposed perpendicularly to a stream may be increased *to any extent* by imparting to the lamina a sufficiently high velocity *in its own plane*. The immense importance of this principle was clearly recognized by Mr Wenham in his valuable paper upon flight*, and a few years later the whole subject was discussed by the greatest authority upon such matters, the late Mr W. Froude, with characteristic insight and lucidity†.

The theoretical problem of determining the resistance from the first principles of hydrodynamics is not free from difficulty, even in the case of two dimensions, where a long rectangular lamina is exposed obliquely to a stream whose direction is perpendicular to the longer sides. The formula‡ resulting from the theory of Kirchhoff, viz.

$$\frac{\pi \sin \alpha}{4 + \pi \sin \alpha} \rho V^2, \qquad \text{(1)}$$

where ρ is the density of the fluid, and V is the total velocity of the stream flowing at the angle α with the plane of the lamina, shows that when α is small the resistance is nearly proportional to $\sin \alpha$. Moreover, (1) agrees with the experiments of Vince§.

It will be seen that the laws of resistance were fairly well established many years ago, at least in their main outlines. Nevertheless, there was ample room for the systematic and highly elaborate experiments recorded in the memoir whose title stands at the head of this article. The work appears to have been executed with the skill and thoroughness which would naturally be expected of the author, and will doubtless prove of great service to those engaged upon these matters. The scanty reference to previous knowledge, which Prof. Langley holds out some promise of extending in subsequent publications, makes it rather difficult to pick out the points of greatest novelty. The main problem is, of course, the law of obliquity, and this is attacked with two distinct forms of apparatus. The general character of the results, exhibited graphically on p. 62, will be made apparent from the accompanying reproduction, in which are added a curve D, corresponding to (1) and E, representing the law of $\sin^2 \alpha$. In each case the

* *Report of Aëronautical Society for* 1866.

† *Proc. Inst. Civ. Eng.* 1871 (discussion upon a paper by Sir F. Knowles). [See Vol. I. p. 290.]

‡ See *Phil. Mag.* December, 1876 [Vol. I. p. 291]. Also Basset's *Hydrodynamics*, Vol. I. p. 131.

§ *Phil. Trans.* 1798.

abscissa is the angle α and the ordinate is the normal pressure, expressed as a percentage of that experienced when $\alpha = 90°$. Of Prof. Langley's curves,

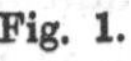
Fig. 1.

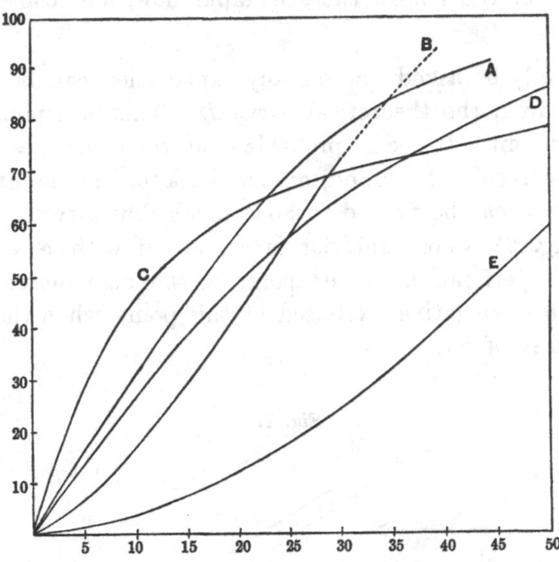

A relates to a *square* plane 12 inches × 12 inches, *B* to a rectangle 6 × 24 inches, and *C* to a rectangle 30 × 4·8 inches, the leading edge (perpendicular to the stream) being in each case specified first, so that the theoretical curve *D* corresponds most nearly to *C*. It will be seen at a glance that at small angles the pressure is enormously greater than according to the law of $\sin^2 \alpha$. The differences between *A*, *B*, *C*, anticipated in a general manner by Wenham and Froude, are of great interest. They demonstrate that in proportion to area a long narrow wing is more efficient as a support than a short wide one, and that in a very marked degree.

Up to a certain point there is no difficulty in giving a theoretical account of these features. When a rectangular lamina is exposed perpendicularly, there is one point, *i.e.* the centre, at which the velocity of the stream is annulled. At this point the pressure attains the full amount, $\frac{1}{2}\rho V^2$, due to the velocity of the stream, while at every other point the pressure is less, and falls to zero at the boundary. If the lamina is sloped to the stream as in *B* and *C*, there is still a median plane of symmetry; and at one point in this plane, but now in advance of the centre, the full pressure is experienced. In strictness, there is only one point of maximum pressure, whatever may be the proportions of the lamina. But if the rectangle be very elongated, there is practically a great difference in this respect according to the manner of presentation, although the small angle α be preserved unchanged. For when the long edges are perpendicular to the stream (*C*), the motion is nearly in two dimensions, and the region of nearly maximum pressure extends over most of the length. But the case is obviously quite different when it is the short dimension that is perpendicular to the stream, for then along the greater part of the length there is rapid flow, and consequently small pressure.

It will naturally be asked whether any explanation can be offered of the divergence of C from the theoretical curve D. This is a point well worthy of further experiment. It seems probable that the cause lies in the suction operative, as the result of friction, at the back of the lamina. That the suction is a reality may be proved without much difficulty by using a hollow lamina, AB (Fig. 2), whose interior is connected with a manometer. If there be a small perforation at any point C, the manometer indicates the pressure, positive or negative, exercised at this point, when the apparatus is exposed to a blast of air.

Fig. 2.

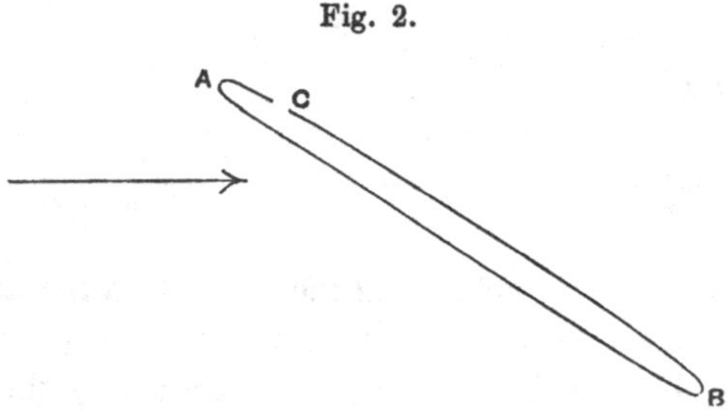

When once the law ot obliquities is known, the problem of aërial maintenance presents no further theoretical difficulty. It was successfully treated many years ago by Penaud*, and somewhat later by Froude, whose interesting letters, written shortly before his death, have recently been published†. In perhaps the simplest form of the question the level is supposed to be maintained with the aid, *e.g.*, of screw propulsion, the necessary maintenance being secured by an aëroplane slightly tilted (α) upwards in front. The work required to be expended in order to maintain a given weight depends upon the area of plane, the inclination, and the speed. Penaud's results show that, *if skin friction could be neglected*, the necessary work might be diminished indefinitely, even with a given area of wing. For this purpose, it would only be necessary to increase the speed and correspondingly to diminish α. But when skin friction is taken into account, the work can only be reduced to a minimum, and to do this with a given area of wing requires a definite (large) velocity, and a definite (small) inclination. The accurate determi-

* See *Report of Aëronautical Society for* 1876.

† *Edinburgh Proceedings.* R. E. Froude, 1891.

nation of the tangential, as well as of the normal, force experienced by an inclined plane is thus of essential importance in the question of flight.

The work of Penaud seems to be so little known that it has been thought desirable to recapitulate some of his theoretical conclusions. But we owe to Penaud not merely sound theory, but the actual construction of a successful flying machine, in which horizontal flight is maintained by a screw propeller. In these models the energy is stored by means of stretched india-rubber, a method available only upon a small scale. It is probable that the principle of the rocket might be employed with advantage; and even upon a large scale, the abolition of all machinery would allow of considerable extravagance in the use of explosive material. This method is especially adapted to the very high speeds which on other grounds are most suitable.

In the chapter on "The Plane Dropper" some striking experiments are described, illustrating the effect of a forward movement in retarding the fall of a horizontal plane. Prof. Langley seems hardly to recognize that there is nothing really distinctive in this arrangement when he says:—

"It is, of course, an entirely familiar observation that we can support an inclined plane by moving it laterally, deriving our support in this case from the upward component of pressure derived from the wind of advance; but, so far as I am aware, this problem of the velocity of fall of a horizontal plane moving horizontally in the air has never been worked out theoretically or determined experimentally, and I believe that the experimental investigation whose results I am now to present is new."

But, apart from the complications which attend the establishment of a uniform *régime*, there is no essential difference between the two cases. The hydrodynamical forces depend only upon the magnitude of the relative velocity and upon the inclination of this relative velocity to the plane. All else is a question merely of ordinary elementary mechanics.

It is interesting to note that Prof. Langley's experience has led him to take a favourable view of the practicability of flight upon a large scale. Such was also the opinion of Penaud, who (in 1876) expresses his conviction "that, in the future more or less distant, science will construct a light motor that will enable us to solve the problem of aviation." But sufficient maintaining power is not the only requisite; and it is probable that difficulties connected with stability, and with safe alighting at the termination of the adventure, will exercise to the utmost the skill of our inventors.

[1901. Some of the problems here referred to are further discussed in the Wilde Lecture on the *Mechanical Principles of Flight* (*Manchester Proceedings*, Vol. XLIV. Part II. pp. 1—26, 1899.]

185.

ON REFLEXION FROM LIQUID SURFACES IN THE NEIGHBOURHOOD OF THE POLARIZING ANGLE*.

[*Philosophical Magazine,* XXXIII. pp. 1—19, Jan. 1892.]

By the experiments of Jamin and others it has been abundantly proved that in the neighbourhood of the polarizing angle the reflexion of light from ordinary transparent liquids and solids deviates sensibly from the laws of Fresnel, according to which the reflexion of light polarized perpendicularly to the plane of incidence should vanish when the incidence takes place at the Brewsterian angle. It is found, on the contrary, that in most cases the residual light is sensible at all angles, and that the change of phase by 180°, which, according to Fresnel's formula, should occur suddenly, in reality enters by degrees, so that in general plane-polarized light acquires after reflexion a certain amount of ellipticity. Although Jamin describes the non-evanescence at the polarizing angle and the ellipticity in its neighbourhood as "deux ordres de phénomènes de nature différente," it is clear that they are really inseparable parts of one phenomenon. If we suppose the incident light polarized perpendicularly to the plane of incidence to be given, the vibration which determines the reflected light at various angles may be represented in amplitude and phase by the situation of points relatively to an origin and coordinate axes. Thus, according to Fresnel's formula, the locus of these points is the axis of abscissæ XX' itself, the point O corresponding to the polarizing angle, at which the reflexion vanishes, and in passing which there is a sudden change of phase of 180°. If the reflexion remains finite at all angles, the curve in question meets the axis YY' at some point P, not coincident with O, and the corresponding phase differs by a quarter-period from the phases met with at a distance from this angle. So

* [1901. A preliminary account of the experiments here detailed was given before the British Association (see *Report*, Aug. 21, 1891, p. 563). The principal observation had been recorded still earlier (*Phil. Mag.* Nov. 1890; Vol. III. p. 396).]

far as experiment can yet show, this curve may be a straight line parallel to XX', and at a short distance from it. If it lie above XX', the reflexion is what Jamin characterizes as positive; if below, the reflexion is negative.

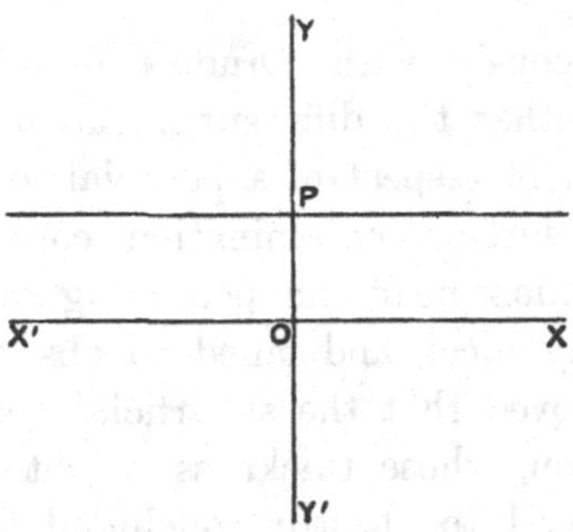

To this order of approximation the behaviour of a transparent body reflecting light of given wave-length is determined by two constants, (1) the refractive index μ, and (2) the intensity of reflexion at the angle $\tan^{-1}\mu$ when the light is polarized perpendicularly to the plane of incidence. The most convenient form of the second constant for experimental purposes is the ratio of reflected amplitudes for the two principal planes when the light, incident at the angle $\tan^{-1}\mu$, is polarized at 45° to these planes. It may be called the ellipticity, and, after Jamin, will be denoted by k. According to Fresnel $k=0$; but Jamin found for water $k=-\cdot 00577$, and for absolute alcohol $k=+\cdot 00208$. Contrasting liquids with solids, he remarks*, "on vient de voir que leur polarisation est elliptique, et qu'il est impossible d'en trouver la cause dans une constitution moléculaire anormale." And, again:—"Il est jusqu'à présent impossible de constater une relation simple entre la valeur du coefficient k et l'indice de réfraction; tout porte à croire, au contraire, que ces deux constantes sont indépendantes, l'une de l'autre. Mais, à défaut de loi précise, on peut remarquer une tendance du coefficient k à diminuer avec l'indice, et quand celui-ci est approximativement égal à 1·45, k est nul: l'indice continuant à décroître, le coefficient k reprend des valeurs sensibles et croissantes."

Since the time of Jamin many valuable observations upon reflexion have been made by Quincke and others, which it is not necessary for my purpose further to refer to. In 1889† Drude made the very important observation that the ellipticity of a freshly split surface of rock-salt is very small, but that it rapidly increases on standing. And he concludes generally that solid bodies reflect from natural cleavage surfaces according to the laws of Fresnel.

* "Mémoire sur la réflexion de la Lumière à la surface des Liquides." *Ann. Chim.* XXXI. p. 165 (1851).

† *Wied. Ann.* XXXVI. p. 532.

It is remarkable that fluids are excluded from this statement. Indeed Drude expressly remarks that in fluids the natural surface-sheet, which forms the transition from one medium to another, exercises an essential influence upon the reflexion.

Nearly contemporaneously with Drude's observation I made a first attempt to examine whether the different behaviour which Plateau found for water and for alcohol in respect of superficial viscosity, and which I was disposed to attribute to surface contamination, corresponded to anything in the phenomena of reflexion near the polarizing angle; but at that time I was misled by a faulty nicol, and failed to observe anything distinctive. Subsequently, having proved that the superficial viscosity of water was due to a greasy contamination, whose thickness might be much less than one millionth of a millimetre, I too hastily concluded that films of such extraordinary tenuity were unlikely to be of optical importance, until, prompted by a remark of Sir G. Stokes, I made an actual estimate of the effect to be expected. The thickness required to stop camphor movements, viz. 2×10^{-7} centim., is about $\frac{1}{300}$ of λ_D. This will be one factor in the expression for the amplitude of reflexion due to grease. There will be another somewhat small factor expressive of the difference of optical quality between grease and water; so that the intensity of reflexion at the polarizing angle might on this view be of the order 10^{-6}, or 10^{-7}, a quantity quite appreciable, when the incident light is from the sun. Thus encouraged, I returned to the attack, and on October 2nd, 1890, examined the image of the sun as reflected from water at the polarizing angle. The arrangements were very simple. Sunlight, reflected horizontally into the dark room from a heliostat outside, was again reflected downwards by an adjustable mirror. The water was contained in a large porcelain dish to the depth of about an inch, and at the bottom of the dish was placed a piece of darkly-coloured glass, not quite horizontal, with the view of annulling the diffuse reflexion. The reflected image was observed with a nicol, from which the glass covering disks had been removed, simply held in the hand. The appearance of the dark spot, brought to the centre of the sun's disk, was at once recognized to be dependent upon the condition of the water-surface. When the surface was clean, the spot was dark and its width (in the plane of incidence) was about $\frac{1}{3}$ or $\frac{1}{4}$ of the solar diameter. Moreover there was a strong appearance of colour, brown above and blue below, evidently due to the dependence of the polarizing angle upon the wave-length of the light. But when the surface was greasy, *even although camphor fragments still rotated briskly*, the band lost its darkness, became diffuse, and showed but little colour. When the greasy film was about sufficient to stop the camphor movements, the contrast with the effect of clean water was very marked*.

* *Phil. Mag.* November 1890, p. 400. [Vol. III. p. 396.]

The surface was cleansed by the aid of the expansible hoop employed in my former experiments. This is made of very thin sheet brass, about 2 inches wide. It is placed upon the water, already reasonably clean, in its contracted condition, so that the area enclosed is but small. When it is opened out, say to a circle of about 10 inches diameter, the internal surface of the water is rendered more clean, and the external less clean, than before. To get the best result it is desirable to go through the operation of expanding two or three times, probably because the cleaned water-surface acquires grease from the internal surface of the brass hoop. It will be evident that the action depends upon the hoop not being completely wetted*. Otherwise the grease could repass from the outside back into the interior. For this reason the hoop cannot be expected to succeed with liquids like alcohol.

By taking advantage of the apparent motion of the sun in altitude, these observations may be repeated in summer without any heliostat, or reflexion, other than that from the water itself. Thus on June 26 the dish was placed on a table below the window of an undarkened room, and the passage of the spot across the sun's disk was watched. The spot was central at about $4^h\ 0^m$, and the instant of centrality could be determined to within 10^s, and probably to within 5^s. On August 15, when the sun's motion in altitude was slower, centrality occurred at about $3^h\ 10^m$, and the precise instant was less well determined.

To see the band at its best requires an unusually good nicol. Whether on account of residual defects in the nicols, or in the lenses of my eyes, vision was improved by the use of a horizontal slit, about $\frac{1}{16}$ inch wide, cut out of black paper, and attached to the cork mounting of the nicol on the side next the eye. Under these conditions the band seen from clean water looks black and well defined, and of width amounting to $\frac{1}{4}$ or $\frac{1}{5}$ of the solar diameter. A still further improvement sometimes attends the use of a second nicol, held parallel to the first, through which the light passes before reflexion from the water. With these arrangements the band is visibly deteriorated by quantities of grease far less than is required to check the camphor movements.

It has been mentioned that the dark band from clean water was fairly narrow; and it will be of interest to inquire what is to be expected upon the assumption that Fresnel's formulæ really express the facts of the case. We will write

$$S = \frac{\sin(\theta - \theta_1)}{\sin(\theta + \theta_1)}, \qquad T = \frac{\tan(\theta - \theta_1)}{\tan(\theta + \theta_1)}; \qquad \text{..............(1)}$$

* Since imperfect wetting must be attributed to residual grease, it would appear that the operation of the hoop is incomplete at best. Nevertheless, it is a very useful and convenient appliance.

so that the ratio of amplitudes of the two polarized components, corresponding to a primitive polarization at 45°, is

$$\frac{T}{S} = \frac{\cos(\theta + \theta_1)}{\cos(\theta - \theta_1)}, \quad \text{.................................(2)}$$

vanishing when $\theta + \theta_1 = \frac{1}{2}\pi$, that is when $\theta = \tan^{-1}\mu$. We will suppose that the angle of incidence has approximately this value, and write $\theta + \delta\theta$, $\theta_1 + \delta\theta_1$ for θ, θ_1 respectively. Thus in the neighbourhood of the polarizing angle the ratio is

$$-\frac{\delta\theta + \delta\theta_1}{\cos(\theta - \theta_1)} \text{ approximately.}$$

Now

$$\sin\theta = \mu \sin\theta_1, \qquad \cos\theta\, \delta\theta = \mu \cos\theta_1\, \delta\theta_1,$$

so that

$$\delta\theta_1 = \frac{\cos\theta\, \delta\theta}{\mu \cos\theta_1} = \frac{\sin\theta_1\, \delta\theta}{\mu \sin\theta} = \frac{\delta\theta}{\mu^2}.$$

Hence

$$\frac{T}{S} = -\frac{(\mu^2 + 1)\,\delta\theta}{\mu^2 \cos^2(\theta - \theta_1)}. \quad \text{.............................(3)}$$

For water

$$\mu_D = 1{\cdot}3336, \quad \theta = \tan^{-1}\mu = 53°\ 8', \quad \theta_1 = 36°\ 52';$$

and

$$T/S = 1{\cdot}627\,\delta\theta, \quad \text{................................(4)}$$

$$T^2/S^2 = 2{\cdot}649\,(\delta\theta)^2. \quad \text{............................(5)}$$

Let us calculate the ratio corresponding to the upper or lower limb of the sun when the spot is central; that is, let $\delta\theta$ be the angular radius of the sun, whose value in minutes is 16. Thus

$$\delta\theta = \frac{16\pi}{10800},$$

and corresponding thereto from (5)

$$T^2/S^2 = 5{\cdot}74 \times 10^{-5}. \quad \text{............................(6)}$$

The width of the band actually observed had been estimated at about $\frac{1}{5}$ of the solar diameter, so that at its limits

$$T^2/S^2 = 2 \times 10^{-6}.$$

The band was thus about as narrow as Fresnel's formulæ would lead one to expect, and its deterioration by a film of grease might be anticipated as at least probable, from the rough estimate above given of the effect of such a film.

The results so far obtained were already sufficient to show that Jamin's value of k, viz. $-{\cdot}00577$, must be (numerically) much in excess of the truth. For according to it, since $k^2 = 3{\cdot}33 \times 10^{-5}$, the minimum illumination at the centre of the spot should be half as great as Fresnel's formulæ make it at the limb of the sun, so that the whole diameter of the sun would be almost equally obscured. The observed narrowness of the band, even in the absence of all precise measures, thus constitutes a proof that Jamin's k is several times too great, and suffices to render it almost certain that the water-surface with which he worked was highly contaminated.

It has already been mentioned that a well-formed band was attended with a marked appearance of colour. The account of this rendered by Fresnel's formulæ is quite satisfactory. Let us calculate the illumination at the centre of the band corresponding to μ, due to a change from μ to $\mu + \delta\mu$, comparing it, as usual, with S^2. In the differentiation θ is to be treated as constant, and the change in θ_1, viz. $\delta\theta_1$, is due to $\delta\mu$. From (2),

$$\frac{T}{S} = \frac{\cos(\frac{1}{2}\pi + \delta\theta_1)}{\cos(\theta - \theta_1 - \delta\theta_1)} = \frac{-\delta\theta_1}{\cos(\theta - \theta_1)};$$

and the relation between $\delta\theta_1$ and $\delta\mu$ is

$$\cos\theta_1\,\delta\theta_1 = -\sin\theta\,\delta\mu/\mu^2,$$

or

$$\delta\theta_1 = -\delta\mu/\mu^2,$$

since $\cos\theta_1 = \sin\theta$. Thus

$$\frac{T}{S} = \frac{\delta\mu}{\mu^2\cos(\theta - \theta_1)}. \qquad\qquad (7)$$

In the case of water,

$$\mu_G = 1{\cdot}341, \quad \mu_B = 1{\cdot}331, \quad \delta\mu = {\cdot}010.$$

From these data,

$$T^2/S^2 = 3{\cdot}46 \times 10^{-5};$$

showing that if the spot is central for Fraunhofer's line B, the illumination at the centre for G is more than half as great as is found (6) for B at the upper and lower limbs of the sun. A considerable development of colour is thus to be expected, when the band is well formed.

The band may be achromatized with the aid of a suitable prism, held between the nicol and the eye, but of course at the expense of introducing colour at the upper and lower limbs of the sun. I had at my disposal a glass prism of 10°. This diminished, but could not annul, the colour when held nearly in the position of minimum deviation; but by sufficient sloping the band was practically achromatized. When more dispersive materials, *e.g.* benzole or bisulphide of carbon, were substituted for water, the development of colour is very great, and in the case of the latter made it impossible to judge of the perfection of the band. The above-mentioned glass prism was of course quite insufficient for compensation.

The magnitude of these chromatic effects is given at once by Brewster's law, which we may write in the form

$$\tan(\theta + \delta\theta) = \mu + \delta\mu.$$

Thus

$$\delta\theta = \cos^2\theta\,\delta\mu = \frac{\delta\mu}{1 + \mu^2}, \qquad\qquad (8)$$

which gives the angular displacement of the centre of the dark band, due to the change from μ to $\mu + \delta\mu$. Let us inquire what small angle (i) must be given to a prism *of the same material* in order that it may be capable of compensating the colour. The deviation D is equal to $(\mu - 1)\,i$, so that $\delta D = \delta\mu \,.\, i$. Hence, if $\delta D = \delta\theta$,

$$i = \frac{1}{1+\mu^2}, \qquad\qquad(9)$$

The necessary angle is thus independent of the dispersive power, and does not vary rapidly with the refractive power, of the substance. For water, $i = 9/25$ in circular measure, or about 22°. For glass ($\mu = 1{\cdot}5$) we should have $i = 18°$.

An attempt was made to achromatize the band from bisulphide of carbon with a 15° prism of that material. So far as could be judged the colour was compensated, but the observation was imperfect on account of the insufficient angular magnitude of the solar disk.

These experiments on the achromatization of the band had been made in the hope of thereby reducing its apparent width, seeing that according to (8) the difference of position for the lines B and G amounts, in the case of water, to 13′, much more than the apparent width of the band. But the width of the achromatized band could not be set at much less than $\frac{1}{5}$ of the sun's diameter*. It seems that in estimating the dimensions of the uncorrected band the eye instinctively allows for the influence of colour.

In experimenting upon water various kinds were tried. Usually the tap-water (from an open cistern) behaved after expansion as well as did distilled water. The brass hoop, judiciously applied, appears to be capable of removing ordinary surface-contamination; but the appearance of the band is liable to be deteriorated by suspended matter, which detracts from the central darkness. So far as could be judged by this method of observation, the best bands were sensibly perfect. There was no evidence of any departure from the law of Fresnel.

Similar results were obtained from other liquids, *e.g.* strong alcohol, sulphuric acid, benzole. Special interest attached to an observation upon a saturated solution of camphor, of which the superficial tension is much lower (·72) than that of pure water. The band was sensibly perfect.

Oleate of soda ($\frac{1}{40}$) was troublesome on account of the difficulty of avoiding scum. A pretty good band could be obtained, certainly inferior to the best, possibly owing to residual scum, but much better than from water greased with olive-oil to the point at which the camphor motions are just stopped.

* A coloured glass is still less effective than the compensating prism. A reduction in the intensity of the light necessarily broadens the band. A similar effect occurs if the sun is not quite clear.

The results last recorded prove that the optical effect is not determined by surface-tension, for the tension of the oleate solution is much less than that of any merely greased surface. A similar conclusion was suggested by the observed difference of behaviour of various parts of the same surface. A surface, originally clean, and then greased with olive-oil carried upon a previously ignited platinum wire, frequently showed streakiness, when the eye, observing through the nicol, as usual, was focused upon the surface.

Except perhaps in the case of oleate, none of these experiments, many times repeated, gave any evidence of a real departure of properly skimmed surfaces from the laws of Fresnel; and it looked very much as if all the results enunciated for liquids by Jamin were vitiated by the presence of greasy films. That a film of extreme tenuity would suffice was certain. The band from water was very obviously deteriorated by a film of olive-oil, which needed to be condensed four or five times in order to stop the camphor movements.

But it was impossible to rest here. It was necessary actually to measure, or, if that were not possible, to find limits for, the ellipticity of the various surfaces. And for this purpose a much more elaborate apparatus had to be installed.

Sunlight, reflected horizontally from the heliostat, passed through a diaphragm in the shutter of about $\frac{1}{4}$ inch diameter, and thence to a collimating lens of 23 inches focus. It was next reflected in the required oblique direction by an adjustable mirror, and caused to traverse the polarizing nicol, mounted in a circle that allowed the orientation of the nicol to be read to a minute of angle. After reflexion from the surface under examination the light traversed in succession a quarter-wave-plate and the analysing nicol, and was then received into the eye, either directly, or with the intervention of a small telescope magnifying about twice. In either case the eye was focused upon the diaphragm, which was provided with cross wires; so that the rays which fell upon any part of the retina constituted a parallel pencil, not only at the surface of the liquid, but also in their passage through the nicols and quarter-wave-plate. The latter was of mica, and both it and the analysing nicol were mounted so as to be capable of rotation about the direction of the reflected ray.

The adjustments were made as follows. The analyser and quarter-wave-plate being removed, the mirror and polarizer were adjusted until the dark spot was central in relation to the cross wires. A rotation of the mirror, altering the angle of incidence, moves the spot vertically, while a rotation of the polarizer moves it horizontally. The zero for the eye-nicol could have been found by rotating the polarizer and then recovering the dark spot; but in order to avoid risks of displacement, which might be fatal in such a delicate inquiry, I preferred to leave the first nicol untouched, and to depolarize the light by the introduction of a parallel plate of quartz. With the aid of this the analysing nicol could be set, and then the mica. If, with the quartz plate in action, the spot is dark and central, all is well adjusted. On removal of the quartz, the band is now seen in full perfection.

One of the difficulties in these experiments lay in the extreme sensitiveness of the liquid surfaces to tremor, a sensitiveness aggravated by the perfect cleanliness required. It had been thought that it would suffice to mount the apparatus upon a shelf attached to the walls of the building, and isolated from the floor. But it appeared that the slightest touch upon the tangent-screw of the divided circle, such as it is necessary to make at the moment of observation, entailed a most distracting tremor. A remedy was found in suspending the dish containing the liquid under examination independently from the roof.

The work has been greatly retarded by want of sunshine. In order to be more independent, I tried to work at the Royal Institution by the electric light. But it appeared impossible to make any observations of value on account of the tremor by which London is pervaded. Moreover the arc-light is very inferior to sunshine for such a purpose.

The theory of the experiment is as follows. According to Fresnel's formulæ the ratios of the reflected to the incident vibrations are, for the two planes of polarization, T and S; in which the reality of T and S indicates that there is no change of phase in reflexion (other than 180°). The ellipticity is represented by the addition to T of iM, where M is small and $i=\sqrt{(-1)}$. Thus if the incident light be polarized in the plane making an angle α with the principal planes, the reflected vibrations may be represented by

$$(T+iM)\cos\alpha, \quad S\sin\alpha.$$

By the action of the mica, or other compensator, a relative change of phase γ is introduced. This is represented by writing for $S\sin\alpha$,

$$S\sin\alpha\,(\cos\gamma+i\sin\gamma).$$

Thus the vibration transmitted by the analyser, set at angle β, is

$$\cos\alpha\cos\beta\,(T+iM)+S\sin\alpha\sin\beta\,(\cos\gamma+i\sin\gamma);$$

and the intensity of this is

$$(T\cos\alpha\cos\beta+S\sin\alpha\sin\beta\cos\gamma)^2$$
$$+(M\cos\alpha\cos\beta+S\sin\alpha\sin\beta\sin\gamma)^2. \quad\text{............(10)}$$

In order that the light may vanish we must have both

$$T+S\tan\alpha\tan\beta\cos\gamma=0, \quad\text{....................(11)}$$

$$M+S\tan\alpha\tan\beta\sin\gamma=0. \quad\text{....................(12)}$$

In the neighbourhood of the polarizing angle, M, S vary slowly, but T varies rapidly. Hence, if γ be given, we may regard (12) as determining $\tan\alpha\tan\beta$, while (11) gives T, and thence the precise angle of reflexion for the dark spot. If there be no ellipticity, $M=0$; whence $\tan\alpha\tan\beta=0$, $T=0$, indicating, as was to be expected, that the dark spot occurs at the Brewsterian angle.

But this law is not universal. For if there be no compensator, $\gamma=0$, and we have as the expression for the intensity,

$$(T\cos\alpha\cos\beta+S\sin\alpha\sin\beta)^2+M^2\cos^2\alpha\cos^2\beta.$$

Hence, if α and β are small, the second term cannot be made to vanish, and the brightness is a minimum when

$$T=-S\tan\alpha\tan\beta.$$

The position of the nearly dark spot is thus dependent upon α, β, and assumes the Brewsterian position only when either α or β vanishes.

In the case of a quarter-wave-plate, $\gamma=\pm\frac{1}{2}\pi$, and the equations become

$$T=0,\quad k=\tan\alpha\tan\beta=\mp M/S. \qquad \text{(13)}$$

The dark spot thus occurs at the Brewsterian angle, while $\tan\alpha\tan\beta$ gives the value of M/S, viz. the k of Jamin. Accordingly if β be set to any convenient angle*, and α be then adjusted so as to bring the dark spot to the central position, the product of the tangents of α and β, each measured from the zeros obtained in the preliminary adjustments, gives k.

But the following procedure not only affords greater delicacy, but makes us comparatively independent of the positions of the zeros. Set β, *e.g.*, to $+30°$, and find α; then reset β to $-30°$. The new value of α would coincide with the old one if there were no ellipticity; and the difference of values measures α upon a doubled scale. If α' be the second value, so that the difference is $\alpha'-\alpha$, then

$$k=\tan 30°\tan\tfrac{1}{2}(\alpha'-\alpha),$$

or as would suffice for all the purposes of the present investigation

$$k=\tfrac{1}{2}\tan 30°\,(\alpha'-\alpha). \qquad \text{(14)}$$

In practice several readings for α would be taken as quickly as possible, β being reversed between each. In this way there is the best chance of distinguishing casual errors of observation from the results of progressive changes in the condition of the surface under examination. For greater security against error due to maladjustments, readings were often taken in all four positions, differing from one another by 90°, of the quarter-wave mica. The observed differences of α should be reversed in adjacent positions of the

* In my apparatus it was convenient to throw the fine adjustment upon α.

mica, and should be identical in the opposite positions, [*i.e.* those] obtained from one another by rotation through 180°.

In the above reasoning γ has been regarded as independent of λ, but this is of course only roughly true. If we neglect the dispersion of the mica, we may take $\gamma = \gamma_0 + \delta\gamma$, where γ_0 relates to the mean ray λ_0, while

$$\delta\gamma / \gamma_0 = - \delta\lambda / \lambda_0. \quad \text{.........................(15)}$$

If the mica be suitably chosen, $\gamma_0 = \pm \frac{1}{2}\pi$.

On this principle of the variability of γ may be explained an effect which was puzzling when first observed. When the water-surface was rather highly contaminated, it was found that the appearance of the spot varied according to the choice of positions for the mica. In one position and its opposite the spot was nearly free from colour*, while in the other two positions, differing from the former by 90°, the coloration was intense. It was evident that some cause was at work, in one case compensating, and in the other doubling, the usual Brewsterian coloration.

If M_0 be the mean value of M, the setting of the nicols will give, as before,

$$\tan\alpha \tan\beta = \mp M_0 / S; \quad \text{.......................(16)}$$

while from (11),

$$T = \pm M_0 \cos\gamma. \quad \text{............................(17)}$$

The angle of reflexion corresponding to darkness is determined by (17), and both sides of the equation are functions of λ. For the mean ray $\gamma = \pm \frac{1}{2}\pi$, and at the correct angle $T = 0$. For a neighbouring ray at the same angle of reflexion we have for T,

$$\frac{dT}{d\lambda_0}\delta\lambda;$$

and for $\cos\gamma$,

$$\cos\tfrac{1}{2}\pi(1 + \delta\gamma/\gamma_0) = -\tfrac{1}{2}\pi d\gamma/\gamma_0.$$

Hence the condition of achromatism is

$$\frac{dT}{d\lambda_0}\delta\lambda \pm \tfrac{1}{2}\pi M_0 \delta\gamma/\gamma_0 = 0;$$

or by (15),

$$\frac{dT}{d\lambda_0} \mp \frac{\pi M_0}{2\lambda_0} = 0. \quad \text{............................(18)}$$

Thus if M_0 be of the right magnitude, the colour will be compensated when $\gamma = \frac{1}{2}\pi$, and doubled when $\gamma = -\frac{1}{2}\pi$, or *vice versâ*.

When the colours were but little dispersed in the plane of incidence, there could usually be observed on sufficiently contaminated surfaces a dispersion laterally, indicating a variation of M with λ. It was to be

* Attention is here fixed upon the central plane of incidence, colour on the right and left of the spot being disregarded.

expected that M should be proportional to λ^{-1}. Not much more could be done experimentally than to verify the direction and order of magnitude of the effect. Thus it appeared that on a greasy surface the difference of readings corresponding to $\beta = \pm 30°$ was greater when the settings were made for the brown than for the blue side of the spot. Of these the former, due to the *absence* of blue, represents the setting proper to blue light.

The angles $\pm 30°$ were found suitable for β. It was at first supposed that advantage would accompany a smaller β; but in this case the spot was too diffused in a horizontal direction to suit the dimensions of the bright field employed. The adjustment of the spot to centrality (right and left) by variation of α was then less certain. On the other hand, a too great increase of β throws excessive stress upon the readings of α.

The delicacy of the apparatus may be measured by the smallest error of α visible on simple inspection. When the light was bright and the reflecting surface steady, a setting for $\beta = +30°$ was visibly wrong on going over to $\beta = -30°$, when the change afterwards found necessary in the setting of α exceeded about 2′. Less than this could hardly be recognized on simple inspection; but the error of a single setting, arrived at by trials backwards and forwards, appeared to be less than 1′. Thus the same readings, taken to the nearest minute, were often recovered many times in succession; but on other occasions larger differences were met with, and it was often difficult to judge whether they were due to imperfect observation or to real changes in the condition of the reflecting surface. In any case it will be a modest estimate to suppose that a difference of one minute in α can be detected on repetition. From this we should get, by (14), as the least observable value of k,

$$k = \tfrac{1}{2} \tan 30° \times \tan 1' = \cdot 00009.$$

Jamin's arrangements do not appear to have allowed of the determination of values of k less than ·001.

The first systematic experiments upon cleaned water-surfaces showed that the ellipticity, if real, was pretty close to the limit of observation. At this stage I expected to find the marked ellipticity of greasy water gradually diminishing to zero as the purifying process was carried further, but remaining always of the same sign, so long as it could be observed at all. This anticipation was not completely verified. The larger differences of α, found with ordinary water upon which camphor fragments were fully active, amounting say to 40′, rapidly diminished under the skimming process, so that the final difference on the purest surfaces seemed just to escape direct observation. It frequently happened that no displacement of the dark spot relatively to the cross wires could be detected on the reversal of β. But when, in order to the highest accuracy, many sets of alternate readings were taken, the difference would come out sometimes in one direction and sometimes in the other. A small difference of 2′, or more, on the side of the contaminated water was easily accepted as due to incomplete cleaning, but I was for a time sceptical as to the significance of similar small differences in the opposite direction. That these differences were not errors of observation was soon apparent; but I thought that they might be of instrumental origin, due perhaps to some maladjustment.

The outstanding question was so small that it might perhaps have been dismissed, but I was unwilling to stop without a determined attempt to get to the bottom of it. The minute reversal of ellipticity stood its ground in spite of repeated remountings of the apparatus; but I feared that it might possibly be due to some, though I was unable to conjecture what, defect in the optical parts themselves. When, however, the nicols at first used were replaced by beautiful prisms made by Steeg and Reuter and the effect still remained, it had to be accepted as genuine, and the conclusion was forced upon me that with some water-surfaces, and those presumably the cleanest, there is a minute ellipticity in the opposite direction to that of ordinary water, and such that the difference of settings for α amounted to about 2′. This corresponds to $k = +\cdot0002$. It will be understood that this is a very minute quantity, but it is not without interest from a theoretical point of view. The fact that k can be positive as well as negative implies of course its possible evanescence. It is, I think, safe to say that some samples of water-surfaces polarize light to perfection.

It will now be desirable to give a specimen of actual observations. The one selected, principally on account of its completeness in respect to the positions of the quarter-wave mica, is dated April 1, 1891, and relates to a surface of tap-water, freshly drawn, and skimmed with the aid of the brass hoop already described. The operation of skimming was repeated after each readjustment of the mica. In the first column the direction of the arrow

Mica	Analyser, β	Polarizer. Separate Readings of α (minutes)	Means	Difference
↓	+30°	31, 32, 32, 31	183° 31½′	+2¾′
	−30	28, 29, 29, 29	183 28¾	
→	+30	28, 31, 29, 30	183 29½	−2¼
	−30	32, 31, 32, 32	183 31¾	
↑	+30	33, 33, 33, 33	183 33	+1½
	−30	30, 32, 32, 32	183 31½	
←	+30	28, 29, 29, 29	183 28¾	−2
	−30	31, 32, 31, 29	183 30¾	

indicates the position of the mica. The second gives the readings of β, the third the individual readings of α, the minutes only being entered*. It will be understood that the readings for $\beta = \pm 30°$ were taken alternately; the first reading under ↓ being 31′, the second 28′, the third 32′, and so on. The fourth column gives the means, and the fifth the difference of these means, which represents ellipticity. The second and fourth differences, corresponding to positions of the mica differing by a right angle from those of the first and third, must have their signs reversed before combination for a final mean difference. We get

$$\tfrac{1}{4}(2\tfrac{3}{4} + 2\tfrac{1}{4} + 1\tfrac{1}{2} + 2) = +2\tfrac{1}{8}'.$$

After the last set the hoop was lifted, so as to allow the return of the contamination. The readings then became

←	+30°	37, 44, 43
	−30°	17, 14, 12

It would appear that the first of these were premature, insufficient time having been allowed for the contamination to spread. The difference, reckoned as before, may now be taken to be about −30′, and is in the *opposite* direction to the small effect of the clean surface. For the contaminated surface $k = -\cdot 0026$, and for the clean $k = +\cdot 00018$†.

Although the above results, and others of a similar nature, obtained both with tap-water and with distilled water, render it practically certain that k is positive for pure water, I do not regard with the same confidence the numerical value above recorded. It is difficult to feel sure that the cleansing was sufficient. A theoretical objection to the hoop method has already been alluded to; and the more perfect methods depending upon the use of convection currents‡ are scarcely applicable here. Attempts were indeed made to work with a surface cleaned by an ascending column of fluid, the column being expanded by heat communicated to it from an immersed platinum spiral, itself warmed by an electric current. But the readings were not accordant; and it appeared that the observations were prejudiced by the deformations of the surface which are the necessary accompaniment of such a flow. Doubts as to the perfection of the cleanliness actually attained lead me to think it possible that the true value of k for an ideal water-surface may be even twice as great as that actually found.

* In almost all the observations the settings were made by myself, and the readings of α at the vernier by Mr Gordon. Without two observers the difficulties would have been much increased.

† The observations so far did not of themselves determine which of the two surfaces has the positive k according to Jamin's convention. It was evident, however, that it must be the contaminated and not the clean surface which corresponds to Jamin's determination of a negative k. Subsequent observations upon reflexion from glass verified this assumption.

‡ See *Roy. Soc. Proc.* "On the Superficial Viscosity of Water," Vol. XLVIII. p. 133. [Vol. III. p. 363.]

Opportunities for useful work upon clean surfaces have been very few, for it is hopeless to attempt observations without a prospect of at least an hour's almost uninterrupted sunshine. But shorter and more uncertain periods may be utilized for observations upon contaminated surfaces, as these do not demand the same care or amount of repetition. As an example of such I will record the readings of June 6, from a water-surface slightly greased with oil of cassia.

↑	$+30°$ $-30°$	$180° 2'$ $180° 54'$	Band achromatic.
←	$+30°$ $-30°$	$180° 47'$ $178° 59'$	Band strongly coloured, red above.

The difference of readings is here about $-50'$, giving $k = -\cdot 0043$. On trial it was found that camphor fragments would just move. The above is an example of the effect of the position of the mica upon the coloration of the band, a subject already discussed.

Experiments were made with the object of comparing different kinds of oil as to their relative effects, optically and upon camphor. It was found, as had been expected, that cassia was more powerful optically than olive-oil. Thus when camphor was nearly dead the difference of readings for olive-oil was about $-30'$ and for cassia about $-48'$.

Interest was felt in the behaviour of a saturated solution of camphor, whose surface-tension is much lower than that of clean water. Observations upon this liquid proved especially difficult, for the dark spot frequently shifted laterally while under inspection, indicating temporary changes in the ellipticity of the particular part of the surface in use. There is little doubt that this complication is due to local evaporation under the influence of light currents of air. As the camphor evaporates from any part of the surface, the tension is momentarily raised, and the surface contracts. If the camphor only were in question, there would probably be no attending optical disturbance, but the local expansions and contractions of the surface lead to attenuation and concentration of the greasy matter present. Under favourable circumstances the difference in the readings of α might be as low (numerically) as $-6'$, and was perhaps due after all to residual greasy matter, other than camphor. In any case the optical effect of the camphor is much less than that of an oily film giving the same surface-tension.

With a strong solution of oleate of soda the difference of α could not be reduced below $-25'$. It is difficult to suppose that this can be due to a film of foreign matter removable by skimming. But the amount of the ellipticity is very low in relation to the surface-tension, which is only about one-third of that of clean water. The value of k corresponding to the above readings is $-\cdot 0021$, only about double of the smallest quantity appreciated by Jamin.

That the surface-tension has no definite relation to the ellipticity is abundantly evident. For example, camphor was quite active upon a surface which gave a difference of readings of $-80'$, corresponding to $k = -\cdot007$. On this occasion the surface had stood for some time without much protection, and it is possible that the effect may have been partly due to dust.

The last example that I will mention of aqueous solutions is a strong brine. This gave a somewhat variable difference of about $-5'$, corresponding to $k = -\cdot00042$. In this case there seemed to be unusual difficulty in getting the surface clean, so that the difference between the brine and pure water is not improbably due to some secondary cause.

Most of the available time was spent upon water in its various states, not only on account of its intrinsic importance, but also because of the presumably greater simplicity of a clean water-surface. The observations are made in an atmosphere which contains no very small proportion of aqueous vapour. When the liquid under examination has an affinity for water, *e.g.* alcohol, it is difficult to form a precise idea as to what may be the condition of the surface. Besides, the arrangements for skimming are less easily applied. On the other hand, the liquids of lower tension are less likely to acquire a film of grease. For alcohol, and also for petroleum, the value of k is about $+\cdot0010$.

The general conclusion to be drawn from these investigations is that the ellipticity of the liquids examined is very much less than was supposed by Jamin, whose results for water and aqueous solutions were almost certainly vitiated by the presence of greasy contamination. Thus the intensity of reflexion from clean water is not much more than $\frac{1}{1000}$ part of that given by Jamin. Moreover, the value of k is positive, and not negative. It is even possible that there would be no sensible ellipticity for the surface of a chemically pure body in contact only with its own vapour. But the surfaces of bodies are the field of very powerful forces of whose action we know but little; and even if there be nothing that could be called chemical change, the mere want of abruptness in the transition would of itself entail a complication. There is thus no experimental evidence against the rigorous applicability of Fresnel's formulæ to the ideal case of an abrupt transition between two uniform transparent media.

September 19.

POSTSCRIPT (*October* 11).

Solutions of saponine and gelatine, substances which confer the foaming property, have been examined. With very small quantities the difference of readings may amount to a degree, *not to be diminished by repeated skimming*. The value of k is thus $-\cdot005$.

The suspicion above suggested that the true value of k for clean water may be numerically higher than is indicated by the results obtained with the aid of the brass hoop has been verified by some observations upon surfaces cleansed by heat. The water, as clean as possible, was contained in a large shallow tin tray. By the application of gentle heat to the part of the tray under optical examination any residual grease is driven off, in consequence of the smaller tension of the warmer surface. If the whole surface is fairly clean to begin with, a very moderate difference of temperature suffices to keep the grease at bay. The difficulties of the experiment have so far prevented a complete series of readings; but the following, obtained on October 2, seem sufficient to establish the fact:—

→	+30°	43,	42,	42,	42	
	−30°	47,	47,	48,	47	
↑	+30°		52,	51,	51,	50
	−30°	47,	46,	47,	45,	48

At the conclusion of the second set the contamination was evidently returning. It would seem that on the cleanest surfaces the difference of readings may amount to 5′, the necessity of readjustment on passing between $\beta = \pm 30°$ being obvious on simple inspection. Corresponding to this

$$k = + \cdot 00042.$$

186.

ON THE THEORY OF SURFACE FORCES. II. COMPRESSIBLE FLUIDS.

[*Philosophical Magazine*, XXXIII. pp. 209—220, 1892.]

In the first part of the paper published under the above title (*Phil. Mag.* Oct. and Dec. 1890 [Vol. III. p. 397]) the theory of Young and Laplace was considered, and further developed in certain directions. The two leading assumptions of this theory are (1) that the range of the cohesive forces, though very small in comparison with the dimensions of ordinary bodies, is nevertheless large in comparison with molecular distances, so that matter may be treated as continuous; and (2) that the fluids considered are incompressible. So far as I am aware, there is at present no reason to suppose that the applicability of the results to actual matter is greatly prejudiced by imperfect fulfilment of (1); but, on the other hand, the assumption of incompressibility is a somewhat violent one, even in the cases of liquids, and altogether precludes the application of the theory to gases and vapours. In the present communication an attempt is made to extend the theory to compressible fluids, and especially to the case of a liquid in contact with its own vapour, retaining the first assumption of continuity, or rather of ultimate homogeneity. There will not be two opinions as to the advantage of the extension to compressible fluids; but some may perhaps be inclined to ask whether it is worth while to spend labour upon a theory which ignores the accumulated evidence before us in favour of molecular structure. To this the answer is that molecular theories are extremely difficult, and that the phenomenon of a change of state from vapour to liquid is of such extreme importance as to be worthy of all the light that can be thrown upon it. We shall see, I think, that a sufficient account can be given without introducing the consideration of molecules, which on this view belongs to another stage of the theory.

If p denote the ordinary hydrostatical pressure at any point in the interior of a self-attracting fluid, ρ the density, and V the potential, the equation of equilibrium is

$$dp = \rho dV. \quad \text{.................................(1)}$$

If, as we shall here suppose, the matter be arranged in plane strata, the expression for the potential at any point is

$$V = 2\pi \int_{-\infty}^{+\infty} \rho' \psi(z)\, dz, \quad \text{..........................(2)}$$

where ρ' is the density at a distance z from the point in question. Expanding in series, we may write

$$\rho' = \rho + z\frac{d\rho}{dz} + \frac{z^2}{1\,.\,2}\frac{d^2\rho}{dz^2} + \ldots,$$

so that

$$V = 2K\,.\,\rho + 2L\frac{d^2\rho}{dz^2} + \ldots\ldots, \quad \text{........................(3)}$$

where

$$K = 2\pi \int_0^\infty \psi(z)\, dz, \qquad L = \pi \int_0^\infty z^2 \psi(z)\, dz. \quad \text{...............(4)}$$

The integrals involving odd powers of z disappear in virtue of the relation

$$\psi(-z) = \psi(z).$$

We may use (3) to form an expression for the pressure, applicable to regions of *uniform* density (and potential). Thus, integrating (1) from a place where $\rho = \rho_1$ to one where $\rho = \rho_2$, we have

$$\begin{aligned} p_2 - p_1 &= \int \rho dV = [\rho V] - \int V d\rho \\ &= 2K(\rho_2^2 - \rho_1^2) - \int d\rho\, \{2K\rho + 2L d^2\rho/dz^2 + \ldots\} \\ &= K(\rho_2^2 - \rho_1^2) - \int d\rho\, \{2L d^2\rho/dz^2 + \ldots\}. \end{aligned}$$

In the latter integral each term vanishes. For example,

$$\int \frac{d^2\rho}{dz^2} d\rho = \tfrac{1}{2}\int d\left(\frac{d\rho}{dz}\right)^2 = \tfrac{1}{2}\left(\frac{d\rho}{dz}\right)_2^2 - \tfrac{1}{2}\left(\frac{d\rho}{dz}\right)_1^2,$$

and at the limits all the differential coefficients of ρ vanish by supposition. Thus, in the application to regions of uniform density—uniform, that is, through a space exceeding the range of the attractive forces—

$$p_2 - p_1 = K(\rho_2^2 - \rho_1^2)\,; \quad \text{..........................(5)}$$

or, as we may also write it,

$$p = \varpi + K\rho^2, \quad \text{................................(6)}$$

where ϖ is a constant, denoting what the value of p would be in a region where $\rho = 0$. We may regard ϖ as the *external* pressure operative upon the fluid. Equation (5) may also be obtained, less analytically, by the argument employed upon a former occasion*, and still more simply perhaps by consideration of the forces operative upon the entire mass of fluid included between the two strata in question regarded as a rigid body. It is very important to remember that *it ceases to apply at places where ρ is varying*, and that unless the strata are plane it requires correction even in its application to regions of uniform density.

In the case of a uniform medium, (6) gives the relation between the external pressure ϖ, measured in experiments, and the total internal pressure p, found by adding to the former the intrinsic pressure $K\rho^2$. By the constitution of the medium, independently of the self-attracting property, there is a relation between p and ρ, and thence, by (6), between ϖ and ρ. If we suppose that the medium, freed from self-attraction, would obey Boyle's law, $p = k\rho$, and

$$\varpi = k\rho - K\rho^2. \qquad (7)$$

According to (7), when ρ is very small, ϖ varies as ρ. As ρ increases, ϖ increases with it, until $\rho = k/2K$, when ϖ reaches a maximum. Beyond this point ϖ diminishes as ρ increases, and this without limit. The curve which represents the relationship of ϖ and ρ is a parabola; and it is evident that all beyond the vertex represents unstable conditions. For at any point on this portion the pressure diminishes as ρ increases. If, therefore, the original uniformity were slightly disturbed, without change of total volume, one part of the fluid becoming denser and the other rarer than before, the latter would tend still further to expand and the former to contract. And according to our equations the collapse would have no limit.

Fig. 1.

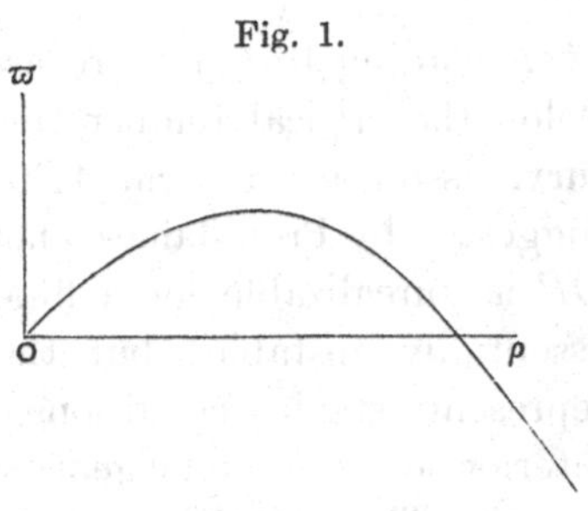

Points on the parabola between O and the vertex represent conditions which are stable so far as the interior of the fluid is concerned, but it may be necessary to consider the action of the walls upon the fluid situated in their neighbourhood. The simplest case is when the containing vessel, which may be a cylinder and piston, exercises no attraction upon the fluid. The fluid may then be compressed up to the vertex of the parabola without losing its uniformity or becoming unstable. If, however, there be sufficient attraction between the walls of the vessel and the fluid, instability leading to total collapse will set in before the vertex is reached.

It will be seen that condensation to a denser state is easily explained, without any reference to molecules, as a direct consequence of self-attraction in a medium otherwise obeying Boyle's law. The objection that may be

* "On Laplace's Theory of Capillarity," *Phil. Mag.* Oct. 1883. [Vol. II. p. 231.]

raised at this point is rather that the explanation is too good, inasmuch as it points to indefinite collapse, instead of to a high, but finite, contraction in the condensed part.

A simple and well-known modification provides an escape from a conclusion which follows inevitably from a rigorous application of Boyle's law. A provision is required to prevent extreme collapse, and this we may find in the assumption that a constant must be subtracted from the volume in order to obtain the quantity to which the pressure is proportional. In this case it is usual and convenient to express the relation by the volume v of the unit mass, rather than by the density. We have

$$p(v-b) = \text{constant},$$

or

$$(\varpi + K/v^2)(v-b) = \text{constant}, \quad \text{.....................(8)}$$

the well-known equation of Van der Waals. Here b is the smallest volume to which the fluid can be compressed; and under this law the collapse of the fluid is arrested at a certain stage, equilibrium being attained when the values of ϖ are again equal for the condensed and uncondensed parts of the fluid.

According to (8), there are three values of v corresponding to a given ϖ. Below the critical temperature the three values are real, and the isothermal curve assumes the form *ABCDEFGH* (Fig. 2) suggested by Prof. James Thomson. The part *DF* is unrealizable for a fluid in mass, being essentially unstable; but the parts *AD*, *FH* represent stable conditions, so far as the interior of the homogeneous fluid is concerned. The line *CG* represents the (external) pressure at which the vapour can exist in contact with the liquid in mass, and the isothermal found by experiment is usually said to be *HGECBA*. This statement can hardly be defended. If a vapour be compressed from *H* through *G*, it can only travel along the straight line from *G* towards *E* under very peculiar conditions. Apart from the action of the walls of the containing vessel, and of suspended nuclei, the path from *G* to *F* must be followed. The path from *G* to *E* implies that the vapour at *G* is in contact with the liquid in mass. This is by supposition not the case; and the passage in question could only be the result of foreign matter whose properties happened to coincide with those of the liquid. If the walls attract the vapour less than the vapour attracts itself, they cannot promote condensation, and the path *HGF* must be pursued. In the contrary case condensation must begin before *G* is reached, although it may be to only a limited extent.

Fig. 2.

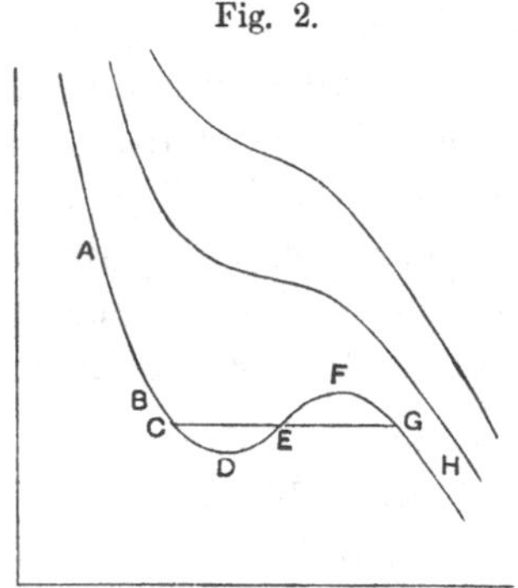

Probably the latter is the state of things usually met with in practice. So soon as the walls are covered with a certain thickness of liquid, the path coincides with a portion of GEC, and the angle at G is only slightly rounded off.

Similar considerations apply at the other end of the straight course. If the liquid be expanded through C, it will not, in general, pass along CE, but will continue to pursue the curve CD, and will even attain the limit D, if the attraction of the walls upon the liquid be not less than that of the liquid upon itself. In the contrary case separation will suddenly occur at a point upon the wall, a bubble of vapour will be formed, and a point on the straight line CE will be attained. It is thus scarcely conceivable that a fluid should follow the broken course $ABCEGH$ without some rounding of the corners, or else of overshooting the points C, G, with subsequent precipitation upon the line CEG.

A very important question is the position of the line CG. Maxwell* showed that inasmuch as the area of the curve represents work performed at a constant temperature, it must be the same for the complete course as for the broken one. The line CG is therefore so situated as to cut off equal areas above and below.

This discussion is of course quite independent of the precise form of the relation between p and v. All that is necessary is such a modification of Boyle's law at great densities as will secure the fluid against indefinite collapse under the influence of its self-attraction.

We will now pass to the question of the transition from liquid to vapour, still supposing the strata to be plane. This is a problem considered by Maxwell in his article upon "Capillary Action" in the *Encyclopædia Britannica*†; but his solution appears to me to be vitiated by more than one oversight. By differentiation of (6) he obtains (with A written for K)

$$dp = 2A\rho d\rho,$$

and thence, by (1),

$$2A\rho d\rho = \rho dV;$$

so that

$$V = 2A\rho + \text{constant}.$$

In the subsequent argument the identity of A with K is overlooked; and the whole process is vitiated by the illegitimate differentiation of (6), which is only applicable at places where ρ is not varying. The final result, which appears to be arrived at without any assumption as to the physical connexion between p and ρ, is thus devoid of significance.

* *Nature*, Vol. XI. p. 358, 1875; *Scientific Papers*, Vol. II. p. 418.

† *Scientific Papers*, Vol. II. p. 560.

Let us integrate (1) from a place in the vapour round which the density has the uniform value ρ_1 to a place in the liquid where the uniform density is ρ_2. Thus

$$\int_{(1)}^{(2)} \frac{dp}{\rho} = V_2 - V_1 = 2K(\rho_2 - \rho_1), \quad \text{.....................(9)}$$

by (3). The external pressure is uniform throughout, and may be denoted by ϖ'; and by (6),

$$\varpi' = p_1 - K\rho_1^2 = p_2 - K\rho_2^2. \quad \text{.....................(10)}$$

At places where ρ is varying, that is in the transitional layer, ϖ, as given by (6), does not represent the external pressure; but we will still regard it as defined analytically by (6). Thus

$$\int_{(1)}^{(2)} \frac{dp}{\rho} = \int_{(1)}^{(2)} \frac{1}{\rho}\left(\frac{d\varpi}{d\rho} + 2K\rho\right) d\rho = \int_{(1)}^{(2)} \frac{1}{\rho}\frac{d\varpi}{d\rho}\, d\rho + 2K(\rho_2 - \rho_1). \quad \text{......(11)}$$

By comparison of (9) and (11),

$$\int_{(1)}^{(2)} \frac{1}{\rho}\frac{d\varpi}{d\rho}\, d\rho = 0\,; \quad \text{.............................(12)}$$

or on integration by parts,

$$\left[\frac{\varpi}{\rho}\right]_{(1)}^{(2)} + \int_{(1)}^{(2)} \frac{\varpi}{\rho^2}\, d\rho = 0.$$

The values of ϖ at the limits are the same, and have been denoted by Hence

$$\int_{(1)}^{(2)} \frac{\varpi - \varpi'}{\rho^2}\, d\rho = 0. \quad \text{.........................(13)}$$

Since $d\rho/\rho^2 \propto dv$, this equation, obtained by purely hydrostatical methods applied to the liquid and vapour and the layer of transition between them, has precisely the same significance as Maxwell's theorem upon the position of the line *CG* in J. Thomson's diagram. In that theorem ϖ represents the external pressure that would be exerted by the fluid in various states of uniform density, some of which are not realizable. In the subject of the present investigation all the densities intermediate between those of the vapour and liquid actually occur; but, except at the extremities, ϖ no longer represents external pressure.

The explanation of the stable existence in the transitional layer of certain densities, which would be unstable in mass, depends of course upon the fact that in the transitional layer the complete self-attraction due to the density is not developed in consequence of the rapid variation of density in the neighbourhood.

The distribution of density in the transitional layer, and the tension of the surface, can only be calculated upon the basis of a knowledge of the physical constitution of the fluid as expressed by the relation between p and

ρ, and by the law of self-attraction. Poisson's contention that the surface-tension cannot be found upon the supposition of an abrupt transition from the liquid to its vapour is evidently justified; and since the thickness of the layer of transition is necessarily of the order of the range of the attraction, it follows that the correction for gradual transition is not likely to be small. A complete calculation of a particular case would be of interest, even on rather forced suppositions; but the mathematical difficulties are considerable. An approximate investigation might be conducted as follows:—

From (1) and (3),

$$\int \frac{dp}{\rho} = V = \rho \,.\, 2K + \frac{d^2\rho}{dz^2} 2L + \ldots.$$

If we neglect the terms in $d^4\rho/dz^4$, &c., this becomes

$$2L\frac{d^2\rho}{dz^2} = \int \frac{dp}{\rho} - 2K \,.\, \rho = f(\rho) - 2K \,.\, \rho, \qquad \text{..............(14)}$$

where $f(\rho) = \int dp/\rho$ is a function of ρ given by the constitution of the medium.

Equation (14) may now be integrated by quadratures.

$$L\left(\frac{d\rho}{dz}\right)^2 = \int f(\rho)\, d\rho - K\rho^2,$$

and

$$z = L^{\frac{1}{2}} \int \left\{\int f(\rho)\, d\rho - K\rho^2\right\}^{-\frac{1}{2}} d\rho. \qquad \text{..................(15)}$$

It is possible that a graphical process would be found suitable. Equation (14) determines the curvature at any point of the curve, representing the relation between ρ and z, in terms of the coordinates and the slope.

When the relation between ρ and z is known, the calculation of the surface-tension is a matter of quadratures. Probably the simplest way of considering the question is to regard the free surface as spherical (liquid within and vapour without), and to calculate the difference of pressures.

We have from (1),

$$p_2 - p_1 = [\rho V] - \int_{(1)}^{(2)} V d\rho = 2K(\rho_2{}^2 - \rho_1{}^2) - \int_{(1)}^{(2)} V \frac{d\rho}{dz} dz, \qquad \text{......(16)}$$

z being measured outwards along the radius. The question is thus reduced to the determination of V at the various points of the layer of transition, for all of which $z = R$ approximately. Let P (Fig. 3) be a point at which V is to be estimated, so that $OP = z$, and let AQB be a spherical shell of radius $z - \zeta$, of thickness $d\zeta$, and of density ρ'. We have first to estimate the potential dV of this shell at P.

Fig. 3.

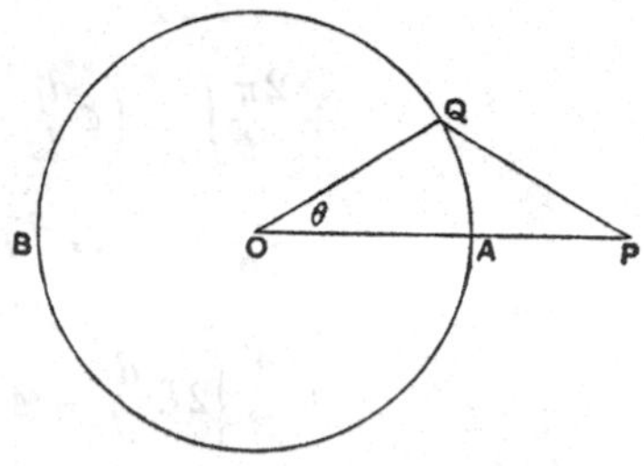

The element of mass at Q is

$$\rho' . 2\pi \sin\theta \, d\theta \, (z-\zeta)^2 \, d\zeta.$$

If, as before, $\phi(f)$ express the ultimate law of attraction, and

$$\Pi(f) = \int_f^\infty \phi(f) \, df,$$

we have to multiply the above element of mass by $\Pi(f)$. Now

$$f^2 = PQ^2 = z^2 + (z-\zeta)^2 - 2z(z-\zeta)\cos\theta,$$

so that

$$-d\cos\theta = \frac{f\,df}{z(z-\zeta)}.$$

The element of the potential is therefore

$$\frac{2\pi\rho'(z-\zeta)\,d\zeta}{z} . \Pi(f) f \, df.$$

In the integration the limits of f are AP and BP. The former is denoted by ζ, and the latter may be identified with ∞, since z or R is supposed to be a very large multiple of the range of the forces. Accordingly for the potential at P of the whole shell, we have

$$dV = \frac{2\pi\rho'(z-\zeta)\,d\zeta\,\psi(\zeta)}{z}, \quad \text{.....................(17)}$$

where, as usual,

$$\psi(\zeta) = \int_\zeta^\infty \Pi(f) f \, df. \quad \text{..........................(18)}$$

To find the whole potential at P, (17) must be integrated with respect to ζ from $-\infty$ to $+\infty$, ρ' being treated as a function of ζ. As we need only consider P near the layer of transition, z in (17) may be identified with R.

If the transition is continuous, we may expand ρ' in the series

$$\rho' = \rho - \zeta\frac{d\rho}{dz} + \frac{\zeta^2}{1.2}\frac{d^2\rho}{dz^2} - \ldots;$$

and then at the point P,

$$V = 2\pi\int_{-\infty}^{+\infty}\left(\rho + \frac{\zeta^2}{1.2}\frac{d^2\rho}{dz^2} + \ldots\right)\zeta\psi(\zeta)\,d\zeta$$

$$+ \frac{2\pi}{R}\int_{-\infty}^{+\infty}\left(\zeta\frac{d\rho}{dz} + \frac{\zeta^3}{1.2.3}\frac{d^3\rho}{dz^3} + \ldots\right)\zeta\psi(\zeta)\,d\zeta$$

$$= 2K.\rho + 2L\frac{d^2\rho}{dz^2} + 2M\frac{d^4\rho}{dz^4} + 2N\frac{d^6\rho}{dz^6} + \ldots$$

$$+ \frac{2}{R}\left\{2L\frac{d\rho}{dz} + 4M\frac{d^3\rho}{dz^3} + 6N\frac{d^5\rho}{dz^5} + \ldots\right\}, \quad \text{.................(19)}$$

where (as in Maxwell's "Capillary Action")

$$K=\pi\int_{-\infty}^{+\infty}\psi(\zeta)\,d\zeta,\qquad L=\tfrac{1}{2}\pi\int_{-\infty}^{+\infty}\zeta^2\psi(\zeta)\,d\zeta,$$

$$M=\frac{\pi}{4\,!}\int_{-\infty}^{+\infty}\zeta^4\psi(\zeta)\,d\zeta,\qquad N=\frac{\pi}{6\,!}\int_{-\infty}^{+\infty}\zeta^6\psi(\zeta)\,d\zeta.\quad\text{......(20)}$$

When (19) is multiplied by $d\rho/dz$ and integrated across the whole layer of transition, we get for the part independent of R,

$$2K\int_{(1)}^{(2)}\rho\,\frac{d\rho}{dz}\,dz=K(\rho_2^2-\rho_1^2)$$

simply, all the other terms in $L, M, \ldots$ vanishing. Hence by (16), with integration by parts,

$$p_2-p_1=K(\rho_2^2-\rho_1^2)$$
$$-\frac{2}{R}\left\{2L\int_{(1)}^{(2)}\left(\frac{d\rho}{dz}\right)^2dz-4M\int_{(1)}^{(2)}\left(\frac{d^2\rho}{dz^2}\right)^2dz+6N\int_{(1)}^{(2)}\left(\frac{d^3\rho}{dz^3}\right)^2dz-\ldots\right\}\text{....(21)}$$

The first term upon the right in (21) is the same as when the strata are plane. The second gives the capillary tension (T), and we conclude that when the transition is continuous

$$T=2L\int_{(2)}^{(1)}\left(\frac{d\rho}{dz}\right)^2dz-4M\int_{(2)}^{(1)}\left(\frac{d^2\rho}{dz^2}\right)^2dz+\ldots\ldots\ldots\ldots\text{(22)}$$

From these results we see that "the existence of a capillary force is connected with suddenness of transition from one medium to another, and that it may disappear altogether when the transition is sufficiently gradual*."

The series (22) would probably suffice for the calculation of surface-tension between liquid and vapour when once the law connecting ρ and z is known. It is possible, however, that its convergence would be inadequate, and in this respect it must certainly fail to give the result for an abrupt transition. In the latter case, where the whole variation of density occurs at one place, (16) becomes

$$p_2-p_1=2K(\rho_2^2-\rho_1^2)-(\rho_2-\rho_1)V,\ \ldots\ldots\ldots\ldots\ldots\ldots\text{(23)}$$

V relating to the place in question. And by (17)

$$V=\int_{-\infty}^{+\infty}2\pi\rho'(1-\zeta/R)\psi(\zeta)\,d\zeta$$
$$=2\pi(\rho_2+\rho_1)\int_0^{\infty}\psi(\zeta)\,d\zeta-\frac{2\pi}{R}(\rho_2-\rho_1)\int_0^{\infty}\zeta\psi(\zeta)\,d\zeta.$$

Thus

$$p_2-p_1=K(\rho_2^2-\rho_1^2)+2T/R,\ \ldots\ldots\ldots\ldots\ldots\ldots\text{(24)}$$

if

$$T=\pi\int_0^{\infty}\zeta\psi(\zeta)\,d\zeta\,.\,(\rho_2-\rho_1)^2,\ \ldots\ldots\ldots\ldots\ldots\ldots\text{(25)}$$

* "On Laplace's Theory of Capillarity," *Phil. Mag.* October 1883. [Vol. II. p. 234.]

where (25) agrees with the value of the tension found for this case by Laplace.

In the application to a sphere of liquid surrounded by an atmosphere of vapour, equations (9), (11), (12) remain unchanged, in spite of the curvature of the surface. If ϖ'' denote the external pressure acting upon the vapour,

$$p_1 = \varpi'' + K\rho_1^2, \qquad \text{......(26)}$$

$$p_2 = \varpi'' + K\rho_2^2 + 2T/R. \qquad \text{......(27)}$$

The symbol ϖ is still regarded as defined algebraically by (6), so that

$$\varpi_1 = \varpi'', \qquad \varpi_2 = \varpi'' + 2T/R. \qquad \text{......(28)}$$

Integrating (12) by parts, we find

$$\frac{\varpi_2}{\rho_2} - \frac{\varpi_1}{\rho_1} + \int_{(1)}^{(2)} \frac{\varpi}{\rho^2} d\rho = 0\,;$$

or by (28),

$$\int_{(1)}^{(2)} \frac{\varpi - \varpi''}{\rho^2} d\rho + \frac{2T}{R\rho^2} = 0. \qquad \text{......(29)}$$

In this equation ϖ is a known function of ρ. If we compare it with (13), where ϖ' represents the external pressure of the vapour in contact with a *plane* surface of liquid, we shall be able to estimate the effect of the curvature. It is to be observed that the limits of integration are not the same in the two cases. If we retain ρ_1, ρ_2 for the plane surface, and for the curved surface write $\rho_1 + \delta\rho_1$, $\rho_2 + \delta\rho_2$, we have from (29)

$$\frac{\varpi_2 - \varpi''}{\rho_2^2} \delta\rho_2 - \frac{\varpi_1 - \varpi''}{\rho_1^2} \delta\rho_1 + \int_{\rho_1}^{\rho_2} \frac{\varpi - \varpi''}{\rho^2} d\rho + \frac{2T}{R(\rho_2 + \delta\rho_2)} = 0\,;$$

or by (28),

$$\int_{\rho_1}^{\rho_2} \frac{\varpi - \varpi''}{\rho^2} d\rho + \frac{2T}{R\rho_2} = 0. \qquad \text{......(30)}$$

The limits of integration are now the same as in (13), so that by subtraction

$$(\varpi' - \varpi'')\left(\frac{1}{\rho_2} - \frac{1}{\rho_1}\right) = \frac{2T}{R\rho_2},$$

or

$$\varpi'' = \varpi' + \frac{2T\rho_1}{\rho_2 - \rho_1}. \qquad \text{......(31)}$$

This is the value for the excess of vapour-pressure in equilibrium with a convex surface that is given in Maxwell's "Heat" as a deduction from Sir W. Thomson's principle.

The application of this principle may be extended in another direction. When liquid rises in a capillary tube open above, the more attenuated vapour at the upper level is in equilibrium with the concave surface, and the more dense vapour below is in equilibrium with the plane surface of the liquid.

But, as was pointed out in the former paper, the rise of liquid is not limited to the height of the meniscus. Above that point the walls of the tube are coated with a layer of fluid, of gradually diminishing thickness, less than the range of forces, and extending to an immense height. *At every point the layer of fluid must be in equilibrium with the vapour to be found at the same level.* The data scarcely exist for anything like a precise estimate of the effect to be expected, but the argument suffices to show that a solid body brought into contact with vapour at a density which may be much below the so-called point of saturation will cover itself with a layer of fluid, and that this layer may be retained in some degree even in what passes for a good vacuum. The fluid composing the layer, though denser than the surrounding atmosphere of vapour, cannot properly be described as either liquid or gaseous.

In our atmosphere fresh surfaces, *e.g.* of split mica or of mercury, attract to themselves at once a coating of moisture. In a few hours this is replaced, or supplemented, by a layer of grease, which gives rise to a large variety of curious phenomena. In the case of mica the fresh surface conducts electricity, while an old surface, in which presumably the moisture has been replaced by grease, insulates well.

187.

ON THE RELATIVE DENSITIES OF HYDROGEN AND OXYGEN. II.

[*Proceedings of the Royal Society*, L. pp. 448—463, 1892.]

IN a preliminary notice upon this subject*, I explained the procedure by which I found as the ratio of densities 15·884. The hydrogen was prepared from zinc and sulphuric, or from zinc and hydrochloric, acid, and was liberated upon a platinum plate, the generator being in fact a Smee cell, enclosed in a vessel capable of sustaining a vacuum, and set in action by closing the electric circuit at an external contact. The hydrogen thus prepared was purified by corrosive sublimate and potash, and desiccated by passage through a long tube packed with phosphoric anhydride. The oxygen was from chlorate of potash, or from mixed chlorates of potash and soda.

In a subsequent paper on the "Composition of Water†," I attacked the problem by a direct synthesis of water from weighed quantities of the two component gases. The ratio of atomic weights thus obtained was 15·89.

At the time when these researches were commenced, the latest work bearing upon the subject dated from 1845, and the number then accepted was 15·96. There was, however, nothing to show that the true ratio really deviated from the 16 : 1 of Prout's law, and the main object of my work was to ascertain whether or not such deviation existed. About the year 1888, however, a revival of interest in this question manifested itself, especially in the United States, and several results of importance have been published. Thus, Professor Cooke and Mr T. W. Richards found a number which, when corrected for an error of weighing that had at first been overlooked, became 15·869.

The substantial agreement of this number with those obtained by myself seemed at first to settle the question, but almost immediately afterwards

* *Roy. Soc. Proc.* Vol. XLIII. p. 356, February, 1888. [Vol. III. p. 37.]

† *Roy. Soc. Proc.* Vol. XLV. p. 425, February, 1889. [Vol. III. p. 233.]

there appeared an account of a research by Mr Keiser, who used a method presenting some excellent features, and whose result was as high as 15·949. The discrepancy has not been fully explained, but subsequent numbers agree more nearly with the lower value. Thus, Noyes obtains 15·896, and Dittmar and Henderson give 15·866.

I had intended further to elaborate and extend my observations on the synthesis of water from weighed quantities of oxygen and hydrogen, but the publication of Professor E. W. Morley's masterly researches upon the "Volumetric Composition of Water*" led me to the conclusion that the best contribution that I could now make to the subject would be by the further determination of the relative densities of the two gases. The combination of this with the number 2·0002†, obtained by Morley as the mean of astonishingly concordant individual experiments, would give a better result for the atomic weights than any I could hope to obtain directly.

In all work of this sort, the errors to be contended with may be classed as either systematic or casual. The latter are eliminated by repetition, and are usually of no importance in the final mean. It is systematic errors that are most to be dreaded. But although directly of but little account, casual errors greatly embarrass a research by rendering difficult and tedious the detection of systematic errors. Thus, in the present case, almost the only source of error that can prejudice the final result is impurity in the gases, especially in the hydrogen. The better the hydrogen, the lighter it will prove; but the discrimination is blunted by the inevitable errors of weighing. After perhaps a week's work it may become clear that the hydrogen is a little at fault, as happened in one case from penetration of nitrogen between the sealed-in platinum electrodes and the glass of the generator.

Another difficulty, which affects the presentation of results, turns upon the one-sided character of the errors most to be feared. As has been said, impure hydrogen can only be too heavy, and another important source of error, depending upon imperfect establishment of equilibrium of pressure between the contents of the globe and the external atmosphere, also works one-sidedly in the same direction. The latter source of error is most to be feared immediately after a re-greasing of the tap of the globe. The superfluous grease finds its way into the perforation of the plug, and partially blocks the passage, so that the six minutes usually allowed for the escape of the initial excess of pressure in the globe may become inadequate. Partly

* *Amer. Journ. Sci.* March, 1891.

† It should not be overlooked that this number is difficult to reconcile with views generally held as to the applicability of Avogadro's law to very rare gases. From what we know of the behaviour of oxygen and hydrogen gases under compression, it seems improbable that volumes which are as 2·0002 : 1 under atmospheric conditions would remain as 2 : 1 upon indefinite expansion. According to the formula of Van der Waals, a greater change than this in the ratio of volumes is to be expected. [1901. In later experiments Morley obtained 2·0027.]

from this cause and partly from incomplete washing out of nitrogen from the generator, the first filling of a set was so often found abnormally heavy that it became a rule in all cases to reject it. From these and other causes, such as accidental leakages not discovered at the time, it was difficult to secure a set of determinations in which the mean really represented the most probable value. At the same time, any arbitrary rejection of individual results must be avoided as far as possible.

In the present work two objects have been especially kept in view. The first is simplicity upon the chemical side, and the second the use of materials in such a form that the elimination of impurities goes forward in the normal working of the process. When, as in the former determinations, the hydrogen is made from zinc, any impurity which that material may contain and communicate to the gas cannot be eliminated from the generator; for each experiment brings into play a fresh quantity of zinc, with its accompanying contamination. Moreover, the supply of acid that can be included in one charge of the generator is inadequate, and good results are only obtained as the charge is becoming exhausted. These difficulties are avoided when zinc is discarded. The only material consumed during the experiments is then the water, of which a large quantity can be included from the first. On the other hand, the hydrogen liberated is necessarily contaminated with oxygen, and this must be removed by copper contained in a red-hot tube. In the experiments to be described the generator was charged with potash*, and the gases were liberated at platinum electrodes. In the case of a hydrogen filling the oxygen blew off on one side from a mercury seal, and on the other the hydrogen was conveyed through hot tubes containing copper. The bulk of the aqueous vapour was deposited in a small flask containing strong solution of potash, and the gas then passed over solid potash to a long tube packed with phosphoric anhydride. Of this only a very short length showed signs of being affected at the close of all operations.

With respect to impurities, other than oxygen and oxides of hydrogen, which may contaminate the gas, we have the following alternatives. Either the impurity is evolved much more rapidly than in proportion to the consumption of water in the generator, or it is not. If the rate of evolution of the impurity, reckoned as a fraction of the quantity originally present, is not much more rapid than the correspondingly reckoned consumption of water, the presence of the impurity will be of little importance. If on the other hand, as is probable, the rate of evolution is much more rapid than the consumption of water, the impurity is soon eliminated from the residue, and the gas subsequently generated becomes practically pure. A similar argument holds good if the source of the impurity be in the copper, or even

* At the suggestion of Professor Morley, the solution was freed from carbonate or nearly so, by the use of baryta, of which it contained a slight excess.

in the phosphoric anhydride; and it applies with increased force when at the close of one set of operations the generator is replenished by the mere addition of water. It is, however, here assumed that the apparatus itself is perfectly tight.

Except for the reversal of the electric current, the action of the apparatus is almost the same whether oxygen or hydrogen is to be collected. In the latter case the copper in the hot tubes is in the reduced, and in the former case in the oxidised, state. For the sake of distinctness we will suppose that the globe is to be filled with hydrogen.

The generator itself (Fig. 1) is of the U-form, with unusually long branches, and it is supplied from Grove cells with about 3 ampères of electric current. Since on one side the oxygen blows off into the air, the pressure in the generator is always nearly atmospheric. Some trouble has been caused by leakage between the platinum electrodes and the glass. In the later experiments to be here recorded these joints were drowned with mercury. On leaving the generator the hydrogen traverses a red-hot tube of hard glass charged with copper*, then a flask containing a strong solution of potash, and afterwards a second similar hot tube. The additional tube was introduced with the idea that the action of the hot copper in promoting the union of the hydrogen with its oxygen contamination might be more complete after removal of the greater part of the oxygen, whether in the combined or in the uncombined state. From this point onward the gas was nearly dry. In the earlier experiments the junctions of the hard furnace tubes with the soft glass of the remainder of the apparatus were effected by fusion. One of these joints remained in use, but the others were replaced by india-rubber connexions *drowned in mercury.* It is believed that no leakage occurred at these joints; but as an additional security a tap was provided between the generator and the furnace, and was kept closed whenever there was no forward current of hydrogen. In this way the liquid in the generator would be protected from any possible infiltration of nitrogen. Any that might find its way into the furnace tubes could easily be removed before the commencement of a filling.

Fig. 1.

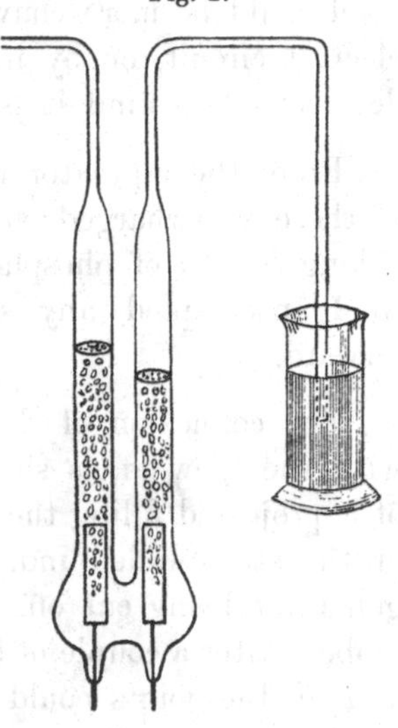

Almost immediately upon leaving the furnace tubes the gas arrives at a tap which for distinctness may be called the regulator. In the generator

* The copper must be free from sulphur; otherwise the contamination with sulphuretted hydrogen is somewhat persistent.

and in the furnace tubes the pressure must be nearly atmospheric, but in the globe there is (at the commencement) a vacuum. The transition from the one pressure to the other takes place at the regulator, which must be so adjusted that the flow through it is approximately equal to the production of gas. At first the manipulation of the regulator was a source of trouble and required almost constant attention, but a very simple addition gave the desired control. This was merely a long wooden arm, attached to the plug, which served both as a lever and as an indicator. Underneath the pointed extremity was a small table to which its motions could be referred. During the first two-thirds of a filling very little readjustment was needed, and the apparatus could be left for half an hour with but little fear of displacing too much the liquid in the generator. Towards the close, as the motive force fell off, the tap required to be opened more widely. Sometimes the recovery of level could be more conveniently effected by insertion of resistance into the electric circuit, or by interrupting it altogether for a few minutes. Into details of this kind it is hardly necessary to go further.

From the regulator the gas passed to the desiccating tubes. The first of these was charged with fragments of solid potash, and the second with a long length of phosphoric anhydride. Finally, a tube stuffed with glass wool intercepted any suspended matter that might have been carried forward.

The connexion of the globe with the generator, with the Töppler, and with the blow-off, is shown in the accompanying Fig. 2. On the morning of a projected filling the vacuous globe would be connected with the free end of the stout-walled india-rubber tube, and secured by binding wire. The generator being cut off, a high vacuum would be made up to the tap of the globe. After a couple of hours' standing the leakage through the india-rubber and at the joints could be measured. The amount of the leakage found in the first two hours was usually negligible, considered as an addition to a globeful of hydrogen, and the leakage that would occur in the hours following would (in the absence of accidents) be still smaller. If the test were satisfactory, the filling would proceed as follows:—

The electric current through the generator being established and the furnace being heated, any oxygen that might have percolated into the drying tubes had first to be washed out. In order to do this more effectively, a moderate vacuum (of pressure equal to about 1 inch of mercury) was maintained in the tubes and up to the regulator by the action of the pump. In this way the current of gas is made very rapid, and the half-hour allowed must have been more than sufficient for the purpose. The generator was then temporarily cut off, and a high vacuum produced in the globe connexion and in the blow-off tube, which, being out of the main current of gas, might be supposed to harbour impurities. After this the pump would be cut off, the connexion with the generator re-established, and, finally, the tap of the globe cautiously opened.

Fig. 2.

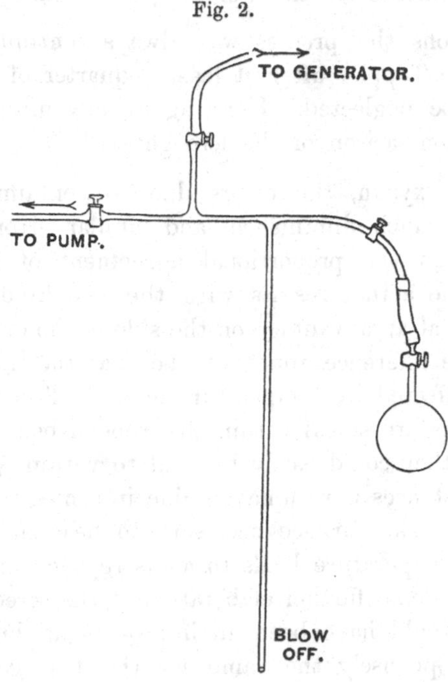

The operation of filling usually occupied from two to three hours. When the gas began to blow off under an excess of pressure represented by about half an inch of mercury, the blow-off cistern was lowered so as to leave the extremity of the tube free. For two minutes the current of gas from the generator was allowed to flow through, after which the generator was cut off, and the globe left in simple communication with the atmosphere, until it was supposed that equilibrium of pressure had been sufficiently established. Doubts have at various times been felt as to the interval required for this purpose. If too little time is allowed, there will remain an excess of pressure in the globe, and the calculated weight of the filling will come out too high. On the other hand, an undue prolongation of the time might lead to a diffusion of air back into the globe. In a special experiment no abnormal weight was detected after half-an-hour's communication, so that the danger on this side appeared to be small. When the passages through the taps were free from grease, one or two minutes sufficed for the establishment of equilibrium, but there was always a possibility of a partial obstruction. In the results to be presently given four minutes were allowed after the separation from the generator. It may be remarked that a part of any minute error that may arise from this source will be eliminated in the comparison with oxygen, which was collected under like conditions.

The reading of the barometers and thermometers at the moment when the tap of the globe was turned off took place as described in the former paper. The arrangements for the weighings were also the same.

In the evacuations the process was always continued until, as tested by the gauge of the Töppler after at least a quarter of an hour's standing, the residue could be neglected. Here, again, any minute error would be eliminated in the comparison of the two gases.

In the case of oxygen, the errors due to contamination (even with hydrogen) are very much diminished, and similar errors of weighing tell very much less upon the proportional agreement of the final numbers. A comparison of the actual results with the two kinds of gas does not, however, show so great an advantage on the side of the oxygen as might have been expected. The inference appears to be that the individual results are somewhat largely affected by temperature errors. Two thermometers were, indeed, used (on opposite sides) within the wooden box by which the globe is surrounded, and they could easily be read to within $\frac{1}{20}$° C. But in other respects, the circumstances were unfavourable in consequence of the presence in the same room of the furnace necessary to heat the copper. An error of $\pm 0{\cdot}1$° C. in the temperature leads to a discrepancy of 1 part in 1500 in the final numbers. Some further elaboration of the screening arrangements actually employed would have been an improvement, but inasmuch as the circumstances were precisely the same for the two gases, no systematic error can here arise. The thermometers were, of course, the same in the two cases.

The experiments are grouped in five sets, two for oxygen and three for hydrogen. In each set the work was usually continued until the tap of the globe required re-greasing, or until, owing to a breakage or to some other accident, operations had to be suspended.

Oxygen.

1891	Weight	Bar. temp., F.	Globe temp., C.
	grams	°	°
June 29	2·5182	70	20·85
July 2	2·5173	69	20·60
July 4	2·5172	67½	19·75
July 6	2·5193	70½	21·40
July 9	2·5174	64	17·60
July 10	2·5177	65½	19·05
Mean	2·51785	68	20°

The six fillings were all independent, except that of July 6, when the bulk of the oxygen remaining from the previous filling was not removed. It so happens that this case shows the greatest discrepancy, but there seems to be no sufficient reason for rejecting it.

Hydrogen.

1891	Weight	Bar. temp., F.	Globe temp., C.
	gram	°	°
July 31	0·15807	60½	15·90
August 4	0·15816	65	18·00
August 6	0·15811	66½	19·20
August 8	0·15803	65	18·15
August 11	0·15801	66	19·15
August 13	0·15809	68½	20·10
Mean	0·15808	65	18°

Hydrogen.

1891	Weight	Bar. temp., F.	Globe temp., C.
	gram	°	°
September 22.........	0·15800	58	14·5
September 24.........	0·15820	61½	16·3
September 28.........	0·15792	62	17·6
September 30.........	0·15788	63½	18·1
October 2	0·15783	62	17·3
Mean	0·15797	61	17°

Hydrogen.

1891	Weight	Bar. temp., F.	Globe temp., C.
	gram	°	°
October 26	0·15807	55	13·30
October 28	0·15801	56	14·00
October 31	0·15817	50	10·95
November 3	0·15790	53½	12·10
November 5	0·15810	55	12·00
November 7	0·15798	50	10·70
November 10	0·15802	48	9·30
November 13	0·15807	55½	12·70
Mean	0·15804	53	12°

Oxygen.

1891	Weight	Bar. temp., F.	Globe temp., C.
	grams	°	°
November 30	2·5183	53	12·15
December 3	2·5168	56	13·55
December 5	2·5172	56½	14·15
December 7	2·5181	58½	14·70
December 8	2·5156	51	11·15
Mean	2·5172	55	13°

In almost every case the weight of the globe *full* is compared with the mean of the immediately preceding and following weights *empty*. The numbers recorded in the second column are derived from the readings of the balance by the introduction of corrections—

(1) For the errors of the weights themselves, found by a systematic comparison, only relative values uncorrected for buoyancy being required.

(2) For the deviation of the mean* barometric reading at the time of filling from 30 inches (as read upon the vernier).

(3) For the deviation of the temperature of the barometers (Column 3) from 60° F.

(4) For the deviation of the temperature of the gas (as read upon the thermometers) from 12° C.

As an example, I will take in detail the calculation for the hydrogen filling of October 26. After the evacuation of October 24, the working globe (14) with its compensating volume piece and 0·4778 gm. stood on the left of the balance with globe (11) on the right. The position of equilibrium of the pointer, as determined after four different releasements, each observed in the usual manner, was 19·02 scale divisions. In like manner, after the evacuation of October 27, with the same weights in use, the equilibrium position of the pointer was 18·46. After the filling of October 26, the weights associated with (14) were 0·3220 gm. instead of 0·4778; and the pointer reading was 20·08. So far as the weights are concerned, the value of the hydrogen would be 0·4778 − 0·3220, or 0·1558 gm.; but to this we must add a correction corresponding to 1·34 scale divisions, being the difference between 20·08 and ½ (19·02 + 18·46). At the time in question, the value of a scale division was 0·00020 gm., so that we obtain—

$$0{\cdot}1558 + 0{\cdot}00027 = 0{\cdot}15607.$$

* There were two barometers.

The particular weights in use on this occasion were such that no correction is necessary in order to allow for their errors.

The mean barometer reading at the time of filling was 29·742, so that the factor required on this account is 30 : 29·742. The correction for temperature of gas is from 13·3 to 12°.

$$\begin{array}{lll} \text{Log } 0{\cdot}15607 \ldots\ldots\ldots\ldots & = & \bar{1}{\cdot}19332 \\ \text{For barometer } \ldots\ldots\ldots & & 0{\cdot}00375 \\ \text{For temperature } \ldots\ldots & & 0{\cdot}00198 \\ \hline \text{Log } 0{\cdot}15814 \ldots\ldots\ldots\ldots & = & \bar{1}{\cdot}19905 \end{array}$$

To this a correction for the temperature of the *barometer* has still to be applied. For 1° F. the correcting factor is (1 − 0·000089), or for 5° F. (1 − 0·000445). From 0·15814 we are thus to subtract 0·00007, giving the tabular number 0·15807.

A further minute correction to the mean of each set may be made for the temperature of the glass. A warm globe is larger than a cold one, and consequently holds more gas. If we suppose that the volume expansion of the glass per degree C. is 0·000025, we find, corrected to 12° C.—

Hydrogen.

1891	Weight	Bar. temp., F.	Globe temp., C.	Corrected to 12°
	gram	°	°	gram
July	0·15808	65	18	0·158056
September.........	0·15797	61	17	0·157950
October	0·15804	53	12	0·158040
Mean.........		60	16	0·158015

Oxygen.

1891	Weight	Bar. temp., F.	Globe temp., C.	Corrected to 12°
	grams	°	°	grams
June	2·51785	68	20	2·51735
November.........	2·51720	55	13	2·51713
Mean.........		61½	16½	2·57124

The means here exhibited give the weights of the two gases as they would be found with the globe at 12° C., and the barometers at 60° F. and at 30 inches. The close agreement of the mean temperatures for the two gases shows how little room there is for systematic error dependent upon imperfections in the barometers and thermometers. But the results still require modification before they can be compared with the view of deducing the relative densities of the gases.

In the first place, there is a systematic, though minute, difference in the pressures hitherto considered as corresponding. The terminal of the blow-off tube is 33 inches below the centre of the globe at the time of filling. In the one case this is occupied by hydrogen, and in the other by oxygen. If we treat the latter as the standard, we must regard the hydrogen fillings as taking place under an excess of pressure equal to $\frac{15}{16}$ of the weight of a column of oxygen 33 inches high; and this must be compared with 30 inches of mercury. Hence, if we take the sp. gr. of oxygen under atmospheric conditions at 0·0014, and that of mercury at 13·6, the excess of pressure under which the hydrogen was collected is as a fraction of the whole pressure

$$\frac{33}{30} \cdot \frac{15}{16} \cdot \frac{0{\cdot}0014}{13{\cdot}6} = 0{\cdot}000106;$$

and $0{\cdot}000106 \times 0{\cdot}158 = 0{\cdot}000017$. This, then, is what we must subtract from the weight of the hydrogen on account of the difference of pressures due to the gas in the blow-off tube. Thus

$$H = 0{\cdot}157998, \qquad O = 2{\cdot}51724.$$

{These numbers are not quite comparable with those given in the former communication, inasmuch as the standard temperature then used *for the barometers* was 55° F. Reduced so as to correspond to 60°, the former numbers become

$$H = 0{\cdot}15797, \qquad O = 2{\cdot}5174.$$

The agreement is satisfactory, especially when it is remembered that both gases were prepared by different methods in the two sets of experiments.—Feb. 17.}

But there is still another and a more important correction to be introduced. In my former paper it was shown that when the weighings are conducted in air the true weight of the gas contained in the globe is not given by merely subtracting the weight of the globe when empty from the weight when full. When the globe is empty, its external volume is less than when full, and thus, in order to obtain the true weight, the apparent weight of the gas must be increased by the weight of air whose volume is equal to the change of volume of the globe.

In order to determine the amount of this change of volume, the globe is filled to the neck with recently boiled distilled water, and the effect is observed upon the level in the stem due to a suction of, say, 20 inches of mercury. It is not advisable to carry the exhaustion much further for fear of approaching too nearly the point at which bubbles of vapour may be formed internally. In the earlier experiments, described in the preliminary note, the upper surface of the liquid was in the stem of the globe itself (below the tap), and the only difficulty lay in the accurate estimation of a change of volume occurring in a wide and somewhat irregular tube. The method employed was to produce, by introduction of a weighed quantity of mercury, a rise of level equal to that caused by the suction.

The advantage of this procedure lay in the avoidance of joints and of the tap itself, but, for the reasons given, the readings were not quite so accurate as might be desired. I wished, therefore, to supplement, if possible, the former determination by one in which the change of volume occurred in a tube narrower and of better shape. With this object in view, the stem of the globe was prolonged by a graduated tubular pipette attached with the aid of india-rubber. The tubes themselves were treated with gutta-percha cement, and brought almost into contact. It had hardly been expected that the joint would prove unyielding under the applied suction, but it was considered that the amount of the yielding could be estimated and allowed for by operations conducted *with tap closed.* The event, however, proved that the yielding at the joint was scarcely, if at all, perceptible.

The pipette, of bore such that 16 cm. corresponded to 1 c.c., was graduated to 0·01, and was read by estimation to 0·001 c.c. In order the better to eliminate the changes due to temperature, readings under atmospheric pressure, and under a suction of 20 inches of mercury, were alternated. On January 28, 1892, a first set gave $0·648 - 0·300 = 0·348$, a second gave $0·6645 - 0·316 = 0·3485$, and a third gave $0·675 - 0·326 = 0·349$. Similar operations with tap closed* gave no visible movement.

The result of the day's experiments was thus 0·3485 for 20 inches, or 0·523 for 30 inches, suction. Similar experiments on January 28, at a different part of the graduation, gave 0·526. On this day the yielding with tap closed was just visible, and was estimated at 0·001. As a mean result, we may adopt 0·524 c.c. The graduation of the pipette was subsequently verified by weighing a thread of mercury that occupied a measured length.

A part of the above-measured volume is due to the expansion of the water when the pressure is relieved. We may take this at 0·000047 of the volume per atmosphere. The volume itself may be derived with sufficient accuracy for the present purpose from the weight of its oxygen contents. It is 2·517/0·00137, or 1837 c.c. The expansion of the water per atmosphere is thus $0·000047 \times 1837$, or 0·087 c.c. This is to be subtracted from 0·524, and leaves 0·437 c.c. This number applies strictly to the volume enclosed

* For greater security the tap was turned while the interior was under suction.

within the glass, but the change in the external volume of the globe will be almost the same*.

The correction now under consideration is thus the weight of 0·437 c.c. of air at the average temperature of the balance room. The density of this air may be estimated at 0·00122; so that the weight of 0·437 c.c. is 0·000533 gm. This is the quantity which must be added to the apparent weights of the gases. The former estimate was 0·00056 gm. The finally corrected weights are thus

$$H = 0{\cdot}158531, \qquad O = 2{\cdot}51777\,;$$

and for the ratio of densities we have

15·882.

This corresponds to a mean atmospheric condition of pressure and temperature.

If we combine the above ratio of densities with Professor Morley's ratio of volumes, viz. 2·0002 : 1, we get, as the ratio of atomic weights, 15·880.

If we refer to the table, we see that the agreement of the first and third series of hydrogen weighings is very good, but that the mean from the second series is decidedly lighter. This may have been in part fortuitous, but it is scarcely probable that it was so altogether. Under the circumstances we can hardly reckon the accuracy of the final results as closer than $\frac{1}{3000}$.

A word should perhaps be said upon a possible source of systematic error, viz. mercury vapour. There is no doubt that hydrogen passed over mercury takes up enough to cause a slow and superficial, but quite distinct, discoloration of sulphur over which it subsequently flows. In the experiments here recorded, the gas did not, indeed, flow over mercury in mass, but, inasmuch as mercury was used to secure the tightness of some of the joints, it is difficult to feel sure of its absence. Again, in evacuations conducted with a mercury pump can the vacuum be regarded as free from mercury vapour, which, it must be remembered, would not show itself upon the gauge of the Töppler? If both the hydrogen and the "vacuum" were saturated with mercury vapour, the result of the weighings would, according to Dalton's law, be free from its influence. The same may be said of any volatile impurity arising from the grease† upon the stopcocks. As the

* For a spherical shell of glass of uniform thickness and with elastic constants following Poisson's law, the ratio of the difference of the internal and external expansion to either of them is $4t/3a$, where t is the thickness of the shell, and a the mean radius. In the present application the value of a/t, deduced from the measured circumference and from the weight of glass, is about 110.

{Perhaps an arrangement in which the *external* volume is directly measured would have been preferable. No allowance for expansion of water would then be needed.—Feb. 17.}

† Composed of vaseline and beeswax.

matter stands, the results must, I think, be regarded as affected with a possible error amounting to a fraction of the weight of mercury vapour at the temperatures employed. But this is probably a very small quantity.

According to Hertz*, the vapour-pressure of mercury at 15° C. would be about 0·001 mm. If this be correct, the weight of mercury vapour in an atmosphere of hydrogen would be as a fraction of the latter†

$$\frac{0{\cdot}001}{760} \times [100] = \frac{1}{[7600]}.$$

It appears that in an investigation of hydrogen aiming at an accuracy of 1/10,000 the question of mercury vapour requires very careful consideration.

The accompanying table of results found by various experimenters may be useful for comparison:—

Name	Date	Atomic weights	Densities
Dumas	1842	15·96	—
Regnault	1845	—	15·96
Rayleigh	1888	—	15·884
Cooke and Richards	1888	15·869	—
Keiser	1888	15·949	—
Rayleigh	1889	15·89	—
Noyes	1890	15·896	—
Dittmar	1890	15·866	—
Morley	1891	15·879	—
Leduc	1891	—	15·905
Rayleigh	1892	—	15·882

In conclusion, I must express my obligations to Mr Gordon, who has assisted me throughout. The work has been unusually tedious, partly from its inherent nature, requiring as it does a certainty of 0·1 milligram in the weighings, and still more from the constant liability to accidents, which may render nugatory a large amount of preparatory work.

[1901. For further investigations respecting the density of gases see *Proc. Roy. Soc.* LIII. p. 481, 1892; LV. p. 340, 1894; LVII. p. 265, 1895. These papers will be included in Vol. IV.]

* *Wied. Ann.* Vol. XVII. p. 199.

† [1901. In the original the numbers were 200 and 3800. I owe the correction to Prof. Japp.]

188.

SUPERHEATED STEAM.

[*Nature*, XLV. pp. 375, 376, 438, 512, 1892.]

I HAVE noticed a curious misapprehension, even on the part of high authorities, with respect to the application of Carnot's law to an engine in which the steam is superheated after leaving the boiler. Thus, in his generally excellent work on the steam-engine*, Prof. Cotterill, after explaining that in the ordinary engine the superior temperature is that of the boiler, and the inferior temperature that of the condenser, proceeds (p. 141): "When a superheater is used, the superior temperature will of course be that of the superheater, which will not then correspond to the boiler pressure."

This statement appears to me to involve two errors, one of great importance. When the question is raised, it must surely be evident that, in consideration of the high latent heat of water, by far the greater part of the heat is received at the temperature of the boiler, and not at that of the superheater, and that, of the relatively small part received in the latter stage, the effective temperature is not that of the superheater, but rather the mean between this temperature and that of the boiler. An estimate of the possible efficiency founded upon the temperature of the superheater is thus immensely too favourable. Superheating does not seem to meet with much favour in practice; and I suppose that the advantages which might attend its judicious use would be connected rather with the prevention of cylinder condensation than with an extension of the range of temperature contemplated in Carnot's rule.

If we wish effectively to raise the superior limit of temperature in a vapour-engine, we must make the boiler hotter. In a steam-engine this means pressures that would soon become excessive. The only escape lies in

* Second edition (Spon: London, 1890). [1900. Prof. Cotterill explained subsequently (*Nature*, Vol. XLV. p. 414, 1892) that I had misunderstood the passage in question.]

the substitution for water of another and less volatile fluid. But, of liquids capable of distillation without change, it is not easy to find one suitable for the purpose. There is, however, another direction in which we may look. The volatility of water may be restrained by the addition of saline matters, such as chloride of calcium or acetate of soda. In this way the boiling temperature may be raised without encountering excessive pressures, and the possible efficiency, according to Carnot, may be increased.

The complete elaboration of this method would involve the condensation of the steam at a high temperature by reunion with the desiccating agent, and the communication of the heat evolved to pure water boiling at nearly the same temperature, but at a much higher pressure. But it is possible that, even without a duplication of this kind, advantage might arise from the use of a restraining agent. The steam, superheated in a regular manner, would be less liable to premature condensation in the cylinder, and the possibility of obtaining a good vacuum at a higher temperature than usual might be of service where the supply of water is short, or where it is desired to effect the condensation by air.

[1900. The indications given above were, it would seem, too concise. See a further correspondence (*Nature*, Vol. XLV. pp. 413—414, 438, 486, 510). The proposal was to condense the pure steam reversibly by reunion with the desiccating agent. To this end it would be necessary to have a supply of solution of the same strength as in the boiler, but at the temperature and pressure of condensation. Theoretically, the amount of this supply would need to be an indefinite multiple of the water evaporated and to be condensed, but in practice a ratio of 4 or 5 times would suffice. To effect, without serious dissipation of energy, the necessary changes of pressure and temperature as the solution circulates between the boiler and the condenser, pumps and regenerators would be needed. The former would do work in restoring the solution (with the condensed water) to the boiler, but this would be largely compensated by the solution flowing the other way from the boiler to the condenser. In the regenerator the heat would be passed from the outgoing to the returning solution with but a small difference of temperature and, therefore, with a small dissipation of energy. A similar use of the principle of regeneration is made in the air liquefying machines of Linde and Hampson.]

Heat Engines and Saline Solutions.

Mr Macfarlane Gray (p. 414) appears to call in question my assertion that in a vapour-engine a saline solution may take the place of a simple liquid when it is desired to replace water by a substance of less volatility, and that the advantage which Carnot proved to attend a high temperature can thus be attained without encountering an unduly high pressure. He contends that "the saline mixture is not the working substance. Carnot's law refers to the working substance only, and not to anything left in the boiler."

Perhaps the simplest way of meeting this objection is to point out that Maxwell's exposition of Carnot's engine (*Theory of Heat*, chapter viii.) applies *without the change of a single word*, whether the substance in the cylinder be water, mercury, or an aqueous solution of chloride of calcium. In each case there is a definite relation between pressure and temperature; and (so far as the substance is concerned) all that is necessary for the reversible operation of the engine is that the various parts of the working substance should be in equilibrium with one another throughout.

Let us compare the behaviour of water in Carnot's engine before and after the addition of chloride of calcium, supposing that the maximum and minimum pressures are the same in the two cases. The only effect of the addition is to raise both the superior and the inferior temperatures. The heat rejected at the inferior temperature may still be available for the convenient operation of an engine working with pure water. At the upper limit, *all* the heat is received at the highest point of temperature—a state of things strongly contrasted with that which obtains when vapour rising from pure water is afterwards superheated.

Heat Engines and Saline Solutions.

I have nothing to modify in what I have written under the above heading (p. 438); but to deal completely with all the questions raised by Mr Macfarlane Gray (p. 486) would require half a treatise. I will limit myself to a few brief remarks.

(1) In Carnot's engine there is no (separate) boiler or condenser.

(2) When I spoke of the various parts of the working substance being in equilibrium with one another, I referred to complete equilibrium, thermal as well as mechanical. If the temperature varies from one part to another there is no equilibrium.

(3) On the above understanding the pressure of vapour in equilibrium with a saline solution of a given strength is a definite function of the temperature.

(4) Let me suggest that the origin of the difficulty may lie in the phrase "superheated vapour" which has not so definite a meaning as Mr Gray seems to ascribe to it. Whether vapour be superheated or no, depends, not only

upon the condition of the vapour itself, *but also upon the bodies with which it is in contact.* Vapour which is merely saturated in contact with a saline solution must be regarded as superheated when contact with the solution is cut off. In the first situation it would condense upon compression, and in the second situation it would not.

In conclusion, I will hazard the prediction that, if the heat-engines of the distant future are at all analogous to our present steam-engines, either the water (as the substance first heated) will be replaced by a fluid of less inherent volatility, or else the volatility of the water will be restrained by the addition to it of some body held in solution.

189.

ABERRATION*.

[*Nature*, XLV. pp. 499—502, 1892.]

UNDER this head may conveniently be considered not only the apparent displacement of the stars discovered by Bradley, but other kindred phenomena dependent upon the velocity of light bearing but a finite ratio to that of the earth in its orbit round the sun, and to other astronomical velocities.

The explanation of stellar aberration, as usually given, proceeds rather upon the basis of the corpuscular than of the wave theory. In order to adapt it to the principles of the latter theory, Fresnel found it necessary to follow Young in assuming that the æther in any vacuous space connected with the earth (and therefore practically in the atmosphere) is undisturbed by the earth's motion of 19 miles per second. Consider for simplicity the case in which the direction of the star is at right angles to that of the earth's motion, and replace the telescope, which would be used in practice, by a pair of perforated screens, on which the light falls perpendicularly. We may further imagine the luminous disturbance to consist of a single plane pulse. When this reaches the anterior screen, so much of it as coincides with the momentary position of the aperture is transmitted, and the remainder is stopped. The part transmitted proceeds upon its course through the æther independently of the motion of the screens. In order, therefore, that the pulse may be transmitted by the aperture in the posterior screen, it is evident that the line joining the centres of the apertures must not be perpendicular to the screens and to the wave front, as would have been necessary in the case of rest. For in consequence of the motion of the posterior screen in its own

* This paper was written in 1887, when I was occupied with my article upon "Wave Theory" for the *Encyclopædia Britannica*, and at a time when a more extensive treatment was contemplated than was afterwards found practicable. Friends upon whom I can rely are of opinion that its publication may be useful; and, as I am not able to give it a complete revision, I prefer to let it stand under its original date, merely warning the reader that very important work has since been published by Michelson.—January, 1892.

plane the aperture will be carried forward during the time of passage of the light. By the amount of this motion the second aperture must be drawn backward, in order that it may be in the place required when the light reaches it. If the velocity of light be V, and that of the earth be v, the line of apertures giving the apparent direction of the star must be directed forwards through an angle equal to v/V. More generally, if the angle between the star and the point of the heavens towards which the earth is moving be α, there will be an apparent displacement towards the latter point, expressed by $\sin\alpha \,.\, v/V$, and independent of the position upon the earth's surface where the observation is made. The ratio v/V is about $\frac{1}{10000}$.

The aperture in the anterior screen corresponds to the object-glass of the telescope with which the observation would actually be made, and which is necessary in order to produce agreement of phase of the various elementary waves at a moderately distant focal point. The introduction of a refracting medium would complicate the problem, and is not really necessary for our present purpose. As has been shown (*Phil. Mag.* March 1881 [Vol. I. p. 513], "On Images formed without Reflection or Refraction"), the only use of an object-glass is to shorten the focal length. Our imaginary screens may be as far apart as we please, and if the distance is sufficient, the definition, and consequently the accuracy of alignment, is as great as could be attained with the most perfect telescope whose aperture is equal to that in the anterior screen.

It appears, then, that stellar aberration in itself need present no particular difficulty on the wave theory, unless the hypothesis of a quiescent æther at the earth's surface be regarded as such. But there are a variety of allied phenomena, mostly of a negative kind, which require consideration before any judgment can be formed as to the degree of success with which the wave theory meets the demands made upon it. In the first place, the question arises whether terrestrial optical phenomena could remain unaffected by the supposed immense relative motion of instruments and of the æther; whether reflection, diffraction, and refraction, as ordinarily observed by us, could be independent of the direction of the rays relatively to the earth's motion. It may be stated at once that no such influence has been detected, even in experiments carefully designed with this object in view.

Another class of experiments, with the results of which theory must be harmonized, are those of Fizeau and Michelson upon the velocity of light in ponderable refracting media which have a rapid motion (relatively to the instruments and other surrounding bodies) in the direction of propagation, or in the opposite direction. These very important researches have proved that in the case of water the velocity of the ponderable medium is not without effect; but that the increment or decrement of the velocity of propagation is very decidedly less than the velocity of the water. On the other hand, the motion of air, even at high velocities, has no perceptible influence upon the propagation of light through it.

Again, it has been found by Airy*, as the result of an experiment originally suggested by Boscovitch, that the constant of stellar aberration is the same, whether determined by means of a telescope of the ordinary kind, or by one of which the tube is filled with water. It is clear that, according to Fresnel's views of the condition of the æther at the earth's surface, this agreement must involve some particular supposition as to the propagation of light in moving refracting media.

The theory of these phenomena must evidently turn upon the question whether the æther at the earth's surface is at rest, absolutely or relatively to the earth†; and this fundamental question has not yet received a certain answer. The independence of terrestrial optical phenomena of the earth's motion in its orbit is, of course, more easily explained upon the latter alternative; or rather no explanation is required. But in that case the difficulty is thrown upon stellar aberration, which follows a more simple law than we should expect to apply in the case of an æther disturbed by the passage of a body in its neighbourhood. Prof. Stokes has, indeed, attempted a theory on these lines‡, by supposing the ætherial motion to be what is called in hydrodynamics irrotational. In strictness there is, however, no such motion possible, subject to the condition of vanishing absolutely at a great distance, and relatively at the earth's surface; and it does not appear that the objection thus arising can be satisfactorily met.

If we start from the experimental facts which have the most direct bearing upon the question under discussion, we are led to regard Fresnel's views (doubtless in some generalized form) as the more plausible. From the results of Fizeau and Michelson relative to air, we may conclude with tolerable confidence that a small mass of ponderable matter, of very low refracting power, moving in space, would not appreciably carry the æther with it. The extension of the argument to a body as large as the earth is not unnatural, though it involves certainly an element of hypothesis. In like manner, if the globe were of water, we should expect the æther to be carried forward, but not to the full amount. The simplest supposition open to us is that, in any kind of ponderable matter, forming part of a complex mass, the æther is carried forward with a velocity dependent upon the local refracting power, but independent of the refracting power and velocity of other parts of the mass. In the earth's atmosphere, where the refracting power is negligible, the æther would be sensibly undisturbed.

* *Proc. Roy. Soc.* xx. 1872, p. 35; xxi. 1873, p. 121.

† An accusation of crudeness might fairly be brought against this phraseology; but an attempt to express the argument in more general language would probably fail, and would in any case be tedious.

‡ *Phil. Mag.* xxviii. 1846, p. 76; xxix. 1846, p. 6.

If we agree to adopt this point of view provisionally, we have next to consider the relation between the velocity of luminous propagation in moving ponderable matter and the refractive index. The character of this relation was discovered by Fresnel, whose argument may be thrown into the following form.

Consider the behaviour of the æther when a plate of ponderable matter (index $= \mu$) is carried forward through vacuum with velocity v in a direction perpendicular to its plane. If D be the density of the æther in vacuum, and D_1 the density in the refracting medium, then, according to Fresnel's views as to the cause of refraction, $D_1 = \mu^2 D$. The æther is thus condensed as the plate reaches it; and if we assume that the whole quantity of æther is invariable, this consideration leads to the law giving the velocity (xv) with which the denser æther within the plate must be supposed to be carried forward. For conceive two ideal planes, one in the plate and one in the anterior vacuous region, to move forward with velocity v. The whole amount of æther between the planes must remain unchanged. Now, the quantity entering (per unit area and time) is Dv, and the quantity leaving is $D_1(v - xv)$. Hence

$$x = 1 - \mu^{-2},$$

so that the velocity with which the æther in the plate is carried forward is $v(1 - \mu^{-2})$, tending to vanish as μ approaches unity. If V be the velocity of light in vacuum and V/μ the velocity in the medium at rest, then the absolute velocity of light in the moving medium is

$$V/\mu \pm v(1 - \mu^{-2}). \quad\text{.............................(1)}$$

Whatever may be thought of the means by which it is obtained, it is not a little remarkable that this formula, and no other, is consistent with the facts of terrestrial refraction, if we once admit that the æther in the atmosphere is at absolute rest. It is not probable that the æther, in moving refracting bodies, can properly be regarded as itself in motion; but if we knew more about the matter we might come to see that the objection is verbal rather than real. Perhaps the following illustration may assist the imagination. Compare the æther in vacuum to a stretched string, the transverse vibrations of which represent light. If the string is loaded, the velocity of propagation of waves is diminished. This represents the passage of light through stationary refracting matter. If now the loads be imagined to run along the string with a velocity not insensible in comparison with that of the waves, the velocity of the latter is modified. The substitution of a membrane for a string will allow of a still closer parallel. It appears that the suggested model would lead to a somewhat different law of velocity from that of Fresnel; but in bringing it forward the object is merely to show that we need not interpret Fresnel's language too literally.

We will now consider a few examples of the application of the law of velocity in a moving medium; and first to the experiment of Boscovitch, in which stellar aberration is observed with a telescope filled with water. We have only to suppose the space between the two screens of our former explanation to be occupied by water, which is at rest relatively to the screens. In consequence of the movement of the water, the wave, after traversing the first aperture, is carried laterally with the velocity $v(1-\mu^{-2})$, and this is to be subtracted from the actual velocity v of the aperture of the posterior screen. The difference is $\mu^{-2}v$. The ratio of this to the velocity of light in water (V/μ) gives the angular displacement of the second aperture necessary to compensate for the motion. We thus obtain $\mu^{-1}v/V$. This angle, being measured in water, corresponds to v/V in air; so that the result of the motion is to make the star appear as if it were in advance of its real place by the angle v/V, precisely as would have happened had the telescope contained air or vacuum instead of water.

We will now calculate the effect of the motion of a plate perpendicular to its own plane upon the retardation of luminous waves moving in the same (or in the opposite) direction. The velocity of the plate is v, its index is μ, and its thickness is d. Denoting, as before, the velocity of the æther within the plate by xv, and supposing, in the first place, that the signs of v and V are the same, we have, for the absolute velocity of the wave in the plate,

$$V/\mu + xv.$$

We have now to express the time (t) occupied by the wave in traversing the plate. This is not to be found by simply dividing d by the above written velocity; for during the time t the anterior face of the plate (which the wave reaches last) is carried forward through the distance vt. Thus, to determine t we have

$$(V/\mu + xv)\,t = d + vt,$$

whence

$$\frac{Vt}{d} = \frac{\mu}{1+(x-1)\,\mu v/V}. \qquad \text{.........................(2)}$$

The time, t_0, which would have been occupied in traversing the same distance $(d+vt)$, had the plate been away is given by

$$Vt_0 = d + vt;$$

so that

$$\frac{Vt_0}{d} = 1 + \frac{\mu v/V}{1+(x-1)\,\mu v/V}. \qquad \text{.....................(3)}$$

Thus

$$\frac{V(t-t_0)}{d} = \frac{\mu(1-v/V)}{1+(x-1)\,\mu v/V} - 1. \qquad \text{.....................(4)}$$

Substituting in this Fresnel's value of x, viz. $(1-\mu^{-2})$, and neglecting as insensible the square of v/V, we find

$$V(t-t_0) = (\mu-1)\,d\,(1-v/V). \qquad \text{.....................(5)}$$

If we suppose that part of the original wave traverses the plate, and that part passes alongside, (5) gives the relative retardation—that is, the distance between the wave-fronts which were originally in one plane. It would appear at first sight that this result would give us the means of rendering v evident. For the retardation, depending upon the sign of v/V, will be altered when the direction of the light is reversed, and this we have it in our power to bring about by simply turning our apparatus through 180°. A more careful examination will, however, lead us to a different conclusion.

The most obvious way of examining the retardation would be to use homogeneous light, and, by producing regular interference of the two portions, to observe the positions of the fringes, and any displacement that might result from a shift of the apparatus relatively to the direction of the earth's motion. But if we employ for this purpose a terrestrial flame, *e.g.* that of a Bunsen's burner containing sodium, we have to take into account the fact that the source is itself in motion. For it is evident that the waves which pass in a given time through any point towards which the source is moving are more numerous than had the source been at rest, and that the wave-lengths are correspondingly shortened. If v be the velocity of the source, the wave-length is changed from λ to $\lambda(1 - v/V)$. At a point behind, from which the source is retreating, the wave-length is $\lambda(1 + v/V)$. We shall have occasion to refer again to this principle, named after Döppler, as applied by Huggins and others to the investigation of the motion of the heavenly bodies in the line of sight.

Referring now to (5), we see that, although the *absolute* retardation is affected by v, yet that the retardation *as measured in wave-lengths* remains unaffected. If, then, there be, in the absence of v, an agreement of phase between the two interfering beams, the introduction of v will cause no disturbance. Consequently no shifting of the interference bands is to be expected when the apparatus is turned so that the direction of propagation makes in succession all possible angles with that of the earth's motion.

The experiment has been modified by Hoek*, who so arranged matters as to eliminate the part of the retardation independent of v. As before, of two parallel beams A and B, one, A, passes through a plate of refracting medium; the other, B, through air. The beams are then collected by a lens, at the principal focus of which is placed a mirror. After reflection by this mirror, the beams exchange paths, B returning through the plate, and A through air. Apart, therefore, from a possible effect of the motion, there would be complete compensation and no final difference of path. As to the effect of the motion, it would appear at first sight that it ought to be sensible. During the first passage, A is (on account of v) accelerated: on the return, B is retarded; and thus we might expect, upon the whole, a relative

* *Archives Néerlandaises*, t. III. p. 180 (1868), t. IV. p. 443 (1869).

acceleration of A equal to $(\mu - 1) d . 2v/V$. But here, again, we have to consider the fact that another part of the apparatus, viz. the mirror, partakes of the motion. In the act of reflection the original retardation of A is increased by twice the distance through which the mirror retreats in the interval between the arrival of the two waves. This distance is (with sufficient approximation) $(\mu - 1) d . v/V$; so that the influence of the movement of the mirror just compensates the acceleration of A which would have resulted in the case of a fixed mirror. On the whole, then, and so long as the square of v/V may be neglected, no displacement of fringes is to be expected when the apparatus is turned. The fact that no displacement was observed by Hoek, nor in an analogous experiment by Mascart*, proves that if the stationary condition of the æther in terrestrial vacuous spaces be admitted, we are driven to accept Fresnel's law of the rate of propagation in moving refracting media.

What is virtually another form of the same experiment was tried by Maxwell†, with like negative results. In this case, prisms were used instead of plates; and the effect, if existent, would have shown itself by the displacement of the image of a spider-line when the instrument was turned into various azimuths.

On the basis of Fresnel's views it may, in fact, be proved generally that, so far as the first power of v/V is concerned, the earth's motion would not reveal itself in any phenomenon of terrestrial refraction, diffraction, or ordinary refraction. The more important special cases were examined by Fresnel himself, and the demonstration has been completed by Stokes‡. Space will not allow of the reproduction of these investigations here, and this is the less necessary, as the experiment of Hoek, already examined, seems to raise the principal question at issue in the most direct manner.

Another point remains to be touched upon. We have hitherto neglected *dispersion*, treating μ as constant. In stationary dispersing media, μ may be regarded indifferently as a function of the wave-length or of the periodic time. When, however, the medium is in motion, the distinction acquires significance; and the question arises, What value of μ are we to understand in the principal term V/μ of (1)? Mascart points out that the entirely negative results of such experiments as those above described indicate that, in spite of the difference of wave-length due to the motion, we must take the same value of μ as if the medium and the source had been at rest, or that μ is to be regarded as a function of the *period*.

Mascart has experimented also upon the influence of the earth's motion upon double refraction with results which are entirely negative. The theo-

* *Ann. de l'École Normale*, t. III. (1874).

† *Phil. Trans.* 1863, p. 532.

‡ *Phil. Mag.* XXVIII. p. 76 (1846). See also Mascart, *Ann. de l'École Normale*, t. I. (1872), t. III. (1874); and Verdet, *Œuvres*, t. IV., deuxième partie.

retical interpretation must remain somewhat ambiguous, so long as we remain in ignorance of the mechanical cause of double refraction.

Reference has already been made to the important experiments of Fizeau and Michelson upon the velocity of light in moving media. The method, in its main features, is due to the former*, and is very ingeniously contrived for its purpose. Light issuing from a slit is rendered parallel by a collimating lens, and is then divided into two portions, which traverse tubes containing running water. After passing the tubes, the light falls upon a focussing lens and mirror (as in Hoek's experiment), the effect of which is to interchange the paths. Both rays traverse both tubes; and, consequently, when ultimately brought together, they are in a condition to produce interference bands. If now the water is allowed to flow through the tubes in opposite directions, one ray propagates itself throughout *with* the motion of the water, and the other *against* the motion of the water; and thus, if the motion has any effect upon the velocity of light, a shift of the bands is to be expected. This shift may be doubled by reversing the flow of water in the tubes.

Fizeau's investigation has recently been repeated in an improved form by Michelson†.

"Light from a source at *a* falls on a half-silvered surface, *b*, where it divides: one part following the path *bcdefbg*, and the other the path *bfedcbg*. This arrangement has the following advantages: (1) it permits the use of an extended source of light, as a gas flame; (2) it allows any distance between

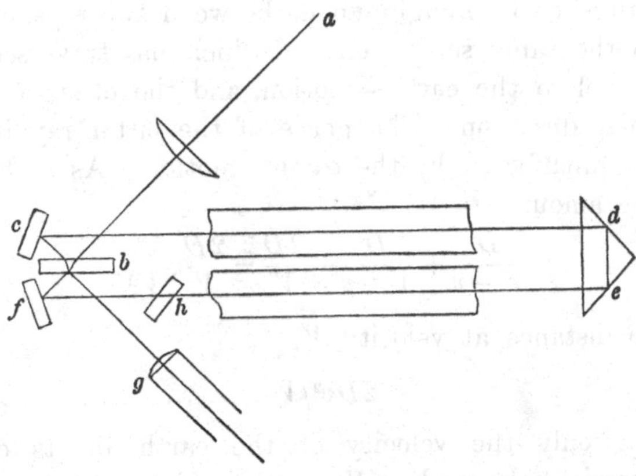

the tubes which may be desired; (3) it was tried by a preliminary experiment, by placing an inclined plate of glass at *h*. The only effect was either to alter the width of the fringes, or to alter their inclination; *but in no case*

* *Ann. de Chimie*, III. lvii. (1859).

† *American Journal*, Vol. XXXI. p. 377 (1886).

was the centre of the white fringe affected. Even holding a lighted match in the path had no effect on this point.

"The tubes containing the fluid were of brass, 28 mm. internal diameter; and in the first series of experiments, a little over 3 metres in length, and in the second series a little more than 6 metres."

Even with the longer tubes and the full velocity (about 8 metres per second) the displacement on reversal amounted to less than the width of a fringe. Nevertheless, fairly concordant results were arrived at; and they showed that the fraction (x) of the velocity of the water (v) by which the velocity of light is altered is ·434, with a possible error of $\pm$·02. The numerical value of the theoretical expression is

$$x = 1 - \mu^{-2} = \cdot 437,$$

in very close accordance.

"The experiment was also tried with air moving with a velocity of 25 metres per second. The displacement was about ·01 of a fringe; a quantity smaller than the probable error of observation. The value calculated from $(1 - \mu^{-2})$ would be ·0036."

We have seen that, so far as the first power of v/V is concerned, Fresnel's theory agrees with all the facts of the case. The question whether it is possible to contrive an experiment in which v^2/V^2 shall be sensible, has been considered by Michelson*, who, having arrived at an affirmative conclusion, proceeded to attack this very difficult experimental problem. In Michelson's apparatus interference is brought about between two rays, coming of course originally from the same source, one of which has traversed, to and fro, a distance D parallel to the earth's motion, and the other a like distance in the perpendicular direction. The phase of the latter ray is considered by Michelson to be unaffected by the earth's motion. As to the former, it is retarded by the amount

$$\frac{D}{V-v} + \frac{D}{V+v} - \frac{2D}{V} = \frac{2D}{V} \cdot \frac{v^2}{V^2},$$

or, reckoned in distance at velocity V,

$$2Dv^2/V^2 \text{..................................} (6)$$

"Considering only the velocity of the earth in its orbit, the ratio $v/V = 10^{-4}$ approximately, and $v^2/V^2 = 10^{-8}$. If $D = 1200$ mm., or in wave-lengths of yellow light, 2,000,000, then in terms of the same unit,

$$2Dv^2/V^2 = \cdot 04.$$

"If, therefore, an apparatus is so constructed as to permit two pencils of light, which have travelled over paths at right angles to each other, to

* *American Journal*, XXII. p. 120 (1881).

interfere, the pencil which has travelled in the direction of the earth's motion will in reality travel ·04 of a wave-length further than it would have done were the earth at rest. The other pencil, being at right angles to the motion, would not be affected.

"If now the apparatus be revolved through 90°, so that the second pencil is brought into the direction of the earth's motion, its path will be lengthened ·04 wave-length. The total change in the position of the interference bands would be ·08 of the distance between the bands, a quantity easily measurable."

In the actual experiment, the earth's velocity was not available to the full extent, and the displacement to be expected on this account was reduced to ·048; but Michelson considers that some addition to it should be made on account of the motion of the solar system as a whole. The displacement actually found was ·022: and when the apparatus was employed in such azimuths that the rotation should have had no effect in any case, ·034. These results are very small, and Michelson gives reasons for regarding them as partially systematic errors of experiment. He concludes that there is no real displacement of the bands, and that the hypothesis of a stationary æther is thus shown to be inconsistent with fact.

It has, however, been recently pointed out by Lorentz* that Michelson has overestimated the effect to be expected according to Fresnel's views. The ray which travels perpendicularly to the earth's motion is not unaffected thereby, but is retarded to an amount represented by Dv^2/V^2. The outstanding relative retardation is thus only Dv^2/V^2, instead of the double of this quantity. Accepting this correction, we have to expect, according to Fresnel's views, a shift of only ·024 of a band in Michelson's experiment.

Under these circumstances Michelson's results can hardly be regarded as weighing heavily in the scale. It is much to be wished that the experiment should be repeated with such improvements as experience suggests. In observations spread over a year, the effects, if any, due to the earth's motion in its orbit, and to that of the solar system through space, would be separated.

On the whole, Fresnel's hypothesis of a stationary æther appears to be at the present time the more probable; but the question must be considered to be an open one. Further evidence would be most important: but it is difficult to see from what quarter anything essentially new can be expected. It might be worth while for astronomers to inquire whether it is really true, as is generally assumed, that stellar aberration is independent of the position upon the earth's surface from which the observation is made. Another question that might, perhaps, be submitted with advantage to an experimental

* "Over den invloed dien de beweging der aarde of de licht vershijnnselen uitoefent." (Amsterdam, 1886.)

examination is whether the propagation of light in air is affected by the rapid motion of heavy masses parallel to, and in the immediate neighbourhood of, the ray.

If we once admit the principle that, whatever the explanation may be, no ordinary* terrestrial observation is affected by the earth's motion, it is easy to give an account of what must happen when the light comes from an external source which may have a motion in the line of sight. Imagine, for example, a spectroscopic examination of a soda flame situated on a star and vibrating in identical periods with those of terrestrial soda flames. In accordance with Döppler's principle, the wave-lengths are altered by a relative motion in the line of sight, and the fact may be rendered evident by a comparison between the spectra of the star and of the terrestrial flame, held so as to be seen in the same direction. The simplest case is when the flame is entirely external to the apparatus, so that both lights are treated in precisely the same way. It is evident that, under these circumstances, the difference between the two cannot fail to become apparent: and this way of regarding the matter shows also that the apparent displacement of the bright lines in the stellar spectrum is dependent upon the relative, and not further upon the absolute, motions of the star and of the earth. The mean of observations, equally distributed over the year, would thus give data for determining the relative motion in the line of sight of the star and of the solar system.

If the external source be the sun itself, it might be thought that the spectra must agree almost perfectly, the eccentricity of the earth's orbit being so very small. But the sun is a revolving body, and consequently a distinction must be made according to the part of the sun from which the light proceeds. It is found, in fact, that a very sensible shift takes place in the positions of the dark lines according as the light under observation comes from the advancing or from the retreating limb. This circumstance has been successfully employed by Thollon and Cornu to distinguish between lines having a solar and a terrestrial origin. In the latter case it is a matter of indifference from which part of the sun the light proceeds.

In general optical theory the finiteness of the velocity of light is usually disregarded. Velocities at least ten times greater than that of the earth in its orbit are, however, known to astronomers: and such must begin to exercise a sensible influence upon radiation. Moreover, in so wide a generalization as the theory of exchanges, the neglect of even a small quantity is unsatisfactory. Prof. Balfour Stewart has discussed the influence of the motion of a plate exercising selective absorption upon the equilibrium of radiation within an enclosure. He argues that a disturbance will ensue, involving a violation of

* This qualification is inserted in order to exclude such an experiment as that of Michelson, just described, in which an attempt is made to render sensible an effect depending on v^2/V^2.

the second law of thermodynamics, unless compensated by some other effect not hitherto recognised. It appears, however, more probable that the whole radiation coming *from* and *through* a plate would not be altered by its motion. Whatever effect (in accordance with Döppler's law) the motion has upon the radiation from the plate, a similar effect would be produced upon the absorbing power. On this view the only result of the motion would be to change the wave-length of the rays most powerfully emitted and absorbed, but without disturbing the balance required by the theory of exchanges. The moving plate would in fact be equivalent to a stationary one of slightly different quality.

190.

REMARKS ON MAXWELL'S INVESTIGATION RESPECTING BOLTZMANN'S THEOREM.

[*Philosophical Magazine*, XXXIII. pp. 356—359, 1892.]

THE investigation in question, which was published by Maxwell in the 12th volume of the *Cambridge Philosophical Transactions** only a short time before his death, has been the subject of some adverse criticism at the hands of Sir W. Thomson† and of Mr Bryan‡. The question is indeed a very difficult one; and I do not pretend to feel complete confidence in the correctness of the view now to be put forward. Nevertheless, it seems desirable that at the present stage of the discussion some reply to the above-mentioned criticisms should be hazarded, if only in order to keep the question open.

The argument to which most exception has been taken is that by which Maxwell (*Scientific Papers*, II. p. 722) seeks to prove that the mean kinetic energy corresponding to every variable is the same. In the course of it, the expression (T) for the kinetic energy is supposed to be reduced to a sum of squares of the component momenta, an assumption which Mr Bryan characterizes as fallacious. But here it seems to be overlooked that Maxwell is limiting his attention to systems *in a given configuration*, and that no dynamics are founded upon the reduced expression for T. The reduction can be effected in an infinite number of ways. We may imagine the configuration in question rendered one of stable equilibrium by the introduction of suitable forces proportional to displacements. The principal modes of isochronous vibration thus resulting will serve the required purpose. I do not see the applicability to this argument of the warning quoted from Routh's *Rigid Dynamics*. Perhaps the objection is felt that the conclusion

* *Scientific Papers*, Vol. II. p. 713.

† *Proc. Roy. Soc.* June, 1891.

‡ "On the Present State of our Knowledge of Thermodynamics," *Brit. Assoc. Report*, 1891. I am indebted to the author for an advance copy of this valuable report.

cannot be true in the absence of a complete specification of the variables. This is a point that may require further examination. I admit that the argument seems to imply that the conclusion possesses something of an invariantic character.

The nature of the question may be illustrated by an example approximately realized in the billiard-table, viz. the path of an elastic particle moving in a plane without loss of energy and limited within an enclosure. The fundamental assumption is that, apart from exceptional cases, the particle, starting from a given point, will sooner or later traverse that point in *every direction*; and the conclusion founded upon this assumption is that in the long run all directions through the point are equally favoured. I do not see that there is here anything to be specially surprised at. If the premises are admitted, the conclusion seems natural enough.

In another part of his investigation Maxwell puts forward under the same reserves the more general hypothesis that not merely does the system pass through a given configuration with every possible system of velocities consistent with the energy condition, but also through every configuration which can be reached without violation of the same condition. In the billiard-table example this means that every part of the table is reached sooner or later; and, as we have seen, every part that is reached is traversed as much in one direction as in another. In this case, where there is no potential energy, we may indeed go further*. Maxwell's equation (41) shows that any part of the table is occupied in the long run as much as any other; so that all points, as well as all directions, are equally probable.

To my mind the difficulty of Maxwell's investigation lies more in the premises than in the deductions†. It is easy to propose particular cases for which the hypothesis is manifestly untrue. For example, if the table be circular, a particle projected otherwise than along a diameter will leave a central circular area uninvaded, and in the outer zone will not pass through a given point in every direction, even when the projection is such that the path is not re-entrant. The question is how far the considerations advanced by Maxwell justify us in putting aside these cases as too exceptional to interfere with the general proposition, which, at any rate in its application to physics, is essentially one of probability.

Having found Maxwell's demonstration of the fundamental theorem

$$dq_1'\ldots dq_n'\,dp_1'\ldots dp_n' = dq_1\ldots dq_n\,dp_1\ldots dp_n$$

* [1901. Since $n=2$ in Maxwell's equation, this conclusion follows, even if there be potential energy.]

† The particular case for which Burnside obtained a result inconsistent with Maxwell's conclusions is emphasized by Mr Bryan. But Mr Burbury is of opinion that the discordant result depends upon an error of calculation, and that when this is set right the discrepancy disappears (*Proc. Roy. Soc.* November 19, 1891, p. 176).

difficult to follow, I have sought to simplify it by an arrangement such that the initial and final times t' and t may be considered as absolutely fixed throughout the discussion. The following, dependent upon the substitution for the "action" A of Hamilton's "principal function" S, seems to meet the requirements of the case. By definition,

$$S = \int_{t'}^{t} (T - V)\, dt = \tfrac{1}{2}A - \int_{t'}^{t} V dt;$$

and, as in Thomson and Tait's *Natural Philosophy*, § 319,

$$\begin{aligned}\delta S &= \tfrac{1}{2}\delta A - \int_{t'}^{t} \delta V\, dt \\ &= \tfrac{1}{2}\{\Sigma m (\dot{x}\,\delta x + \ldots)\} - \tfrac{1}{2}[\Sigma m (\dot{x}\,\delta x + \ldots)] + \tfrac{1}{2}\int_{t'}^{t} dt\, [\delta T + \delta V - 2\delta V];\end{aligned}$$

so that

$$\delta S = \{\Sigma m (\dot{x}\,\delta x + \ldots)\} - [\Sigma m (\dot{x}\,\delta x + \ldots)],$$

or in generalized coordinates

$$\delta S = \Sigma p\, \delta q - \Sigma p'\, \delta q'. \quad \ldots\ldots\ldots\ldots\ldots\ldots(1)$$

In this equation all the motions contemplated are unconstrained, and occupy the fixed time $t - t'$. The total energy E is variable from one motion to another, and S is to be regarded as a function of the q's and q''s.

The initial and final momenta are thus expressed by means of S in the form

$$p_r' = -\frac{dS}{dq_r'}, \qquad p_r = \frac{dS}{dq_r}; \quad \ldots\ldots\ldots\ldots\ldots\ldots(2)$$

so that

$$\frac{dp_r'}{dq_s} = -\frac{d^2S}{dq_r'dq_s} = -\frac{dp_s}{dq_r'}. \quad \ldots\ldots\ldots\ldots\ldots\ldots(3)^*$$

Thus, using S with $t - t'$ constant, instead of (as in Maxwell's investigation) A with E constant, we get

$$dq_1' \ldots dq_n'\, dp_1' \ldots dp_n' = dq_1' \ldots dq_n'\, dq_1 \ldots dq_n$$

$$\times \begin{vmatrix} \dfrac{dp_1'}{dq_1} & \cdots\cdots & \dfrac{dp_n'}{dq_1} \\ \cdots & \cdots & \cdots \\ \cdots & \cdots & \cdots \\ \dfrac{dp_1'}{dq_n} & \cdots\cdots & \dfrac{dp_n'}{dq_n} \end{vmatrix}. \quad \ldots\ldots\ldots\ldots\ldots\ldots(4)$$

* As an example the motion of a particle in two dimensions about a centre of force may be considered. q_r, q_s are then the rectangular coordinates of the particle at a fixed time t; q_r', q_s' the coordinates at the fixed time t', while p_r, p_s and p_r', p_s' are the component velocities at the same moments.

In equation (3) r and s may be identical.

On the left side the motion is defined by the initial q's and p's at time t'; on the right by the initial and final q's and by $t-t'$ (not E, which is a dependent variable).

In like manner

$$dq_1 \dots dq_n \, dp_1 \dots dp_n = dq_1' \dots dq_n' \, dq_1 \dots dq_n$$

$$\times \begin{vmatrix} \frac{dp_1}{dq_1'} & \cdots\cdots & \frac{dp_n}{dq_1'} \\ \cdots & \cdots & \cdots \\ \cdots & \cdots & \cdots \\ \frac{dp_1}{dq_n'} & \cdots\cdots & \frac{dp_n}{dq_n'} \end{vmatrix} \quad \text{.} \quad \ldots\ldots\ldots\ldots\ldots\ldots(5)$$

By the relation (3) proved above the two determinants in (4) and (5) are equal, and thus

$$dq_1' \dots dq_n' \, dp_1' \dots dp_n' = dq_1 \dots dq_n \, dp_1 \dots dp_n, \ldots\ldots\ldots\ldots(5)$$

the required conclusion.

[1901. For a further discussion of this subject the reader is referred to a paper on the "Law of Partition of Kinetic Energy" (*Phil. Mag.* XLIX. p. 98, 1900).]

191.

ON THE PHYSICS OF MEDIA THAT ARE COMPOSED OF FREE AND PERFECTLY ELASTIC MOLECULES IN A STATE OF MOTION*.

[*Phil. Trans.* 183 A, pp. 1—5, 1892.]

THE publication of this paper after nearly half a century demands a word of explanation; and the opportunity may be taken to point out in what respects the received theory of gases had been anticipated by Waterston, and to offer some suggestions as to the origin of certain errors and deficiencies in his views.

So far as I am aware, the paper, though always accessible in the Archives of the Royal Society, has remained absolutely unnoticed. Most unfortunately the abstract printed at the time (*Roy. Soc. Proc.* 1846, Vol. v. p. 604; here reprinted as Appendix I.) gave no adequate idea of the scope of the memoir, and still less of the nature of the results arrived at. The deficiency was in some degree supplied by a short account in the *Report of the British Association* for 1851 (here reprinted as Appendix II.), where is distinctly stated the law, which was afterwards to become so famous, of the equality of the kinetic energies of different molecules at the same temperature.

My own attention was attracted in the first instance to Waterston's work upon the connection between molecular forces and the latent heat of evaporation, and thence to a paper in the *Philosophical Magazine* for 1858, "On the Theory of Sound." He there alludes to the theory of gases under consideration as having been started by Herapath in 1821, and he proceeds:—

"Mr Herapath unfortunately assumed heat or temperature to be represented by the simple ratio of the velocity instead of the square of the velocity—being in this apparently led astray by the definition of motion

* [From an Introduction to a Memoir, entitled as above, by J. J. Waterston, received Dec. 11, 1845, read March 5, 1846.]

generally received—and thus was baffled in his attempts to reconcile his theory with observation. If we make this change in Mr Herapath's definition of heat or temperature, viz., that it is proportional to the *vis viva*, or square velocity of the moving particle, not to the momentum, or simple ratio of the velocity, we can without much difficulty deduce, not only the primary laws of elastic fluids, but also the other physical properties of gases enumerated above in the third objection to Newton's hypothesis. In the Archives of the Royal Society for 1845—1846, there is paper 'On the Physics of Media that consists of perfectly Elastic Molecules in a State of Motion,' which contains the synthetical reasoning upon which the demonstration of these matters rests. The velocity of sound is therein deduced to be equal to the velocity acquired in falling through three-fourths of a uniform atmosphere. This theory does not take account of the size of the molecules. It assumes that no time is lost at the impact, and that if the impacts produce rotatory motion, the *vis viva* thus invested bears a constant ratio to the rectilineal *vis viva*, so as not to require separate consideration. It also does not take account of the probable internal motion of composite molecules; yet the results so closely accord with observation in every part of the subject as to leave no doubt that Mr Herapath's idea of the physical constitution of gases approximates closely to the truth. M. Krönig appears to have entered upon the subject in an independent manner, and arrives at the same result; M. Clausius, too, as we learn from his paper 'On the Nature of the Motion we call Heat' (*Phil. Mag.* Vol. XIV. 1857, p. 108)."

Impressed with the above passage and with the general ingenuity and soundness of Waterston's views, I took the first opportunity of consulting the Archives, and saw at once that the memoir justified the large claims made for it, and that it marks an immense advance in the direction of the now generally received theory. The omission to publish it at the time was a misfortune, which probably retarded the development of the subject by ten or fifteen years. It is singular that Waterston appears to have advanced no claim for subsequent publication, whether in the Transactions of the Society, or through some other channel. At any time since 1860 reference would naturally have been made to Maxwell, and it cannot be doubted that he would have at once recommended that everything possible should be done to atone for the original failure of appreciation.

It is difficult to put oneself in imagination into the position of the reader of 1845, and one can understand that the substance of the memoir should have appeared speculative and that its mathematical style should have failed to attract. But it is startling to find a referee expressing the opinion that "the paper is nothing but nonsense, unfit even for reading before the Society." Another remarks "that the whole investigation is confessedly founded on a principle entirely hypothetical, from which it is the object to deduce a mathematical representation of the phenomena of elastic media.

It exhibits much skill and many remarkable accordances with the general facts, as well as numerical values furnished by observation.......The original principle itself involves an assumption which seems to me very difficult to admit, and by no means a satisfactory basis for a mathematical theory, viz., that the elasticity of a medium is to be measured by supposing its molecules in vertical motion, and making a succession of impacts against an elastic gravitating plane." These remarks are not here quoted with the idea of reflecting upon the judgment of the referee, who was one of the best qualified authorities of the day, and evidently devoted to a most difficult task his careful attention; but rather with the view of throwing light upon the attitude then assumed by men of science in regard to this question, and in order to point a moral. The history of this paper suggests that highly speculative investigations, especially by an unknown author, are best brought before the world through some other channel than a scientific society, which naturally hesitates to admit into its printed records matter of uncertain value. Perhaps one may go further and say that a young author who believes himself capable of great things would usually do well to secure the favourable recognition of the scientific world by work whose scope is limited, and whose value is easily judged, before embarking upon higher flights.

One circumstance which may have told unfavourably upon the reception of Waterston's paper is that he mentions no predecessors. Had he put forward his investigation as a development of the theory of D. Bernoulli, a referee might have hesitated to call it nonsense. It is probable, however, that Waterston was unacquainted with Bernoulli's work, and doubtful whether at that time he knew that Herapath had to some extent foreshadowed similar views.

At the present time the interest of Waterston's paper can, of course, be little more than historical. What strikes one most is the marvellous courage with which he attacked questions, some of which even now present serious difficulties. To say that he was not always successful is only to deny his claim to rank among the very foremost theorists of all ages. The character of the advance to be dated from this paper will be at once understood when it is realised that Waterston was the first to introduce into the theory the conception that heat and temperature are to be measured by *vis viva*. This enabled him at a stroke to complete Bernoulli's explanation of pressure by showing the accordance of the hypothetical medium with the law of Dalton and Gay-Lussac. In the second section the great feature is the statement (VII.), that "in mixed media the mean square molecular velocity is inversely proportional to the specific weight of the molecules." The proof which Waterston gave is doubtless not satisfactory; but the same may be said of that advanced by Maxwell fifteen years later. The law of Avogadro follows at once, as well as that of Graham relative to diffusion. Since the law of equal energies was actually published in 1851, there can be no hesitation, I think, in attaching Waterston's name to it. The attainment of correct results in the third section, dealing with adiabatic expansion, was only prevented by a slip of calculation.

In a few important respects Waterston stopped short. There is no indication, so far as I can see, that he recognised any other form of motion, or energy, than the translatory motion, though this is sometimes spoken of as vibratory. In this matter the priority in a wider view rests with Clausius. According to Waterston the ratio of specific heats should be (as for mercury vapour) 1·67 in all cases. Again, although he was well aware that the molecular velocity cannot be constant, there is no anticipation of the law of distribution of velocities established by Maxwell.

A large part of the paper deals with chemistry, and shows that his views upon that subject also were much in advance of those generally held at the time..........

With the exception of some corrections relating merely to stops and spelling the paper is here reproduced exactly as it stands in the author's manuscript.—Dec. 1891.

[1901. It may be added that Waterston's memoir contains the first calculation of the molecular velocity, and further that it points out the relation of this velocity to the velocity of sound. The earliest actual *publication* of such a calculation is that of Joule, who gives for the velocity of hydrogen molecules at 0° C. 6055 feet per second (*Manchester Memoirs,* Vol. IX. p. 107, Oct. 1848; *Phil. Mag.* Ser. 4, Vol. XIV. p. 211; Joule's *Scientific Papers,* Vol. I. p. 295), thus anticipating by eight or nine years the first paper of Clausius (*Pogg. Ann.* 1857), to whom priority is often erroneously ascribed.]

192.

EXPERIMENTS UPON SURFACE-FILMS.

[*Philosophical Magazine*, XXXIII. pp. 363—373, 1892.]

THE experiments here described are rather miscellaneous in character, but seem of sufficient interest to be worthy of record. The greater number of them have been exhibited in the course of lectures at the Royal Institution.

The Behaviour of Clean Mercury.

According to Marangoni's rule, water, which has the lower surface-tension, should spread upon the surface of mercury; whereas the universal experience of the laboratory is that drops of water standing upon mercury retain their compact form without the least tendency to spread. To Quincke belongs the credit of dissipating the apparent exception. He found that mercury specially prepared behaves quite differently from ordinary mercury, and that a drop of water deposited thereon spreads over the whole surface. The ordinary behaviour is evidently the result of a film of grease, which adheres with great obstinacy.

The process described by Quincke is somewhat elaborate; but my experience with water suggested that success might not be so difficult, if only the mistake were avoided of pouring the liquid to be tried from an ordinary bottle. In the early experiments upon the camphor movements difficulty seems to have been experienced in securing sufficiently clean water surfaces. The explanation is probably to be found in the desire to use distilled water, and to the fact that the liquid would usually be simply poured from a stock bottle into the experimental vessel. No worse procedure could be devised; for the free surface in the bottle is almost sure to be dirty, and is transferred in great part to the vessel. In my experience water from the dirtiest cistern will exhibit the camphor movements, provided that it be drawn in the usual manner from a tap, and that the precaution be taken to give the vessel a preliminary rinsing.

In order to carry out the idea of drawing the liquid from underneath, an arrangement was provided like an ordinary wash-bottle, and was filled with tolerably clean mercury. As experimental vessels watch-glasses are convenient. They may be dipped into strong sulphuric acid, rinsed in distilled water, and dried over a Bunsen flame. When the glasses are cool they may be charged with mercury, of which the first portion is rejected. Operating in this way there was no difficulty in obtaining surfaces upon which a drop of water would spread, although, from causes that could not always be traced, a certain proportion of failures was met with.

Exposure of the glasses to the atmosphere soon tells upon the success of the experiment, although on one occasion spreading occurred after a glass had stood (with protection from dust) for 20 hours. Even so short an exposure as 10 minutes was found to prejudice the condition of the mercury surface. Although something here may have depended upon the special character of the sample of mercury, it will be advisable in repeating the experiment to pour the mercury at the last moment.

As might be expected, the grease which produces these effects is largely volatile. In many cases a very moderate preliminary warming of the watch-glass makes all the difference in the behaviour of the drop.

So far as I have observed, the spreading of the drop takes place always in a leisurely fashion. If a little powder of recently ignited magnesia be dusted over the mercury, there is no violent repulsion of the dust before the advancing water. But if a small drop of oil be substituted for the water, the powder is flashed away so quickly that the eye cannot follow the operation. The difference between the two cases appears to depend upon the atmospheric moisture. As soon as the mercury is poured, it coats itself with an aqueous film, and the subsequent spreading of the drop takes place upon a surface whose affinity for water is already largely satisfied. A drop of water that has spread and then partially gathered up again (as usually happens after a short interval) shows an interesting behaviour when breathed upon. The disk contracts somewhat, and then as the breath, which need hardly be visible, passes off, expands again; and thus a number of times. The temporary character of the effect indicates that it is due rather to the moisture of the breath than to any greasy contamination—a view confirmed by subsequent experiments, in which the breath was replaced by a current of pure air which had passed through warm water.

In the experiment with a powdered surface, the dust may be driven from the neighbourhood of a drop of petroleum by the action of vapour, without actual contact of the liquids.

Drops of Bisulphide of Carbon upon Water.

The behaviour of a drop of CS_2 placed upon clean water is also at first sight an exception to Marangoni's rule. So far from spreading over the surface, as according to its lower tension it ought to do, it remains suspended in the form of a lens. And dust which may be lying upon the surface is not driven away to the edge upon the deposit of the drop, as would happen in the case of oil. A simple modification of the experiment suffices, however, to clear up the difficulty. If *after* the deposit of the drop a little lycopodium be scattered over the surface, it is seen that a circular space surrounding the drop, of perhaps the size of a shilling, remains bare, and this however often the dusting be repeated, as long as any of the CS_2 remains. The interpretation can hardly be doubtful. The bisulphide is really spreading all the while, but on account of its volatility is unable to reach any considerable distance. Immediately surrounding the drop there is a film moving outwards at a high speed, and this carries away almost instantaneously any dust that may fall upon it. The phenomenon above described requires that the water surface be clean. If a very little grease be present, there is no outward flow and dust remains undisturbed in the immediate neighbourhood of the drop. With the aid of the vertical lantern, and a shallow dish whose bottom is formed of plate glass, these experiments are easily shown to an audience.

Movements of Dust.

When dust of sulphur or lycopodium is scattered upon the surface of water contained in a partially filled vessel, it is found that after a few seconds the dust leaves the edge and that a clear ring is formed of perhaps a centimetre in width. Two explanations suggest themselves. The action may be due to grease communicated to the surface from the edge of the vessel; or, secondly, it may be the effect of gravity upon those particles of the dust which lie within the limits of the capillary meniscus. The first explanation is rendered improbable by the non-progressive, or at least but very slowly progressive, character of the effect; and it is negatived by a repetition of the experiment in a varied form. It is found that if the vessel, whether of glass or metal, be filled *over the brim*, so that the capillary meniscus is convex, then, although as before a bare margin is formed, the effect is due to a motion of the dust outwards (instead of inwards, as in the former case), and is therefore not to be attributed to grease.

A similar movement of dust was to be observed in the experiment above recorded, where magnesia was scattered upon a pool of mercury, and is undoubtedly due to gravity; but the full explanation is not so simple as might appear at first sight.

Even in the interior parts of the surface at a distance from the edge the sulphur particles do not retain their initial positions, but form aggregates into which continually increasing numbers are attracted. This is also due to gravity, neighbours tending, as it were, to fall into the depression by which every particle is surrounded.

Camphor Movements a Test of Surface-Tension.

The theory of these movements, due to Van der Mensbrugghe, implies that they will take place with greater or less vigour so long as the tension of the surface, which may be in some degree contaminated, is greater than that of a saturated solution of camphor. If, however, the contamination be so great that the tension falls below this point, the solution of camphor can no longer spread upon the surface, and the movements cease. Thus, according to this theory and to observations* upon a saturated solution of camphor, the movements are an indication that the actual tension does not fall below ·71 of that of pure water.

Although there appeared to be no reason for distrusting this view, it was thought desirable to examine specially whether the cessation of the movements was really a question of surface-tension only, without regard to the character of the contamination. The readiest method of ensuring the equality of the tensions of two surfaces contaminated with different materials is to make the two surfaces parts of one surface, for two parts of the same surface cannot be at rest unless they have the same tension. The method of experiment was therefore to divide a surface of clean water contained in a large dish into two parts by a line of dust, and to communicate different kinds of grease to the surfaces on the two sides of the indicating line. If, for example, a small chip of wood, slightly greased with olive-oil, be allowed to touch one part of the surface, the line of dust is repelled by the expansion of that part, but the effect may be compensated by a slight greasing of the other part with oil of cassia. By careful alternate additions the line of dust may be kept central, while the two halves become increasingly greased with the two kinds of oil. At every stage of this process, so long as the surface is at rest, the tension of all parts is necessarily the same.

A large number of substances have thus been tried in pairs, of which may be mentioned oils of olive, cassia, turpentine, lavender, cinnamon, anise, petroleum, pseudocumene. In no case could any difference be detected in the behaviour of camphor fragments on the two sides. Whenever possible, the quantities of oil were adjusted to the point at which the movements were just ceasing. In case of overshooting the mark, the excess of oil could be easily removed by strips of paper, partially immersed and then withdrawn, the action being equivalent to an expansion of the surface. In several cases the volatility of the substance with which the surface was contaminated led to a subsequent retraction of the line of dust. Thus freshly distilled oil of turpentine, even at first barely capable of arresting the movements, soon passes off.

* *Phil. Mag.* November 1890. [Vol. III. p. 394.]

As was shown by Tomlinson, oil of anise is incapable of arresting the camphor movements. In the experiment with a partition of dust, olive-oil will drive oil of anise into a very small space, whose area is doubtless dependent upon the amount of other impurities present. In this case, as in all others, the behaviour of camphor is the same on the various parts of the surface.

It may thus be taken as established that the relation of a contaminated surface to the camphor movements is one of surface-tension only.

A similar method of experimenting may be applied to a rough determination of the degree of purity of cleansed surfaces. The whole of the surface under test is lightly dusted over, and olive-oil is applied at several places close to the circumference until camphor movements are nearly arrested. After each addition of oil the dusted area contracts, and at the close of operations it gives a measure of the extent to which the original contamination must be concentrated in order to stop camphor.

A few numbers may be given as examples, although in all probability the result is influenced by a variety of circumstances. A circular area of 10 inches diameter, occupied by tap-water, and cleansed by the flexible hoop described in former papers, was tested on July 28, 1891. The application of oil, just sufficient to stop the camphor movements, drove the dust into a central circular patch of $2\frac{1}{2}$ inches diameter. When the surface was in its natural condition, unpurified by the action of the hoop, the central patch was of about 5 inches diameter. These numbers, approximately verified on repetition, show that the natural surface was about 4 times, and the purified about 16 times better than according to the camphor standard. The difference between the two cases is less than was expected, and would perhaps have been greater had distilled water been employed. It must be remembered also that contact with dust (sulphur) is unfavourable to the purity of a water surface. In a very good light a special dusting might probably be dispensed with, the motion of the surface being evidenced by inevitable motes.

If the dust be applied in the first instance to a small central patch, which is then touched internally with a very small quantity of oil, the expansion of the dust in the form of a ring is followed by a slight but unmistakable rebound. The effect appears to take place when the surface is very clean to begin with, and is then somewhat difficult of explanation. I am disposed to think that it must be attributed in all cases to initial contamination. This is concentrated in front of the rapidly advancing ring, and has not time to diffuse itself equally over the whole external area. Under the influence of inertia the expansion of the central area may then proceed so far that its tension becomes greater than that of the parts immediately surrounding.

Influence of Heat.

For a lecture experiment the effect of heat is best shown by holding a hot body near the surface of water contained in a shallow vessel with a glass bottom. The hot body may be the end of a glass rod heated by a flame, or more conveniently a small spiral of platinum wire, rendered incandescent at will by an electric current. The immediate effect of the heat is to lower the tension of the part of the surface affected; but the visible result depends entirely upon whether the surface be clean or otherwise. In the former case the heated surface expands, and an outward current is generated. This is rendered evident by the clearing away of dust. But if the original contamination exceed a very small quantity, a moderate expansion of the heated area brings the tension again up to equality with that of the surrounding surface, and there is no further action. In this case there is no visible clearing away of dust under the hot body.

Under favourable circumstances a very slight elevation of temperature suffices. On July 28 a shallow tin vessel 8 × 5 inches, the lid of a biscuit-box, was levelled and filled with tap-water from a rubber hose, after a thorough preliminary rinsing *in situ*. A little dust (sulphur) was then scattered over, and the finger was brought underneath into contact with the bottom of the dish. After about 20 seconds the dust opened out, and a bare spot was formed over the finger of about 1½ inches diameter. A spirit-flame, applied for a few seconds under one end of the dish, cleared away the dust from the larger part of the area. If when quiet was nearly restored, a little fresh dust was applied, and the experiment with the finger repeated, the effect was more pronounced than before, and the bared space much larger, showing that the treatment with the spirit-flame had driven away most of the residual contamination.

The best effects were obtained with a dish somewhat larger than that above mentioned; and in subsequent experiments the difference of temperature between different parts was more readily maintained by the use of a

Fig. 1.

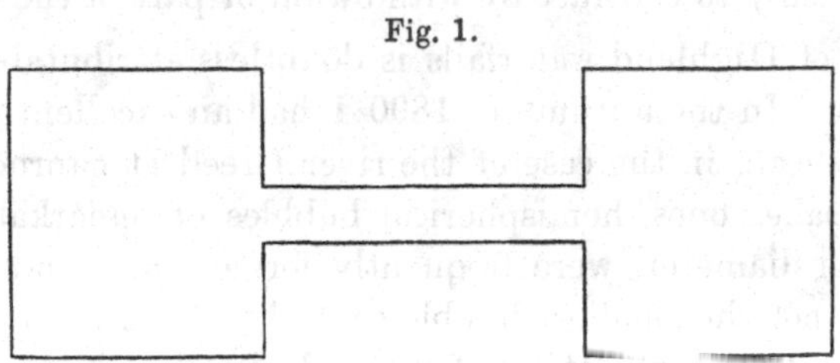

vessel in which the main portions were connected by a comparatively narrow channel. In this way the tensions of surfaces, contaminated in different degrees, may be equalized, the warmer purer surface in one compartment balancing the colder but greasier surface in the other. And the actual temperature-difference necessary for equilibrium gives a measure of the small difference of tensions to be compensated*.

When the surface of the liquid in the tin vessel is but very slightly greased, a spot can no longer be cleared by the warmth of the finger held underneath. Indeed the spirit-flame itself soon becomes ineffective. And yet the greasing may be so slight that camphor fragments move with apparently unabated vigour.

It is of interest to compare the behaviour of saturated solution of camphor with that of greasy water. The former can scarcely be brought to rest, unless covered up. This is doubtless due to evaporation of camphor, aided by local draughts. A spirit-flame drives away dust in a manner impossible in the case of a merely greasy surface, whose tension may nevertheless be decidedly higher than that of the camphorated water.

It may here be mentioned that the lowering of tension by camphor follows a different law from the lowering caused by soap. In the latter case the fall of tension requires time, and at the first moment of its formation a free surface has almost the tension of pure water. Similar experiments to those formerly recorded† with soapy water have shown that the ratio of tensions for pure water and for solution of camphor are the same at the first moment of the formation of a free surface as when the measures are conducted statically.

Saponine and Soap.

A strong infusion of horse-chestnuts allowed excellent bubbles to be blown, up to 4 inches [10 cm.] or more in diameter. When the interiors of equal bubbles of soap and of saponine were brought into communication, the latter contracted and the former expanded, showing that the tension of the saponine film was the greater. In order to obtain equilibrium, the diameter of the saponine bubble required to be about half as great again as that of the soap bubble. These saponine bubbles exhibited the characteristic wrinkling, when caused suddenly to contract by withdrawal of part of the contained air.

The foaming of Highland waterfalls is doubtless attributable to dissolved vegetable matter. In the autumn of 1890 I had an excellent opportunity of observing these effects in the case of the river Creed at Stornoway. By the coalescence of smaller ones, hemispherical bubbles of remarkable size, up to a foot or more in diameter, were frequently formed, and endured for a few seconds; and yet not the smallest bubble could be blown from a tobacco-pipe. However, by collecting some of the foam and allowing it to subside, which took a good while, I obtained liquid from which bubbles could be blown with

* The lowering of tension per degree Cent. is said to be ·0018 of the total value.

† *Proc. Roy. Soc.* March 1890. [Vol. III. p. 341.]

a pipe up to 4 inches diameter. But these bubbles behaved like soap, and not as had been rather expected, like saponine, remaining perfectly tight and smooth when the included air was rapidly withdrawn.

Separation of Motes.

In the course of some experiments last year, in illustration of Sir G. Stokes's theory of ternary mixtures, I had prepared an association* of water, alcohol, and ether, in which the quantity of alcohol was so adjusted that the tendency to divide into two parts was almost lost. As it was, division took place after shaking into two nearly equal parts, and these parts were of almost identical composition. On placing the bottle containing the liquids in the concentrated light from an arc lamp, I was struck with the contrast between the appearance of the two parts. The lower, more aqueous, layer was charged with motes, while the upper, more ethereal, layer was almost perfectly free from them. Some years ago I had attempted the elimination of motes by repeated distillation of liquid in vacuum, conducted without actual ebullition, but I had never witnessed as the result of this process anything so clear as the ethereal mixture above described.

The observation with the ternary association, which happened to be the first examined, is interesting, because the approximate equality of the liquids suggests that the explanation has nothing directly to do with gravitation. But the presence of alcohol is not necessary. Ether and water alone shaken together exhibit the same phenomenon. It would appear that when the two liquids are mixed together in a finely divided condition, the motes attach themselves by preference to the more aqueous one, and thus when separation into two distinct layers follows, the motes are all to be found below†.

An obvious explanation, which, however, stands in need of confirmation, is that under the play of the capillary forces the energy is least when the motes, which may be presumed to be denser than either liquid, are in contact with the denser rather than with the rarer of the two. The density here referred to is that which occurs in Laplace's theory of capillarity, and may need to be distinguished from ordinary mechanical density.

I have lately endeavoured to obtain some confirmation of the views above expressed by the use of other liquids. It would evidently be satisfactory to exhibit the selection of motes by the upper, instead of by the lower, layer. Experiments with bisulphide of carbon and water, and also associations of these two bodies with alcohol, which acts as a solvent to both, gave no

* *Association* is here employed as a general term denoting the juxtaposition of two or more fluids. Whether the result is a *mixture* depends upon circumstances.

† [1901. The clearness of the upper layer, after a mixture of ether and alcohol has been shaken up with dust, had already been observed and explained much as above by Barus (*Am. Journ.* XXXVII. p. 122, 1889).]

definite result, perhaps in consequence of a tendency to the formation of a solid pellicle at the common surfaces. But with chloroform and water, and with associations of chloroform, water, and acetic acid (acting as a common solvent), the experiment succeeded. The motes were always collected in the *upper*, more aqueous, layer, even when the composition of the two layers into which the liquid separated was so nearly the same that a few additional drops of acetic acid sufficed to prevent separation altogether.

In this and similar cases a marked tendency to foaming may be observed when the composition is such that separation just fails to take place.

The Lowering of Tension by the Condensation of Ether Vapour.

The suspension of water in an inverted tube of small bore is familiar to all. The limit of diameter was investigated some years ago by Duprez*. A glass tube, such as that shown in Fig. 2, is ground true at the lower end, and at the upper end is connected to an india-rubber tube provided with a pinch-cock. Water is sucked up from a vessel of moderate size, the rubber is nipped, and by a *quick* motion the tube and the vessel are separated, preferably by a downward movement of the latter. In this way of working Duprez found that the liquid might remain suspended in tubes of diameter up to 16 millim., and with the aid of a sliding plate up to 19·85 millim. The theory is given in Maxwell's article in the *Encyclopædia Britannica* ("On Capillary Action"). For lecture purposes it is well not to attempt too much. The tube employed by me had an internal diameter of 14½ millim., and there was no difficulty in obtaining suspension. The experiment on the effect of ether-vapour was then as follows:—The inverted tube, with its suspended water, being held in a clamp, a beaker containing a few drops of ether was brought up from below until the free surface of the water was in contact with ether vapour. The lowering of tension, which follows the condensation of vapour, is then strikingly shown by the sudden precipitation of the water.

Fig. 2.

Breath Figures and their Projection.

These figures are perhaps most readily prepared upon the plan described in Riess's *Electricity*. The carefully cleaned glass plate upon which the image is to be received is placed upon a flat metallic slab, and upon it again rests the coin to be copied, for example, a shilling. The two conductors form the coatings of a Leyden jar, and are connected by wires to the discharging

* "Sur un cas particulier de l'équilibre des liquides," *Bruxelles Acad. Sci. Mém.* XXVI. 1851; XXVIII. 1854.

terminals of a large Wimshurst machine, the latter being set so as to give sparks about $\frac{1}{4}$ inch long. In my experiments about 20 turns of the handle were found sufficient to impress the latent image.

The projection of the figures, developed upon the glass by breathing, requires a special arrangement, which it is the principal object of this note to describe. For this purpose the light simply transmitted by the undimmed parts of the plate must be intercepted, leaving the image to be formed by the light *diverted* from its path by the condensed breath. The arrangement was as follows:—

The ordinary condenser B (Fig. 3) of the electric lantern was stopped down to an aperture of $\frac{3}{4}$ inch, and provided a somewhat divergent beam

Fig. 3.

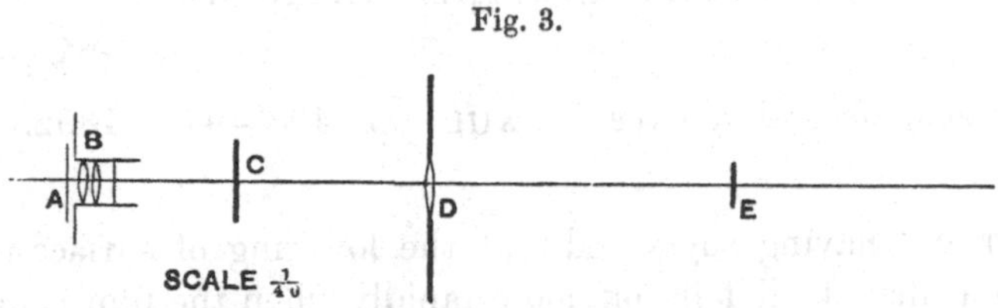

of light of corresponding diameter. At a distance of $15\frac{1}{2}$ inches from the condenser was placed the slide C upon which a figure had been impressed. The focusing lens D was of plate-glass, 6 inches in diameter and 25 inches focus, and was of course distant from the breath figure by an amount slightly exceeding its own focal length. Any light that might pass outside was intercepted by a suitable mounting. So far there was nothing peculiar, except in respect to the dimensions of the focusing lens. But now between the latter and the screen was inserted a disk E of black card 2 inches in diameter, at such a distance (40 inches) from the lens as to receive a well-defined image of the hot carbons A. By this disk all regularly refracted light would be stopped, so that the screen would appear dark. If, however, any part of the prepared glass be dimmed by the breath, light is there diverted from its path, and thus escaping the stop proceeds to form an image of the part in question upon the screen. The dewed parts of the breath-figure are accordingly seen bright upon a dark ground; and with the arrangement described, in which the large diameter of the focusing lens is a leading feature, the projected images are very beautiful. A similar method would probably be adequate to the projection of smoke-jets.

In conclusion I may mention that the latent images can be developed in a more durable manner by a deposit of *silver*, the arrangements being such as are adopted for the silvering of mirrors, except that the action is stopped at an earlier stage. The washed and dried deposit may then be protected from mechanical injury by a coat of varnish.

193.

ON THE THEORY OF SURFACE FORCES. III.—EFFECT OF SLIGHT CONTAMINATIONS.

[*Philosophical Magazine*, XXXIII. pp. 468—471, 1892.]

OBSERVATION* having suggested that the lowering of surface-tension of water due to a film of oil falls off more rapidly when the film is attenuated than the thickness of the film itself can be supposed to do, I was led to examine the question theoretically; and the result shows that, according to the principles of Young and Laplace, the lowering of tension due to a very thin film should be in proportion, not to the thickness, but to the *square* of the thickness of the film. In the calculations which follow the fluids are supposed to be incompressible, a layer of density ρ and thickness α being interposed between fluids of densities ρ_2 and ρ_1 (Fig. 1). The thickness α, as well as the range of the forces, is supposed to be negligible in comparison with the radius of curvature R of the surfaces of separation.

Fig. 1.

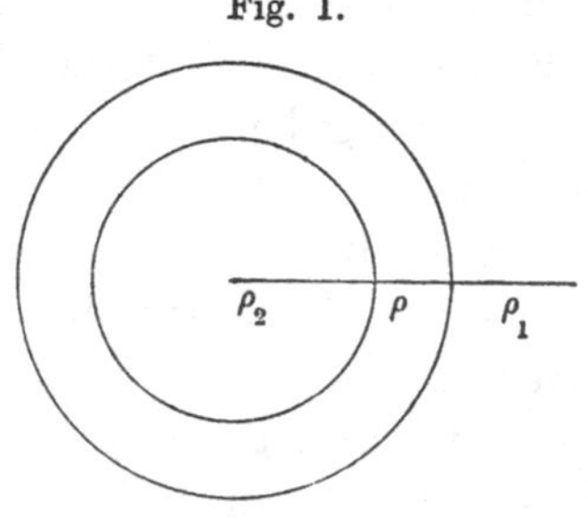

By II. (16)† we have for the difference of pressures in the inner and outer liquids,

$$p_2 - p_1 = 2K(\rho_2^2 - \rho_1^2) - \int_{(1)}^{(2)} V d\rho$$

$$= 2K(\rho_2^2 - \rho_1^2) - (\rho - \rho_1) . V(\rho, \rho_1) - (\rho_2 - \rho) . V(\rho_2, \rho), \quad \text{.........(1)}$$

where $V(\rho, \rho_1)$, $V(\rho_2, \rho)$ denote the potentials at the surfaces of separation.

* See, for example, Miss Pockels on "Surface Tension," *Nature*, Vol. XLIII. p. 437 (1891).

† *Phil. Mag.* Feb. 1892. [Vol. III. p. 519.]

Again, by II. (17),

$$V(\rho, \rho_1) = \int_{-\infty}^{+\infty} 2\pi\rho'(1 - \zeta/R)\,\psi(\zeta)\,d\zeta$$

$$= 2\pi(\rho_2 + \rho_1)\int_0^\infty \psi(\zeta)\,d\zeta - 2\pi(\rho_2 - \rho)\int_0^\alpha \psi(\zeta)\,d\zeta$$

$$- \frac{2\pi}{R}\left\{(\rho_2 - \rho_1)\int_0^\infty \psi(\zeta)\,\zeta\,d\zeta - (\rho_2 - \rho)\int_0^\alpha \psi(\zeta)\,\zeta\,d\zeta\right\},$$

and $$V(\rho_2, \rho) = 2\pi(\rho_2 + \rho_1)\int_0^\infty \psi(\zeta)\,d\zeta + 2\pi(\rho - \rho_1)\int_0^\alpha \psi(\zeta)\,d\zeta$$

$$- \frac{2\pi}{R}\left\{(\rho_2 - \rho_1)\int_0^\infty \psi(\zeta)\,\zeta\,d\zeta - (\rho - \rho_1)\int_0^\alpha \psi(\zeta)\,\zeta\,d\zeta\right\}.$$

Consider now

$$(\rho - \rho_1).V(\rho, \rho_1) + (\rho_2 - \rho).V(\rho_2, \rho), \quad\text{..............(2)}$$

and collect separately the part independent of R, and that proportional to R^{-1}. For the first we have

$$2\pi(\rho_2^2 - \rho_1^2)\int_0^\infty \psi(\zeta)\,d\zeta, \quad\text{......................(3)}$$

the same as if $\alpha = 0$.

For the second, omitting the factor $-2\pi/R$, we get

$$(\rho - \rho_1)\left\{(\rho_2 - \rho_1)\int_0^\infty \psi(\zeta)\,\zeta\,d\zeta - (\rho_2 - \rho)\int_0^\alpha \psi(\zeta)\,\zeta\,d\zeta\right\}$$

$$+ (\rho_2 - \rho)\left\{(\rho_2 - \rho_1)\int_0^\infty \psi(\zeta)\,\zeta\,d\zeta - (\rho - \rho_1)\int_0^\alpha \psi(\zeta)\,\zeta\,d\zeta\right\},$$

or $$(\rho_2 - \rho_1)^2\int_0^\infty \psi(\zeta)\,\zeta\,d\zeta - 2(\rho - \rho_1)(\rho_2 - \rho)\int_0^\alpha \psi(\zeta)\,\zeta\,d\zeta. \quad\text{.......(4)}$$

Now $$K = 2\pi\int_0^\infty \psi(\zeta)\,d\zeta,$$

so that $$p_2 - p_1 = K(\rho_2^2 - \rho_1^2) + 2T/R, \quad\text{....................(5)}$$

where $$T = \pi(\rho_2 - \rho_1)^2\int_0^\infty \psi(\zeta)\,\zeta\,d\zeta - 2\pi(\rho - \rho_1)(\rho_2 - \rho)\int_0^\alpha \psi(\zeta)\,\zeta\,d\zeta. \quad\text{...(6)}$$

The tension of the composite surface is thus given by (6).

If $\alpha = 0$, we fall back upon the case of a simple sudden transition from ρ_2 to ρ_1, and we get as before

$$T = \pi(\rho_2 - \rho_1)^2\int_0^\infty \psi(\zeta)\,\zeta\,d\zeta. \quad\text{......................(7)}$$

Again, if $\alpha = \infty$,

$$T = \pi\left\{(\rho_2 - \rho)^2 + (\rho - \rho_1)^2\right\}\int_0^\infty \psi(\zeta)\,\zeta\,d\zeta. \quad\text{.............(8)}$$

This corresponds to the formation of two independently acting tensions between the two pairs of liquids.

To pass from these verifications to circumstances of novelty, let us now suppose that α is small compared with the range of the forces. When ζ is small, $\psi(\zeta)$ may be identified with $\psi(0)$, and we have

$$\delta T = -\pi(\rho-\rho_1)(\rho_2-\rho)\,.\,\psi(0)\,.\,\alpha^2, \quad\ldots\ldots\ldots\ldots\ldots\ldots\ldots(9)$$

showing that in the limit δT is proportional to the *square* of the thickness α.

According to Young's supposition I. (19)* of a constant attraction within the range a,

$$\psi(\zeta) = \tfrac{1}{2}a(a^2-\zeta^2) - \tfrac{1}{3}(a^3-\zeta^3),$$

so that $\psi(0) = \frac{1}{6}a^3$; and more generally whether α be great or small,

$$\int_0^\alpha \psi(\zeta)\,\zeta\,d\zeta = \alpha^2(\tfrac{1}{12}a^3 - \tfrac{1}{8}a\,\alpha^2 + \tfrac{1}{15}\alpha^3). \quad\ldots\ldots\ldots\ldots\ldots(10)$$

The general formula (6) may be applied also to the case of a thin lamina by supposing that $\rho_2 = \rho_1 = \rho_0$. Thus

$$T = 2\pi(\rho-\rho_0)^2\int_0^\alpha \psi(\zeta)\,\zeta\,d\zeta \quad\ldots\ldots\ldots\ldots\ldots\ldots\ldots(11)$$

gives the tension of a lamina of density ρ and thickness α surrounded by fluid of density ρ_0†. Here again, if α be very small, the integral reduces to $\frac{1}{2}\alpha^2\psi(0)$, so that the tension varies as the square of α.

It must be understood that the lamina is here supposed to be of uniform constitution, and that thus the result is probably inapplicable to soap-films.

The law that the effect of a film when very thin is proportional to the square of the thickness is of considerable interest. It is here deduced upon the basis of Laplace's more special hypothesis that the specific differences between various bodies in relation to capillary properties may be represented by a simple coefficient in the expression for the law of force. But it is not difficult to see that the conclusion is really independent of this restriction, and that it holds good even though the law of force for every pair of bodies is entirely arbitrary.

[1901. Subsequent experiments by Miss Pockels and by the author (*Phil. Mag.* XLVIII. p. 331, 1899) show that the actual falling off is more sudden than that above calculated, probably in consequence of the discontinuity which enters when the layer is only one or two molecules thick.]

* *Phil. Mag.* Oct. 1890. [Vol. III. p. 404.]

† In Maxwell's solution of this problem, Art. "Capillary Action," *Enc. Brit.*, the tension of the lamina is given at double the above value.

194.

ON THE QUESTION OF THE STABILITY OF THE FLOW OF FLUIDS.

[*Philosophical Magazine*, XXXIV. pp. 59—70, 1892.]

IT is well known that while Sir G. Stokes's theory of viscous flow gives a completely satisfactory account of what is observed in the case of capillary tubes, no theory at present exists to explain the complete change in the laws of flow which supervenes when the tubes are of larger diameter and the velocities not very small. Prof. Osborne Reynolds* has applied the theory of dynamical similarity to this question, and has shown both by theory and experiment that the change in the law of resistance occurs when $c\rho w/\mu$ has a certain value, where c is a linear parameter such as the diameter of the tube, w is the velocity, μ the coefficient of friction, and ρ the density. The conclusion is perhaps most easily reached by applying the method of dimensions to the expression for the ratio (P) of the difference of pressures at two points along the length of the tube to the distance between the points. The dimensions of this ratio are those of a force divided by a volume; and if we assume that it may be expressed in terms of ν† (equal to μ/ρ), c, ρ, and w in the form

$$c^x \nu^y \rho^z w^n,$$

we have the three relations

$$-2 = x + 2y - 3z + n, \quad -2 = -y - n, \quad 1 = z,$$

so that

$$x = n - 3, \quad y = 2 - n, \quad z = 1,$$

and

$$P \propto \nu^2 c^{-3} \rho \,.\, (cw/\nu)^n. \quad \text{..........................(1)}$$

Since n is here indeterminate, all we can infer from dynamical similarity is that

$$P = \nu^2 c^{-3} \rho\, f(cw/\nu), \quad \text{..........................(2)}$$

where f is an arbitrary function.

* *Phil. Trans.* CLXXIV. p. 935 (1883).

† Of which the dimensions are 2 in space and −1 in time.

For capillary tubes and moderate velocities P varies as the first power of w, so that in (1) $n = 1$. In this case

$$P = A\,\nu\,c^{-2}\rho\,w, \qquad\qquad (3)$$

A being an arbitrary constant. When, on the other hand, cw/ν is great, experiment shows that $n = 2$ nearly. If this law be exact, (1) gives

$$P = B\,c^{-1}\rho\,w^2, \qquad\qquad (4)$$

independent of ν. The second power of the velocity and independence of viscosity are thus inseparably connected.

In the above theory no account is taken of any variation in the walls of the tubes. Either they must be perfectly smooth or else the *irregularities must be in proportion to the diameters.* Under this limitation (2) would appear to hold good, at least if there be no finite slip at the walls.

The proportionality to ρ, expressed in (4), has probably not been tested experimentally. Neither is there any complete theoretical deduction of (4). But a comparison with Torricelli's law of efflux is significant. The resistance is the same as if it were necessary to renew continually the velocity of the liquid at intervals which are proportional to the diameters of the pipes.

The connexion between the alteration in the law of resistance and the transition from regularly stratified to eddying motion has been successfully traced by Reynolds. The question is, Why do eddies arise and take possession? From the description and drawings given by Reynolds it is natural to suppose that in the absence of viscosity the stratified motion would be unstable, and that it is stable in small tubes and at low velocities only in consequence of the steadying effect of viscosity then acting at an advantage. It was with this idea that (at an earlier date*) I attempted an investigation of the stability of stratified flow in two dimensions, fully expecting to find it unstable. The result, however, was to show that in the absence of viscosity the stratified flow between two parallel walls was not unstable, provided that the law of flow were such that the curve representing the velocities in the various strata was of one curvature throughout, a condition satisfied in the case in question. To be more precise, it was proved that if the deviation from the regularly stratified motion were, as a function of the time, proportional to e^{int}, then n could have no imaginary part.

On the other hand, if the condition as to the curvature of the velocity curve be violated, n may acquire an imaginary part, and the resulting disturbance of the steady motion is exponentially unstable, as was shown by several examples in the paper referred to, and in a later one† in which the subject was further pursued.

* *Proc. Math. Soc.* February 12, 1880. [Vol. I. p. 474.]

† *Ibid.* November 1887. [Vol. III. p. 17.]

We are thus confronted with a difficulty. For if the investigation in question can be applied to a fluid of infinitely small viscosity, how are we to explain the observed instability which occurs with moderate viscosities? It seems very unlikely that the first effect of increasing viscosity should be to introduce an instability not previously existent, while, as observation shows, a large viscosity makes for stability.

Several suggestions towards an explanation of the discrepancy present themselves. In the first place, irregularities in the walls, not included in the theoretical investigation, may play an essential part. Again, according to the view of Lord Kelvin, the theoretical stability for infinitely small disturbances at all viscosities may not extend beyond very narrow limits; so that in practice and under finite disturbances the motion would be unstable, unless the viscosity exceeded a certain value. Two other suggestions which occurred to me at the time of writing my first paper as perhaps pointing to an explanation may now be mentioned. It is possible that there may be an essential difference between the motion in two dimensions to which the calculations related, and that in a tube of circular section on which observations are made. And, secondly, it is possible that, after all, the investigation in which viscosity is altogether ignored is inapplicable to the limiting case of a viscous fluid when the viscosity is supposed infinitely small. There is more to be said in favour of this view than would at first be supposed. In the calculated motion there is a finite slip at the walls, and this is inconsistent with even the smallest viscosity. And, further, there are kindred problems relating to the behaviour of a viscous fluid in contact with fixed walls for which it can actually be proved* that certain features of the motion which could not enter into the solutions, were the viscosity ignored from the first, are nevertheless independent of the magnitude of the viscosity, and therefore not to be eliminated by supposing the viscosity to be infinitely small. Another case that may be instanced is that of a large stream of viscous fluid flowing past a spherical obstacle. As Sir G. Stokes has shown, the steady motion is the same whatever be the degree of viscosity; and yet it is entirely different from the flow of an inviscid fluid in which no rotation can be generated. Considerations such as this raise doubts as to the interpretation of much that has been written on the subject of the motion of inviscid fluids in the neighbourhood of solid obstacles.

The principal object of the present communication is to test the first of the two latter suggestions. It will appear that, as in the case of motion between parallel plane walls, so also for the case of a tube of circular section, no disturbance of the steady motion is exponentially unstable, provided viscosity be altogether ignored.

* "On the Circulation of Air in Kundt's Tubes," *Phil. Trans.* November 1883. [Vol. II. p. 239.]

Referring the motion to cylindrical coordinates z, r, θ, parallel to which the component velocities are w, u, v, we have*

$$\frac{\partial u}{\partial t} - \frac{v^2}{r} = \frac{dQ}{dr}, \quad \frac{\partial v}{\partial t} + \frac{uv}{r} = \frac{1}{r}\frac{dQ}{d\theta}, \quad \frac{\partial w}{\partial t} = \frac{dQ}{dz},$$

$$\frac{\partial}{\partial t} = \frac{d}{dt} + u\frac{d}{dr} + \frac{v}{r}\frac{d}{d\theta} + w\frac{d}{dz},$$

where $-Q = V + p/\rho$, [and V is the potential of the impressed forces].

These are the general equations. In order to apply them to the present problem of small disturbances from a steady motion represented by

$$u = 0, \quad v = 0, \quad w = W,$$

where W is a function of r only, we will regard the complete motion as expressed by u, v, $W + w$, and neglect the squares of the small quantities u, v, w, which express the disturbance.

Thus,

$$\frac{du}{dt} + W\frac{du}{dz} = \frac{dQ}{dr}, \quad \text{.........................(1)}$$

$$\frac{dv}{dt} + W\frac{dv}{dz} = \frac{dQ}{rd\theta}, \quad \text{.........................(2)}$$

$$u\frac{dW}{dr} + \frac{dw}{dt} + W\frac{dw}{dz} = \frac{dQ}{dz}, \quad \text{.........................(3)}$$

which, with the "equation of continuity,"

$$\frac{d(ru)}{dr} + \frac{dv}{d\theta} + r\frac{dw}{dz} = 0, \quad \text{.........................(4)}$$

determine the motion.

The next step is to introduce the supposition that as functions of t, z, θ, the variables u, v, w, and Q are proportional to $e^{i(nt+kz+s\theta)}$.

We get

$$i(n + kW)u = \frac{dQ}{dr}, \qquad (n + kW)v = \frac{s}{r}Q, \quad \text{..............(5)}$$

$$u\frac{dW}{dr} + i(n + kW)w = ikQ, \quad \text{....................(6)}$$

$$\frac{d}{dr}(ru) + isv + ikrw = 0. \quad \text{.......................(7)}$$

From these equations three of the variables may be eliminated, so as to obtain an equation in which the fourth is isolated. The simplest result is that in which Q is retained. It is

$$\frac{d^2Q}{dr^2} + \frac{1}{r}\frac{dQ}{dr} - Q\left(\frac{s^2}{r^2} + k^2\right) - \frac{2k}{n + kW}\frac{dW}{dr}\frac{dQ}{dr} = 0. \quad \text{...........(8)}$$

* Basset's *Hydrodynamics*, § 470.

But the equation in u lends itself more readily to the imposition of boundary conditions. If $s=0$, that is in the case of symmetrical disturbances, the equation in u is obtained at once by differentiation of (8), and substitution of u from (5). After reduction it becomes

$$(n+kW)\left\{\frac{d^2u}{dr^2}+\frac{1}{r}\frac{du}{dr}-\frac{u}{r^2}-k^2u\right\}-ku\left\{\frac{d^2W}{dr^2}-\frac{1}{r}\frac{dW}{dr}\right\}=0. \quad \text{....(9)}$$

If the undisturbed motion be that of a highly viscous fluid in a circular tube, W is of the form $A+Br^2$, and the second part of (9) disappears. There can then be admitted no values of n, except such as make $n+kW=0$ for some value of r included within the tube. For the equation

$$\frac{d^2u}{dr^2}+\frac{1}{r}\frac{du}{dr}+\frac{u}{r^2}-k^2u=0, \quad \text{.......................(10)}$$

being that of the Bessel's function of the first order with a purely imaginary argument, admits of no solution consistent with the conditions that $u=0$ when r vanishes, and also when r has the finite value appropriate to the wall of the tube. But any value assumed by $-kW$ is an admissible solution for n. At the place where $n+kW=0$, (10) need not be satisfied, and under this exemption the required solution may be obtained consistently with the boundary conditions. It is included in the above statement that no admissible value of n can include an imaginary part.

If s be not zero, we have in transforming to u to include also terms arising from the differentiation in (8) of $-Qs^2/r^2$, that is

$$2s^2r^{-3}Q-\frac{s^2}{r^2}\frac{dQ}{dr},$$

for the second of which we substitute from (5), and for the first from (8) itself. The result is

$$(n+kW)\left[\frac{d^2u}{dr^2}+\frac{1}{r}\frac{du}{dr}\frac{3s^2+k^2r^2}{s^2+k^2r^2}\right.$$
$$\left.+\frac{u}{r^2}\left\{-1-k^2r^2-s^2+\frac{2s^2}{s^2+k^2r^2}\right\}\right]$$
$$=ku\left\{\frac{d^2W}{dr^2}-\frac{1}{r}\frac{dW}{dr}\frac{k^2r^2-s^2}{k^2r^2+s^2}\right\}. \quad \text{..................(11)}$$

From (11) we may fall back on the case of two dimensions by supposing r to be infinite. But, in order not to lose generality, we must at the same time allow s to be infinite, so that, for example, $s=k'r$. Thus, writing x for r, and y for $r\theta$, we find for the differential equation applicable to the solution in which all the quantities are proportional to $e^{i(nt+kz+k'y)}$,

$$(n+kW)\left\{\frac{d^2u}{dx^2}-k^2u-k'^2u\right\}=ku\frac{d^2W}{dx^2}, \quad \text{..................(12)}$$

agreeing with that formerly discussed except for a slight difference of notation.

We will now consider (11) in the abbreviated form,

$$(n+kW)\left\{\frac{d^2u}{dr^2}+\frac{a}{r}\frac{du}{dr}+b\frac{u}{r^2}\right\}=W_1ku,$$

where a is a positive number not less than unity; or, again,

$$\frac{d}{dr}\left(r^a\frac{du}{dr}\right)+br^{a-2}u=\frac{kur^aW_1}{n+kW}. \quad \text{......................(13)}$$

The question proposed for consideration is whether (13) admits of a solution with a complex value of n, subject to the conditions that for two values of r, say r_1 and r_2, u shall vanish. This represents the flow of fluid through a channel bounded by two coaxal cylinders.

Suppose, then, that n is of the form $p+iq$, and u of the form $\alpha+i\beta$, where p, q, α, β are real. Separating the real and imaginary parts in (13), we get

$$\frac{d}{dr}\left(r^a\frac{d\alpha}{dr}\right)+br^{a-2}\alpha=\frac{kr^aW_1}{(p+kW)^2+q^2}\{(p+kW)\alpha+q\beta\}, \quad \text{....(14)}$$

$$\frac{d}{dr}\left(r^a\frac{d\beta}{dr}\right)+br^{a-2}\beta=\frac{kr^aW_1}{(p+kW)^2+q^2}\{(p+kW)\beta-q\alpha\}; \quad \text{....(15)}$$

and thence

$$\beta\frac{d}{dr}\left(r^a\frac{d\alpha}{dr}\right)-\alpha\frac{d}{dr}\left(r^a\frac{d\beta}{dr}\right)=\frac{kr^aW_1(\alpha^2+\beta^2).q}{(p+kW)^2+q^2}. \quad \text{..........(16)}$$

We now integrate this equation with respect to r over the space between the walls, viz. from r_1 to r_2. The integral of the left-hand member is

$$\beta r^a\frac{d\alpha}{dr}-\alpha r^a\frac{d\beta}{dr}, \quad \text{............................(17)}$$

and this vanishes at both limits, β and α being there zero. The integral of the right-hand member of (16) is accordingly zero, from which it follows *that if W_1 be of one sign throughout*, q must vanish—that is to say, no complex value of n is admissible.

The general value of W_1, viz.

$$\frac{d^2W}{dr^2}-\frac{1}{r}\frac{dW}{dr}\frac{k^2r^2-s^2}{k^2r^2+s^2}, \quad \text{..........................(18)}$$

reduces in the case of two dimensions to d^2W/dr^2, or, as we may then write it, d^2W/dx^2. Instability, at any rate of the full-blown exponential sort, is thus excluded, provided d^2W/dx^2 is of one sign throughout the entire region of flow limited by the two parallel plane walls.

Commenting upon this argument, Lord Kelvin* remarks that the disturbing infinity, which arises in (13) when n has a value such that $n+kW$ vanishes at some point in the field of motion, "vitiates the seeming

* *Phil. Mag.* Aug. 1887, p. 275.

proof of stability." Perhaps I went too far in asserting that the motion was thoroughly stable; but it is to be observed that if n be complex, there is no "disturbing infinity." The argument, therefore, does not fail, regarded as one for excluding complex values of n. What happens when n has a real value such that $n + kW$ vanishes at an interior point, is a subject for further examination.

The condition for two dimensions that d^2W/dx^2 be of one sign throughout is satisfied for a law of flow such as that of a viscous fluid, and we shall see that the corresponding condition for (17) in the more general problem is also satisfied in the case of the steady flow of a viscous fluid between cylindrical walls at r_1 and r_2. The most general form of W for steady motion symmetrical about the axis is*

$$W = Ar^2 + B\log r + C, \quad \text{.......................(19)}$$

in which the constants A, B, C are related by the conditions

$$0 = Ar_1^2 + B\log r_1 + C,$$

$$0 = Ar_2^2 + B\log r_2 + C.$$

From the last two equations we derive

$$A(r_2^2 - r_1^2) + B\log r_2/r_1 = 0, \quad \text{....................(20)}$$

so that A and B have opposite signs. Introducing the value of W from (18), we obtain as the special form here applicable

$$\frac{4s^2A - 2k^2B}{k^2r^2 + s^2},$$

which is thus of one sign throughout the range. A small disturbance from the steady motion expressed by (19) is therefore not exponentially unstable.

The result now obtained is applicable however small may be the inner radius r_1 of the annular channel. But the extension to the case of the ordinary pipe of unobstructed circular section may be thought precarious, when it is remembered that provision must be made for a possible finite value of u when $r = 0$. But although α and β may be finite at the lower limit, the annulment of (17) is secured by the factor r^α; so that complex values of n are still excluded, provided W_1 be of unchangeable sign. In the present case the B of (19) vanishes, and we have

$$d^2W/dr^2 = 0, \quad r^{-1}dW/dr = 2A;$$

so that (18) gives

$$W_1 = \frac{4s^2A}{k^2r^2 + s^2},$$

satisfying the prescribed condition.

* Basset's *Hydrodynamics*, § 514.

The difficulty in reconciling calculation and experiment is accordingly not to be explained by any peculiarity of the two-dimensional motion to which calculation was first applied. It may indeed be argued that the instabilities excluded are only those of the exponential type, and that there may remain others on the borderland of the form $t \cos t$, &c. But if the above calculations are really applicable to the limiting case of a viscous fluid when the viscosity is infinitely small, we should naturally expect to find that the smallest sensible viscosity would convert the feebly unstable disturbance into one distinctly stable, and if so the difficulty remains. Speculations on such a subject in advance of definite arguments are not worth much; but the impression upon my mind is that the motions calculated above for an absolutely inviscid liquid may be found inapplicable to a viscid liquid of vanishing viscosity, and that a more complete treatment might even yet indicate instability, perhaps of a local character, in the immediate neighbourhood of the walls, when the viscosity is very small.

It is on the basis of such a complete treatment, in which the terms representing viscosity in the general equations are retained, that Lord Kelvin* arrives at the conclusion that the flow of viscous fluid between two parallel walls is fully stable for infinitesimal disturbances, however small the amount of the viscosity may be. Naturally, it is with diffidence that I hesitate to follow so great an authority, but I must confess that the argument does not appear to me demonstrative. No attempt is made to determine whether in free disturbances of the type e^{int} (in his notation $e^{\iota\omega t}$) the imaginary part of n is finite, and if so whether it is positive or negative. If I rightly understand it, the process consists in an investigation of forced vibrations of arbitrary (real) frequency, and the conclusion depends upon a tacit assumption that if these forced vibrations can be expressed in a periodic form, the steady motion from which they are deviations cannot be unstable. A very simple case suffices to prove that such a principle cannot be admitted. The equation to the motion of the bob of a pendulum situated near the highest point of its orbit is

$$d^2x/dt^2 - m^2x = X, \quad \text{.............................(21)}$$

where X is an impressed force. If $X = \cos pt$, the corresponding part of x is

$$x = -\frac{\cos pt}{p^2 + m^2}; \quad \text{.............................(22)}$$

but this gives no indication of the inherent instability of the situation expressed by the free "vibrations,"

$$x = Ae^{mt} + Be^{-mt}. \quad \text{.............................(23)}$$

As a preliminary to a more complete investigation, it may be worth while to indicate the solution of the problem for the two-dimensional motion of viscous liquid between two parallel planes, in the relatively very simple case

* *Phil. Mag.* Aug. and Sept. 1887.

where there is no foundation of steady motion. The equation, given in Lord Kelvin's paper, for the motion of type $e^{i(nt+kz)}$ is

$$i\mu\left(\frac{d^4u}{dx^4} - 2k^2\frac{d^2u}{dx^2} + k^4u\right) + n\left(\frac{d^2u}{dx^2} - k^2u\right) = 0. \qquad (24)$$

The boundary conditions, say at $x = \pm a$, are that u, (v), and w shall there vanish, or by (7) that

$$u = 0, \quad du/dx = 0.$$

The following would then be the proof from the differential equation that for all the admissible values of n, p is zero and q is positive.

Writing as before, $u = \alpha + i\beta$, and separating the real and imaginary parts, we find

$$-\mu\left(\frac{d^2}{dx^2} - k^2\right)^2\beta + p\left(\frac{d^2\alpha}{dx^2} - k^2\alpha\right) - q\left(\frac{d^2\beta}{dx^2} - k^2\beta\right) = 0, \qquad (25)$$

$$\mu\left(\frac{d^2}{dx^2} - k^2\right)^2\alpha + p\left(\frac{d^2\beta}{dx^2} - k^2\beta\right) + q\left(\frac{d^2\alpha}{dx^2} - k^2\alpha\right) = 0. \qquad (26)$$

Multiply (25), (26) by α, β respectively, add, and integrate with respect to x over the range of the motion. The coefficient of q is

$$\int\left\{\beta\frac{d^2\alpha}{dx^2} - \alpha\frac{d^2\beta}{dx^2}\right\}dx,$$

and this is equal to zero in virtue of the conditions at the limits. In like manner the coefficient of μ is zero, as appears on successive integrations by parts. The coefficient of p is

$$-\int\left\{\left(\frac{d\alpha}{dx}\right)^2 + \left(\frac{d\beta}{dx}\right)^2 + k^2\alpha^2 + k^2\beta^2\right\}dx;$$

so that $p = 0$.

Again, multiply (25) by β, (26) by α, and subtract. On integration as before the coefficient of q is

$$\int\left\{\left(\frac{d\alpha}{dx}\right)^2 + \left(\frac{d\beta}{dx}\right)^2 + k^2\alpha^2 + k^2\beta^2\right\}dx,$$

and that of μ is

$$-\int\left\{\left(\frac{d^2\alpha}{dx^2}\right)^2 + \left(\frac{d^2\beta}{dx^2}\right)^2 + 2k^2\left(\frac{d\alpha}{dx}\right)^2 + 2k^2\left(\frac{d\beta}{dx}\right)^2 + k^4\alpha^2 + k^4\beta^2\right\}dx.$$

Hence q has the same sign as μ, that is to say, q is positive. That n in e^{int} is a pure positive imaginary is no more than might have been inferred from general principles, seeing that the problem is one of the small motions about equilibrium of a system devoid of potential energy.

Since (24) is an equation with constant coefficients, the normal functions in this case are readily expressed. Writing it in the form

$$\left\{\frac{d}{dx^2} - k^2 - \frac{in}{\mu}\right\}\left\{\frac{d^2}{dx^2} - k^2\right\}u = 0, \qquad (27)$$

we see that the four types of solution are

$$e^{kx},\quad e^{-kx},\quad e^{ik'x},\quad e^{-ik'x},$$

where

$$-k'^2 = k^2 + in/\mu\,; \quad\text{.............................(28)}$$

or, if we take advantage of what has just been proved,

$$k'^2 = q/\mu - k^2, \quad\text{.............................(29)}$$

where q and μ are positive. It will be seen that the odd and even parts of the solution may be treated separately. Thus, for the first,

$$u = A\sinh kx + B\sin k'x, \quad\text{.......................(30)}$$

and the conditions to be satisfied at $x = \pm a$ give

$$0 = A\sinh ka + B\sin k'a,\quad 0 = kA\cosh ka + k'B\cos k'a\,; \quad\text{......(31)}$$

so that the equation for k' is

$$\frac{\tan k'a}{k'a} = \frac{\tanh ka}{ka}. \quad\text{..........................(32)}$$

Again, for the solution involving the even functions,

$$u = C\cosh kx + D\cos k'x, \quad\text{.......................(33)}$$

where

$$\frac{\cot k'a}{k'a} = -\frac{\coth ka}{ka}. \quad\text{..........................(34)}$$

Equations (32), (34) give an infinite number of real values for k', and when these are known q and n follow from (29).

The most persistent motion (for which q is smallest) corresponds to a small value of k, and to the even functions of (33). In this case from (34)

$$k'a = \pi,\ 2\pi,\ 3\pi,\ \text{\&c.},$$

the first of which gives as the smallest value of q

$$q = \mu\pi^2/a^2. \quad\text{................................(35)}$$

The corresponding form for u is

$$u = e^{ikz-qt}\{1 + \cos(\pi x/a)\}. \quad\text{......................(36)}$$

This type of motion is represented by the arrows in the following diagram:—

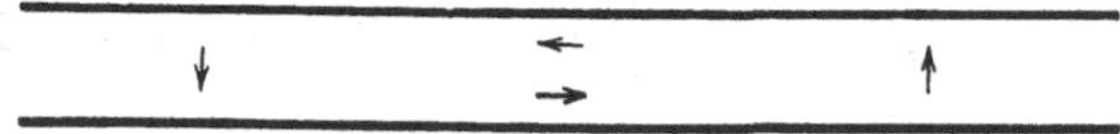

On the other hand the smallest value of q under the head of the odd functions is

$$q = \mu\pi^2(1\cdot4303)^2/a^2, \quad\text{.........................(37)}$$

and the motion is of the type

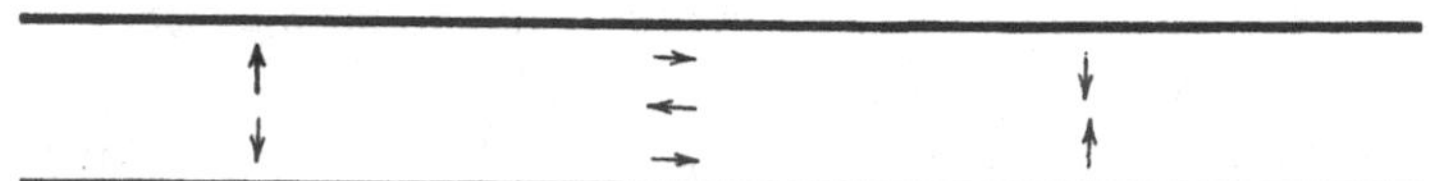

195.

ON THE INSTABILITY OF A CYLINDER OF VISCOUS LIQUID UNDER CAPILLARY FORCE.

[*Philosophical Magazine*, XXXIV. pp. 145—154, 1892.]

THE main outline of the theory of the instability of a long cylinder of liquid is due to Plateau, who showed that if the equilibrium surface $r = a$ be slightly deformed so as to become

$$r = a + \alpha \cos kz, \quad \text{.............................(1)}$$

in which z is measured parallel to the axis, the deformation is stable or unstable according as ka is greater or less than unity; that is, according as the wave-length λ of the varicosity is less or greater than $2\pi a$, the circumference of the cylinder. The solution of the merely statical question is, however, insufficient for the application to the important problem of the disintegration of a jet of liquid. A deformation of any wave-length exceeding $2\pi a$ increases exponentially with the time (e^{qt}); and what we require to know is the relation between q and λ. A value of λ, if any, for which q is a maximum, determines the mode of maximum instability; and this in general tends more and more to be the actual mode of disintegration as the initial disturbances are reduced without limit.

As resisting the capillary force, Plateau seems to have had in view only viscosity. "Par suite des viscosités intérieure et superficielle, le rapport entre la longueur normale des divisions et le diamètre du cylindre surpasse toujours la limite de la stabilité. Il se réduirait sans doute à cette limite même si le liquide était exempt de toute viscosité*." In the case of jets composed of liquids such as water, the influence of viscosity might be expected to be small, but it would be erroneous to conclude that the limit $ka = 1$ would then correspond to the mode of maximum instability. *Inertia*

* *Statique expérimentale et théorique des liquides soumis aux seules forces moléculaires*, 1873 Vol. II. p. 231.

would necessarily play a leading part, and from a complete investigation* it appears that the mode of maximum instability corresponds with $\lambda = 4{\cdot}51 \times 2a$, exceeding very considerably the circumference of the cylinder.

The importance attached by Plateau to viscosity suggested an investigation in which this property should be included; and the results would at any rate find application to threads of materials like glass and sealing-wax, in which viscosity would predominate over inertia. Having in my mind some old experiments upon the behaviour of fine threads of treacle deposited upon paper, which slowly resolve themselves into drops having a very similar appearance to those obtained from a jet of water, I rather expected to find that under the influence of viscosity alone the mode of resolution would be nearly the same as under the influence of inertia alone. This anticipation proved to be wide of the mark, the result showing that under viscosity alone the value of λ for maximum instability would be very great. And a little consideration shows that the retarding forces exerted by the paper support may well be of quite a different character from those due to mere fluid viscosity. In the latter case the gathering together from considerable distances is but little resisted, the motion not differing greatly from that of a solid body, whereas such a mode of resolution would be greatly impeded by the contact with paper. In order better to represent such contact forces, I have considered the problem in the form which it assumes when the resistances are proportional to the absolute velocities of the parts. This admits of easy solution, and the result illustrates the behaviour of the thread of treacle in contact with paper, and shows that there is a marked difference between this case and that of a thread whose disintegration is resisted by true fluid viscosity.

The introduction of resistances proportional to absolute velocities does not interfere with the irrotational character of the motion of otherwise frictionless fluid†. The radial and axial velocities u, w may thus, as usual, be regarded as derived from a velocity-potential according to the equation

$$u = d\phi/dr, \qquad w = d\phi/dz. \qquad \text{.......................(2)}$$

If the resistance is μ' times the velocity, the general equation of pressure, viz.

$$p/\rho = R - d\phi/dt - \tfrac{1}{2}U^2, \qquad \text{.......................(3)}$$

becomes for the present purpose, where U^2 may be neglected,

$$p = -\mu'\phi - \rho\, d\phi/dt. \qquad \text{.........................(4)}$$

The quantities defining the motion are as functions of z proportional to e^{ikz}, and as functions of t proportional to e^{int}, where k is real, but n may be

* *Proc. Math. Soc.* November 1878. [Vol. I. p. 361.] See also below.

† *Theory of Sound*, Vol. II. § 239 (1878).

complex. The general equation for the velocity-potential of an incompressible fluid, viz. $\nabla^2\phi = 0$, thus becomes

$$\frac{d^2\phi}{dr^2} + \frac{1}{r}\frac{d\phi}{dr} - k^2\phi = 0,$$

of which the solution, subject to the condition to be imposed when $r = 0$, is

$$\phi = AJ_0(ikr),$$

or rather

$$\phi = Ae^{i(nt+kz)} J_0(ikr). \qquad (5)$$

At the same time p is given by

$$p = -(\mu' + in\rho)\phi. \qquad (6)$$

We have now to consider the boundary condition, applicable when $r = a$. The displacement ξ at the surface is connected with ϕ by the equation

$$\xi = \int u\,dt = \int \frac{d\phi}{dr}\,dt = \frac{1}{in}\frac{d\phi}{dr}. \qquad (7)$$

The variable part of the pressure is due to the tension T, which is supposed to be constant, as is practically the case in the absence of surface-contamination. The curvature in the plane of the axis is $-d^2\xi/dz^2$, or $k^2\xi$. The curvature in the perpendicular direction is $(a+\xi)^{-1}$, or $1/a - \xi/a^2$. Thus

$$p = \frac{T\xi(k^2a^2-1)}{a^2}; \qquad (8)$$

and the boundary condition is

$$\frac{T(k^2a^2-1)}{ina^2}\frac{d\phi}{dr} = -(\mu' + in\rho)\phi;$$

or by (5),

$$\frac{T}{\rho a^3}\frac{(k^2a^2-1)\,ika\,.\,J_0'(ika)}{J_0(ika)} + in(in + \mu'/\rho) = 0, \qquad (9)$$

a quadratic equation by which n is determined.

If $\mu' = 0$,

$$(in)^2 = \frac{T}{\rho a^3}\frac{(1-k^2a^2)\,ikaJ_0'}{J_0}, \qquad (10)$$

as found in the former paper. In this expression $ikaJ_0'/J_0$ is a real positive quantity for all (real) values of ka; so that the displacement is exponentially unstable if $ka < 1$, and periodic if $ka > 1$, as was to be expected. In the former case the values of in are numerically greatest when $ka = \pi/4{\cdot}5$.

In the other extreme case where inertia may be neglected in comparison with viscosity, we have

$$in = \frac{T}{\rho a^3}\frac{(1-k^2a^2)J_0'}{\mu'/\rho\,.\,J_0}, \qquad (11)$$

so that the instability is greatest when ka has the same value as in the first case.

The general form of the quadratic is

$$(in)^2 + in \,.\, \mu'/\rho + H(k^2a^2 - 1) = 0, \quad\quad (12)$$

where H is positive.

If $ka < 1$, both values of in are real, one being positive and the other negative. The displacement is accordingly unstable, and the greatest instability occurs with the above-defined value of ka. If, on the other hand, $ka > 1$, the values of in may be either real or imaginary. In the former case both values are negative, and in the latter the real parts are negative, so that the deformations are stable.

The investigation applicable to a real viscous liquid of viscosity μ, or $\rho\nu$, is much more complicated than the foregoing, mainly in consequence of the non-existence of a velocity-potential. But inasmuch as the motion is still supposed to be symmetrical about the axis, the equation of continuity gives

$$u = \frac{1}{r}\frac{d\psi}{dz}, \qquad w = -\frac{1}{r}\frac{d\psi}{dr}, \quad\quad (13)$$

where ψ is Stokes's current function. For small motions ψ satisfies the equation*

$$\left(\frac{d^2}{dr^2} - \frac{1}{r}\frac{d}{dr} + \frac{d^2}{dz^2} - \frac{1}{\nu}\frac{d}{dt}\right)\left(\frac{d^2}{dr^2} - \frac{1}{r}\frac{d}{dr} + \frac{d^2}{dz^2}\right)\psi = 0. \quad\quad (14)$$

In the present question ψ as a function of z and t is proportional to $e^{i(nt+kz)}$, and it may be separated into two parts, ψ_1 and ψ_2, of which ψ_1 satisfies

$$\frac{d^2\psi_1}{dr^2} - \frac{1}{r}\frac{d\psi_1}{dr} - k^2\psi_1 = r\frac{d}{dr}\left(\frac{1}{r}\frac{d\psi_1}{dr}\right) - k^2\psi_1 = 0, \quad\quad (15)$$

and ψ_2 satisfies

$$\frac{d^2\psi_2}{dr^2} - \frac{1}{r}\frac{d\psi_2}{dr} - k'^2\psi_2 = r\frac{d}{dr}\left(\frac{1}{r}\frac{d\psi_2}{dr}\right) - k'^2\psi_2 = 0, \quad\quad (16)$$

where

$$k'^2 = k^2 + in/\nu. \quad\quad (17)$$

At the surface we have to consider the normal stress P, and the tangential stresses. Of the latter one vanishes in virtue of the symmetry, and the other is to be made to vanish in conformity with the condition that there is to be no impressed tangential force†. Thus

$$\frac{du}{dz} + \frac{dw}{dr} = 0, \quad\quad (18)$$

or in terms of ψ by (13)

$$\frac{d^2\psi}{dr^2} - \frac{1}{r}\frac{d\psi}{dr} + k^2\psi = 0. \quad\quad (19)$$

* *Camb. Trans.* 1850. See also Basset's *Hydrodynamics*, Vol. II. p. 259.

† It is here assumed that there is no "superficial viscosity."

Introducing ψ_1, ψ_2, and having regard to (15), (16), we may express this condition in the form

$$2k^2\psi_1 + (k'^2 + k^2)\psi_2 = 0, \qquad (20)$$

which is to be satisfied when $r = a$.

Again, for the normal stress,

$$P = -p + 2\mu\frac{du}{dr} = \rho\left(\frac{n}{k}w - \frac{\nu}{ik}\nabla^2 w\right) + 2\mu\frac{du}{dr}$$

$$= \mu\left\{\frac{1}{ik}\nabla^2\left(\frac{1}{r}\frac{d\psi}{dr}\right) + 2ik\frac{d}{dr}\left(\frac{\psi}{r}\right)\right\} - \frac{n\rho}{kr}\frac{d\psi}{dr}. \qquad (21)$$

Herein

$$\nabla^2\left(\frac{1}{r}\frac{d\psi}{dr}\right) = \frac{1}{r}\frac{d}{dr}r\frac{d}{dr}\left(\frac{1}{r}\frac{d\psi}{dr}\right) - \frac{k^2}{r}\frac{d\psi}{dr}.$$

For ψ_1,

$$\frac{1}{r}\frac{d}{dr}\left(r\frac{d}{dr}\frac{1}{r}\frac{d\psi_1}{dr}\right) - \frac{k^2}{r}\frac{d\psi_1}{dr} = 0,$$

for ψ_2,

$$\frac{1}{r}\frac{d}{dr}\left(r\frac{d}{dr}\frac{1}{r}\frac{d\psi_2}{dr}\right) - \frac{k^2}{r}\frac{d\psi_2}{dr} = \frac{k'^2 - k^2}{r}\frac{d\psi_2}{dr};$$

so that

$$P = \mu\left\{\frac{k'^2 - k^2}{ikr}\frac{d\psi_2}{dr} + 2ik\frac{d}{dr}\frac{\psi_1 + \psi_2}{r}\right\} - \frac{n\rho}{kr}\frac{d(\psi_1 + \psi_2)}{dr}. \qquad (22)$$

The variable part of the capillary pressure is, as we have already seen,

$$T\xi(k^2a^2 - 1)/a^2,$$

in which

$$\xi = \int u\,dt = k\psi/na.$$

Thus, the condition to be satisfied when $r = a$ is

$$\frac{T(1 - k^2a^2)}{a^2}\frac{k\psi}{na} = \mu\left\{\frac{k'^2 - k^2}{ika}\frac{d\psi_2}{dr} + 2ik\frac{d}{dr}\frac{\psi}{r}\right\} - \frac{n\rho}{ka}\frac{d\psi}{dr}. \qquad (23)$$

The forms of ψ_1, ψ_2 are to be determined by the equations (15), (16), and by the conditions to be satisfied when $r = 0$. It will be observed that ψ_1 satisfies the condition appropriate to the stream-function when there is a velocity-potential. This would be of the form

$$\phi = e^{ikz}J_0(ikr), \qquad (24)$$

so that

$$\psi_1 = \int (ru)\,dz = \frac{r}{ik}\frac{d\phi}{dr} = re^{ikz}J_0'(ikr).$$

Thus

$$\psi_1 = ArJ_0'(ikr) \qquad (25)$$

is the most general form admissible, as may be verified by differentiation. In this $J_0(ikr)$ satisfies the equation

$$J_0''(ikr) + \frac{1}{ikr}J_0'(ikr) + J_0(ikr) = 0. \qquad (26)$$

Since (16) differ from (15) only by the substitution of k' for k, the general form for ψ_2 is

$$\psi_2 = BrJ_0'(ik'r). \quad\text{.............................(27)}$$

By use of these values the first boundary condition (20) becomes

$$2k^2AJ_0'(ika) + (k'^2 + k^2)\,BJ_0'(ik'a) = 0. \quad\text{..............(28)}$$

We have next to introduce the same values into the second boundary condition (23). In this

$$\frac{d\psi_1}{dr} = ikrA\left[J_0''(ikr) + \frac{1}{ikr}J_0'(ikr)\right] = -AikaJ_0(ika)$$

by (26). In like manner,

$$d\psi_2/dr = -Bik'aJ_0(ik'a).$$

Thus

$$\frac{T(1-k^2a^2)}{\rho a^3}\frac{ka}{n}[AJ_0'(ika) + BJ_0'(ik'a)]$$

$$= -\nu\left[B\frac{k'(k'^2-k^2)}{k}J_0(ik'a) + 2k^2AJ_0''(ika) + 2kk'BJ_0''(ik'a)\right]$$

$$+\frac{n}{ka}[AikaJ_0(ika) + Bik'aJ_0(ik'a)]. \quad\text{..............................(29)}$$

Between (28) and (29) we now eliminate the ratio A/B, and thus obtain as the equation by which [in conjunction with (17)] the value of n is to be determined

$$\frac{T(1-k^2a^2)}{\rho a^3}\frac{ka}{n}\frac{k'^2-k^2}{k'^2+k^2}J_0'(ika)$$

$$= -2k^2\nu\left\{J_0''(ika) - \frac{2kk'}{k'^2+k^2}\frac{J_0'(ika)}{J_0'(ik'a)}J_0''(ik'a)\right.$$

$$\left. - \frac{k'(k'^2-k^2)}{k(k'^2+k^2)}\frac{J_0'(ika)}{J_0'(ik'a)}J_0(ik'a)\right\}$$

$$+\frac{n}{ka}\left\{ikaJ_0(ika) - \frac{2k^2}{k'^2+k^2}\frac{J_0'(ika)}{J_0'(ik'a)}ik'aJ_0(ik'a)\right\}. \quad\text{..........(30)}$$

We shall now apply this result to the particular case where the viscosity is very great in comparison with the inertia. The third part of (30) may then be omitted, and we have to seek the limiting form of the remainder when k' is nearly equal to k, as we see must happen by (17). In the first part,

$$\frac{k'^2-k^2}{k'^2+k^2} = \frac{\delta k}{k}.$$

In the second,

$$J_0''(ika) - \frac{2kk'}{k'^2+k^2}\frac{J_0'(ika)}{J_0'(ik'a)}J_0''(ik'a) = \frac{ika\,\{J_0''^2 - J_0'J_0'''\}\,\delta k}{kJ_0'},$$

and

$$-\frac{k'(k'^2-k^2)}{k(k'^2+k^2)}\frac{J_0'(ika)}{J_0'(ik'a)}J_0(ik'a) = -\frac{J_0\delta k}{k}.$$

Thus the limiting form is

$$\frac{T(1-k^2a^2)}{\mu a \,.\, n} = -\frac{2ka \,.\, ika}{J_0'^2}\left\{J_0''^2 - J_0'J_0''' - \frac{J_0J_0'}{ika}\right\},$$

in which, however, we may effect further simplifications by means of the properties of J_0. We find by use of (26)

$$J_0''^2 - J_0'J_0''' - \frac{J_0J_0'}{ika} = J_0^2 + J_0'^2\left(1+\frac{1}{k^2a^2}\right),$$

so that, finally,

$$in = -\frac{T(1-k^2a^2)}{2\mu a \,.\, k^2a^2\{J_0^2/J_0'^2 + 1 + 1/k^2a^2\}}. \quad \text{.............(31)}$$

In (31) the argument of J_0, J_0' is ika, or z as we will call it for brevity. And by a known property $J_0' = -J_1$. Now

$$J_0(z) = 1 - \frac{z^2}{2^2} + \frac{z^4}{2^2 \,.\, 4^2} - \ldots,$$

$$J_1(z) = \frac{z}{2}\left\{1 - \frac{z^2}{2 \,.\, 4} + \frac{z^4}{2 \,.\, 4^2 \,.\, 6} - \ldots\right\};$$

so that if $x = ka$

$$J_0(ix) = 1 + \frac{x^2}{2^2} + \frac{x^4}{2^2 \,.\, 4^2} - \ldots,$$

$$J_1(ix) = \frac{ix}{2}\left\{1 + \frac{x^2}{2 \,.\, 4} + \frac{x^4}{2 \,.\, 4^2 \,.\, 6} + \ldots\right\}.$$

These functions have been tabulated by Prof. A. Lodge * under the notation $I_0(x)$, $I_1(x)$, where

$$I_0(x) = J_0(ix) = 1 + \frac{x^2}{2^2} + \frac{x^4}{2^2 \,.\, 4^2} + \ldots, \quad \text{.......................(32)}$$

$$I_1(x) = -iJ_1(ix) = \frac{x}{2}\left\{1 + \frac{x^2}{2 \,.\, 4} + \frac{x^4}{2 \,.\, 4^2 \,.\, 6} + \ldots\right\}. \quad \text{.........(33)}$$

In this notation

$$x^2\{J_0^2(ix)/J_1^2(ix) + 1 + 1/x^2\} = x^2 + 1 - x^2I_0^2(x)/I_1^2(x), \quad \text{......(34)}$$

and we have to consider the march of (34) as a function of x.

When x is very small,

$$I_0(x) = 1 - \tfrac{1}{4}x^2, \qquad I_1(x) = \tfrac{1}{2}x + \tfrac{1}{16}x^3,$$

so that

$$(34) = -3 + \text{terms in } x^4;$$

and then from (31)

$$in = \frac{T}{6\mu a}. \quad \text{...............................(35)}$$

* *Brit. Ass. Report*, 1889, p. 28.

We shall see that this corresponds to the maximum instability, and it occurs when the wave-length of the varicosity is very large in comparison with the diameter of the cylinder. The following table gives the values of (34) for specified values of x:—

x	(34)	x	(34)
0·0	−3·0000	1·0	−3·0188
0·2	−3·0000	2·0	−3·2160
0·4	−3·0004	4·0	−4·458
0·6	−3·0023	6·0	−6·247

It will be seen that the numerical value of (34) is least when $x=0$, which is also the value of x for which the numerator of (31) is greatest. On both accounts, therefore, in is greatest when x or $ka=0$. But over the whole range of the instability from $ka=0$ to $ka=1$, (34) differs but little from -3, so that we may take as approximately applicable

$$in=\frac{T(1-k^2a^2)}{6\mu a}. \qquad (36)$$

The result of the investigation is to show that when viscosity is paramount long threads do not tend to divide themselves into drops at mutual distances comparable with the diameter of the cylinder, but rather to give way by attenuation at few and distant places. This is, I think, in agreement with the observed behaviour of highly viscous threads of glass or treacle when supported only at the terminals. A separation into numerous drops, or a varicosity pointing to such a resolution, may thus be taken as evidence that the fluidity has been sufficient to bring inertia into play.

The application of (31) to the case of stability ($ka>1$) is of less interest, but it may be worth while to refer to the extreme case where the wave-length of the varicosity is *very small* in comparison with the diameter. We then fall upon the particular case of a plane surface disturbed by waves of length λ. The result, applicable when the viscosity is so great that inertia may be left out of account, is the limit of (31) when a, or x, is infinite, while k remains constant, or

$$in=\frac{Tk}{2\mu}\div \text{Lim}\, x\,\{J_0^2(ix)/J_1^2(ix)+1\}.$$

By means of the expressions appropriate when the argument is large, the limit in question may be proved to be -1; so that

$$in=-\frac{Tk}{2\mu}. \qquad (37)$$

If gravity be supposed operative in aid of the restoration of equilibrium, we should have to include in the boundary condition relative to pressure a term

$g\rho\xi$ in addition to $Tk^2\xi$; so that the more general result is obtainable by adding $g\rho/k^2$ to T. Thus

$$in = -\frac{k}{2\mu}\left(T + \frac{g\rho}{k^2}\right), \quad \text{.........................(38)}$$

giving the rate of subsidence of waves upon the surface of a highly viscous material. It could of course be more readily obtained directly.

When gravity operates alone,

$$in = -\frac{g\rho}{2\mu k} = -\frac{g}{2\nu k}, \quad \text{.........................(39)}$$

which agrees with a conclusion of Prof. Darwin*. A like result may be obtained from equations given by Mr Basset†.

* *Phil. Trans.* 1879, p. 10. In equation (12) write $i/a = k$, and make $i = \infty$.

† *Hydrodynamics*, Vol. II. § 520, equations (21), (27). See also Tait, *Edinb. Proc.* 1890, p. 110.

196.

ON THE INSTABILITY OF CYLINDRICAL FLUID SURFACES.

[*Philosophical Magazine*, XXXIV. pp. 177—180, 1892.]

IN former papers* I have investigated the character of the equilibrium of a cylindrical fluid column under the action of capillary force. If the column become varicose with wave-length λ, the equilibrium is unstable, provided λ exceed the circumference ($2\pi a$) of the cylinder; and the degree of instability, as indicated by the value of q in the exponential e^{qt} to which the motion is proportional, depends upon the value of λ, reaching a maximum when $\lambda = 4{\cdot}51 \times 2a$. In these investigations the external pressure is supposed to be constant; and this is tantamount to neglecting the inertia of the surrounding fluid.

When a column of liquid is surrounded by air, the neglect of the inertia of the latter will be of small importance; but there are cases where the situation is reversed, and where it is the inertia of the fluid outside rather than of the fluid inside the cylinder which is important. The phenomenon of the disruption of a jet of air delivered under water, easily illustrated by instantaneous photography [Vol. III. p. 443], suggests the consideration of the case where the inside inertia may be neglected; and to this the present paper is specially directed. For the sake of comparison the results of the former problem are also exhibited.

Since the fluid is supposed to be inviscid, there is a velocity-potential, proportional to e^{ikz} as well as to e^{qt}, and satisfying the usual equation

$$\frac{d^2\phi}{dr^2} + \frac{1}{r}\frac{d\phi}{dr} - k^2\phi = 0. \quad \text{.........................(1)}$$

If the fluid under consideration is inside the cylinder, the appropriate solution of (1) is

$$\phi = J_0(ikr) = I_0(kr); \quad \text{.........................(2)}$$

and the final result for q^2 is

$$q^2 = \frac{T}{\rho a^3}\frac{(1-k^2a^2)\,ika\,.\,J_0'(ika)}{J_0(ika)} = \frac{T}{\rho a^3}\frac{(1-k^2a^2)\,ka\,I_0'(ka)}{I_0(ka)}, \quad \text{......(3)}$$

* (1) "On the Instability of Jets," *Math. Soc. Proc.* November 1878. [Vol. I. p. 361.]
(2) "On the Capillary Phenomena of Jets," *Proc. Roy. Soc.* May 1879. [Vol. I. p. 377.]
(3) "On the Instability of a Cylinder of Viscous Liquid under Capillary Force," *suprà*, p. 145. [Vol. III. p. 585.]

in which T represents the capillary tension, ρ the density, and, as usual,

$$I_0(x) = J_0(ix) = 1 + \frac{x^2}{2^2} + \frac{x^4}{2^2 . 4^2} + \frac{x^6}{2^2 . 4^2 . 6^2} + \dots, \qquad (4)$$

$$I_1(x) = iJ_0'(ix) = \frac{x}{2} + \frac{x^3}{2^2 . 4} + \frac{x^5}{2^2 . 4^2 . 6} + \dots . \qquad (5)$$

But if the fluid be outside the cylinder, we have to use that solution of (1) for which the motion remains finite when $r = \infty$. This may be expressed in two ways*. When r is great we have the semi-convergent form

$$\phi = -\left(\frac{\pi}{2kr}\right)^{\frac{1}{2}} e^{-kr} \left\{1 - \frac{1^2}{1 . 8kr} + \frac{1^2 . 3^2}{1 . 2 . (8kr)^2} - \frac{1^2 . 3^2 . 5^2}{1 . 2 . 3 . (8kr)^3} + \dots\right\}, \qquad (6)$$

and for all values of r the fully convergent series

$$\phi = (\gamma + \log \tfrac{1}{2}kr) I_0(kr) - \frac{k^2 r^2}{2^2} S_1 - \frac{k^4 r^4}{2^2 . 4^2} S_2 - \dots, \qquad (7)$$

in which γ is Euler's constant, equal to ·5772..., and

$$S_n = 1 + \tfrac{1}{2} + \tfrac{1}{3} + \dots + 1/n. \qquad (8)$$

In this case the solution of the problem becomes

$$q^2 = \frac{T}{\rho a^3} \frac{(k^2 a^2 - 1) ka \phi'(ka)}{\phi(ka)}, \qquad (9)$$

ϕ being defined by (7). In (9) ρ represents the inertia of the external fluid, that of the internal fluid being neglected, while in the corresponding formula (3) ρ is the inertia of the internal fluid, that of the external fluid being neglected. There would be no difficulty in writing down the analytical solution applicable to the more general case where both densities are regarded as finite.

The accompanying Table gives the values of

$$\left\{\frac{(1 - x^2) x I_1(x)}{I_0(x)}\right\}^{\frac{1}{2}}, \qquad (10)$$

to which q in (3) is proportional, and of

$$\left\{\frac{(x^2 - 1) x \phi'(x)}{\phi(x)}\right\}^{\frac{1}{2}}, \qquad (11)$$

corresponding in a similar manner to (9). In the second case we have

$$\phi(x) = (\gamma + \log \tfrac{1}{2}x) I_0(x) - \frac{x^2}{2^2} S_1 - \frac{x^4}{2^2 . 4^2} S_2 - \frac{x^6}{2^2 . 4^2 . 6^2} S_3 - \dots, \qquad (12)$$

$$x\phi'(x) = I_0(x) + (\gamma + \log \tfrac{1}{2}x) x I_1(x) - \frac{x^2}{2} S_1 - \frac{x^4}{2^2 . 4} S_2 - \frac{x^6}{2^2 . 4^2 . 6} S_3 - \dots . \qquad (13)$$

On account of the factor $(1 - x^2)$ both (10) and (11) vanish when $x = 0$ and when $x = 1$. Beyond $x = 1$, (10), (11) become imaginary, indicating stability. It will be seen that when the fluid is internal the instability is a maximum between $x = ·6$ and $x = ·7$; and when the fluid is external, between $x = ·4$ and $x = ·5$. That the maximum instability would correspond to a longer wave-length in the case of the external fluid might have been expected, in view of the greater room available for the flow. The same consideration also explains the higher maximum attained by (11) than by (10).

* See the writings of Sir G. Stokes; or *Theory of Sound*, § 341.

x	$I_0(x)$	$x I_1(x)$	(10)	$-\phi(x)$	$x\phi'(x)$	(11)
0·0	1·0000	·0000	·0000	∞	1·0000	·0000
0·1	1·0025	·0050	·0703	2·4270	·9854	·6339
0·2	1·0100	·0201	·1382	1·7527	·9551	·7233
0·3	1·0226	·0455	·2012	1·3724	·9169	·7795
0·4	1·0404	·0816	·2567	1·1146	·8738	·8113
0·5	1·0635	·1289	·3015	·9244	·8283	·8198
0·6	1·0920	·1882	·3321	·7774	·7817	·8022
0·7	1·1264	·2603	·3433	·6607	·7353	·7535
0·8	1·1665	·3463	·3269	·5654	·6894	·6625
0·9	1·2130	·4474	·2647	·4869	·6449	·5017
1·0	......	......	·0000	......	......	·0000

In order the better to study the region of the maximum, the following additional values have been calculated by the usual bisection formula

$$\tfrac{1}{2}(q+r)+\tfrac{1}{16}\{q+r-(p+s)\}.$$

x	(10)	x	(11)
·65	·3406	·45	·8186
·70	·3433	·50	·8198
·75	·3397	·55	·8147

The value of x for which (10) is a maximum may now be found from Lagrange's interpolation formula. It is

$$x = {\cdot}696,$$

corresponding to $\lambda = 2a \times \pi/x = 4{\cdot}51 \times 2a,$ ....(14)

and agreeing with the value formerly obtained by a different procedure.

In like manner we get for the value of x giving maximum instability in the case of the external fluid,

$$x = {\cdot}485,$$

and $\lambda = 6{\cdot}48 \times 2a.$ ....(15)

Some numerical examples applicable to the case of water were given in a former paper. It appeared that for a diameter of one millimetre the disturbance of maximum instability is multiplied one-thousand-fold in about one-fortieth of a second of time. This is for the case of internal fluid. If the fluid were external, the amplification in the same time would be more than one-million-fold.

END OF VOL. III.

CAMBRIDGE: PRINTED BY J. AND C. F. CLAY, AT THE UNIVERSITY PRESS.

Printed in the United States
By Bookmasters